Advances in Plant Ecophysiology Techniques

AF694572

Adela M. Sánchez-Moreiras • Manuel J. Reigosa
Editors

Advances in Plant Ecophysiology Techniques

Springer

Editors
Adela M. Sánchez-Moreiras
Department of Plant Biology and Soil Science
University of Vigo
Vigo, Spain

Manuel J. Reigosa
Department of Plant Biology and Soil Science
University of Vigo
Vigo, Spain

ISBN 978-3-030-06617-8 ISBN 978-3-319-93233-0 (eBook)
https://doi.org/10.1007/978-3-319-93233-0

© Springer International Publishing AG, part of Springer Nature 2018
Softcover re-print of the Hardcover 1st edition 2018
This work is subject to copyright. All rights are reserved by the Publisher, whether the whole or part of the material is concerned, specifically the rights of translation, reprinting, reuse of illustrations, recitation, broadcasting, reproduction on microfilms or in any other physical way, and transmission or information storage and retrieval, electronic adaptation, computer software, or by similar or dissimilar methodology now known or hereafter developed.
The use of general descriptive names, registered names, trademarks, service marks, etc. in this publication does not imply, even in the absence of a specific statement, that such names are exempt from the relevant protective laws and regulations and therefore free for general use.
The publisher, the authors and the editors are safe to assume that the advice and information in this book are believed to be true and accurate at the date of publication. Neither the publisher nor the authors or the editors give a warranty, express or implied, with respect to the material contained herein or for any errors or omissions that may have been made. The publisher remains neutral with regard to jurisdictional claims in published maps and institutional affiliations.

Printed on acid-free paper

This Springer imprint is published by the registered company Springer Nature Switzerland AG
The registered company address is: Gewerbestrasse 11, 6330 Cham, Switzerland

Preface

The book *Advances in Plant Ecophysiology Techniques* has its origin on the successful launching of the *Handbook of Plant Ecophysiology Techniques* released by Kluwer Academic Publishers in 2001. As with any book written about a rapidly growing discipline, the techniques included in that handbook considerably evolved and changed during the last 17 years. Particularly, molecular and microscopic methods have been highly improved during these years, which made necessary an update on the techniques used to measure ecophysiological parameters.

This book will be useful, we hope, for expert and non-initiated scientists and graduate and post-graduate students starting their research studies or looking for new techniques to complement the parameters they already measure. Researchers interested in Plant Ecophysiology, Agronomy, Forestry, Environmental Sciences, and other related disciplines will find here a plethora of techniques easy to apply and reproduce.

Advances in Plant Ecophysiology Techniques consists of 28 chapters, written by a variety of scientists all over the world, which cover the most relevant parameters necessary to be measured when plant response to biotic/abiotic factors is under study. Every chapter included in this handbook has a theoretical part, reviewing the analyzed parameter, the techniques used for its measurement, and the meaning it has in ecophysiology, and a more empirical part focusing on the selected technique, with hand-by-hand procedures as used in the laboratory, which will allow the end users to easily reproduce the method in their labs.

Although every chapter can be read independently, we have made an effort to avoid repetitions and to group the different chapters from the most general to the most complex techniques.

The first chapter written by Mercedes Verdeguer focuses on the bioassays used to measure the impact of biotic/abiotic factors on the growth and germination of seedlings and adult plants. This chapter is well complemented by the second chapter studying root morphology, written by Maria Rosa Abenavoli and colleagues.

After this morphological approach, there are four chapters related to the study of photosynthesis into the plant: photosynthetic pigment content by Beatriz Fernández-Marín and collaborators, net photosynthetic rate and respiration by Jaume Flexas and colleagues, and fluorescence and thermoluminescence by Amarendra Mishra cover all aspects of the photosynthetic status of the plant. In addition, water status is very well covered by Ricardo Aroca and his group and Hamlyn Jones, who does a nice review on the key biophysical equations involved in the use of thermal sensing for the study of plant water relations. In this group of chapters are also included two chapters related to oxidative and photooxidative stress markers, one of them written by Erola Fenollosa and Sergi Munné-Bosch and focused on the use of stress markers to unravel plant invasion success, and the other written by Adele Muscolo and colleagues and focused on the role that reactive oxygen species (ROS) have on plant response and the way to measure them.

The next two chapters, written by Teodoro Coba de la Peña, Adela M. Sánchez-Moreiras, and Fabrizio Araniti, are focused on the use of flow cytometry to measure cell cycle and other physiological parameters, such as pH, calcium content, glutathione level, etc., including the use of mutants to take all the advantages of this technique.

The following five chapters are focused on different microscopic techniques ranging from optical to confocal and transmission electronic microscopy and their use for the measurement of a variety of different parameters, such as mitotic index and plant microtubules (by Elisa Graña), programmed cell death (by John Conway and Paul McCabe), structural and ultrastructural morphology (by Adela M. Sánchez-Moreiras and colleagues), and many physiological parameters measured by fluorescence (by Elfrieda Fodor and Ferhan Ayaydin) or confocal (by Leonardo Bruno and collaborators) microscopy. The advances we have seen in these techniques during the last years make these chapters especially useful for plant scientists.

The next four chapters are devoted to –omics and SAR/QSAR techniques. Leonardo Bruno and colleagues addressed multi-probe whole mount mRNA in situ hybridization for the quick and simultaneous measurement of the expression of several genes, while metabolomics significance and measurement has been addressed by Leslie Weston and her group, proteomics was reviewed by Jesus Jorrín and colleagues, and SAR/QSAR techniques for activity studies have been written by Marta Teijeira and María Celeiro.

After these chapters, Adela M. Sánchez-Moreiras and collaborators present an approach for the testing and the study of the mode of action of natural compounds, including morphological, physiological, biochemical, as well as molecular techniques. This chapter is a good example how successful can be mixing different scientific expertise when approaching the response of the plant to external factors. The next chapter focused on allelochemical detoxification in plants and written by Margot Schulz and colleagues is highly related to this one, as plants have evolved different strategies to cope with the presence of these potentially toxic natural compounds. Moreover, Chap. 24 by Fabrizio Araniti and collaborators is also related to plant natural compounds/secondary metabolites, as it focuses on the chemical characterization of volatile organic compounds (VOCs) using two different techniques.

The use of carbon radiochemicals and stable isotopes for plant (Geneviève Chiapusio and collaborators) and soil (Adrià Barbeta and colleagues) studies gives a nice environmental perspective and has been addressed in the following two chapters. Also with an eye on the soils, the next chapter written by Paula Castro and colleagues focuses on different approaches to structurally and functionally characterize key soil microorganisms.

Finally, a nice chapter about computational approach to study ecophysiology has been written by Amarendra Mishra.

All these chapters together represent a useful compendium of the most interesting parameters on plant ecophysiology, which we hope will become a personal guide for plant scientists.

We want to show our warmest gratitude to all the authors for their involvement in this project and also to Springer for their support and help during the period of editing this handbook. And finally, we want to thank all readers who will use these chapters to update their knowledge and to improve their research in Plant Ecophysiology and other related sciences.

Vigo, Spain

Adela M. Sánchez-Moreiras
Manuel J. Reigosa

Contents

Chapter 1
In Vitro and *In Vivo* Bioassays

Mercedes Verdeguer

1 Introduction

The plant's environment is determined by all the physical and chemical factors that characterize the habitats, and also by the effects that other co-occurring organisms induce on them. The functional study of plant's behaviour in their environment linked ecology and physiology in a new discipline, the ecophysiology (Pardos 2005). Ecophysiologists, or physiological ecologists deal with ecological questions like the mechanisms that regulate and control growth, reproduction, survival, abundance, and geographical distribution of plants, as these processes are affected by interactions of plants with their physical, chemical, and biotic environment. The knowledge of these ecophysiological patterns and mechanisms helps to understand the functional significance of specific plant traits and their evolutionary heritage. Ecophysiological techniques have greatly advanced understanding of photosynthesis, respiration, plant water relations, and plant responses to abiotic and biotic stress, from immediate to evolutionary timescales. Many important societal issues, as agriculture, climate change, or nature conservation, benefit from the application of an ecophysiological perspective (Lambers et al. 2008). Plants have adapted to an incredible range of environments, and the fields of ecological and environmental plant physiology have provided tools for understanding the survival, distribution, productivity, and abundance of plant species across the diverse climates of our planet (Ainsworth et al. 2016).

One of the branches of Ecophysiology focuses on the study of the physiological interactions of plants with other plants, animals and microorganisms. Some plants have the capacity of inhibiting the growth or development of surrounding plants by releasing chemical compounds known as allelopathic compounds or allelochemicals. Plants introduce allelochemicals into the environment through foliar leaching,

M. Verdeguer (✉)
Instituto Agroforestal Mediterráneo, Universitat Politècnica de València, Valencia, Spain
e-mail: merversa@doctor.upv.es

© Springer International Publishing AG, part of Springer Nature 2018

A. M. Sánchez-Moreiras, M. J. Reigosa (eds.), *Advances in Plant Ecophysiology Techniques*, https://doi.org/10.1007/978-3-319-93233-0_1

root exudation, residue decomposition, volatilization and debris incorporation into soil (Inderjit and Keating 1999). Most allelochemicals are secondary metabolites, which are obtained through branching of the main metabolic pathways of carbohydrates, fats and aminoacids (Lotina-Hennsen et al. 2006). Secondary metabolites act sometimes as allelochemicals; however, the terms 'allelochemical' and 'secondary metabolite' should not be used as synonyms (Whittaker and Feeny 1971).

In this chapter, *in vitro* and *in vivo* assays for the study of plant interactions, especially for the study of the effects of potential phytotoxic secondary metabolites (allelochemicals), are in detail described.

2 *In Vitro* Bioassays

The research of phytotoxic chemicals with potential utilization as herbicides is one of the areas in which allelopathic studies are more promising nowadays (Macías et al. 2001; Benvenuti et al. 2017). These studies need to have into account the following focus points: (1) the knowledge of the organism that release the chemicals ('donor plant'); (2) the knowledge of the chemicals involved in the interaction; (3) the observation of the plant that acts as the receiver ('target plant') to discover the effects of the chemicals and their mechanism/s of action; and (4) the way the chemicals go from the donor to the target plant, as well as the transformations, induced by biotic and/or abiotic factors, that the allelochemicals could undergo in between, as this could essentially modify their effects (Macías et al. 2008). Analytical techniques and bioassays are basic tools for the detection and understanding of the effects of these compounds.

The most widely used bioassays are seed germination assays, carried out on Petri dishes with filter paper as the most common support (Leather and Einhellig 1988), or agar as an alternative; and seedling growth tests (Lotina-Hennsen et al. 2006).

2.1 Seed Germination Bioassays

Although seed germination bioassays seem simple, some decisions are important in order to have representative and useful results.

First, depending on the objectives of the study, a mixture of compounds or a single compound could be tested. This is sometimes a point of controversy.

Allelopathic activity in field situations is thought to be often due to the joint action of mixtures of allelochemicals rather than to the action of a single allelochemical (Einhellig 1995). For example, Lydon et al. (1997) reported that soil amended with pure artemisinin was less inhibitory to the growth of redroot pigweed than soil amended with a chemically more complex annual wormwood leaf extract. There are many more examples supporting the synergistic action of mixture compounds (Barney et al. 2005; Koroch et al. 2007). Araniti et al. (2013) found that the inhibitory effects caused by *Calamintha nepeta* methanolic extract on *Arabidopsis thaliana* depended on the combined action of different molecules. Inderjit et al. (2002) argued that the understanding of the joint action of phytotoxins in allelopathy research is mainly hindered due to the lack of a well-defined reference model and to methodological problems. Determining which compounds are involved in causing phytotoxicity will be important in developing predictive models for toxicity. Although plant physiologists have successfully determined the mode of action of several individual allelochemicals, more research is needed to evaluate the mode of action of compounds in a mixture belonging to different chemical classes (Inderjit and Duke 2003).

Essential oils or aqueous or solvent extracts contain a mixture of compounds. Therefore, it is necessary to test them as a whole to know the effects that they could exert, but complementary tests with isolated compounds that are part of the mixture could also be done to better understand their activities and mechanism/s of action (Araniti et al. 2013). When more components are part of the mixture more difficult is to comprehend how they are acting and more complicated trials need to be performed.

The second step when planning bioassays is the selection of the target species. Lettuce seeds (*Lactuca sativa* L.) have been widely used because of their fast germination and high sensitivity. Some crop species have been also used for these purposes. There is concern that these species do not reflect what is occurring on natural ecosystems but they offer researchers a starting point from which to learn more about phytotoxicity and allelochemicals. There are some criteria that seeds must satisfy to be used in phytotoxic bioassays: being readily available; being affordable; germinating quickly, completely, and uniformly; and producing repeatable and reliable results. They must also be sensitive enough to respond to a variety of chemicals with different biochemical effects and to offer researchers a means by which to help identifying the mechanisms of action of active compounds. *Arabidopsis thaliana* (L.) has been also used as model species, as it is sensitive to a variety of potent allelochemicals and satisfies all of the selection criteria for target species (Pennacchio et al. 2005).

The most important consideration when choosing the species for developing a bioassay for an allelopathic study is to select the target species from both mono and dicotyledons (Lotina-Hennsen et al. 2006), because they have different metabolism and different responses to the phytotoxic compounds. It has been demonstrated that phytotoxic compounds cause different effects depending on the species they are acting against (Reigosa et al. 1999; Verdeguer et al. 2009a; Graña et al. 2013). The selected target species also depend on the objectives of the research. For studies of allelopathic compounds action in natural ecosystems it is necessary to select the species that are receiving the compounds in nature (Herranz et al. 2006). When searching new bioherbicides it is interesting to select different important weeds and crops as target species (Benvenuti et al. 2017) while when focusing in the mechanism of action is better selecting *A. thaliana* as target species (Reigosa and Malvido-Pazos 2007).

In vitro seed germination bioassays can be performed placing the seeds on filter paper (Dudai et al. 1999), agar (Pederson 1986) or other inert substrates, as sand or vermiculite (Dayan and Duke 2006). Filter paper is the most used substrate for many reasons, like its easy availability and management, low cost and the fact that usually limited quantities of natural compounds are available for testing. Depending on the thickness of the filter paper employed the requirements of water or water plus solvent solution of the compound(s) studied will be different. For example, Whatman n° 1 paper, which is one of the most used (Barnes and Putnam 1987; Reigosa and Malvido-Pazos 2007) is 87 g/m^2 and Whatman n° 3 paper, which has also been used for germination bioassays (Dudai et al. 1999), is 185 g/m^2. It is important that the filter paper is wet enough to allow seed germination but without excessive water to prevent fungal growth and to avoid anaerobic conditions that could difficult seed germination. Previous assays could be done with control seeds to determine the adequate water or solution supply.

Most allelochemicals or their mixtures (e.g. essential oils) are not soluble in water. In some studies, they are applied directly to the filter paper (Dudai et al. 1999; Verdeguer et al. 2009a), while in others, different solvents are used based on the compound's solubility to obtain aqueous solutions, like ethanol, dimethyl sulfoxide (Reigosa and Malvido-Pazos 2007) or Tween 20 (Angelini et al. 2003). When using solvents, it is necessary to include a control with the concentration employed to verify that they are not causing undesirable phytotoxic effects. It is important to have a good solution of the compound(s) studied, otherwise the results could be not as expected. For testing essential oils, Fitoil, a biological adjuvant containing 40% of soybean oil can be used to solubilize them in water (Verdeguer 2011).

Another aspect to consider when planning seed germination tests is the number of seeds and replications used. Normally, at least five repetitions are performed (Dudai et al. 1999; Angelini et al. 2003; Verdeguer et al. 2009a; Graña et al. 2013; Benvenuti et al. 2017). The number of seeds included can vary depending on the size of the seed and the purpose of the experiment, and also on the diameter of the Petri dish employed. The inhibitory potential of essential oils against different crops and weeds was tested in Petri dishes of 6 cm with 20 seeds each and 5 replications (Dudai et al. 1999). Benvenuti et al. (2017) used Petri dishes of 15 cm diameter and 50 seeds each with 3 repetitions, repeated twice (6 replications in total) to evaluate the effect of Compositae essential oils on *Amaranthus retroflexus* L. and *Setaria viridis* (L.) P. Beauv. Petri dishes of 9 cm diameter are commonly used for germination bioassays. Depending on the size of the seeds tested is recommendable to change the number of seeds placed in each Petri dish. Eight replicates with 24 seeds each were used for large-seeded species, whereas five replicates with 50 seeds each were used for small-seeded species for testing the phytotoxic potential of citral on weeds germination (Graña et al. 2013; Díaz-Tielas et al. 2014).

If the germination tests are being maintained for more than 5–7 days, to study the effect of natural products in the development of the seedlings after germination, it also affects the number of seeds to be used, for example, *Portulaca oleracea* L., which is an important weed in Mediterranean summer crops has very small seeds and 20 or more seeds can be easily placed in a Petri dish of 9 cm diameter

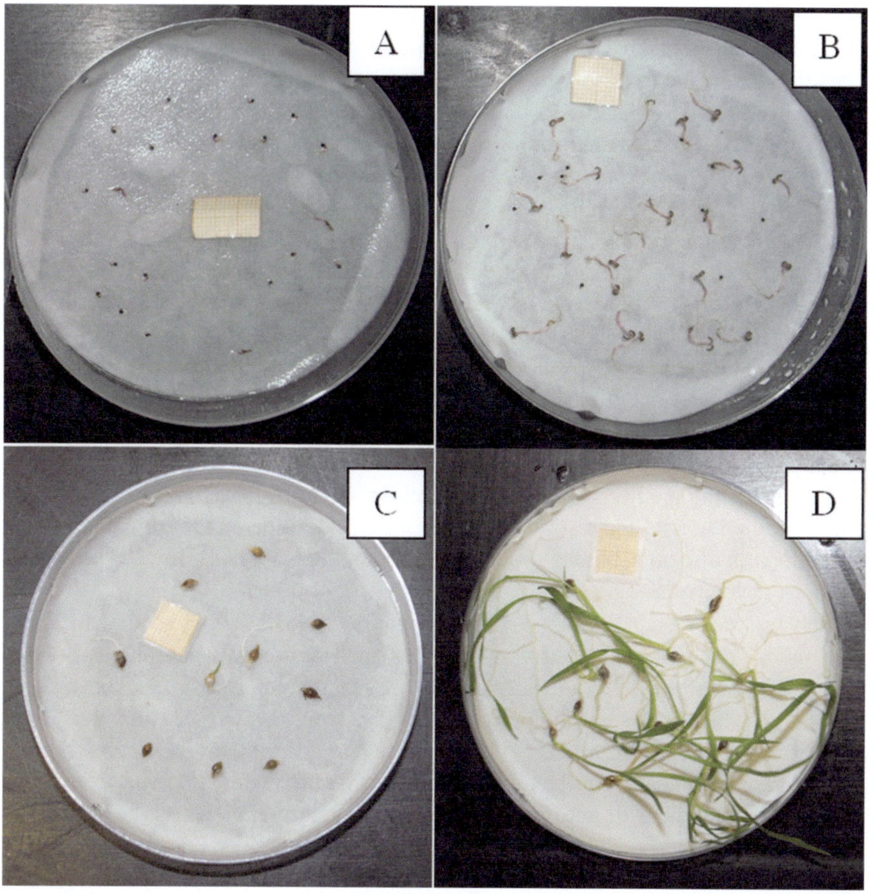

Fig. 1.1 *Portulaca oleracea* seeds after 3 days of incubation (**a**) (30 °C 12 h light/20 °C 8 h dark) and after 14 days (**b**). *Echinochloa crus-galli* seeds after 3 days of incubation (**c**) and after 14 days (**d**) in the same conditions

(Fig. 1.1a, b). *Echinochloa crus-galli* (L.) P. Beauv., which is an important weed in rice crops has greater seeds and is not recommendable to use more than 10 seeds in each Petri dish if the seedlings development has to be analysed, because when germination starts seedlings develop very quickly and take up a lot of space (Fig. 1.1a, b). In this case 5 repetitions of 20 seeds each could be used for *P. oleracea* and 10 repetitions of 10 seeds each could be used for *E. crus-galli*. It is important that the concentrations of compound(s) tested are the same in both cases. There is also a bit controversy with this issue because some authors stated that the doses received by seeds are different in both cases (when placing different seed number) but also the seeds requirements are different. As the doses of herbicides in field conditions are measured by surface area, we could use the same idea for the doses used in Petri dishes.

In a given soil volume containing a specific amount of phytotoxin, plants growing at low densities have a larger amount of toxin available per plant than at high densities, where the toxin is shared (and thus diluted) among many plants, receiving each plant a proportionately smaller dose. Lower phytotoxin concentrations could produce equivalent or greater inhibitory effects than higher concentrations when the amount available per plant is greater (Weidenhamer 1996). The degree of inhibition caused by an allelopathic compound will depend upon both phytotoxin concentration and the total amount of phytotoxin available (Weidenhamer et al. 1987).

Some species can present autotoxicity, like alfalfa, *Medicago sativa* L., (Chon and Nelson 2001) or tobacco, *Nicotiana tabacum* L. (Deng et al. 2017) or *Prosopis juliflora* (Sw.) DC. (Warrag 1995), among others. When testing these species, it is recommended to use well plate systems in order to separate the seeds and do not mask the effects of the allelochemicals tested with the autotoxins released by these seeds or seedlings. It is important to note that the International Seed Testing Association (ISTA) has developed the International Rules for Seed Testing, to uniform seed testing methods. The 2018 edition is available at the ISTA website. They are recommendations and testing methods for seeds designated for growing of crops or production of plants, but for spontaneous and medicinal plants or weed seeds had not been developed.

The last decision regarding seed germination tests is to define the duration. Depending on the objective, it can be established normally when all seeds have germinated or when there are not changes in germination rates after consecutive counts. Previous assays can be performed to determine how much time needs the majority of seeds to germinate.

Different parameters can be evaluated in germination tests. It is very important to use suitable indices of germination because the interpretation and the biological meaning of the assays depend of them (Chiapusio et al. 1997). For example, if it is considered only the total final seed germination rate (percentage of seeds germinated at the end of the experiment) delays on germination cannot be measured neither evaluated, while this is an important effect of allelochemicals on germination. The speed of germination (proportion of germinated seeds obtained in n days or hours) reflects better what happens during the germination process and is a good indicator for delays in germination. The speed of accumulated germination (cumulative number of seeds that germinate on time N since set up of the experiment) and the coefficient of the rate of germination (number of seeds germinated on time T) are less sensitive than the speed of germination to show delays on germination and their interpretation must be done carefully. Control and treated seeds must be compared at each exposure time for better conclusions. Other parameters that can be evaluated are the abnormal symptoms observed in seeds, and the radicle length, among others. If the tests are performed to investigate the phytotoxicity of some products on seed germination could be interesting to determine the concentration of product that causes 50% inhibition of germination rate (IC_{50}) and the concentration of product that causes 80% inhibition (IC_{80}). It is also important to verify if the phytotoxic effects on seed germination are reversible or irreversible. For this purpose, the seeds that had been in contact with the phytotoxic product are transferred to water (Petri dishes are prepared as they were controls) after the germination tests, and their germination is evaluated again. If they germinate the phytotoxic effects are reversible. This is important for biodiversity conservation purposes.

A protocol for testing natural products phytotoxic effects on seed germination is here described. Seeds of different weeds, monocotyledons and dicotyledons are used in order to study different responses depending on the type of weed. For small-sized seed weeds, 20 seeds are placed in Petri dishes of 9 cm diameter between two layers of filter paper (50 g/m^2) wetted with 4 mL of the corresponding treatment, and 5 replicates are performed. For large-sized weed, 10 seeds and 10 replicates are used. The treatments applied are control (distilled water) and 4 different concentrations of the product assayed. The Petri dishes are incubated in a chamber with 60% HR, 30 °C 12 h light and 20 °C 8 h dark. Photos of Petri dishes are registered after 3, 5, 7, 10 and 14 incubation days and are processed with Digimizer and ImageJ to determine different parameters: number of germinated seeds per day, abnormal seeds, radicle length and seedling length, to calculate germination speed and rates. At the end of the assay also fresh and dry weight of treated plants can be registered.

2.2 *Seedling Growth Bioassays*

Seedling growth of the developed seedlings can be evaluated in order to determine the effects of one compound or a mixture of compounds on seedling growth after germination, if the seeds were able to germinate after the treatment, and this is performed as described in the previous part. Photos are registered from the Petri dishes after 3, 5, 7, 10 and 14 incubation days and processed with Digimizer and ImageJ to determine radicle and seedling length and other parameters of interest.

There are also specific assays performed to study the effects of natural products applied in post emergence, once the seeds have been germinated. In this assays seedlings of the target plants are produced following the protocols described in the previous section for the controls and then ten plants with uniform radicle length (5 mm) are selected and transferred to Petri dishes with the treatment to analyse the effects (Fig. 1.2). Five replicates are prepared for each treatment and the Petri dishes are incubated in the same conditions used for germination tests (Verdeguer 2011). Photos are registered after 3, 5, 7, 10 and 14 days since the plants are transferred and are then processed with Digimizer and ImageJ to measure radicle and seedling length. Other studies follow similar protocols using seedlings of 1–2 mm radicle length that are transferred to Petri dishes with the treatments and the growth is measured for 48 h (Dudai et al. 1999; Araniti et al. 2017).

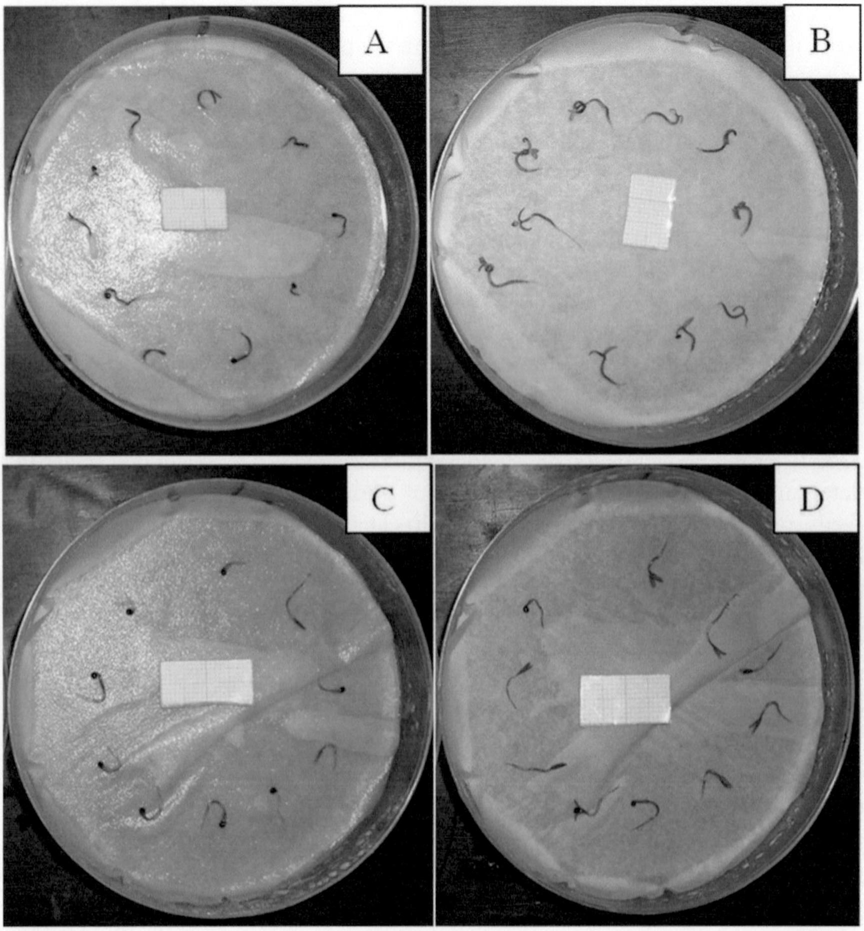

Fig. 1.2 *Chenopodium album* seedlings of control (**a**, **b**) and treatment of 100% concentration of *Eucalyptus camaldulensis* aqueous extract (**c**, **d**) after 3 and 14 days incubation (30 °C 12 h light/20 °C 8 h dark)

3 *In Vivo* Assays

There are few studies about the *in vivo* phytotoxic activity of natural products against weeds. One of the most important reasons is because of the lack of the natural products, which normally are laboratory obtained compounds without an adequate formulation for being applied in *in vivo* conditions, which is a limiting factor when planning assays. In the last years, groups working on this research area have made efforts to develop protocols for *in vivo* testing of natural products. In this section, some of these protocols will be described. Natural products can be tested in pre and post-emergence, and they can be applied by irrigation or spraying.

3.1 Pre-emergence In Vivo *Assays*

When testing natural products in pre-emergence against weeds there are two different possibilities: working with soil, and its seed bank, that has not been treated with herbicides, or working with peat substrate and sow specific weed seeds. Other important considerations are how, when and how often to supply the treatments. Assays with soil are better performed in trays than in pots, because soil in small pots can cause compaction problems, it is necessary a careful water management. It is recommendable to inventory the weeds that are growing in the field from which the soil is collected, and is imperative to correctly homogenize the soil collected before using it for the experiments.

In a greenhouse experiment, trays of 56.5 × 36.5 × 12 cm were filled with 5 cm perlite at the basis, and 5 cm soil from a citrus orchard where herbicides had not been applied for the last 5 years (Fig. 1.3). Three replications were prepared for each treatment. Treatments were applied in pre-emergence, and three aqueous extracts were tested: *Lantana camara*, *Eucalyptus camaldulensis* and *Eriocephalus africanus* at 100% dose. Controls were irrigated with water and the other trays were irrigated with the respective treatments (4 L per tray). Treatments were applied only once, and trays were irrigated once a week. Emerged plants were counted and

Fig. 1.3 Greenhouse *in vivo* experiments. Filling with perlite the basis of the trays (**a**), trays with soil before being treated (**b**), trays just after treatments application (**c**), trays 4 weeks after treatments application

classified, and the weed density was also calculated. Height, fresh and dry weight of the emerged plants was also registered. The experiment was finished when there were no significant differences in the number of emerged plants between control and treatments (Verdeguer et al. 2009b).

When using peat substrate, the experiments can be performed in seedbeds or pots. It is necessary take into account the same considerations about seeds size and number of repetitions that were exposed for *in vitro* bioassays. For example, in pots of 8 × 8 × 7 cm 20 small-sized weed seeds or 10 big-sized weed seeds can be placed.

3.2 *Post-emergence* In Vivo *Assays*

It is important to consider that treatments can be supplied by irrigation or spraying and their activity is influenced by the applying method (Fig. 1.4) (Graña et al. 2013; Díaz-Tielas et al. 2014; Castañeda 2017). As described in the previous section, assays can be performed with soil or with peat substrate. In assays with soil it is necessary to irrigate it to allow the germination of the seeds that are in the seed bank of the soil and the growth of the seedlings. In the assays with peat substrate, seeds are sown in pots or seedbeds, and when seedlings develop a thinning can be done if it is necessary. When the emerged plants reach the desired development stage, treatments can be applied by both methods, in order to determine which is the most effective.

The phenological stage in which the plant is treated is also an important factor to consider, because the activity of natural, and also synthetic products is very influenced by it. Previous assays could be performed with weeds in different stages to determine the best moment for application. For example, for *Arabidopsis thaliana*, treatments were applied when it had five fully expanded leaves (Graña et al. 2013; Araniti et al. 2017). For *E. crus-galli* it is recommended to apply herbicides for a best control when it has 2–3 true leaves and for *Conyza* species at the rosette stage. The treatments can be applied once (Verdeguer et al. 2009b; Castañeda 2017) or repeated times (Graña et al. 2013; Araniti et al. 2017). After the treatments are applied, phytotoxic effects can be evaluated. Mortality, observed damages (classified with a phytotoxicity scale), height, fresh and dry weight of the plants can be recorded to determine the herbicidal activity of the tested products.

Fig. 1.4 Greenhouse *in vivo* experiment. Treatments with *Eucalyptus camaldulensis* essential oil in post emergence applied by irrigating (**a**, **c**) and spraying (**b**, **d**), *Conyza bonariensis* controls (**a**, **b**) and treated plants (**c**, **d**)

References

Ainsworth EA, Bernacchi CJ, Dohleman FG (2016) Focus on ecophysiology. Plant Physiol 172:619–621

Angelini LG, Carpanese G, Cioni PL, Morelli I, Macchia M, Flamini G (2003) Essential oils from Mediterranean lamiaceae as weed germination inhibitors. J Agric Food Chem 51:6158–6164

Araniti F, Graña E, Reigosa MJ, Sánchez-Moreiras AM, Abenavoli MR (2013) Individual and joint activity of terpenoids, isolated from *Calamintha nepeta* extract, on *Arabidopsis thaliana*. Nat Prod Res 27:2297–2303

Araniti F, Sánchez-Moreiras AM, Graña E, Reigosa MJ, Abenavoli MR (2017) Terpenoid trans-caryophyllene inhibits weed germination and induces plant water status alteration and oxidative damage in adult arabidopsis. Plant Biol 19:79–89

Barnes JP, Putnam AR (1987) Role of benzoxazinones in allelopathy by rye (*Secale cereale* L.). J Chem Ecol 13:889–906

Barney JN, Hay AG, Weston LA (2005) Isolation and characterization of allelopathic volatiles from mugwort *Artemisia vulgaris*. J Chem Ecol 31:247–265

Benvenuti S, Cioni PL, Flamini G, Pardossi A (2017) Weeds for weed control: Asteraceae essential oils as natural herbicides. Weed Res 57:342–353

Castañeda LG (2017) Control de *Conyza bonariensis* con aceites esenciales de plantas mediterráneas *Thymbra capitata, Mentha x piperita, Santolina chamaecyparissus* y *Eucalyptus camaldulensis*. Trabajo Final de Máster. Universitat Politècnica de València

Chiapusio G, Sánchez AM, Reigosa MJ, González L, Pellissier F (1997) Do germination indices adequately reflect allelochemical effects on the germination process? J Chem Ecol 23:2445–2453

Chon SU, Nelson CJ (2001) Effects of experimental procedures and conditions on bioassay sensitivity of alfalfa autotoxicity. Commun Soil Sci Plant Anal 32:1607–1619

Dayan F, Duke S (2006) Clues in the search for new herbicides. In: Reigosa M, Pedrol N, González L (eds) Allelopathy. Springer, Dordrecht, pp 63–83

Deng J, Zhang Y, Hu J, Jiao J, Hu F, Li H, Zhang S (2017) Autotoxicity of phthalate esters in tobacco root exudates: effects on seed germination and seedling growth. Pedosphere 27:1073–1082

Díaz-Tielas C, Sotelo T, Graña E, Reigosa MJ, Sánchez-Moreiras AM (2014) Phytotoxic potential of trans-chalcone on crop plants and model species. J Plant Growth Regul 33:181–194

Dudai N, Poljakoff-Mayber A, Mayer AM, Putievsky E, Lerner HR (1999) Essential oils as allelochemicals and their potential use as bioherbicides. J Chem Ecol 25:1079–1089

Einhellig FA (1995) Allelopathy: current status and future goals. In: Inderjit, Dakshini KMM, Einhellig FA (eds) Allelopathy: organisms, processes and applications. American Chemical Society, Washington, DC, pp 1–24

Graña E, Sotelo T, Díaz-Tielas C, Reigosa M, Sánchez-Moreiras AM (2013) The phytotoxic potential of the terpenoid citral on seedlings and adult plants. Weed Sci 61:469–481

Herranz JM, Ferrandis P, Copete MA, Duro EM, Zalacain A (2006) Effect of allelopathic compounds produced by *Cistus ladanifer* on germination of 20 Mediterranean taxa. Plant Ecol 184:259–272

Inderjit, Duke SO (2003) Ecophysiological aspects of allelopathy. Planta 217:529–539

Inderjit, Keating KI (1999) Allelopathy: principles, procedures, processes, and promises for biological control. Adv Agron 67:141–231

Inderjit, Streibig J, Olofsdotter M (2002) Joint action of phenolic acid mixtures and its significance in allelopathy research. Physiol Plant 114:422–428

Koroch AR, Rodolfo Juliani H, Zygadlo JA (2007) Bioactivity of essential oils and their components. In: Berger RG (ed) Flavours and fragrances: chemistry, bioprocessing and sustainability. Springer, Berlin/Heidelberg, pp 87–115

Lambers H, Pons TL, Chapin FS (2008) Plant physiological ecology. Springer, Berlin/Heidelberg

Leather GR, Einhellig FA (1988) Bioassay of naturally occurring allelochemicals for toxicity. J Chem Ecol 14:1821–1828

Lotina-Hensen B, King-Díaz B, Aguilar MI, Hernández Terrones MG (2006) Plant secondary metabolites. Targets and mechanisms of allelopathy. In: Reigosa MJ, Pedrol N, González L (eds) Allelopathy: a physiological process with ecological implications. Springer, Dordrecht, pp 229–265

Lydon J, Teasdale JR, Chen PK (1997) Allelopathic activity of annual wormwood (*Artemisia annua*) and the role of artemisinin. Weed Sci 45:807–811

Macías FA, Molinillo JMG, Galindo JCG, Varela RM, Simonet AM, Castellano D (2001) The use of allelopathic studies in the search for natural herbicides. J Crop Prod 4:237–255

Macías FA, Oliveros-Bastidas A, Marin D, Carrera C, Chinchilla N, Molinillo JMG (2008) Plant biocommunicators: their phytotoxicity, degradation studies and potential use as herbicide models. Phytochem Rev 7:179–194

Pardos JA (2005) Ecophysiology, a meeting point between function and management of forest ecosystems. Investigación Agraria: Sistemas y Recursos Forestales 14:277–291

Pederson GA (1986) White clover seed germination in agar containing tall fescue leaf extracts. Crop Sci 26:1248–1249

Pennacchio M, Jefferson L, Havens K (2005) *Arabidopsis thaliana*: a new test species for phytotoxic bioassays. J Chem Ecol 31:1877–1885

Reigosa MJ, Malvido-Pazos E (2007) Phytotoxic effects of 21 plant secondary metabolites on *Arabidopsis thaliana* germination and root growth. J Chem Ecol 33:1456–1466

Reigosa MJ, Souto XC, González L (1999) Effect of phenolic compounds on the germination of six weeds species. Plant Growth Regul 28:83–89

Verdeguer M (2011) Fitotoxicidad de aceites esenciales y extractos acuosos de plantas. mediterráneas para el control de arvenses. Tesis doctoral, Universitat Politècnica de València

Verdeguer M, Blazquez MA, Boira H (2009a) Phytotoxic effects of *Lantana camara*, *Eucalyptus camaldulensis* and *Eriocephalus africanus* essential oils in weeds of Mediterranean summer crops. Biochem Syst Ecol 37:362–369

Verdeguer M, Garcia D, Blázquez MA, Boira H (2009b) Potencial alelopático de extractos acuosos de *Lantana camara, Eucalyptus camaldulensis* y *Eriocephalus africanus* y posible uso como herbicidas naturales. In: Herbologia e Biodiversidade numa Agricultura Sustentável ,vol 1, pp 403–406

Warrag MOA (1995) Autotoxic potential of foliage on seed germination and early growth of mesquite (*Prosopis juliflora*). J Arid Environ 31:415–421

Weidenhamer JD (1996) Distinguishing resource competition and chemical interference: overcoming the methodological impasse. Agron J 88:866–875

Weidenhamer JD, Morton TC, Romeo JT (1987) Solution volume and seed number: often overlooked factors in allelopathic bioassay. J Chem Ecol 13:1481–1491

Whittaker RH, Feeny PP (1971) Allelochemics: chemical interactions between species. Science 171:757–770

Chapter 2
Root Morphology

Antonio Lupini, Fabrizio Araniti, Antonio Mauceri, Maria Princi, Antonino Di Iorio, Agostino Sorgonà, and Maria Rosa Abenavoli

1 Introduction

Root system defined as the "*Hidden Half*" of plant, has not attracted a great deal of attention for a long time from plant biologists. In recent years, through the new innovative techniques, root system has been deeply studied allowing all to reveal its structure, function, but also its genetic potential, which could be manipulated to improve crop yield and plant survival in stressful environments. Plant root system has three major functions: site of water and nutrients acquisition from the soil, essential support for plant anchoring and sensor of abiotic and biotic stresses. It also points out secondary functions such as photoassimilates storage, phytohormones synthesis and clonal propagation.

Root system growth and development depend on endogenous genetic program as well as biotic and abiotic environmental factors (Forde and Lorenzo 2001). For example, nutrient status of the plant and/or photosynthate supplies (intrinsic factors) and nutrients distribution in the soil and/or soil texture and/or water potential and/or microorganisms and/or salinity (extrinsic factors) highly modified root system (Walch-Liu et al. 2006). This considerable root plasticity provides to plants to maximize both reproductive success and survival, especially in severe environments.

Root system of monocots is more complex than that of dicotyledoneous plants. Monocots display a "fibrous root system", with several moderately branching roots growing from the stem. In particular, fibrous root system of monocots is characterized by having a mass of similarly sized roots, consisting of a primary root

A. Lupini · F. Araniti · A. Mauceri · M. Princi · A. Sorgonà · M. R. Abenavoli (✉)
Department Agraria, Mediterranea University, Reggio Calabria, Italy
e-mail: mrabenavoli@unirc.it

A. Di Iorio
Department of Biotechnology and Life Science, University of Insubria, Varese, Italy

© Springer International Publishing AG, part of Springer Nature 2018

A. M. Sánchez-Moreiras, M. J. Reigosa (eds.), *Advances in Plant Ecophysiology Techniques*, https://doi.org/10.1007/978-3-319-93233-0_2

emerging from embryo, a variable number of seminal roots that originate post-embryonically and crown roots emerging from the stem. All major root types form lateral roots and undergo higher-order branching (Hochholdinger et al. 2004; Nibau et al. 2008). In cereals such as maize, rice and wheat, lateral roots arise from the phloem pole pericycle with a contribution of endodermis (Hochholdinger and Zimmermann 2008). In contrast, dicots have a "taproot system", which is characterized by having a primary root (the taproot) emerging from the embryo and growing downward. As roots mature, quiescent cells within pericycle begin competent to sustain a set of divisions to form lateral root primordia (Malamy 2009). Lateral roots elongate and are further undergone to reiterative branching to form secondary and tertiary laterals. Ultimately, adventitious roots at the shoot-root junction can be formed (Hochholdinger et al. 2004; Nibau et al. 2008). Newly emerged roots are highly sensitive to environmental signals such as gravity, which drives root growth towards moisture and nutrients in the soil (Muday and Rahman 2007).

Fibrous roots are excellent for soil erosion control because the mass of roots cling to soil particles, while taproots anchor plants deeply helping to prevent wind damages and stabilize plants especially in beach and sand dune areas. Taproots are often modified to store starches and sugars as shown in sweet potatoes, sugar beet or carrot. Taproot systems are also important adaptations for searching for water from deeper sources, conferring great drought resistance compared to fibrous ones. Whereas, fibrous roots near the soil surface, may absorb quickly water also from even a light rainfall.

1.1 Root System Analysis

Root system analysis can be approached in terms of structure, morphology and architecture. Root structure takes in account all anatomical aspects of the individual roots; while root morphology deals with the surface traits of the individual root axis, such as length, biomass, diameter, surface area, root specific length, fineness and root density. It also includes epidermal features such as root hairs, root headset, root axis ripple, and cortical senescence. Finally, root architecture defines the spatial configuration of root system in the soil (Lynch 1995). Next paragraphs deal with root morphology and architecture whose analysis is carried out by WinRhizo system, the image analysis system described in this chapter.

1.2 Root Morphology and Its Parameters

Root morphology was described through the estimation of several parameters, which allow better understanding of root functional aspects, especially those related to resources acquisition from soil. Root length (LR) is a morphological parameter that best describes root ability to explore the soil (Ryser 1998) and to acquire water and nutrients. In particular, LR is positively correlated with ions absorption with low mobility such as phosphate (Newman and Andrews 1973). Change in root length (RL) is determined by two morphological components: root biomass (RB, g) and specific root length i.e. root length per unit root dry weight (SRL, cm g^{-1}) as described by the following relationship:

$$RL = RB^{*}SRL\left(cm\right)$$

To better emphasize the importance of root length in resources acquisition, it has been proposed the root length ratio (RLR) parameter, i.e. root length per unit of the plant's dry biomass. The RLR defines "how much plant biomass is transformed into root length", or better, "how much plant invests in root system". Higher RLR value allows the plant a greater soil exploration and, consequently, an increased resources acquisition. Therefore, this parameter is a good index to describe the plant's ability to acquire soil resources. For this reason, it is strongly modulated by changes in nutrient and water availability: it increases under lower nitrate concentrations (Ryser and Lambers 1995; Sorgonà et al. 2005) and drought stress in drought-tolerant bean landrace (Abenavoli et al. 2016).

Change in RLR is determined by different morphological components: root mass ratio i.e. relative biomass allocation to root per unit of plant's dry biomass (RMR, g g^{-1}) and specific root length, i.e. root length per unit root dry weight, (SRL, cm g^{-1}), which, in turn, depends on root fineness, i.e. root length per unit root volume (RF, cm cm^{-3}) and tissue density, i.e. root dry mass per unit root volume (RDT, g cm^{-3}) (Ryser 1998). The relationships among these parameters are the following:

$$RLR = RL / PB = RMR^{*}SRL\left(cm\,g^{-1}\right)$$

$$RMR = RB / PB\left(g\,g^{-1}\right)$$

$$SRL = RL / RB = RF / RDT\left(cm\,g^{-1}\right)$$

$$RF = RL / RV\left(cm\,cm^{-3}\right)$$

$$RTD = RB / RV\left(g\,cm^{-3}\right)$$

Therefore, plants may produce longer roots either by increasing biomass allocation to roots, as demonstrated under a low supply of nitrogen (Ryser and Lambers 1995; Sorgonà et al. 2005, 2007a) or by increasing root fineness and/or reducing root tissue density, leaving biomass allocation unchanged (Ryser 1998).

In particular, RMR parameter is a relative index, which indicates the biomass allocation towards root that is modified by environmental conditions such as nutrient availability (Ryser and Lambers 1995).

On the other hand, the SRL is a root parameter related to the ability of species or genotypes to effectively spend their photosyntates for the limiting resources acquisition. Then, the SRL is a root trait, which better explains the root system investment in terms of costs (biomass) and benefits (length). Therefore, plants with high SRL achieve the same purpose in terms of elongation but with lower costs (less biomass) than those with low SRL, showing more efficiency in mineral resources acquisition. However, SRL is a "complex parameter" determined both by the root fineness (FR) and the root total density (DTR), which can often nullify the SRL variation in response to environmental conditions (Ryser 1998).

In particular, the fineness (FR) is functionally correlated with the adverse environmental conditions and, in particular, with the habitats. On the other hand, root tissue density (RTD) is the other component of SRL, whose modulation can increase the plant's ability to capture soil resources (high length or root length ratio) without changing biomass allocation or root fineness. For example, higher RLR was maintained by lower root tissue density at different N and P (Ryser and Lambers 1995) and nitrate levels (Sorgonà et al. 2007a). Furthermore, this parameter appears to be negatively correlated with growth rate of the plant species (Wahl and Ryser 2000).

1.3 Root System Architecture

Root system architecture (RSA) defines the spatial arrangement of root axes in soil (Lynch 1995) and varies hugely among species and within species, also matching to different environment where plants grown (Loudet et al. 2005; Osmont et al. 2007). The importance of root architecture relies on its closely relationship with plant yield performances (Lynch 1995) and nutrient acquisition efficiency (Sorgonà et al. 2005, 2007b).

Different techniques are developed for studying root architecture both in lab and field conditions. Several phenotyping systems use 2D imaging captured by cameras (Le Marie et al. 2014) or flatbed scanners (Adu et al. 2014), following 3-D root architecture analysis by topological (Fitter 1991; Sorgonà et al. 2007b), fractal (Eghball et al. 1993; Lynch and Van Beem 1993; Berntson 1994) and computer simulation analysis (Lynch and Nielsen 1996). Further, 3D root architecture has been captured in plants grown on gel by 3D laser scanners (Fang et al. 2009) and digital cameras (Iyer-Pascuzzi et al. 2010), or in pot filled with soil by X-ray computed tomography (Hargreaves et al. 2009; Tracy et al. 2010) and magnetic resonance imaging (Jahnke et al. 2009) and in field by magnetic resonance imaging (Zhou and Luo 2009), magnetic resonance imaging-positron emission tomography (Jahnke et al. 2009), ground-penetrating radar (Hruska et al. 1999; Hagrey 2007), manual (Oppelt et al. 2000) and digital measurements (Danjon et al. 1999; Di Iorio et al. 2005).

Next paragraphs detail on topological and fractal analysis of root architecture by 2-D images captured and analyzed with WinRhizo system.

1.4 Topological Analysis

The topology, which is a non-metric aspect of root architecture, refers to the hierarchical order by which the individual axes are connected to each other in determining the degree or the mode of branching of roots. The topology is analyzed by quantification of different non-metric parameters such as the links, the magnitude (μ) of the entire root system, the altitude (a) and the total exterior pathlength (Pe) (Fitter 1996). The links define root segments between two nodes or between a vertex and a node. The links are divided into those with meristem end, exterior links, and those that combine two branching points, internal links (II). In turn, the exterior links that connect with the interior links are defined exterior-interior links (EI), while if they bind with other exterior links are called exterior-exterior links (EE) (Fig. 2.1).

The altitude is the number of links in the longest path through the system starting from the base; while the Pe is the sum of all the routes from the base toward the exterior links (Fig. 2.2a). The slope of the linear curve obtained by plotting the values of a (Loga) or Pe (LogPe) as a function of magnitude (Logμ) is used as topological index (TI) (Fig. 2.2b). High slope values indicate a herringbone-type branching pattern of root system, while low slope values are typical of the dichotomous model (Fig. 2.2b).

Through the computer simulation, Fitter (1991) pointed out that "the exploitation efficiency" (soil volume within the depletion areas divided the root volume) was positively correlated with TI. In particular, herringbone-type root systems are more

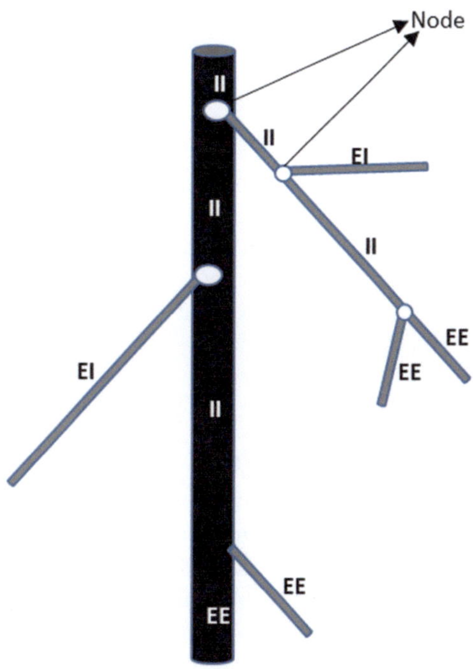

Fig. 2.1 Root system represented by topological model. EE: exterior-exterior link; EI: exterior-interior link; II: interior-interior link

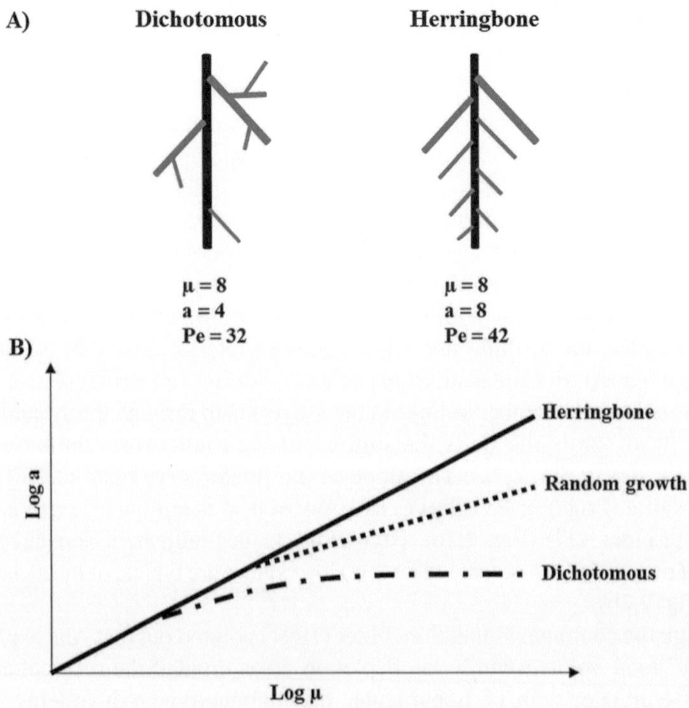

Fig. 2.2 (**a**) Dichotomous- and herringbone-type root systems with the corresponding topological parameter values: μ, magnitude; a, altitude; Pe, total exterior pathlength. (**b**) Loga vs Logμ plot where the herringbone- and dichotomous-type exhibit the lowest and highest slope values, respectively

efficient in exploring and exploiting the edaphic resources while the dichotomous pattern is more efficient in nutrient transport within roots and less expensive for its large number of high magnitude links.

To support the simulation results, several experiments pointed out that root topology exhibited dichotomous traits as mineral nutrients increase in the soil and/or nutrient solution (Fitter and Stickland 1992; Sorgonà and Cacco 2002; Sorgonà et al. 2005).

1.5 Fractal Analysis

Fractal geometry is an interesting geometrical system for the description of very complex and modular natural objects. Root system exhibits a modular growth. Hence, the fractal geometry, by scaling relations, is a promising technique to analyze and quantify the complexity of root architecture both from the structural point of view that functional as, for example, the root system ability to occupy the soil volume.

Fractal analysis for root architecture utilizes box counting method and the eq. $N(L) = K_L^{-D}$ where L is the length of the box side and N(L) is the number of boxes of size L needed to cover the root (Tatsumi 2001; Walk et al. 2004; Eshel 1998). By linear regression of N(L) *vs* L levels are computed the slope (D) and intercept log K. D is the fractal dimension (FD) and log K is associated with fractal abundance (FA). The FD is closely related with root length and biomass (Fitter and Stickland 1992; Lynch and Van Beem 1993), root size (Eghball et al. 1993; Berntson 1994; Lynch and van Beem 1993), root length density (Berntson 1994), root topology (Fitter and Stickland 1992), root length, surface area and branching frequency (Costa et al. 2003), nutrient availability (Eghball et al. 1993), drought tolerance (Wang et al. 2009) and ecophysiological strategies of fruit crops (Oppelt et al. 2000).

2 The WinRhizo System

The WinRhizo system includes:

- a calibrated color optical scanner to acquire images;
- a positioning system to ensure a good arrangement of the root before starting scanner acquisition;
- a work station which allows to store and analyze image by software after acquisition.

The system should use a specific flat-bed scanner as image acquisition devise (Fig. 2.3a), which is characterized by a lighting system placed on both scanner glass parts (top and bottom). Since the system has to acquire macroscopic objects (root and leaf), an optical scanner allows obtaining an image with high resolution (Arsenault et al. 1995).

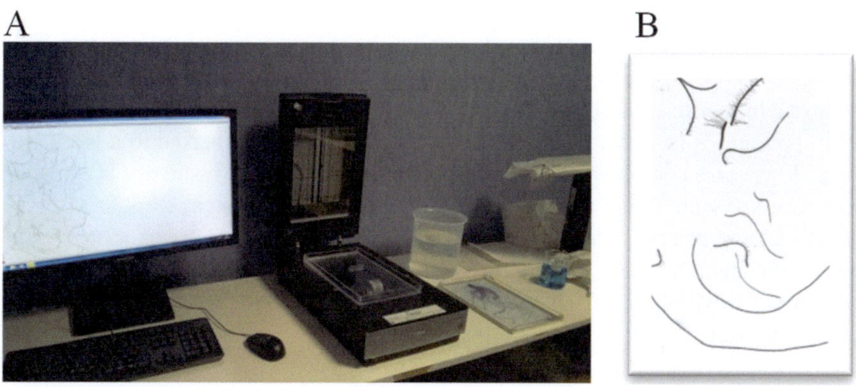

Fig. 2.3 (**a**) Scanning system coupled to Scanner used in lab to acquire root images; (**b**) maize root types acquired (8 days old)

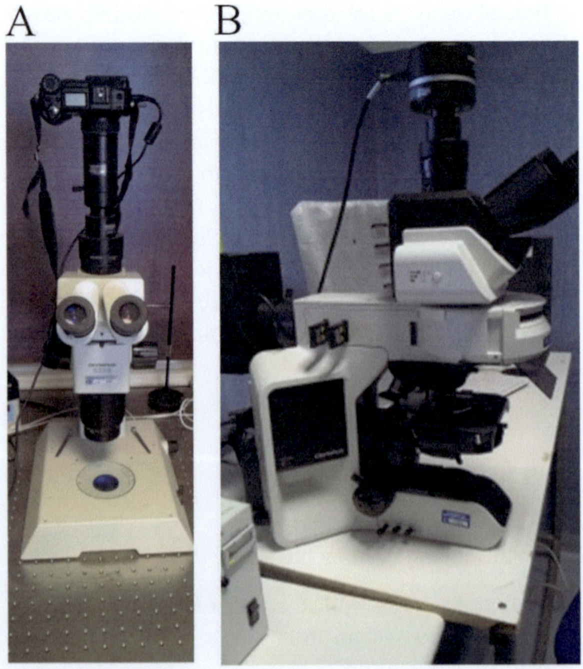

Fig. 2.4 (**a**) Camera connected to a stereomicroscopy; (**b**) microscopy

In addition, excellent optical, electrical and mechanical features are needed to provide a lighting uniformity over entire area thereby avoiding further positions, orientation and light adjustments. For optimizing root image acquisition and analysis, an optical resolution (dots per inch, DPI) at last of 4800 with 0.005 pixel size, fast speed, 16/48 bit per pixel is necessary. Moreover, scanners include a standard calibration (Régent Instruments Inc.) to normalize the data obtained from different model or root position (orientation, lighting, method of acquisition etc.).

On the other hand, the scanning system does not allow to obtain detailed information concerning root tips (e.g. cell organization of root apical meristem, root hair density etc.). Therefore, a camera connected to a stereomicroscope should be useful to have a higher resolution (Fig. 2.4).

In both cases (scanner or camera) these hardware are connected to a work station. However, a computer configuration to image analysis software needs a powerful processor with a high RAM (over 2 GB), an operative system work at 32 or 64 bit and a large screen (at least 24″). The analysis can be carried out on root directly placed on scanning glass or adopting a root position system, which consists in a waterproof tray (Fig. 2.5). This system makes easy the root spatial disposition on the scanner surface improving the image quality.

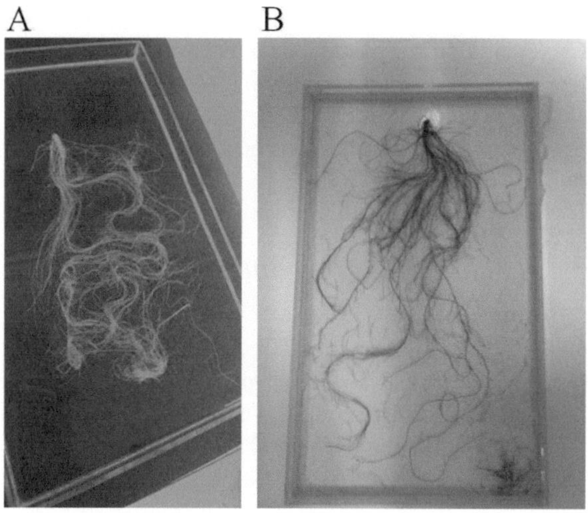

Fig. 2.5 Tomato roots grown in hydroponic system, disposed in water-proof tray, before (**a**) and after (**b**) toluidine staining

3 Image Acquisition

3.1 Root Collection and Stocking

Since there is a huge difference between roots collected from soil and those hydroponically grown, this should be considered during the experimental design. Hydroponic systems or plants cropped on sandy soil can greatly simplify this step, while plant collection from soils rich in clay and/or organic matter can make it extremely difficult and it could increase the loss of roots (especially thinner roots).

WinRhizo allows the automatic acquisition of root parameters. Roots grown in hydroponic systems can be directly acquired, conversely, those collected from soil system need to be accurately washed before image acquisition.

However, root system could be also stocked for a long time without any morphological changes, in a preserving solution, also known as FAA, containing ethanol (96%) (656.25 mL L^{-1}): formaldehyde (37%) (50 mL L^{-1}): glacial acetic acid (99.85%) (50 mL L^{-1}). This stoking method could be useful in experimental setups, with a high number of samples or huge root apparatus. In fact, large roots need to be divided in smaller portions that will be analyzed separately.

3.2 Root Staining

During image acquisition process, roots are digitized through a scanner. To obtain a detailed image of roots, especially the finest ones, could be necessary to increase the contrast of tissue, through a previous staining. It improves both the accuracy of the analysis and the determination of root diameters. Roots are soaked for 5–10 min (depending on root diameter) in 0.1% Toluidine blue solution. After staining, roots have to be washed thoroughly in order to remove the excessive dye (Fig. 2.5). Care should be taken during long dye exposition since it could bring to the production of dye lumps around the root surface, invalidating the analysis.

3.3 Root Positioning and Image Acquisition

Once stained, roots can be directly placed on the water-proof trays filled with distilled water (Fig. 2.5b). Root positioning is helped using tweezers and very fine plastic sticks as tools. Roots should be completely immersed in distilled water, thereby avoiding the presence of shadows as well.

Furthermore, roots should be positioned randomly avoiding overlaps, especially when the analysis aimed to quantify crosses, forks and tips numbers. For roots with large dimensions, the positioning in the trays could be difficult or not possible. Therefore, as before indicated, roots have to be separated in smaller portions (Fig. 2.3b), or alternatively, a representative sample of whole root could be scanned, before determining dry mass. Then, the SRL (Specific Root Length) of sample can be used to estimate the total root length.

All root adjustments should be done on trays directly placed on scanner glass to avoid shifts during acquisition.

As previously introduced, WinRhizo is coupled to a calibrated scanner, which lights roots from above and below while being scanned further reducing shadows on root image. The first step in root acquisition consists in creating a preview of whole root that should be analyzed for improving the final acquisition quality of image in terms of brightness, contrast, exposition, etc. A pivotal parameter that should be considered during image acquisition is the final resolution, which depends on type of samples. A low resolution reduces excessively the image quality causing loss of precision in measurement. For example, *Zea mays* roots may be scanned at 200–400 dpi, while tomato roots should be scanned at 500–600 dpi.

Moreover, in order to reduce the file size, root acquisition could be done in gray-scale images. This option is suggested only for root length measurements and not for other parameters.

At the end of scanner acquisition, images are saved and stored in *.tiff* or *.jpg* format, allowing to acquire a huge number of samples and to carry out the analysis later.

4 Image Analysis

Image analysis starts immediately after root acquisition. One of the most important steps is the choice of the right threshold value (Fig. 2.6). During image analysis, the software uses thresholding to discriminate between root and image aberrations. Based on grayscale values, each pixel is classified as either root or artifact; that's the reason why any shadows should be avoided. Although the software is provided by macros that automatically set the adequate parameters, it is advised to manually set them each time.

Successively, root artifacts, debris or unattractive image regions should be excluded using the dedicated function. Then, the user selects the region/s of interest to run the analysis. Once started, the software collect all the desired parameters at once, then the results are displayed on the monitor and roots are skeletonized by different color traits, which represent the classes of the parameters selected by the user (e.g. diameter, area, volume etc.) (Fig. 2.7). At the end of the analysis the software directly save all the data as *.txt* file, which could be opened in standard spreadsheet statistic programs for further data meaning interpretation. In Table 2.1 is reported an example of data output.

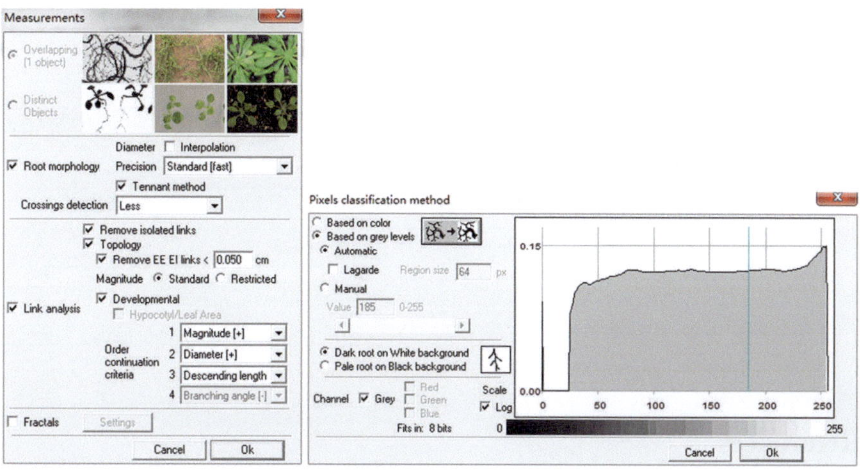

Fig. 2.6 Parameter setting in WinRhizo software

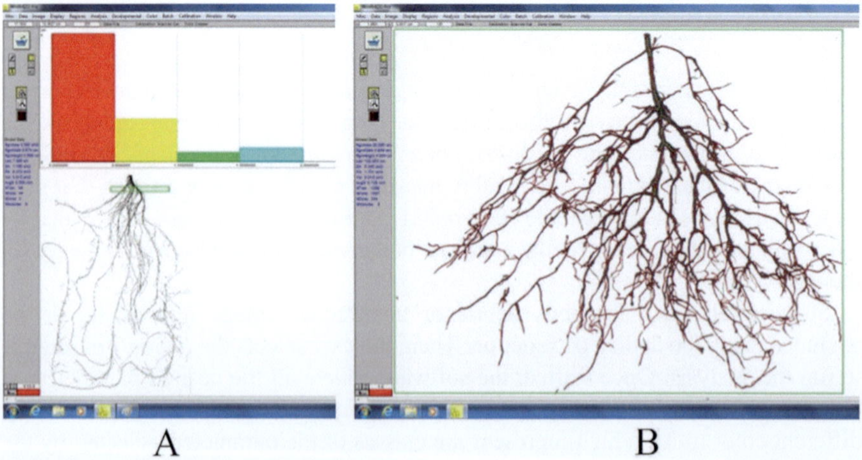

Fig. 2.7 The images reported are from the root image previously shown, but are a selection of a smaller region for clarity. Color traces indicate where roots have been detected. (**a**) Portion of root analyzed (green rectangle on root image) and divided in diameter classes (colored bars); (**b**) root skeletonized after WinRhizo analysis

Table 2.1 The excel files output from .txt generated by WinRhizo software

		Length(cm)	ProjArea(cm^2)	SurfArea(cm^2)	AvgDiam(mm)	LenPerVol(cm/m^3)	RootVolume(cm^3)	Tips	Forks	Crossing
SampleID	COLOR									
SampleID	LINK									
SampleID	DEV									
SampleID	AXIX									
		4.572	2.34	7.36	0.5506	42.47	0.101	62	26	1
Root Length by Diameter Class										
0.0<L<0.10	0.1<L<0.20	0.2<L<0.30	0.30<L<0.40	0.40<L<0.50	0.50<L<0.60	0.60<L<0.70	0.70<L<0.80	0.80<L<0.90	0.90<L<1.0	1.0<L<1.10
0.9370	0.3684	1.08976	10.4566	19.6748	1.7872	0.6394	0.7377	1.2343	0.9545	0.213
Surface Area by Diameter Class										
0.0<SA<0.10	0.1<SA<0.20	0.2<SA<0.30	0.30<SA<0.40	0.40<SA<0.50	0.50<SA<0.60	0.60<SA<0.70	0.70<SA<0.80	0.80<SA<0.90	0.90<SA<1.0	1.0<SA<1.10
0.0212	0.0167	0.0962	1.1866	2.678	0.2908	0.1308	0.4572	0.1986	0.3635	0.3452
Projected Area by Diameter Class										
0.0<PA<0.10	0.1<PA<0.20	0.2<PA<0.30	0.30<PA<0.40	0.40<PA<0.50	0.50<PA<0.60	0.60<PA<0.70	0.70<PA<0.80	0.80<PA<0.90	0.90<PA<1.0	1.0<PA<1.10
0.0067	0.0053	0.0308	0.3775	0.8532	0.0932	0.0416	0.1455	0.0602	0.1145	0.0953
Root Volume by Diameter Class										
0.0<V<0.10	0.1<V<0.20	0.2<V<0.30	0.30<V<0.40	0.40<V<0.50	0.50<V<0.60	0.60<V<0.70	0.70<V<0.80	0.80<V<0.90	0.90<V<1.0	1.0<V<1.10
0	0.0001	0.0007	0.0291	0.0018	0.0021	0.0086	0.0043	0.0084	0.001	0.0006

References

Abenavoli MR, Leone M, Sunseri F, Bacchi M, Sorgonà A (2016) Root phenotyping for drought tolerance in bean landraces from Calabria (Italy). J Agric Crop Sci 202:1–12

Adu MO, Chatot A, Wiesel L, Bennett MJ, Broadley MR, White PJ, Dupuy LX (2014) A scanner system for high-resolution quantification of variation in root growth dynamics of *Brassica rapa* genotypes. J Exp Bot 65:2039–2048

Arsenault JL, Pouleur S, Messier C, Guay R (1995) WinRhizo, a root measuring system with a unique overlap correction method. Hortic Sci 30:906

Berntson GM (1994) Root systems and fractals: how reliable are calculations of fractal dimension? Ann Bot 73:281–284

Costa C, Dwyer LM, Dutilleul P, Foroutan-pour K, Liu A, Hamel C, Smith DL (2003) Morphology and fractal dimension of root systems of maize hybrids bearing the leafy trait. Can J Bot 81:706–713
Danjon F, Sinoquet H, Godin C, Colin F, Drexhage M (1999) Characterisation of structural tree root architecture using 3D digitising and AMAPmod software. Plant Soil 211:241–258
Di Iorio A, Lasserre B, Scippa GS, Chiatante D (2005) Root system architecture of *Quercus pubescens* trees growing on different sloping conditions. Ann Bot 95:351–361
Eghball B, Settimi JR, Maranville JW, Parkhurst AM (1993) Fractal analysis for morphological description of corn roots under nitrogen stress. Agron J 85:287–289
Eshel A (1998) On the fractal dimensions of a root system. Plant Cell Environ 21:247–251
Fang S, Yan X, Liao H (2009) 3D reconstruction and dynamic modeling of root architecture in situ and its application to crop phosphorus research. Plant J 60:1096–1108
Fitter AH (1991) The ecological significance of root system architecture: an economic approach. In: Atkinson D (ed) Plant root growth: an ecological perspective. Blackwell Scientific, Oxford, pp 229–243
Fitter AH (1996) Characteristics and functions of root system. In: Waisel Y, Eshel A, Kafkafi U (eds) Plant roots, the hidden half. Marcel Dekker, New York, pp 1–20
Fitter AH, Stickland TR (1992) Fractal characterization of root system architecture. Funct Ecol 6:632–635
Forde BG, Lorenzo H (2001) The nutritional control of root development. Plant Soil 232:51–68
Hagrey SA (2007) Geophysical imaging of root-zone, trunk, and moisture heterogeneity. J Exp Bot 58:839–854
Hargreaves C, Gregory P, Bengough A (2009) Measuring root traits in barley (*Hordeum vulgare* ssp. *vulgare* and ssp. *spontaneum*) seedlings using gel chambers, soil sacs and X-ray microtomography. Plant Soil 316:285–297
Hochholdinger F, Zimmermann R (2008) Conserved and diverse mechanisms in root development. Curr Opin Plant Biol 11:70–74
Hochholdinger F, Woll K, Sauer M, Dembinsky D (2004) Genetic dissection of root formation in maize (*Zea mays* L.) reveals root-type specific developmental programmes. Ann Bot 93:359–369
Hruska J, Cermak J, Sustek S (1999) Mapping of tree root systems by means of the ground penetrating radar. Tree Physiol 19:125–130
Iyer-Pascuzzi AS, Symonova O, Mileyko Y, Yueling H, Belcher H, Harer J, Weitz JS, Benfey PN (2010) Imaging and analysis platform for automatic phenotyping and trait ranking of plant root systems. Plant Physiol 152:1148–1157
Jahnke S, Menzel MI, Van Dusschoten D, Roeb GW, Buhler J, Minwuyelet S, Blumler P, Temperton VM, Hombach T, Streun M (2009) Combined MRI–PET dissects dynamic changes in plant structures and functions. Plant J 59:634–644
Le Marie C, Kirchgessner N, Marschall D, Walter A, Hund A (2014) Rhizoslides: paper-based growth system for non-destructive, high throughput phenotyping of root development by means of image analysis. Plant Methods 10:10–13
Loudet O, Gaudon V, Trubuil A, Daniel-Vedele F (2005) Quantitative trait loci controlling root growth and architecture in *Arabidopsis thaliana* confirmed by heterogeneous inbred family. Theor Appl Genet 110:742–753
Lynch JP (1995) Root architecture and plant productivity. Plant Physiol 109:7–13
Lynch JP, Nielsen KL (1996) Simulation of root system architecture. In: Waisel Y, Eshel A, Kafkafi U (eds) Plant roots, the hidden half. Marcel Dekker, New York, pp 247–257
Lynch JP, van Beem JJ (1993) Growth and architecture of seedling root of common bean genotypes. Crop Sci 33:1253–1257
Malamy JE (2009) Lateral root formation. In: Beeckman T (ed) Root development. Wiley Online Library, pp 83–126
Muday GK, Rahman A (2007) Auxin transport and the integration of gravitropic growth. In: Gilroy S, Masson P (eds) Plant tropisms. Blackwell Publishing, Oxford, pp 47–78

Newman EI, Andrews RE (1973) Uptake of phosphorus and potassium in relation to root growth and root density. Plant Soil 38:49–69

Nibau C, Gibbs D, Coates J (2008) Branching out in new directions: the control of root architecture by lateral root formation. New Phytol 179:595–614

Oppelt AL, Kurth W, Dzierzon H, Jentschke G, Godbold DL (2000) Structure and fractal dimensions of root systems of four co-occurring fruit tree species from Botswana. Ann For Sci 57:463–475

Osmont KS, Sibout R, Hardtke CS (2007) Hidden branches: developments in root system architecture. Annu Rev Plant Biol 58:93–113

Ryser P (1998) Intra- and interspecific variation in root length, root turn-over and the underlying parameters. In: Lambers H, Poorter H, MMI VV (eds) Inherent variation in plant growth. Physiological mechanism and ecological consequences. Backhuys, Leiden, pp 441–465

Ryser P, Lambers H (1995) Root and leaf attributes accounting for the performance of fast- and slow-growing grasses at different nutrient supply. Plant Soil 170:251–265

Sorgonà A, Cacco G (2002) Linking the physiological parameters of nitrate uptake with root morphology and topology in wheat (*Triticum durum* Desf.) and in citrus rootstock (*Citrus volkameriana* Ten & Pasq). Can J Bot 80:494–503

Sorgonà A, Abenavoli MR, Cacco G (2005) A comparative study between two citrus rootstocks: effect of nitrate on the root morpho-topology and net nitrate uptake. Plant Soil 270:257–267

Sorgonà A, Abenavoli MR, Gringeri PG, Cacco G (2007a) Comparing morphological plasticity of root orders in slow- and fast-growing citrus rootstocks supplied with different nitrate levels. Ann Bot 100:1287–1296

Sorgonà A, Abenavoli MR, Gringeri PG, Lupini A, Cacco G (2007b) Root architecture plasticity of citrus rootstocks in response to nitrate availability. J Plant Nutr 30:1921–1932

Tatsumi J (2001) Fractal geometry of root system morphology: fractal dimension and multifractals. In: Proceedings 6th symposium international society root research, Nagoya, Japan, pp 24–25

Tracy SR, Roberts JA, Black CR, McNeill A, Davidson R, Mooney SJ (2010) The X-factor: visualizing undisturbed root architecture in soils using X-ray computed tomography. J Exp Bot 61:311–313

Wahl S, Ryser P (2000) Root tissue structure is linked to ecological strategies of grasses. New Phytol 148:459–471

Walch-Liu P, Ivanov II, Filleur S, Gan Y, Remans T, Forde BG (2006) Nitrogen regulation of root branching. Ann Bot 97:875–881

Walk TC, Van Erp E, Lynch JP (2004) Modelling applicability of fractal analysis to efficiency of soil exploration by roots. Ann Bot 94:119–128

Wang H, Siopongco J, Wade LJ, Yamauchi A (2009) Fractal analysis on root systems of rice plants in response to drought stress. Environ Exp Bot 65:338–344

Zhou XC, Luo XW (2009) Advances in non-destructive measurement and 3D visualization methods for plant root based on machine vision. In: Proceedings of the 2nd international conference on biomedical engineering and informatics. Tianjin, BMEI'09, pp 1–5

Chapter 3
Plant Photosynthetic Pigments: Methods and Tricks for Correct Quantification and Identification

Beatriz Fernández-Marín, José Ignacio García-Plazaola, Antonio Hernández, and Raquel Esteban

1 Introduction to Photosynthetic Pigments

1.1 Leaf Composition of Photosynthetic Pigments

Chloroplast of green photosynthetic tissues in the Viridiplantae (monophyletic group that includes green algae and terrestrial plants) is characterised by a relatively conserved composition of pigments (Esteban et al. 2015). Leaves of virtually all plant species invariably contain chlorophyll (Chl) *a* and Chl *b*, and six carotenoids. Five of them are xanthophylls (carotenoids that contain oxygen): neoxanthin (Neo), lutein (Lut), violaxanthin (Vio), antheraxanthin (Ant) and zeaxanthin (Zea). The remaining carotenoid is a carotene (no oxygen in the molecule): β-carotene (β-Car). Additionally, certain taxa contain a second carotene: α-Car, which partially substitutes β-Car in some species under low light environment (Young and Britton 1989; Esteban and García-Plazaola 2016). Some species phylogenetically unrelated also include lutein epoxide (Lx), a xanthophyll likewise related to shade acclimation (Matsubara et al. 2005; Esteban et al. 2009b). Regarding pigment concentration, in agreement with the relatively conserved composition of pigments across green photosynthetic organisms, and because the maximum Chl concentration per leaf is limited by specific and physiological constrains (Niinemets 2007), photosynthetic pigments are restricted within specific ranges of concentrations. Highly reliable ranges of pigment content for non-stressed plants that were obtained from two databases comprising more than 800 species can be found in (Esteban et al. 2015; Fernández-Marín et al. 2017) (summarized as reference in Table 3.1).

B. Fernández-Marín (✉) · J. I. García-Plazaola · A. Hernández · R. Esteban
Department of Plant Biology and Ecology, University of the Basque Country (UPV/EHU), Bilbao, Spain
e-mail: beatriz.fernandezm@ehu.es

© Springer International Publishing AG, part of Springer Nature 2018

A. M. Sánchez-Moreiras, M. J. Reigosa (eds.), *Advances in Plant Ecophysiology Techniques*, https://doi.org/10.1007/978-3-319-93233-0_3

Table 3.1 Reliable ranges for chlorophyll contents and photosynthetic pigment ratios of non-stressed plants and their expected trends under high light or stress conditions

	Chl *a* + *b* (μmol m^{-2})	Chl *a* + *b* (μmol g^{-1} DW)	Chl *a*/*b* (mol mol^{-1})	Neo/Chl (mmol mol^{-1})
Non-stressed	145–800	0.2–15.8	2.2–4.2	24–65
Trend under high light or stress	Decrease	Decrease	Rise	Quite stable

	VAZ/Chl (mmol mol^{-1})	AZ/VAZ (mol mol^{-1})	Lut/Chl (mmol mol^{-1})	β-Car/Chl (mmol mol^{-1})
Non-stressed	22–177	0.05–0.35	68–283	27–157
Trend under high light or stress	Rise	Rise	Quite stable	Rise

Derived from (Esteban et al. 2015) and (Fernández-Marín et al. 2017)

Table 3.2 Spectral maxima of main photosynthetic pigments from green plants

	Chl *a*	Chl *b*	Neo	Vio	Lx	Ant	Lut	Zea	α-Car	β-Car
MaxAbs	435	469	437	441	441	446	446	451	446	451
λ(nm)	666	656	466	471	471	476	476	481	471	480

Wavelengths correspond to extracts in acetone obtained by the Photo-Diode Array (PDA) detector of the HPLC (García-Plazaola and Becerril 1999)

1.2 Location and Functions in the Chloroplast

Individual photosynthetic pigments have specific locations and functions within the photosynthetic apparatus. Chlorophyll *a* is located in the reaction centres (RCs) and the antennae (light harvesting complexes, LHCs) of both photosystems (PSI and PSII). Chlorophyll *b*, by contrast, is only bound to LHCs (Croce 2012). Both Chls show slightly different absorption spectra (Table 3.2, see Sect. 2.2) and function as the main light collectors in PSI and PSII. Carotenoids, on the other hand, play multiple roles in photosynthesis: first as light harvesters by broadening the spectrum of light collected by Chls (thanks to their absorbance of blue and blue-green light and to their capacity to transfer the absorbed energy to Chl); and second, as photoprotectants due to their ability to quench singlet oxygen and triplet Chl under excess light conditions. Additionally, they take part in the assembly of photosystems, thereby, altering the structure and function of the photosynthetic apparatus. Moreover, a relatively small fraction of the xanthophyll molecules, is not directly bound to any protein complex (i.e. is free in the lipidic membranes), where directly participate in thylakoid membrane stabilization (Havaux 1998; Dall'Osto et al. 2007b; Polívka and Frank 2010). Such is the remarkable case of some molecules of Zea (Havaux et al. 2004; Gruszecki and Strzalka 2005; Dall'Osto et al. 2010). β-Carotene is mainly found in the core complexes of PSII and PSI and also in LHCI, where it has an important role as quencher of singlet oxygen and triplet Chls (Dall'Osto et al. 2007b; Cazzaniga et al. 2012, 2016), while xanthophylls are mostly bound to antenna complexes (Moradzadeh et al. 2017).

Lutein is the most abundant xanthophyll in the photosynthetic apparatus and is located in LHCs. It can directly quench triplet Chls (Dall'Osto et al. 2006) but additionally plays a crucial role in the stability of LHC trimers (Lokstein et al. 2002) and has a photoprotective role (Dall'Osto et al. 2007a).

Neoxanthin is thought to play mainly a structural role in the assemblage of antenna protein complexes. It is located in the periphery of LHCII where it scavenges singlet oxygen (Dall'Osto et al. 2007a), but additionally, it has been evidenced very recently that Neo competes with Vio in the binding to LHCII, influencing the inter-conversions of Vio to Ant and Zea in the so-called VAZ-cycle (Wang et al. 2017). The conversion of Vio towards Ant and Zea is related to conformational changes in the antennae and with enhanced dissipation of energy as heat, that overall has an important photoprotective role in the photosynthetic apparatus (Demmig-Adams 1998; Johnson et al. 2011).

In parallel to the VAZ-cycle, an inter-conversion from Lx to Lut can also take place in some species (particularly common among some families, as Lauraceae) mainly acclimated to low light environments (i.e. forest understorey) in the so called LxL-cycle (Esteban et al. 2009c; 2010; Esteban and García-Plazaola 2014).

1.3 Dynamics of Photosynthetic Pigments in Response to the Environment

One of the most remarkable features of plant photosynthetic pigments is that their composition and proportion are highly dynamic (in particular, in response to changes in light intensity, and in general, in response to any stress factor), reflecting changes in photosynthetic and photoprotection processes. Environmental stresses (e.g. low temperature, drought, desiccation, salt stress, nutrition deficit, pollutants etc.) depress photosynthesis and consequently, lead to an excess of energy absorbed by Chls that cannot be converted into photochemistry. On a daily scale, cycles of synthesis/degradation of antenna components (including Chls) (Fukushima et al. 2009) and inter-conversion of xanthophylls within the VAZ-cycle (Demmig-Adams et al. 1996) or within the LxL-cycle (Esteban et al. 2009b) occur. The first (present in all the species from the Viridiplantae studied until now) consists on the conversion (de-epoxidation) of Vio into Zea via Ant under stress (i.e. excess irradiance at midday). Under non-stress (i.e. at night), the opposite reaction takes place, giving rise to VAZ-cycle that usually operates following a diurnal rhythm (day/night). This cycle modulates the efficiency of light energy conversion protecting the photosynthetic apparatus from photodamage but also being able to reduce plant productivity up to 20% under fluctuating light conditions (Kromdijk et al. 2016).

A second xanthophyll cycle, the LxL-cycle operates when a sunfleck suddenly increases the irradiation incident on seedling leaves. A fraction of the Lx molecules are de-epoxidised to Lut, which could enhance dissipation and photoprotection in addition to Zea. In contrast to the VAZ-cycle, the re-epoxidation of Lut back to Lx under low light conditions is much slower and it has probably a pre-emptive role in case of repetitive sunflecks (Esteban et al. 2010). The operation of the VAZ-cycle under stressors other than light has been evidenced upon desiccation/rehydration cycles and also under anoxia or heat stress (Fernández-Marín et al. 2009, 2011a, b). Weather the same response is developed by LxL cycle or not, remains still unexplored. Apart from light stress, endogenous circadian rhythms are also among regulatory factors of chlorophyll and carotenoid contents in leaves on a daily basis (García-Plazaola et al. 2017).

Variations in photosynthetic pigments, estimated either as absolute content (per leaf area or per leaf mass) or as ratios (relative amounts of some pigments in comparison to others), can be extremely informative if few considerations are taken into account before data interpretation. First: location of the pigments in the LHCs and/or the RCs; second, if they are usually bound to protein complexes or free in the thylakoid membranes, and third, what their functions are. Thus, differential pigment contents (and proportions) can be found after analysis, regarding leaf acclimation (i.e.: sun *vs* shade exposition) or stress level. Probably due to their large amounts and to its important structural role in the photosynthetic apparatus Neo content is among the most stable under stress (Esteban et al. 2015; Fernández-Marín et al. 2017). By contrast, total Chl content, Chl *a*/*b* ratio, VAZ/Chl and to higher extent AZ/VAZ (de-epoxidated state of VAZ-cycle (Ant+Zea)/(Vio+Ant+Zea)), are among the most variable parameters either in response to high irradiance or to stress. Thus, when comparing sun with shade-exposed leaves, or stressed with non-stressed plants, Chl content is generally lower, while Chl *a*/*b*, VAZ/Chl and AZ/VAZ ratios are higher. Low temperature and drought are among the abiotic stresses inducing more evident changes in the VAZ-cycle. Thus, winter acclimation and also desiccation of photosynthetic tissues can lead to high levels or AZ/VAZ even if stress occurs in the dark (Esteban et al. 2009a; Fernández-Marín et al. 2009, 2011b, 2013; Míguez et al. 2017). Finally, different pigment composition can also reflect inter-specific or inter-ecotypes differences on leaf structure, anatomy and morphology (Camarero et al. 2012; Esteban et al. 2015; Fernández-Marín et al. 2017).

Different methodological approaches allow for the identification and quantification of photosynthetic pigments. Two main types of methodologies can be distinguished: (i) analytical procedures that imply destructive sampling, and (ii) optical methods that allow non-destructive assessment of photosynthetic pigments (Fig. 3.1).

While analytical procedures require the extraction of leaf pigments in an organic solvent, several optical methods can be applied over intact plants at different scales from leaf to landscape. Near-Infrared Reflectance Spectroscopy (NIRS) represents a methodology in between, since it requires destruction of samples, but non-extraction is needed once calibration for a certain sample-type is done. In this chapter, both invasive and non-invasive approaches will be covered with special focus on the analytical (U)HPLC procedure, for which a detailed and reproducible method is specifically provided.

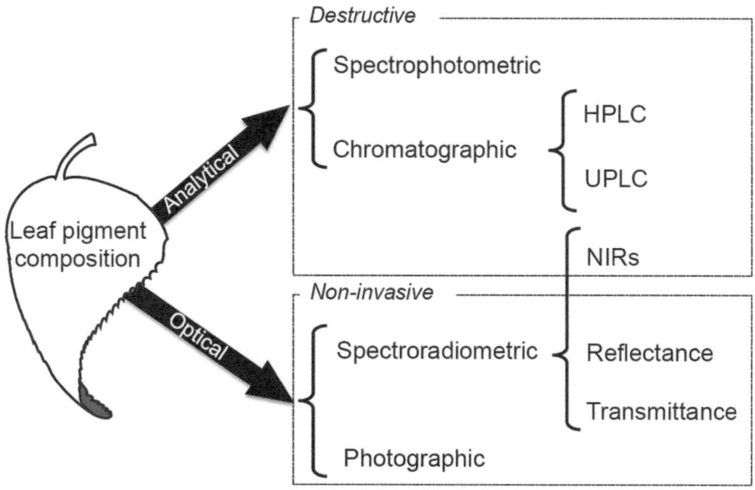

Fig. 3.1 Summary of main procedures currently available for the quantification of photosynthetic pigments: analytical, as spectrophotometric and chromatographic, and optical, as spectroradiometric and photographic

2 Destructive Methods to Quantify Photosynthetic Pigments

2.1 Sample Collection, Storage and Grinding

2.1.1 Sample Collection

When researchers are planning an experiment, in which the determination of chlorophylls and carotenoids is foreseen, special care should be taken at different steps of the procedure. Proper sample collection, (mainly in field studies) is particularly critical since leaf biochemical composition may vary enormously, as a result of individual differences and changing environmental conditions (as explained in the Sect. 1). This step is, therefore, extremely important due to the intrinsic capacity of pigments to quickly respond to the environmental fluctuations (i.e. within seconds (Peguero-Pina et al. 2013)) and the highly variable conditions in field studies. Primarily determinant for changes in the photosynthetic pigments are light, temperature and drought (Esteban et al. 2015; Fernández-Marín et al. 2017). Under this prerequisite, sample collection should be, whenever possible, performed under comparable conditions in order to exclude undesired variations (Tausz et al. 2003) (Fig. 3.2). This means that sampling must preferably be conducted at the same time of the day, sun orientation, relative position of leaf within the crown, etc. As an example, under non-stressful circumstances, north oriented leaves could be sampled at predawn from the lower part of the crown (i.e. low AZ/VAZ will be expected, Fig. 3.2). All these things considered, immediate freezing of leaf sample in liquid nitrogen is strongly recommended to prevent biochemical modifications, whenever available. If liquid nitrogen is not available, rapid desiccation of small leaf samples

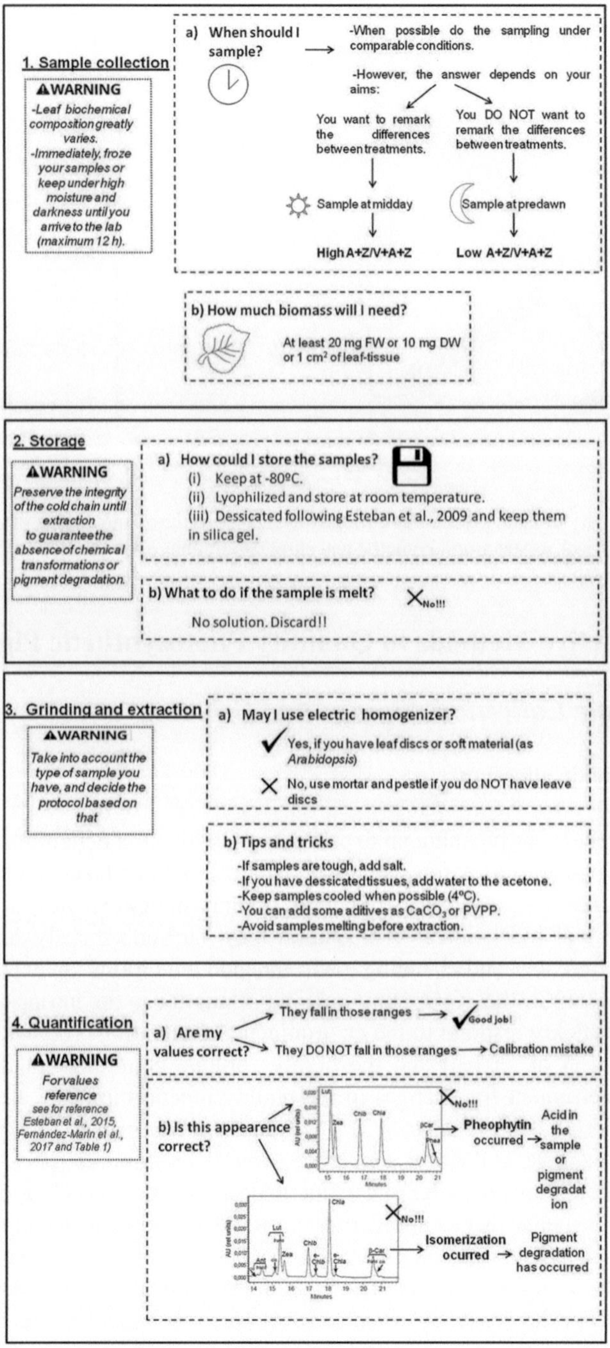

Fig. 3.2 Summary of the procedure for photosynthetic pigment analysis by (U)HPLC, including tricks and tips for a successful assessment

by using silica-gel is a secondly commendable approach, although non-reliable de-epoxidation index will be obtained during the quantification (Esteban et al. 2009a) (Fig. 3.2).

Alternatively, whenever the individual content of Ant, Vio, Zea or the AZ/VAZ are irrelevant for the aim of the experiment, scinded leaves can be collected in the field, keep under saturated atmosphere (to avoid dehydration) overnight and collect samples for pigments in the lab (this procedure is sometimes referred to as "artificial predawn"). When using liquid nitrogen all safety protocols need to be strictly followed.

2.1.2 Storage of Samples

After freezing, plant material should be stored at −80 °C until analyses (Young et al. 1996). Special care has to be taken to prevent sample melting during transfer from Dewar flask to the freezer since keeping the cold chain until the extraction is absolutely crucial to prevent any chemical transformations and/or pigment degradation. If a sample accidentally melts, even for a few seconds, irreversible chemical modifications may occur, altering its chemical composition. This sample should be then discarded for the pigment analyses (Esteban et al. 2009a) (Fig. 3.2). As an alternative to storage at −80 °C, samples can be lyophilized, which implies freezing of the samples and the subsequent sublimation of water under vacuum below the triple point (the temperature and pressure at which the three phases gas, liquid and solid of a substance coexist in thermodynamic equilibrium) (Cherian and Corona 2006). This procedure stabilises and improves the storability of the samples that can be preserved with silica gel at room temperature for several days, and at −20 °C for months. Even so, storage in the dark is recommended to avoid photo-deterioration of the pigments. Finally, when liquid nitrogen is not available (as in many remote areas), samples can be alternatively desiccated (explained in Sect. 2.1.1) and stored in silica according to the procedure of (Esteban et al. 2009a) (Fig. 3.2).

2.1.3 Commendable Amount of Sample and Methods for Grinding

Physical properties of the photosynthetic tissue of interest (e.g. leaf type: broadleaf, needle, etc.; or the phylogenetic group of the species, etc.) represent another major critical point during sampling. Depending on this, the optimal amount of tissue collected for each sample, the grinding procedure (by mortar and pestle or with electronic devices such as homogenizer, mill, or dismembrator) and/or the final pigment expression (per fresh or dry weight, per area or per chlorophyll) may vary (Fig. 3.2). For a typical broad leaf and 1 mL volume of final extraction, approximately 20 mg fresh weight, 10 mg dry weight, or 1 cm^2 of leaf-tissue, are recommended per replicate.

Conservation protocols explained above generate two types of samples: desiccated (and consequently, dehydrated) or frozen (and hydrated), which differ in chemical terms by the presence of water. This water is very important in the extraction procedure to facilitate the extraction of most polar pigments such as some xanthophylls (e.g. Neo, or Vio). For frozen samples, which contain water, pure organic solvent is used, but lyophilized or desiccated samples should be extracted with water-diluted organic solvents 95–98%.

2.2 *Spectrophotometric Assay of Photosynthetic Pigments*

Photosynthetic pigments are among the few plant metabolites that absorb light in the visible range. Furthermore, low polarity of these pigments allows relatively high specificity in the extraction and prevents interference with other coloured metabolites such as anthocyanins or betacyanins, which are hydrophilic. Consequently, its quantitative determination by spectroscopic techniques should be possible in crude pigment extracts. Given the fact that Chl *a* and *b* absorb red light at distinct wavelengths and that the absorbance of carotenoids is negligible in this range (Table 3.2), quantification of both Chls in green leaf extracts containing a mixture of photosynthetic pigments should be easy. In the 1940s, the first spectrophotometric equations for the simultaneous determination of Chl *a* and Chl *b* became available, and in fact those described by (Arnon 1949) are still in use by many scientists. However, several decades later it was evidenced that Arnon equations were inaccurate and were replaced by newer and more precise ones (Porra 2002). Nowadays, a wide range of equations optimized for the spectroscopic properties of chlorophylls and carotenoids extracted in different organic solvents (pure acetone, 80% acetone, chloroform, diethyl ether, dimethyl sulphoxide, methanol, 90% methanol, dimethyl formamide, ethanol and 95% ethanol) are available (see Table 3.3). Furthermore, some of them have been optimized for its use with spectrophotometers of high and low resolution (0.1–0.5 nm or 1–4 nm, respectively (Wellburn 1994).

In a basic protocol, pigment quantification starts with the grinding of plant material and extraction of photosynthetic pigments with an organic solvent (see Sect. 2.1.3). When other non-photosynthetic pigments are present, this step is usually enough to eliminate them from the extract, preventing any potential interference. The protocol simply finishes with the determination of absorbance at several (usually 3) wavelengths that represent the average maximum absorbance of carotenoids (470 nm), Chl *b* (642–653 nm) and Chl *a* (660–666 nm).

This method allows the estimation of the bulk of carotenoids, but the exact composition of each individual carotenoid cannot be estimated in the pigment mixture. The quantification of each carotenoid in principle would require separation analysis by HPLC (see Sect. 2.3). However, some alternative protocols, based on multi-wavelength measurement, have been developed to quantify spectrophotometrically each carotenoid on pigment mixtures (Küpper et al. 2007). Nevertheless, the usefulness of this method is limited by the fact that it does not allow a reliable estimation of carotenoids with identical absorption spectra, such as Zea and β-Car, of great interest for physiological or nutritional studies.

Table 3.3 Equations for the simultaneous determination of Chl *a* and Chl *b* in different organic solvents

Solvent	ABS λ_1	ABS λ_2	Factor A1	Factor A2	Factor B1	Factor B2	Refs
Acetone	661.6	644.8	11.24	2.04	20.13	4.19	*3*
	662	645	11.75	2.35	18.61	3.96	*4*
Acetone 90%	664	647	11.93	1.93	20.36	5.5	*2*
Acetone 80%	663	645	12.7	2.69	22.9	4.68	*1*
	663	646	12.21	2.81	20.13	5.03	*4*
	663.2	646.8	12.25	2.79	21.5	5.1	*7*
	663.6	646.6	12.25	2.55	20.31	4.91	*5*
Chloroform	656.6	647.6	11.47	2	21.85	4.53	*7*
	666	648	10.91	1.2	16.38	4.57	*7*
Diethylether	660.6	642.2	10.05	0.97	16.36	2.43	*3*
	662	644	10.05	0.766	16.37	3.14	*4*
	662	644	10.1	1.01	16.4	2.57	*6*
	662	644	10.05	0.89	16.37	2.68	*8*
Diethylether Water free	660	641.8	9.93	0.75	16.23	2.42	*3*
Diethylether Water sat.	661.6	643.2	10.36	1.28	17.49	2.72	*3*
DMFA	663.8	646.8	12	3.11	20.78	4.88	*5*
	664	647	11.65	2.69	20.81	4.53	*7*
DMSO	665	649	12.19	3.45	21.99	5.32	*7*
	665.1	649.1	12.47	3.62	25.06	6.5	*7*
Ethanol	665	649	13.95	6.88	24.96	7.32	*4*
Ethanol 95%	664.2	648.6	13.36	5.19	27.43	8.12	*3*
Methanol	665.2	652.4	16.72	9.16	34.09	15.28	*7*
	665.2	652	16.29	8.54	30.66	13.58	*5*
	666	653	15.65	7.34	27.05	11.21	*4*

References cited in the table: 1. Arnon (1949), 2. Jeffrey and Humphrey (1975), 3. Lichtenthaler and Buschmann (2001), 4. Lichtenthaler and Wellburn (1983), 5. Porra et al. (1989), 6. Smith and Benitez (1955), 7. Wellburn (1994), 8. Ziegler and Egle (1965). Table shows wavelengths and factors for the two following general equations:

Equation 1: **Chl *a* (μg/mL)** = (Factor A1 × ABS λ_1)–(Factor A2 × ABS λ_2)

Equation 2: **Chl *b* (μg/mL)** = (Factor B1 × ABS λ_2)–(Factor B2 × ABS λ_1)

Overall, the spectrophotometric assay of pigment composition is a simple, fast and reliable when is properly carried out. However, pigment studies are particularly prone to the occurrence of errors (as revised in (Fernández-Marín et al. 2015) caused by the careless use of analytical protocols. In the following lines, some typical problems, frequently encountered during the pigment determinations, are described, together with some alternatives:

- **Low extraction yield**: The extraction yield is not optimal. Differences among solvents, extraction procedures and species properties may account for unrealistic pigment composition in the extract. Optimize the grinding (the finer the powder obtained, the easier the extraction will be) and check always the best extraction medium for each type of plant material.

- **Water in the tissue**: When extracting pigments from fresh or frozen leaves, a certain amount of water present in the tissue is unavoidably added to the extraction medium. This can have effects on the extraction yield of some pigments or in the spectroscopic properties of pigments (and consequently the accuracy of equations). If this issue needs to be solved the best option is to use only freeze-dried plant material that does not contain water. In any case, be aware of the polarity of the solvent (Fig. 3.2).
- **Turbidity**: A plant extract is turbid, and must be centrifuged to remove particles. Once the extract is completely clear, absorbance at 750 nm should be close to zero since photosynthetic pigments do not absorb at this specific wavelength. Nevertheless if, after the clearing process, some turbidity remains, consider that the lower the wavelength is, the higher the scattering of light will be. This effect causes a higher apparent absorbance and leads to an overestimation of Chl *b* respect to Chl *a* and of carotenoids respect to Chls. In some equations, this effect is compensated by subtracting A_{750}. However, the most straightforward way to correct this undesirable effect is to perform a new step of centrifugation and/or filtration.
- **Absorbance out of range**. To obtain realistic measurements, the absorbance range should be 0.3–0.85 (Lichtenthaler and Buschmann 2001). If not, the accuracy or the linearity of the measurement decreases. At this point the solution is to concentrate (by evaporation) or dilute the sample. If some previous measurements can be done, is always better to adjust the extraction volume and the amount of plant tissue to the desired range of absorbance values. Fluxing cold extracts with a gas N_2 gas stream accelerates the evaporation of solvent (concentration of the sample) with minimum risk for sample deterioration.
- **Presence of anthocyanins and other pigments**. The presence of anthocyanins (and other red pigments) is common in leaves from stressed plants. When these pigments cannot be completely excluded from the extract, their quantification equations can be modified by a term that considers and subtracts the residual absorbance of anthocyanins (i.e. A_{537}) (Sims and Gamon 2002).
- **Solvent and equation used do not match each other**. A problem commonly found in the literature is that the solvent used for the extractions is not exactly the same as the one used to derive the equations. Since the spectroscopic properties of pigments, that settle the base for the development of equations, vary (sometimes dramatically) with the nature of solvent, the correspondence between equation and solvent should always be checked directly in the original source, where the method is described. Be critical and avoid using and copying of the method that has been always used in your lab.

2.3 (U)HPLC Method to Assess Photosynthetic Pigment Composition

2.3.1 Generalities About (U)HPLC Methodology for Pigment Quantification

Liquid chromatography is a technique in analytical chemistry used to separate, identify, and quantify each component in a mixture. High performance liquid chromatography (HPLC) is one of the best options to characterise the exact pigment composition in a plant tissue, and it is the standard method to quantify individual carotenoids. An alternative to HPLC is UHPLC that refers to ultra high performance liquid chromatography. This is an evolution of the former HPLC systems that offers several advantages, such as higher peak capacities, smaller peak widths, enhanced sensitivity and higher chromatographic resolution (Maurer et al. 2014). The shorter runtimes also considerably save mobile phase solvents. Nevertheless, UHPLC also has disadvantages in comparison with traditional HPLC: i.e. is more sensitive to complex matrix in the sample, and to buffers of the extraction medium.

An (U)HPLC is a computer-guided instrument that includes at least: mobile phase reservoirs and pumps, injector system, column (usually inserted in a thermo-regulable module) and detector. A few reservoirs hold the solvents (mobile phase) that are fluxed by a high-pressure pump at a specified flow rate. An injector (sample manager or autosampler) is able to introduce the sample into the flowing mobile phase stream. This carries the sample into the HPLC column that contains the chromatographic packing material (called the stationary phase because it is held in place), where the separation of different metabolites from the sample takes place. The compounds will elute from the HPLC column at different times and will be detected by the detector.

Liquid chromatography for measuring photosynthetic pigments commonly uses a photodiode array (PDA) detector with UV–visible absorption detection to measure absorbance of individual pigments, once they have been separated in a reversed-phase column. Reversed phase chromatography uses a hydrophobic stationary phase (polymeric C18 or C30 particles into the column), with a strong affinity for hydrophobic compounds, as carotenoids and chlorophylls; and a polar mobile phase (commonly an aqueous-organic mixture). A gradient system using two different solvents (first the polar one, second the hydrophobic) optimizes the elution of the pigments. As a result, most hydrophobic molecules as β-carotene in the polar mobile phase tend to adsorb to the hydrophobic stationary phase and are retained for longer time during their pass through the column. Consequently, more polar molecules (such as xanthophylls) have higher affinity for the mobile phase and will pass faster through the column. As a result more polar molecules will elute quicker (at shorter retention times) than apolar pigments. Separation of pigments typically last around 30 min in HPLC and about 5 min in UHPLC (although this times are tremendously variable depending on the method, column length, and solvent flow used).

2.3.2 Example of a Reliable Method for Pigment Quantification by HPLC

Several (U)HPLC methods are available in the bibliography to determine carotenoids and Chls. Most of them use reversed phase HPLC systems with a visible light absorption detector and allow reproducible results when conditions (solvents, flow, column characteristics, etc.) are reproduced (Gilmore and Yamamoto 1991; Maurer et al. 2014; Junker and Ensminger 2016b). Here we show a HPLC/UHPLC method derived from (García-Plazaola and Becerril 2001). This method includes a UV-fluorescence detector for the determination of tocopherols in addition to pigments in the same injection. However, because of the scope of this chapter, we will refer only to the pigment analysis. In the next lines, we detail the procedure, some preventive measures and tricks to achieve good results in pigment quantification using HPLC.

- **Extraction**

  1. Main factor that can undesirably alter pigments during extraction are: light, high temperature and acids. So, keeping the extract cold, protected from direct light and from exposure to acids is recommended along the extraction and injection procedure. Extract fresh or frozen leaf sample in ice-bathed 100% acetone (its polarity index provides a good compromise for extracting both the most and the least hydrophobic pigments). In case of freeze-dried samples is commendable to proceed with a double extraction starting with 95% acetone and re-suspending the pellet in a second extraction with pure acetone, what leads to a final extract of 97.5% acetone on average. This will allow good yield-extraction of most polar (in the first extraction) and most apolar (in the second extraction) pigments. In order to prevent pigment degradation due to internal organic acids of the samples, the addition of $CaCO_3$ ($0.5\ g\ L^{-1}$) in the extraction medium is commendable when HPLC is used (but must be avoided in UHPLC systems, since precipitation of the salt can obstruct the tubing). A suitable relationship between amount of plant material and extraction-medium should be used (typically 10–50 mg leaf fresh weight or 0.5–1 leaf cm^2 per mL). Soft fresh or frozen samples (i.e. small pieces of spinach or Arabidopsis leaves) can be directly homogenized with the extraction-medium by using an electric tissue homogeniser, or immediately before the extraction by using mortar and pestle. In this case, chill the mortar with liquid N_2 before adding samples and homogenise to powder just before adding the acetone. Collect the mixture in a 2 mL eppendorf tube. For tough samples you can either add a bit of sand into the mortar, or alternatively, freeze-dry the samples and powder them with a ball-mill or dismembrator before the extraction. In this case, after adding the acetone, vortex the sample vigorously for 10 s. Regardless the chosen extraction method, it is important to keep the samples refrigerated ($\leq$4 °C) and to protect them from direct light (Fig. 3.2).
  2. Extract must be thereafter centrifuged at 4 °C and 16,000 g for 20 min to obtain the supernatant. The pellet can be re-extracted if it contains visible

chlorophylls (Fig. 3.2). If so, repeat the extraction step until the green colour of the pellet had gone and pool it together with the first supernatant.

3. Syringe-filtered the supernatant through a 0.22 μm PTFE filter. The first drops that pass the filter should be discarded to avoid contamination. Fill the HPLC vial and close it immediately with a cap. Extracts can be stored in the freezer (−20 °C) for few days but it is strongly recommended to immediately inject the samples in the HPLC. Long storage of extracts may provoke pigment isomerisation (Esteban et al. 2009a, b, c). The estimated time for the preparation of 24 samples (from extraction to insertion in the HPLC) is of ~2 h.
4. After each sample, wash the mortar, homogenizer, syringes, etc. generously with ethanol absolute or pure acetone.

- **HPLC conditions**

1. The mobile phase in (García-Plazaola and Becerril 2001) consists of two solvents. The solvent A, acetonitrile:methanol:water (84:9:7) v/v/v with TRIS-HCl buffer 10 mM pH 8 and the solvent B, methanol:ethyl acetate (68:32) v/v. Solvents must be HPLC-grade and it is recommended to filter them before using. Vacuum filtering also removes dissolved gases that otherwise could bubble along the chromatographic system. Most modern instruments ultrasonically eliminate gases from solvent when functioning.
2. Photosynthetic pigments are eluted using a linear gradient from 100% A to 100% B for the first 12 min, followed by an isocratic elution of 100% B for the next 6 min. This is followed by 1 min linear gradient from 100% B to 100% A. Finally, an isocratic elution with 100% A is established for a further 6 min, to allow the column to re-equilibrate with solvent A prior to the next injection (see Table 3.4 also for equivalences in case of UHPLC systems).
3. The solvent flow rate is 1.2 mL min^{-1} (0.5 mL/min for UHPLC).
4. Injected sample volume is 15 μL (1 μL for UHPLC).
5. HPLC chromatography is carried out in a Tracer Spherisorb ODS-1 reversed phase column (i.e.: Tracer Spherisorb or Waters® Spheridorb® ODS1). Column is 250 mm long with 4.6 mm diameter and 5 μm of particle size. The use of a guard column preceding the main column is strongly recommended to pro-

Table 3.4 Detailed HPLC gradient for the analysis of photosynthetic pigments after (García-Plazaola and Becerril 2001)

Step	Minutes	Flow (mL/min)	% A	% B
1	0	0	100	0
2	10 (2.4)	1.2 (0.5)	0	100
3	16 (3.6)	1.2 (0.5)	0	100
4	17 (3.8)	1.2 (0.5)	100	0
5	25 (5)	1.2 (0.5)	100	0

The corresponding modifications for its transfer to an UHPLC system are depicted in brackets within the table. Solvent A: acetonitrile 84%, Methanol 9%, H_2O-Tris (10 mM pH 8) 7%. LC-grade water instead of TRIS is recommended in solvent A for UHPLC systems. Solvent B: Methanol 68%, Ethyl acetate 32%

long its life-span and to enhance chromatographic separation (elution) of the pigments. An appropriate guard column would be Nova-Pak C-18 (50 × 3.9 mm; 4 μm). UHPLC column would be Waters® Acquity® UPLC HSS C18 SB (100 × 2.1 mm, 1.8 μm), with a frit filter (0.2 μm, 2.1 mm). Select the temperature of the column into the oven at 30 °C what usually guaranties repeatable separation conditions along the year in non-acclimated laboratories. High temperature accelerates the elution of the pigments but worsens the separation of some of them (i.e. Lut and Zea can be particularly tricky).

6. Each sample is scanned by the PDA in the range 250–700 nm, and peaks are detected and integrated at 445 nm for the quantification of carotenoids and Chls.

- **Pigment identification**

Once the chromatogram is obtained check that all pigments are present and sufficiently resolved (no overlapping peaks). A test of the system can be done by extracting pigments from a green leaf sample (such as spinach or Arabidopsis) and carefully assessing the chromatogram. All green leaves have at least 6 pigments and they should appear in the same order than in the method followed. With the method by García-Plazaola and Becerril (2001) the order of the pigments (with increasing retention times) is Neo, Vio, Lut, Chl *b*, Chl *a*, and β-Car (Fig. 3.3). Retention times and relative order of pigments may vary depending on the method (solvents, extraction medium, etc.). When leaves are illuminated or exposed to a severe stress prior to the sampling, two pigments add to the mentioned list: Ant, eluting between Vio and Lut, and Zea eluting close after its isomer Lut (Fig. 3.3). Peak-pigment identity is confirmed by comparing the visible absorption spectra of each peak (recorded with the PDA) to the literature. Although maximum wavelength of absorption may slightly vary depending on the solvents, most carotenoids and chlorophylls show a characteristic absorption spectrum that allows an almost unequivocal identification (Britton et al. 2004).

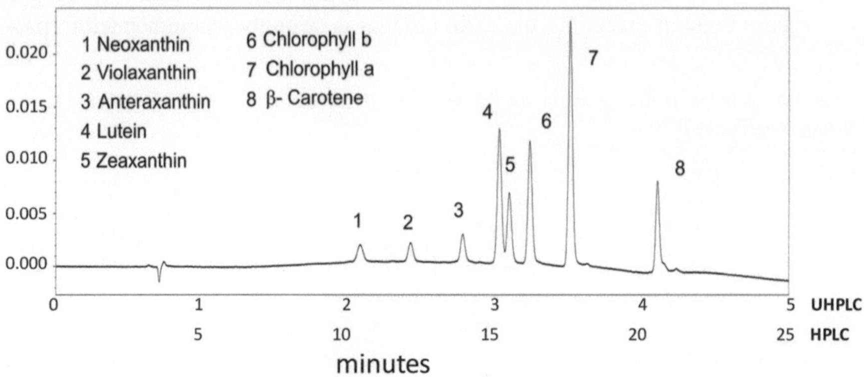

Fig. 3.3 Example of a pigment chromatogram integrated at 445 nm following the method by (García-Plazaola and Becerril 2001)

When a new method is being set up, a useful trick to check the resolution of Zea is to force a high content in the test-sample by placing a leaf under intense illumination for 15 min before sampling. When comparing the chromatograms from a dark-incubated and an illuminated leaf, Vio should decrease and concomitantly Zea should increase. If Zea does not appear or does not increase, it is likely that Zea and Lut (both xanthophylls are isomers) are eluting together. In fact, given the importance of Zea on plant responses, a bad separation of Lut and Zea is the most frequent problem that compromises the usefulness of a chromatographic protocol for eco-physiological studies.

- **Pigment quantification**

Although plant pigments have maximum absorbance at different wavelengths, usually integration of area of different peaks is done at the same wavelength for all compounds (typically 445 nm) where all carotenoids and chlorophylls sensitively absorb. The (U)HPLC software usually integrates all peaks in the chromatogram and gives a table with the retention time of each peak and their area. However, it is almost indispensable to make a manual correction to assure that all peaks areas correspond with the correct pigments. Some pigments like Neo and β-Car can show multiple peaks (*cis* and *trans* isomers) with the same absorption spectra, but this should not represent a problem since peak-areas can be summed. If the noise in the chromatogram is high, is better to increase plant sample amount to solvent ratio during the extraction than increasing the injection volume. Peak area is proportional to pigment amount but the exact relationship between peak area and pigment must be known to make the conversion using equations. For this purpose pigment standards must be used for calibration. The conversion of each peak area to mol or g pigment can be done using a spreadsheet (excel) or using the (U)HPLC software. Although it is possible to program the (U)HPLC software to make automatically all the process (integration, identification, and units conversion) all phases should be thoroughly supervised by the scientist. Final amount of pigment can be expressed in mol or weight units, both per leaf weight or area. Most frequently employed units are: μmol pigment m^{-2}, nmol cm^{-2}, μmol pigment g^{-1} dry weight. Comparison of obtained concentrations with data from the literature are crucial to avoid mistakes in the quantification or in the use of units (Esteban et al. 2015; Fernández-Marín et al. 2015).

- **Signs of pigment degradation**

The occurrence of double peaks indicating presence of isomers/epimers in our sample, as well as the occurrence of unexpected metabolites such as pheophytine can represent signs of sample degradation (Fig. 3.2). Exposition of samples to light, heat or acid (i.e. also intrinsic acids of sample), and also long-storage of samples under inadequate conditions usually explains the formation of isomers and epimers from carotenoids and the degradation of Chls to pheophytine. In Fig. 3.2 several tips and tricks to avoid sample degradation are given at different steps during sampling, storage, processing and analysis.

2.4 Near-Infrared Reflectance Spectroscopy (NIRS) for Pigment Quantification

Near-infrared reflectance spectroscopy (NIRS) analyses the diffuse reflectance of samples and it is based on the absorption of light in the range from 780 to 2500 nm. Each sample produces a unique spectral signature due to specific absorption of bonds such as O–H, C=O, C–N or N–H, characteristic of organic matter. Samples must be grinded and freeze-dried before NIRS analysis, since water has a high absorption in the near-infrared that could invalidate the measurements. Calibration (i.e. comparison of NIRS spectra with HPLC obtained contents) is needed for each individual metabolite (i.e. pigment) and sample type (i.e.: species, leaf developmental stage, etc.). Several recent examples of successful quantification of photosynthetic pigments by NIRS can be found in the literature (Pintó-Marijuan et al. 2013; Fernàndez-Martínez et al. 2017).

3 Non-invasive Analysis of Photosynthetic Pigments

Pigment concentration is undoubtedly most accurately measured by extraction in a solvent followed by its analytical determination. Alternatively, however, it is possible to estimate pigments content using non-destructive and *in situ* optical techniques. When light reaches a leaf most, but not all, of the visible wavelengths (400–700 nm) are absorbed by both Chls and carotenoids. However, these pigments do not capture so efficiently the green light, and as a result (unless there are anthocyanins or betacyanins in the cuticle), leaves display this colour. Although all plants look equally green (to human eye), it is possible to analyse non-absorbed light to derive plant pigment contents and/or relationships between pigments. Both, light reflected by the leaf surface, or light transmitted (light that cross/go through the leaf) can be used to estimate pigments content (Fig. 3.4).

3.1 Measurement of Pigments by Light Transmittance Through the Leaf

Several portable (handheld) models measuring light transmittance are available, such as *CCM-200 (Opti-Science), Spad (Minolta), CL-01 (Hansatech) and Dualex (Dx Force-A)*. All of them are able to estimate Chl content per area by estimating leaf light absorption of red radiation (around 650 nm) and near infrared (NIR) radiation (approximately 850–940 nm). NIR is not absorbed by photoreceptors, so it is used as a reference to correct the detour effect (light scattering) to give a meter output (Shrestha et al. 2012). Furthermore, *Dualex* device combines absorbance and fluorescence measurements in the UV-A band, in the Red and in the NIR to estimate

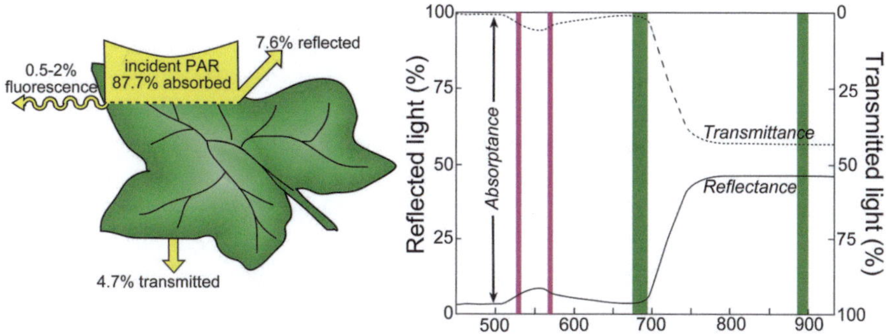

Fig. 3.4 Optical behaviour of incident visible light on a idealised leaf (left) and its spectrum of transmittance and reflectance (right). Pink (wavelengths used for PRI) and green (wavelengths used for NDVI) vertical bars highlight wavelength ranges typically used by non-invasive procedures for quantification of photosynthetic pigments

parameter others than Chl contents as nitrogen balance index, and the polyphenol content (leaf flavonols and anthocyanins indexes) (Cerovic et al. 2012). An important limitation of these devices is that measurements of the chlorophyll content are given in relative units and not in absolute units, and even more, the meter outputs are not totally linear with chlorophyll content (Parry et al. 2014). This means that when the exact chlorophyll content has to be calculated, a calibration with actual chlorophyll contents and relative units has to be done. In addition, the response is species-dependent, which means that a calibration for each plant species is needed. As the chlorophyll content in leaves is closely related with nitrogen content in plant, specifically in crops, the main field for these devices is agronomy where some of them are sold as nitrogen tester for crops. However, they have also been used for research work because they have several advantages: they produce fast, reproducible and non-destructive measurements (Parry et al. 2014), and the devices are also affordable and handheld.

3.2 *Measure of Pigments by Leaf Light Reflectance*

Vegetation reflectance can be measured using spectroradiometers that detect plant reflected light in the visible and NIR spectrum at different spatial and temporal scales, by *in situ* or in remote measurements. At the canopy scale many factors condition light reflectance such as leaf angle, leaf area, illumination, soil optical properties, and consequently a plethora of indexes and correction coefficients developed under different conditions have been widely used to remotely (airborne and satellite) assess changes in vegetation properties (Lu et al. 2015). However, most handy measurements can be done using *in situ* reflectance handheld instruments. Two main indexes, using reflectance at several wavelengths, are used to assess both, chlorophyll contents and VAZ-cycle pigments, the Normalised Difference Vegetation Index (NDVI), and the Photochemical Reflectance Index (PRI), respectively.

The PRI index reflects light use efficiency. Gamon and collaborators (2001) showed that energy dissipation can be monitored by changes in PRI as this values correlated with xanthophyll cycle pigment epoxidation state. This index uses relative reflectance at 531 and 570 nm to assess the VAZ pigment conversion (Gamon et al. 2001). As explained above, these three xanthophylls interconvert each other depending on balance between light use efficiency and light dissipation. The PRI is formulated as follows: [PRI = (R531 − R570)/(R531 + R570)], where R531 and R570 represent the reflectance at 531 and 570 nm respectively. Leaf surface properties (highly reflective cuticle, wax, hair or trichomes presence) can change the pattern of light reflectance, so R570 is used as the reference (Gamon et al. 2001). When light interception exceeds light utilization in photosynthesis Vio converts to Zea to help energy dissipation. Zea, but not Ant, absorb at 531 nm, so R531 and PRI accordingly decrease to zeaxanthin increases. For this reason, this index currently indicates vegetation health (high values) or plant stress (low values).

The NDVI is a vegetation index that increases depending on Chl density per area. Considering that Chls absorb red wavelength (600–700 nm approximately) but not in the NIR, reflectance in the red (R) is inversely related to green biomass. Since the reflectance is also influenced by structural properties in the leaf, this index includes the NIR correction in the calculation (NIR-R/NIR + R). Typically red is measured at 660–670 nm and NIR at 730–740 nm. Several handheld instruments using fixed wavelengths reflectance are available to measure *in situ* NDVI and/or PRI indexes: *PlantPen PRI 200 & NDVI 300 (Photon Systems Instruments), GreenSeeker Handheld (Trimble), RapidSCAN CS-45 Crop (Holland Scientific) and Spectral Reflectance Sensor (Decagon).* Handheld spectroradiometer as *SpectraPen SP 100 and SpectraPen LM 500 (PSI)* and more versatile full-range portable Spectroradiometer as *Apogee Instruments Spectroradiometer, SVC HR-640i (Spectra Vista Corporation) or ASD FieldSpec® spectrometer (Analytical Spectral Devices*) are able to detect reflectance at any wavelength, which allows to calculate custom-made reflectance indexes. Another equipments as *Geenseaker-Crop Sensing System (Trimble) and Crop Circle ACS-430 (Holland Scientific*) are designed for agronomical uses, as they are thought to be used with farming equipment.

It must be advised that PRI, NDVI and, in general, any reflectance index provide relative measurements, but not absolute values of plant pigment contents. Before any buying decision, we strongly recommend visiting the web sites of the manufacturers and comparing the specifications and options of the different instruments to check whether they measure the parameters of interest. Even more important, is to check the literature and applications of these non-invasive measurements.

3.3 Photographic Analysis

Besides spectral analysis of light, ordinary digital photographs can also be used to analyse plant pigment composition by using image programs that discompose photographs into the three RGB (red, green, and blue) channels. A few indexes have been developed from the colour channels information to assess greenness at canopy level and individual leaves (see (Junker and Ensminger 2016a, b) for further details).

References

Arnon DI (1949) Copper enzymes in isolated chloroplasts. Polyphenolxidase in *Beta vulgaris*. Plant Physiol 24:1–15

Britton G, Liaaen-Jensen S, Pfander H (2004) Carotenoids: handbook. Springer, Basel

Camarero JJ, Olano JM, Arroyo Alfaro SJ, Fernández-Marín B, Becerril JM, García-Plazaola JI (2012) Photoprotection mechanisms in *Quercus ilex* under contrasting climatic conditions. Flora 207:557–564

Cazzaniga S, Li Z, Niyogi KK, Bassi R, Dall'Osto L (2012) The Arabidopsis szl1 mutant reveals a critical role of β-carotene in photosystem I photoprotection. Plant Physiol 159:1745–1758. https://doi.org/10.1104/PP.112.201137

Cazzaniga S, Bressan M, Carbonera D, Agostini A, Dall'Osto L (2016) Differential roles of carotenes and xanthophylls in photosystem I photoprotection. Biochemistry 55:3636–3649. https://doi.org/10.1021/acs.biochem.6b00425

Cerovic ZG, Masdoumier G, Ben GN, Latouche G (2012) A new optical leaf-clip meter for simultaneous non-destructive assessment of leaf chlorophyll and epidermal flavonoids. Physiol Plant 146:251–260. https://doi.org/10.1111/j.1399-3054.2012.01639.x

Cherian M, Corona E (2006) Lyophilisation of biologicals. In: Briefings T (ed) Bioprocessing and biopartnering. Touch Briefings, London, pp 20–21

Croce R (2012) Chlorophyll-binding proteins of higher plants and cyanobacteria. In: Eaton-Rye J, Tripathy B, Sharkey TD (eds) Photosynthesis: plastid biology, energy conversion and carbon assimilation, Advances in Photosynthesis and Respiration. Springer, Dordrecht, pp 127–149

Dall'Osto L, Lico C, Alric J, Giuliano G, Havaux M, Bassi R (2006) Lutein is needed for efficient chlorophyll triplet quenching in the major LHCII antenna complex of higher plants and effective photoprotection in vivo under strong light. BMC Plant Biol 6:32. https://doi.org/10.1186/1471-2229-6-32

Dall'Osto L, Cazzaniga S, North H, Marion-Poll A, Bassi R (2007a) The Arabidopsis aba4-1 mutant reveals a specific function for neoxanthin in protection against photooxidative stress. Plant Cell 19:1048–1064. https://doi.org/10.1105/tpc.106.049114

Dall'Osto L, Fiore A, Cazzaniga S, Giuliano G, Bassi R (2007b) Different roles of α- and β-branch xanthophylls in photosystem assembly and photoprotection. J Biol Chem 282:35056–35068. https://doi.org/10.1074/jbc.M704729200

Dall'Osto L, Cazzaniga S, Havaux M, Bassi R (2010) Enhanced photoprotection by protein-bound vs free xanthophyll pools: a comparative analysis of chlorophyll b and xanthophyll biosynthesis mutants. Mol Plant 3:576–593. https://doi.org/10.1093/mp/ssp117

Demmig-Adams B (1998) Survey of thermal energy dissipation and pigment composition in sun and shade leaves. Plant Cell Physiol 39:474–482

Demmig-Adams B, Gilmore AM, Adams WW (1996) In vivo functions of carotenoids in higher plants. FASEB 10:403–412

Esteban R, García-Plazaola JI (2014) Involvement of a second xanthophyll cycle in non-photochemical quenching of chlorophyll fluorescence: the lutein epoxide story. In: Demmig-Adams B, Garab G, Adams WW III, Govindjee (eds) Non-photochemical quenching and energy dissipation in plants, algae and cyanobacteria. Springer, Dordrecht, pp 277–295

Esteban R, García-Plazaola JI (2016) Nonubiquitous carotenoids in higher plants: presence, role in photosynthesis, and guidelines for identification. In: Pessarakli M (ed) Handbook of photosynthesis, 3rd edn. CRC Press, Boca Raton, pp 589–600

Esteban R, Balaguer L, Manrique E, Rubio de Casas R, Ochoa R, Fleck I, Pintó-Marijuan M, Casals I, Morales D, Jiménez MS, Lorenzo R, Artetxe U, Becerril JM, García-Plazaola JI

(2009a) Alternative methods for sampling and preservation of photosynthetic pigments and tocopherols in plant material from remote locations. Photosynth Res 101:77–88. https://doi.org/10.1007/s11120-009-9468-5

Esteban R, Becerril JM, García-Plazaola JI (2009b) Lutein epoxide cycle, more than just a forest tale. Plant Signal Behav 4:342–344. https://doi.org/10.4161/psb.4.4.8197

Esteban R, Olano JM, Castresana J, Fernández-Marín B, Hernández A, Becerril JM, García-Plazaola JI (2009c) Distribution and evolutionary trends of photoprotective isoprenoids (xanthophylls and tocopherols) within the plant kingdom. Physiol Plant 135:379–389. https://doi.org/10.1111/j.1399-3054.2008.01196.x

Esteban R, Matsubara S, Jiménez MS, Morales D, Brito P, Lorenzo R, Fernández-Marín B, Becerril JM, García-Plazaola JI (2010) Operation and regulation of the lutein epoxide cycle in seedlings of *Ocotea foetens*. Funct Plant Biol 37:859–869. https://doi.org/10.1071/FP10014

Esteban R, Barrutia O, Artetxe U, Fernández-Marín B, Hernández A, García-Plazaola JI (2015) Internal and external factors affecting photosynthetic pigment composition in plants: a meta-analytical approach. New Phytol 206:268–280. https://doi.org/10.1111/nph.13186

Fernández-Marín B, Balaguer L, Esteban R, Becerril JM, García-Plazaola JI (2009) Dark induction of the photoprotective xanthophyll cycle in response to dehydration. J Plant Physiol 166:1734–1744. https://doi.org/10.1016/j.jplph.2009.04.019

Fernández-Marín B, Míguez F, Becerril J, García-Plazaola J (2011a) Activation of violaxanthin cycle in darkness is a common response to different abiotic stresses: a case study in *Pelvetia canaliculata*. BMC Plant Biol 11:181. https://doi.org/10.1186/1471-2229-11-181

Fernández-Marín B, Míguez F, Becerril JM, García-Plazaola JI (2011b) Dehydration-mediated activation of the xanthophyll cycle in darkness: is it related to desiccation tolerance? Planta 234:579–588. https://doi.org/10.1007/s00425-011-1420-1

Fernández-Marín B, Kranner I, Sebastián MS, Artetxe U, Laza JM, Vilas JL, Pritchard HW, Nadajaran J, Míguez F, Becerril JM, García-Plazaola JI (2013) Evidence for the absence of enzymatic reactions in the glassy state. A case study of xanthophyll cycle pigments in the desiccation-tolerant moss *Syntrichia ruralis*. J Exp Bot 64:3033–3043. https://doi.org/10.1093/jxb/ert145

Fernández-Marín B, Artetxe U, Barrutia O, Esteban R, Hernández A, García-Plazaola JI (2015) Opening Pandora's box: cause and impact of errors on plant pigment studies. Front Plant Sci 6:148. https://doi.org/10.3389/fpls.2015.00148

Fernández-Marín B, Hernández A, García-Plazaola JI, Esteban R, Míguez F, Artetxe U, Gómez-Sagasti MT (2017) Photoprotective strategies of Mediterranean plants in relation to morphological traits and natural environmental pressure: a meta-analytical approach. Font Plant Sci 19:1051. https://doi.org/10.3389/fpls.2017.01051

Fernàndez-Martínez J, Joffre R, Zacchini M, Fernández-Marín B, García-Plazaola JI, Fleck I (2017) Near-infrared reflectance spectroscopy allows rapid and simultaneous evaluation of chloroplast pigments and antioxidants, carbon isotope discrimination and nitrogen content in *Populus* spp. leaves. For Ecol Manag 399:227–234. https://doi.org/10.1016/j.foreco.2017.05.041

Fukushima A, Kusano M, Nakamichi N, Kobayashi M, Hayashi N, Sakakibara H, Mizuno T, Saito K (2009) Impact of clock-associated Arabidopsis pseudo-response regulators in metabolic coordination. Proc Natl Acad Sci 106:8791–8791. https://doi.org/10.1073/pnas.0904226106

Gamon JA, Field CB, Fredeen AL, Thayer S (2001) Assessing photosynthetic downregulation in sunflower stands with an optically-based model. Photosynth Res 67:113–125. https://doi.org/10.1023/A:1010677605091

García-Plazaola JI, Becerril JM (1999) A rapid high-performance liquid chromatography method to measure lipophilic antioxidants in stressed plants: simultaneous determination of carotenoids and tocopherols. Phytochem Anal 10:307–313

García-Plazaola JI, Becerril JM (2001) Seasonal changes in photosynthetic pigments and antioxidants in beech (*Fagus sylvatica*) in a Mediterranean climate: implications for tree decline diagnosis. Aust J Plant Physiol 28:225–232. https://doi.org/10.1071/PP00119

García-Plazaola JI, Fernández-Marín B, Ferrio JP, Alday JG, Hoch G, Landais D, Milcu A, Tissue DT, Voltas J, Gessler A, Roy J, Resco de Dios V (2017) Endogenous circadian rhythms in pigment composition induce changes in photochemical efficiency in plant canopies. Plant Cell Environ 40:1153–1162. https://doi.org/10.1111/pce.12909

Gilmore AM, Yamamoto HY (1991) Resolution of lutein and zeaxanthin using a non-endcapped, lightly carbon-loaded C18 high-performance liquid chromatographic column. J Chromatogr A 543:137–145. https://doi.org/10.1016/S0021-9673(01)95762-0

Gruszecki WI, Strzalka K (2005) Carotenoids as modulators of lipid membrane physical properties. Biochim Biophys Acta Mol basis Dis 1740:108–115. https://doi.org/10.1016/j.bbadis.2004.11.015

Havaux M (1998) Carotenoids as membrane stabilizers in chloroplasts. Trends Plant Sci 3:147–151. https://doi.org/10.1016/S1360-1385(98)01200-X

Havaux M, Dall'Osto L, Cuiné S, Giuliano G, Bassi R (2004) The effect of zeaxanthin as the only xanthophyll on the structure and function of the photosynthetic apparatus in *Arabidopsis thaliana*. J Biol Chem 279:13878–13888. https://doi.org/10.1074/jbc.M311154200

Jeffrey SW, Humphrey GF (1975) New spectrophotometric equations for determining chlorophylls a, b, c1 and c2 in higher plants, algae and natural phytoplankton. Biochem und Physiol der Pflanz 167:191–194. https://doi.org/10.1016/S0015-3796(17)30778-3

Johnson MP, Goral TK, Duffy CD, Brain AP, Mullineaux CW, Ruban AV (2011) Photoprotective energy dissipation involves the reorganization of photosystem II light-harvesting complexes in the grana membranes of spinach chloroplasts. Plant Cell 23:1468–1479. https://doi.org/10.1105/tpc.110.081646

Junker LV, Ensminger I (2016a) Relationship between leaf optical properties, chlorophyll fluorescence and pigment changes in senescing *Acer saccharum* leaves. Tree Physiol 36:694–711. https://doi.org/10.1093/treephys/tpv148

Junker LV, Ensminger I (2016b) Fast detection of leaf pigments and isoprenoids for ecophysiological studies, plant phenotyping and validating remote-sensing of vegetation. Physiol Plant 158:369–381. https://doi.org/10.1111/ppl.12512

Kromdijk J, Glowacka K, Leonelli L, Gabilly ST, Iwai M, Niyogi KK, Long SP (2016) Improving photosynthesis and crop productivity by accelerating recovery from photoprotection. Science 354(6314):857–861

Küpper H, Seibert S, Parameswaran A (2007) Fast, sensitive, and inexpensive alternative to analytical pigment HPLC: quantification of chlorophylls and carotenoids in crude extracts by fitting with Gauss peak spectra. Anal Chem 79:7611–7627. https://doi.org/10.1021/ac070236m

Lichtenthaler HK, Buschmann C (2001) Chlorophylls and carotenoids: measurement and characterization by UV-VIS spectroscopy. In: Current protocols in food analytical chemistry. Wiley, Hoboken

Lichtenthaler HK, Wellburn AR (1983) Determinations of total carotenoids and chlorophylls a and b of leaf extracts in different solvents. Biochem Soc Trans 11:591–592

Lokstein H, Tian L, Polle JEW, DellaPenna D (2002) Xanthophyll biosynthetic mutants of *Arabidopsis thaliana*: altered nonphotochemical quenching of chlorophyll fluorescence is due to changes in photosystem II antenna size and stability. Biochim Biophys Acta Bioenerg 1553:309–319. https://doi.org/10.1016/S0005-2728(02)00184-6

Lu S, Lu X, Zhao W, Liu Y, Wang Z, Omasa K (2015) Comparing vegetation indices for remote chlorophyll measurement of white poplar and Chinese elm leaves with different adaxial and abaxial surfaces. J Exp Bot 66:5625–5637. https://doi.org/10.1093/jxb/erv270

Matsubara S, Naumann M, Martin R, Nichol C, Rascher U, Morosinotto T, Bassi R, Osmond B (2005) Slowly reversible de-epoxidation of lutein-epoxide in deep shade leaves of a tropical tree legume may "lock-in" lutein-based photoprotection during acclimation to strong light. J Exp Bot 56:461–468. https://doi.org/10.1093/jxb/eri012

Maurer MM, Mein JR, Chaudhuri SK, Constant HL (2014) An improved UHPLC-UV method for separation and quantification of carotenoids in vegetable crops. Food Chem 165:475–482. https://doi.org/10.1016/j.foodchem.2014.05.038

Míguez F, Fernández-Marín B, Becerril JM, García-Plazaola JI (2017) Diversity of winter photoinhibitory responses: a case study in co-occurring lichens, mosses, herbs and woody plants from subalpine environments. Physiol Plant 160:282–296

Moradzadeh M, Sadeghnia HR, Tabarraei A, Sahebkar A (2017) Anti-tumor effects of crocetin and related molecular targets. J Cell Physiol 233:2170–2182. https://doi.org/10.1002/jcp.25953

Niinemets Ü (2007) Photosynthesis and resource distribution through plant canopies. In: Eaton-Rye J, Tripathy B, Sharkey T (eds) Photosynthesis: plastid biology, energy conversion and carbon assimilation, Advances in Photosynthesis and Respiration. Blackwell Publishing Ltd, Dordrecht, pp 1052–1071

Parry C, Blonquist JM, Bugbee B (2014) In situ measurement of leaf chlorophyll concentration: analysis of the optical/absolute relationship. Plant Cell Environ 37:2508–2520. https://doi.org/10.1111/pce.12324

Peguero-Pina JJ, Gil-Pelegrín E, Morales F (2013) Three pools of zeaxanthin in *Quercus coccifera* leaves during light transitions with different roles in rapidly reversible photoprotective energy dissipation and photoprotection. J Exp Bot 64:1649–1661. https://doi.org/10.1093/jxb/ert024

Pintó-Marijuan M, Joffre R, Casals I, De Agazio M, Zacchini M, García-Plazaola JI, Esteban R, Aranda X, Guàrdia M, Fleck I (2013) Antioxidant and photoprotective responses to elevated CO2 and heat stress during holm oak regeneration by resprouting, evaluated with NIRS (near-infrared reflectance spectroscopy). Plant Biol 15:5–17. https://doi.org/10.1111/j.1438-8677.2011.00538.x

Polívka T, Frank HA (2010) Light harvesting by carotenoids. Acc Chem Res 43:1125–1134

Porra RJ (2002) The chequered history of the development and use of simultaneous equations for the accurate determination of chlorophylls a and b. Photosynth Res 73:149–156. https://doi.org/10.1023/A:1020470224740

Porra RJ, Thompson WA, Kriedemann PE (1989) Determination of accurate extinction coefficients and simultaneous equations for assaying chlorophylls a and b extracted with four different solvents: verification of the concentration of chlorophyll standards by atomic absorption spectroscopy. Biochim Biophys Acta Bioenerg 975:384–394. https://doi.org/10.1016/S0005-2728(89)80347-0

Shrestha S, Brueck H, Asch F (2012) Chlorophyll index, photochemical reflectance index and chlorophyll fluorescence measurements of rice leaves supplied with different N levels. J Photochem Photobiol B Biol 113:7–13. https://doi.org/10.1016/j.jphotobiol.2012.04.008

Sims DA, Gamon JA (2002) Relationships between leaf pigment content and spectral reflectance across a wide range of species, leaf structures and developmental stages. Remote Sens Environ 81:337–354. https://doi.org/10.1016/S0034-4257(02)00010-X

Smith JHC, Benitez A (1955) Chlorophylls: analysis in plant materials. In: Modern methods of plant analysis/Moderne Methoden der Pflanzenanalyselant analysis. Springer, Berlin/Heidelberg, pp 142–196

Tausz M, Wonisch A, Grill D, Morales D, Jiménez MS (2003) Measuring antioxidants in tree species in the natural environment: from sampling to data evaluation. J Exp Bot 54:1505–1510. https://doi.org/10.1093/jxb/erg175

Wang K, Tu W, Liu C, Rao Y, Gao Z, Yang C (2017) 9-cis-Neoxanthin in light harvesting complexes of photosystem II regulates the binding of violaxanthin and xanthophyll cycle. Plant Physiol 174:86–96. https://doi.org/10.1104/pp.17.00029

Wellburn AR (1994) The spectral determination of chlorophylls a and b, as well as total carotenoids, using various solvents with spectrophotometers of different resolution. J Plant Physiol 144:307–313. https://doi.org/10.1016/S0176-1617(11)81192-2

Young AJ, Britton G (1989) The distribution of α-carotene in the photosynthetic pigment-protein complexes of higher plants. Plant Sci 64:179–183. https://doi.org/10.1016/0168-9452(89)90022-8

Young A, Phillip D, Savill J (1996) Methods for carotenoid analysis. In: Pessaraki M (ed) Handbook of photosynthesis. Marcel Dekker Inc, New York, pp 597–622

Ziegler R, Egle K (1965) Zur quantitativen Analyse der Chloroplastenpigmente. I. Kritische. berprüfung der spektralphotometrischen Chlorophyll-Bestimmung. Beitr Biol Pflanz 41:11–37

Chapter 4
Measuring Photosynthesis and Respiration with Infrared Gas Analysers

Cyril Douthe, Jorge Gago, Miquel Ribas-Carbó, Rubén Núñez, Nuria Pedrol, and Jaume Flexas

1 Introduction

1.1 Primary Carbon Metabolism and Gas Exchange in Leaves

Earth primary productivity reflects the balance between two important biological processes: photosynthesis and respiration (Atkin et al. 2015; Niinemets 2016). Photosynthesis (*A*) refers to the assimilation of the atmospheric CO_2 and its conversion into sugars, the first basic organic compounds entering the metabolism. This process of CO_2 fixation uses the sun radiation as the energy source, and water as the electron donor, which in turn releases oxygen in the atmosphere. Dark respiration (R_D) or mitochondrial respiration (Atkin and Tjoelker 2003) employs the products of photosynthesis through the glycolysis (cytosol), the tricarboxylic acid cycle (TCA, matrix of mitochondria) and the electron transport rate chain (ETC, inner membrane mitochondria) to produce ATP and carbon skeletons needed for growth, cell maintenance, and other essential cellular processes. During the process of respiration, O_2 is consumed, and CO_2 is released to the atmosphere within the same order of magnitude than photosynthesis (Jansson et al. 2010), which highlights the importance of considering this process in the leaves, whole-plant and global models

Cyril Douthe and Jorge Gago have contributed equally with all other contributors.

C. Douthe · J. Gago · M. Ribas-Carbó · J. Flexas (✉)
Research Group on Plant Biology under Mediterranean Conditions, Universitat de les Illes Balears (UIB)-Instituto de Agroecología y Economía del Agua (INAGEA), Palma de Mallorca, Spain
e-mail: jaume.flexas@uib.es

R. Núñez
Geonica, S.A., Madrid, Spain

N. Pedrol
Research Group on Plant and Soil, Universidade de Vigo, Vigo, Spain

© Springer International Publishing AG, part of Springer Nature 2018
A. M. Sánchez-Moreiras, M. J. Reigosa (eds.), *Advances in Plant Ecophysiology Techniques*, https://doi.org/10.1007/978-3-319-93233-0_4

of carbon, water, and oxygen fluxes (Valentini et al. 2000; Canadell et al. 2007; Atkin et al. 2015). The velocity and extent of both processes can be assessed at the leaf level using infrared-based gas exchange analysers.

In vascular plants, photosynthesis is a complex interaction between biophysical processes and chemical reactions. Leaves are specialized photosynthetic tissues where the CO_2 from the atmosphere can be trapped into the leaf through the stomata to the substomatal cavity, subsequently crossing the mesophyll tissues that comprise several different cell structures (cell wall, plasmalemma, cytosol, chloroplast membrane, stroma), to finally reach the carboxylation sites of the RubisCO (Flexas et al. 2016). RubisCO (ribulose-1,5-bisphosphate carboxylase/oxygenase, EC 4.1.1.39) is the primary enzyme with a central role in photosynthesis, responsible for the CO_2 uptake by photosynthetic organisms. It is important to consider that this enzyme presents two different catalytic activities: carboxylation and oxygenation (i.e., fixation of both CO_2 and O_2). This result in direct competition between O_2 and CO_2 for the final reaction with RuBP (ribulose-1,5-bisphosphate) and the beginning of the Calvin-Benson cycle (Farquhar et al. 1980). While the carboxylase (fixation of CO_2) will end-up in sugar production, the oxygenase activity of Rubisco starts the photorespiration cycle that ends up in the net release of CO_2 (instead of CO_2 fixation). That is why photorespiration is considered as a counter-efficient process for the leaf regarding carbon balance: the oxygenase decreases RubisCO availability for the carboxylase process, it consumes electrons captured from light radiation, and it releases previously fixed carbon to the atmosphere (Galmés et al. 2005, and references therein).

In consequence, the leaf *in vivo* net CO_2 assimilation (A_N), that is measured using an infrared gas analyser (IRGA), is not a *true* photosynthesis rate, but the net balance between the rates of a carbon flux entering the leaf (the gross photosynthesis) and leaving the leaf simultaneously (the photorespiration and the mitochondrial respiration in the light). The combination of these three processes determines the leaf carbon balance (Valentini et al. 1995; Flexas et al. 2002, 2016) that drives primary productivity for any terrestrial ecosystem. This fact highlights the importance of the gas-exchange techniques when considering carbon fluxes in the context of climate change, agriculture, forestry and the understanding of natural ecosystems.

2 Theoretical Approach

2.1 Measuring Leaf Gas-Exchange: Basic Concepts and Measurements

The key point of these measurements is based on the tight relationship between CO_2 assimilation and water losses by transpiration (E) in the leaves through the stomata. By taking profit of this relationship, both specifically responsive infrared wavelengths for CO_2 and water vapour were used to develop sensors as the basis of the infrared gas-exchange analyzers (IRGAs). Simultaneous monitoring of the changes in CO_2 and water vapour across a leaf, provides a precise and integrated *in vivo* measurement of the net photosynthesis and the transpiration in illuminated samples, and also of the mitochondrial respiration rate and the residual transpiration in darkened samples (Field et al. 1989, 2000; Flexas et al. 2012b; Evans and Santiago 2014; Montero et al. 2016).

From the 1980s, gas-exchange analysers have become a common tool for plant ecophysiologists, and especially when the first models of "portable" equipment were developed, opening the possibility to measure plants in field conditions (Field et al. 1989, 2000; Long et al. 1996). There are two main approaches available, open and closed path gas-exchange systems. In closed systems, there is no net flow entrance of air in the chamber, and flux estimations are based on the variation of the gas concentration over time inside the closed circuit, which includes a cuvette with a leaf inside. Instead, the open pathways systems have a net flow of air entering and exiting the system, and the estimations are based on differences of concentration of two split fractions of air, one fraction having flown freely from the entrance to an IRGA, and the other one having passed through the leaf cuvette chamber into a second IRGA (Gallé and Flexas 2010). In this chapter, we will use as an example an open system (Fig. 4.1), the LI-COR 6400XT (LI-COR Inc., NE) portable gas-analyser coupled with chlorophyll fluorescence system.

Basically, the open system of the LI-6400 works as follows (Fig. 4.1): a pump forces the air flow to pass through a circuit, where air is split in two: a fraction goes straight to an IRGA, and the other fraction goes through a second IRGA after passing through the measuring chamber or cuvette with a leaf inside. [CO_2] and [H_2O] are measured in both IRGAs, the reference one reflecting the concentrations entering the chamber (C_e and W_e), and the sample one determining the concentrations after interaction with the leaf and exiting out of the leaf chamber (C_o and W_o).

The differences in [CO_2] and [H_2O] between these two measurements are used to determine the leaf net assimilation and transpiration rates (Fig. 4.2).

CO_2 and water vapour concentrations can be regulated in the equipment. If the user needs a determined concentration of any of the two gases, these can be decreased or even fully removed by passing air through different chemicals. For example, CO_2 is absorbed by soda lime ($Ca(OH)_2$ and NaOH granulates), and water vapour by drierite ($CaSO_4$) or silica gel (caution: before manipulating these chemicals check their safety datasheets carefully!). CO_2 concentration can be automatically regulated using the sensor readings using a mixer that controls the disposable compressed CO_2 gas cylinders to provide the required CO_2 air concentration into the already CO_2-free air (after previous full depletion using soda lime). In the LI-6400 it is not possible to increase the concentration of water vapour automatically using an analogous system; otherwise, with the new equipment LI-COR 6800 (LI-COR inc., NE), and as well the Walz GSF-3000 (Walz, Effeltrich), water vapour can also be automatically controlled regulating the gas concentration employing desiccant and humidifier chemicals integrated into the air pipe system with electro-mechanical valves.

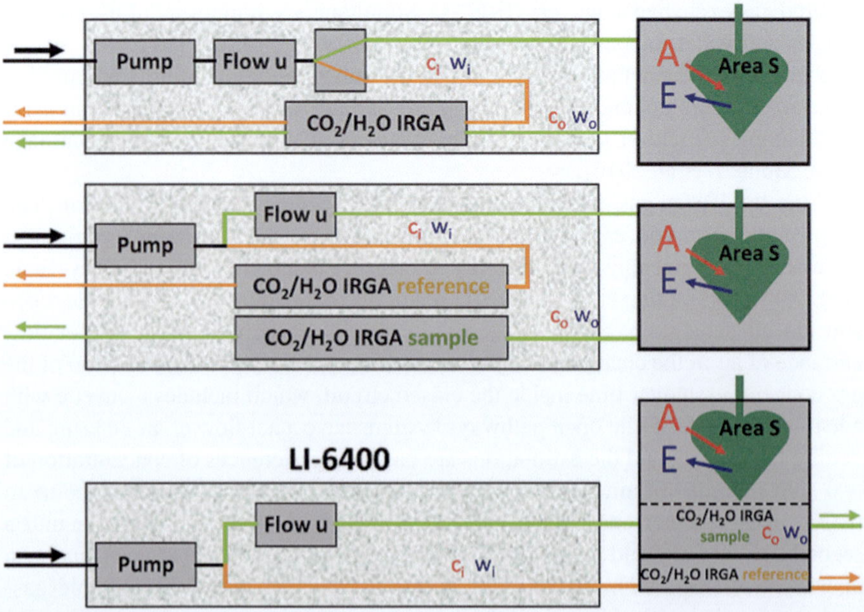

Fig. 4.1 In the LI-6400 Open Flow-Through Systems (bottom), the gas stream is split up into *sample* and reference *flow*, which continuous differential measurements without alternating; moreover, IRGAs are located in the head, so that gas measurements take place in the same space in which leaf is located, thus avoiding delay between response and measurement. For comparison of advantages with respect to previous models: (top) One absolute IRGA, switch between in and out, discontinuous measurements; (middle) two absolute IRGAs, continuous measurements, long tubing; (bottom) two absolute IRGAs, continuous measurements, IRGA in the head, shorter tubing

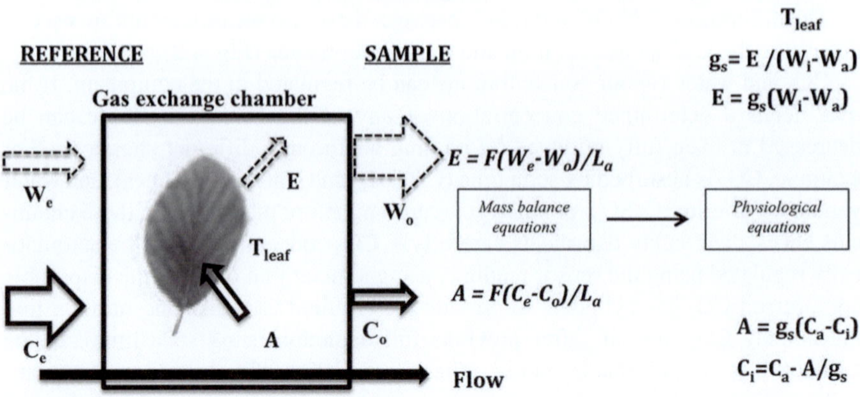

Fig. 4.2 Scheme of the gas-exchange measurement chamber with the main calculations. Photosynthesis (*A*) and transpiration (*E*) are calculated as differences in CO_2 and H_2O concentration, based on the readings of the IRGAs in the Reference and Sample circuits (mass balance equations), with *F* the flow and L_a as leaf area inside the measurement chamber. The mass-balances fluxes are hypothesised to reflect the pure physiological fluxes faithfully. From this basics equations, it can be further calculated the stomatal conductance (g_s) employing T_{leaf} (leaf temperature) and CO_2 concentration at the sub-stomatal cavity (C_i)

CO_2 and water vapour leaf fluxes are measured by the difference between the *reference* and *sample* circuits (Figs. 4.1 and 4.2) as early proposed by Gaastra (1959) and then modified like in von Caemmerer and Farquhar (1981):

$$A_N = u_e \frac{c_e - c_0}{L_a} - c_o E$$

where C_e and C_o are the CO_2 mole fraction at the chamber entrance and output, respectively; u_e is the incoming flow air (mol air s^{-1}), L_a is the leaf area surface (m^2), and E is the transpiration rate (mol H_2O m^{-2} s^{-1}). IRGAs can be used as well to measure leaf dark respiration when the leaf is under darkness conditions: photosynthesis and photorespiration are both suppressed by the absence of light through the chloroplast electron transport chain. When measuring leaf respiration with an IRGA, the "photosynthesis" measured by the device will appear as negative (the system applies the same equation under light or dark conditions). In consequence, that flux must be interpreted as a positive CO_2 flux corresponding to the leaf respiration, driven by the mitochondria in darkness.

Stomatal conductance to water vapour (g_s) can be calculated from E, by using the leaf temperature – wich is measured by a thermocouple placed inside the cuvette (caution: before each use, it should be tested that the thermocouple is working, well calibrated, and in close contact with the leaf to be measured!) – and accounting for the boundary layer effect. This calculation assumes that within the substomatal cavity the relative humidity is around 100%. This allows to calculate the W_i ([H_2O] in the sub-stomatal cavity), that in turn allows the estimation of the conductance of the water pathway using the first Fick's law of diffusion: $g_{sw} = E/(W_i - W_a)$ (with W_a the [H_2O] in the atmosphere – chamber in this case) (Fig. 4.2). Physiologists commonly use g_s more than E, since E is sensible to W_a (a variable atmospheric condition) and this is not a biological process. In turn, g_s is a full biological process mostly reflecting the degree of stomatal aperture (Osmond et al. 1979). Keep in mind that g_s can be affected by external factors, but the leaf itself actively controls it. Stomatal conductance can be expressed in terms of H_2O (g_{sw}) or CO_2 (g_{sc}), with $g_{sw} = 1.6\ g_{sc}$. The 1.6 factor denotes the difference in diffusivity in the air of the two molecules. This allows to calculate the CO_2 concentration at the substomatal cavities (C_i), applying again the first Fick's law of diffusion with $C_i = C_a - A_N/g_{sc}$ where C_a is the atmospheric [CO_2] (inside the chamber in this case) (Fig. 4.2) (Gaastra 1959; von Caemmerer and Farquhar 1981; Gallé and Flexas 2010).

From these measurements, another interesting parameter can be calculated, the water use efficiency (*WUE*), which represents the balance between carbon gains and the associated costs in water. So, employing instantaneous gas exchange measurements, it is easy to directly estimate it using the ratio between A_N and either E (so-called instantaneous WUE) or g_s (intrinsic WUE). This parameter drives plant productivity and the interaction of the plant with a changing environment, becoming highly important to improve irrigation and crop breeding strategies to face with the climatic change challenge threat to agriculture in the semi-arid regions (Gago et al. 2014).

These are the basic measurements that can be performed with a gas-exchange system. It is essential to know how the system works, the theory behind and its practical limitations to guarantee the precision and quality of your data. Moreover, these measurements take more relevance when considering that they are at the basis of many other procedures that are used to characterize the leaf physiology completely. We can also recommend excellent practical protocols for gas-exchange and fluorescence measurements already published as Evans and Santiago (2014) "Gas exchange using a LI-COR 6400", or the Licor LI-6400XT Manual itself.

2.2 Combining Chlorophyll Fluorescence and Gas-Exchange: Opportunities for Deep Photosynthesis Characterization

As mentioned previously, photosynthesis is driven thanks to the energy that comes from the sun. Leaves first capture the photon radiation by the chlorophyll molecules; then, this energy can be transferred through three different main processes: (1) used in photochemistry, where the energy captured is employed in the photosynthetic process; (2) dissipated by an exothermic reaction (heat dissipation); and (3) re-emitted in a longer wavelength (i.e., less energetic radiation than that received), the so-called chlorophyll fluorescence. These processes work in competition, so any decrease in one of them directly imply increases in some of the other two. This theory is employed to estimate the yield of chlorophyll fluorescence, capturing information about photochemistry and heat dissipation.

Currently, theoretical frameworks basically rely on the so-called "Kautsky effect", early observed when a leaf transferred from dark to light has its fluorescence that rapidly increases (within 1 s or so) and then slowly decreases to steady state. This pattern can be explained as follows: In the dark, heat dissipation processes depending on enzyme activity (e.g., those related to the xanthophyll cycle) are disabled (e.g., Murchie and Niyogi 2011; Demmig-Adams et al. 2012). But chlorophyll fluorescence and the early steps of photochemistry (i.e., light capture by antenna chlorophylls, charge separation in the reaction center, and most of the electron transport in the thylakoid) are active because being physical and not enzymatic processes. Photochemistry can absorb a reduced amount of the incoming energy, and therefore all the remaining leads to a rapid large increase of chlorophyll fluorescence from a basal level (F_o) up to its maximum capacity (F_m). Then, as the light is kept on, the RubisCO and other enzymes become activated, as well as the xanthophyll cycle-related heat dissipation. Since these two processes compete with chlorophyll fluorescence for the use of the energy absorbed by chlorophylls, their progressive light-induced activation leads to a subsequent slow decrease of chlorophyll fluorescence that will relax until reaching some steady-state value (F_s). Such effect reflects the competitive balance between the three processes that depend on the light intensity as well as on the physiological status of the leaf. If a short but intense pulse of light is applied now, photochemistry will become rapidly saturated, and chlorophyll fluorescence will rise again within 1 s or so, but still a lower value than Fm (referred as $F_m{}'$). This is because, contrary to darkness, heat dissipation under light condition is competing with chlorophyll fluorescence for the use of the light energy.

Several parameters were defined to determine the PSII photochemistry status, and probably some of the most useful are the following: the maximum efficiency of the PSII ($F_v/F_m = (F_m - F_o)/F_m)$) (dark conditions); the quantum efficiency of the PSII photochemistry (light conditions) ($\Phi_{PSII} = (F_m{}' - F_s)/F_m{}'$); the fraction of open PSII reaction centers, the photochemical quenching ($qP = (F_m{}' - F_s)/(F_m{}' - F_o)$; and the thermal dissipation of energy excess (non photochemical quenching $NPQ = (F_m - F_m{}'/F_m{}')$ (see Demmig-Adams et al. 1996 and Maxwell and Johnson 2000 for further information). In a dark-adapted non-stressed leaf F_v/F_m should range around 0.8, this is the maximum potential fraction of the energy that can be converted in photochemistry. Any decrease of this value would come from either an increase in thermal dissipation (non-photochemical quenching) or photodamage to PSII, indicating different types of photoinhibition process (Genty et al. 1990; Osmond and Förster 2006).

It was reported that the Φ_{PSII} in a light-adapted leaf is a proxy for PSII photochemistry, i.e., for the quantum efficiency of electron transport at the level of PSII. Thus, Φ_{PSII} can be used to estimate the photosynthetic linear electron transport rate (ETR, Genty et al. 1990; Laisk and Oja 2018), with ETR = Φ_{PSII} * PAR_i * α * β, where PAR_i is the incident photosynthetically active light radiation (μmol photons m^{-2} s^{-1}), β the fraction of absorbed light distributed between PSII and PSI, and α the leaf absorbance. Note that Φ_{PSII} and PAR_i can be directly measured by any gas exchange system coupled to a Fluorometer (like a LI-COR 6400 equipped with an LCF). It also gives a direct estimation of the ETR. Using the by-default parameterization (found in the literature), with β = 0.5 and α = 0.87, the α * β product value will be 0.435, but it is highly recommended to perform a direct estimation of this term. First, because ETR estimates are highly sensitive to the α * β value (and thus all the subsequent variables calculated from ETR, like g_m –the mesophyll conductance to CO_2,), and, secondly, because they can largely vary among species and conditions, especially α (see Pons et al. 2009 and Martins et al. 2013 for a detailed method description). Note that leaf absorbance can be measured independently employing a spectroradiometer and an integration sphere; nevertheless, there is no robust easy-to-use method for independent estimation of β. The best way to estimate the value of α * β is measuring the relationship between Φ_{PSII} and Φ_{CO2} through light or CO_2 response curves under non-photorespiratory conditions (Valentini et al. 1995; Martins et al. 2013). This procedure will be described below (Sect. 3.2.4).

2.3 Modeling Gas-Exchange: Going Deeper in the Leaf Photosynthetic Characterization

Combining gas-exchange with fluorescence technologies allows going deeper in the leaf physiology understanding. Some of the most important *in vivo* information that can be retrieved or estimated are: (1) the mesophyll conductance – g_m (that directly restricts the CO_2 available for the RubisCO at the chloroplast stroma site); (2) the rate of photorespiration (that is an important sink of energy and carbon loss for the leaf); (3) relative photosynthetic limitations occurring for a given leaf (stomatal, mesophyll and biochemical limitation; see Grassi and Magnani 2005); and (4) photosynthetic capacity parameters originally established in the model of Farquhar et al. (1980), i.e., the maximum velocity of carboxylation by RubisCO (V_{cmax}), the maximum capacity for electron transport rate and driving the Calvin cycle (J_{max}) and the triose-phospate use (TPU).

1. Mesophyll conductance estimation is based on combined gas-exchange and chlorophyll fluorescence data: this method was established by Harley et al. (1992) and is based on the basic photosynthetic stoichiometry, i.e., that, in the absence of photorespiration, 4 electrons should be processed in the thylakoid electron transport chain to reduce two molecules of NADPH, which are required to fix one CO_2 molecule in a carboxylation event. The idea is to find an estimate of C_c ([CO_2] at the carboxylation site in the chloroplast stroma), and then apply the Fick's law of diffusion again with $g_m = A_N/(C_i - C_c)$. The complete equation to estimate g_m is so:

$$g_m = \frac{A_N}{C_i - \frac{\Gamma^*\left[ETR + 8\left(A_N + R_{day}\right)\right]}{ETR - 4\left(A_N + R_{day}\right)}}$$

where R_{day} is the mitochondrial respiration in light and Γ^* is the CO_2 compensation point in the absence of R_{day}, and it accounts for the fact that measurements are performed under photorespiratory conditions. Keep in mind that even if a model is robust, its correct parameterization (i.e., attributing a value to each parameter in the equation) is crucial to obtain reliable results. So, several methods allow estimating R_{day}, two of them being the Yin et al. (2011) method (requires a light response curve coupled with a Fluorometer), or the more simple Niinemets et al. (2005, 2009) approach, using an empirically-based agreement that R_{day} equals to half R_{dark}. Γ^* can also be estimated *via* several methods: or by gas exchange, that needs two A/Ci curves each performed at 21% and 2% [O_2] (see Yin et al. 2009), or by *in-vitro* estimations of the RubisCO kinetics, from which Γ^* is derived (see Galmés et al. 2017 for an extensive comparison of the methods, and Hermida-Carrera et al. 2016 for RubisCO kinetics database in crops). Note that other methods were developed along the years for those two parameters (like the "Kok" method for R_{day}, ot the "Laisk" method for both R_{day} and Γ^*), but these are now considered non-reliable. Recent literature is now comparing the different methodologies to establish the robustness of each one (see Walker et al. 2016; Galmés et al. 2017; Walker et al. 2017).

2. Estimation the photorespiration: this was established by Epron et al. (1995), based again on the basic stoichiometry of electrons required for a carboxylation or an oxygenation event: $Rp = 1/12[\mathrm{ETR} - 4(AN + \mathrm{Rday})]$.

3. Estimating the relative limitations to photosynthesis: this approach was first proposed by Grassi and Magnani (2005), based on an earlier model by Jones (1985), in which it was not considered the limitation by mesophyll conductance. This kind of analysis can be useful to compare different species (e.g., Carriquí et al. 2015), or compare the photosynthetic performance and limitation under different stressed environments (e.g., Gallé et al. 2009). The aim is to decompose the different factors that limit the photosynthesis at a given moment and to establish a hierarchy of those different limitations. Two of them concern the diffusive limitation: the stomatal (l_s) and the mesophyll (l_m) limitation. They come from the fact that the A_N flow is considered a continuous flow restricted by two resistances (1/conductance) in series, since we assume that $A_N = g_s\,(C_a - C_i) = g_m\,(C_i - Cc)$. The third limitation comes from the carboxylation itself (l_b). So, we can establish the following equations based on Grassi and Magnani (2005):

$$l_s = \left(g_{tot} / g_s \bullet \partial A_N / \partial C_c\right) / \left(g_{tot} + \partial A_N / \partial C_c\right)$$
$$l_m = \left(g_{tot} / g_m \bullet \partial A_N / \partial C_c\right) / \left(g_{tot} + \partial A_N / \partial C_c\right)$$
$$l_b = g_{tot} / \left(g_{tot} + \partial A_N / \partial C_c\right)$$

where g_{tot} is the total conductance to CO_2 between the leaf surface and the carboxylation sites ($1/g_{tot} = 1/g_s + 1/g_m$). Note that this model has been improved by Buckley and Díaz-Espejo (2015), but the complexity of the latter is so that, in many cases, parameterizing this model would not be feasible, for which the Grassi and Magnani approach is still useful.

4. Retrieve the biochemical photosynthetic parameters of a leaf: this approach was used by Farquhar and colleagues at the time to establish their extendedly used model (Farquhar et al. 1980). Their idea consisted in seeing the measured photosynthetic rate as if it was a 'reaction velocity' in response to 'reaction substrate availability' (approached by the C_i estimated during IRGA measurements). In this way, by performing gas exchange meaurements along a CO_2 gradient (i.e., the A-C_i curves), it was possible to apply well known and simple enzyme-reaction equations to estimate the maximum carboxylation by the RubisCO (V_{cmax}, from the portion of the curve where the substrate CO_2 is most limiting, under the rule of the Michaelis-Menten law for the case of inhibitory competition by substrate O_2), the maximum capacity for electron transport (J_{max}, from the CO_2 non-limiting region of the curve, reflecting a limitation by RuBP regeneration and, thus, the activity of photochemistry and the Calvin cycle), and the rate of triose-phosphate utilization (TPU, from the saturated part of the curve at very high [CO_2]). All these parameters can be extracted from the analysis of a complete A-C_i curve performed at ambient O_2 and under saturating light. However, as C_i does not reflect the actual CO_2 concentration at the chloroplast stroma (C_c) it is better to apply the model after considering g_m, i.e., to A-C_c curves (Flexas et al. 2012a). It can be done using gas exchange and chlorophyll fluorescence esti-

mates of g_m and C_c as outlined in previous sections, or directly based on pure gas exchange measurements. For the latter, Ethier and Livingstone (2004) modified the equations use for the fitting of the measured data against the theoretical model (to retrieve V_{cmax}, $J_{max,}$ and TPU). They transformed the Farquhar's model original equation into several non-rectangular hyperbolae, improving the quality of the estimated parameters. They also included g_m into their model (originally considered as infinite in the 1980s Farquhar's model), allowing a g_m estimation without the employment of the chlorophyll fluorescence method (see Ethier and Livingstone 2004). The most interesting point of this method is that it provides a second independent approach to estimate g_m that can be useful to reinforce its estimation through the other methodologies mentioned. Several tools have been developed to facilitate this purpose, like the Excel sheet of Sharkey et al. (2007; Sharkey 2016).

Other useful physiological parameters can be extracted from gas exchange and fluorescence measurements. Other Excel tools exist, like that provided by Bellasio et al. (2016). These authors propose a systematic analysis of light and CO_2 response curves (under ambient and low O_2 concentration) and, in a step-by-step approach, the tool provides: R_{day}, initial PSII (photosystem II) photochemical yield, initial quantum yield for CO_2 fixation (Φ_{CO2}), fraction of incident light harvested by PSII ($\alpha * \beta$ product), initial quantum yield for electron transport, electron transport rate (ETR), photorespiration, stomatal limitation, RubisCO (ribulose 1·5- bisphosphate carboxylase/oxygenase) rate of carboxylation and oxygenation, RubisCO specificity factor, g_m, light and CO_2 compensation point, and RubisCO apparent Michaelis–Menten constant and V_{cmax} (RubisCO CO_2-saturated carboxylation rate).

3 Practical Approach: Hands-on Protocol

3.1 Preparing Your Gas-Exchange Analyser for Precise Measurements

At the beginning of the preparation of your equipment, the most important action is the calibration of the sensors to ensure that your device can reproduce reliable results. For this purpose, regular checks of the different sensors of the gas exchange system are crucial; but all sensors do not require the same checking frequency.

At the beginning of this "hands-on protocol", we split the "checks" of your equipment depending on the frequency that it is recommended to apply them. First, we propose "daily checks" the ones that need absolutely to be done every day before start any measurement, and second "long-term" checks to ensure quality maintenance of the equipment. For this purpose, this "hands-on protocol" employs as an example the LI-COR 6400XT equipped with the fluorescence chamber. All the IRGAs are based on the same concepts, so users of others equipment can also find useful the recommendations that we described below. This protocol is intended to be comple-

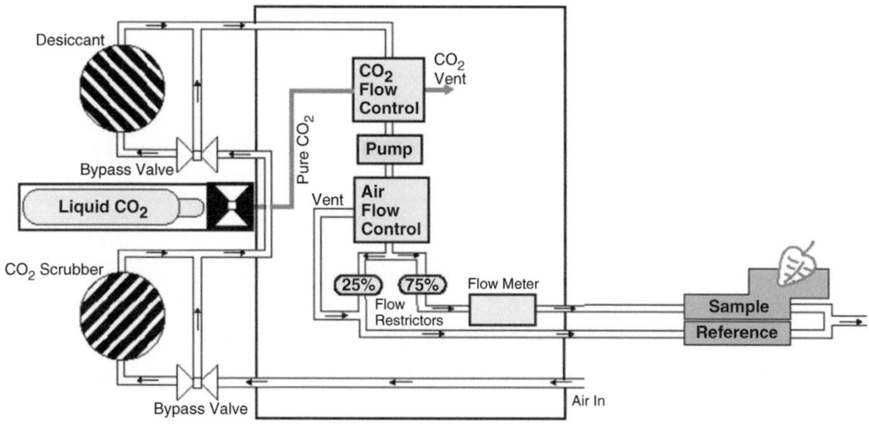

Fig. 4.3 Schematic gas flow and parts of the LI-COR 6400

mentary to the manufacturer's manual (LI-COR 2012), so we strongly recommend a careful reading of the manual to guarantee the proper utilisation of the equipment. Figure 4.3 will help you to understand what you are checking, where, and why.

3.1.1 Daily Checks Before Measuring

These checks consist of evaluating the most important sensors of the gas exchange system to give the best chance to perform good and reliable measurements.

First of All: "Check Around"

- **Ensure to plug every cable and tube in its right position**. For example, the *Reference* and *Sample* tubes on the console side have the same connector, check that the black taped tube (*Sample*) is on its good (*Sample*) position.
- **Check that the exhaust tube** (right-angled semi-transparent 10 cm tube, bottom-side of the chamber) is in place.
- **Check** that **the gaskets of the chamber are in good state and overlap** each other perfectly when you close the leaf measurement chamber.
- When you are sure that everything is at its place, go to next step.

1. **Start the machine**, and then ***Scrub*** **the desiccant and soda lime tubes**. Meanwhile, the system is opening and you are doing the other checks, the pump system will empty the airflow circuit of CO_2 and water. You may save some time at the moment to check the *Zero* of the IRGA. You can also place CO_2 cartridge to **fill the CO_2 mixer if needed**.
2. **Check the "Match Valve" test** visually during the opening sequence of the system (the downside of the head). You can also directly activate it in the measurements menu "Match" button to check its good functioning. This is what happens:

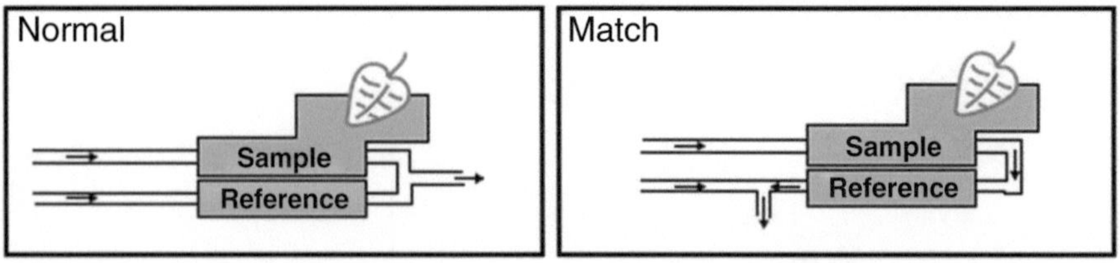

3. **Check the *Zero* of the flow meter**. For that, go to "Calib Menu" and select the "Flow meter zero..." wait for the countdown, then the values in mV should be within 1 mV from zero. Adjust it in consequence, but keep in mind this parameter is not likely to change day by day.
4. **Check the "*Max*" of the flow**. Go to "New Measurements", and fix the flow to 1000 µmol s^{-1}, then the reading value should be >700 µmol s^{-1} (menu *b*). On the contrary, a resistance on the flow is happening through the air circuit: check the air mufflers (white filters) there is two inside the desiccant and two in the soda lime tube. They are very likely to provoke this problem. If ok, fix back the flow to 500 µmol s^{-1} (for 6 cm^2 chamber) or 300 µmol s^{-1} (for 2 cm^2 chamber) as standard measurement flow rates.
5. **Check the "*Zero*" of the thermocouple**. To do that, disconnect the thermocouple from the LI-6400 head (purple 2-pin connector, see the manual for more information), then the *Tleaf* (leaf temperature value) value must be close by ± 0.1 °C from the *Tblock* (block temperature value) value (menu *h*). If not, this can be adjusted with the small screw of the downside of the head, close to the *Reference* and *Sample* tubes connection (see manual for further information). Adjustment of *Tleaf* zero must be made in stable temperature condition and after a warm-up of the system (ideally ~30 min).
6. **Check the sensitivity of the thermocouple**. Gently touch the thermocouple with the tip of your finger and check the proper variation in *Tleaf* (menu *h,* it must increase). Check later also that each *Tblock*, *Tair,* and *Tleaf* gives reasonable values. For your next measurements set the *Tblock* or *Tleaf* as desired.
7. **Set the light "ON"**, usually with 90% red and 10% blue. Check that LEDs are active, and pay attention to the purplish color that ensures that blue LEDs are active. Check that the reading value of PARin (menu *g*) agree with your settings (chamber must be closed).
8. **Check the leaf fan** (or mixing fan). Change its value (Function Key 1, f1) from 5 to 0 to stop it, and then set it up again at 5. If you listen carefully (place the chamber close to your ear because the noise change is not easy to distinguish), you will hear a change of the sound coming from inside the leaf chamber. If not, check the fuses inside the console, some debris that can obstruct the fan. Unfortunately, if the leaf fan is broken, you need to replace it following the manufacturer's manual strictly.
9. **Check the Zero of the IRGA.** This procedure consists of passing CO_2 and H_2O free air thought the IRGA; then check if the sensor reading is close to zero. Look for values within 5 µmol mol^{-1} for CO_2 and 0.5 mmol mol^{-1} for H_2O (fol-

lowing the LI-COR manual, v 6.2, p 4–5). Focus on the *Reference* value only, since *Sample* will take the *Reference* value after a *Match*. Zeroing the IRGA must be done with care, essentially with totally free CO_2 and H_2O air and a perfectly sealed circuit. In most of the cases, non-zero values come from non-fresh chemicals or leaks through the air circuit. The common "guilties" are bad tube connections, the bad seal of the leaf chamber (2 × 3 o-rings at the chamber/head connection), or leaks from the desiccant and soda lime tubes themselves or their connection with the console. An IRGA is unlikely to drift from several CO_2 μmol mol^{-1} between days, as the temperature does not change drastically. **Zeroing the IRGA is encouraged to be performed only in laboratory conditions with calibrated gas tanks (pure N_2, see below).**

10. **Set the desired CO_2 concentration.** You need to close well the chamber, adjust the tight screw and wait for mixer stabilization. Then do a "Match" to get the *Sample* reading be "calibrated" based the *Reference*. Now, you can check possible leaks of the chamber gaskets by a blow-test around the chamber. If there is no increase >1 μmol CO_2 mol^{-1}, **then, after all, you are ready to measure a leaf.**

3.1.2 "Long-Term" Maintenance

"Long-term" maintenance procedures are not likely to be performed every day. They are most likely to be done… let's say, once per month, for example, but this mostly depends on the frequency and intensity employing of your equipment. In general, they can be performed when some problems are detected and can help to solve it. A typical example of these type of maintenance could be the CO_2 Mixer calibration: if the CO_2 Mixer needs too much time to reach the targeted [CO_2] (or the same for Light intensity and the light source), this can be a signal to do the "Mixer Calibration" procedure. However, since "prevention is always better than cure", it is better to check these procedures periodically.

Internal Calibrations (Calibrations that Do Not Require an External Item for the Procedure)

1. Mixer calibration

This routine checks the control signal (mV) of the CO_2 mixer and the [CO_2] delivered by the mixer itself into the circuit. This procedure is not strictly a calibration itself, in the sense that does not adjust the reading value of a sensor, but helps the CO_2 Mixer to reach more rapidly the desired [CO_2]. Go to "Calib Menu" and follow instructions detailed in the manual equipment. When the mixer needs too much time to reach the targeted [CO_2], this routine can solve this problem. Also, keep in mind to frequently change the filter present inside the mixer (do it without CO_2 cartridge inside), because the accumulated oil can also provoke problems of CO_2 regulation.

2. Light Source Calibration

The LEDs source and the LCF (Leaf Chamber Fluorometer) can be calibrated in the same way it is for the Mixer. The light source will associate the different voltage feeding the LEDs with the corresponding reading value of PARin. This association will help to reach faster the desired light intensity. There is also a "Zero PARin" procedure that checks the offset reading that can remain in darkness. Check it monthly can avoid this problem.

3. Fluorometer calibration

The Leaf Chamber Fluorometer (LCF) also needs some specific calibration/checks. We will find the same procedure as for classical LEDs light source: a check of the *Zero*, and a calibration curve mV *versus* measured light. However, other procedures are specific to the LCF. One of the most important is the **"Square Flash Calibration"**, that is impaired with the **"MultiPhase Flash" method** (MPF). It is highly recommended to set the flash method on "MPF" (Function Key 8, f2, type = "Multiphase"), since it will ensure a better estimate of the Fm′ parameter even when saturation values are not easy to reach with your leaf (Loriaux et al. 2013). You will find this in the "Calib menu", then "LCF source". Keep in mind that other procedures can help to determine the "Optimum Flash Intensity" and the "Optimum Measuring Intensity". The first one is less important since using the MPF method avoid the previous commented problems of PSII saturation. The second one helps to determine the ideal intensity for Fo determination (in darkness, without inducing photosynthesis).

External Calibrations (Calibrations that Required an External Item for the Procedure)

1. Calibrating the IRGA: Zero and Span

Zeroing the IRGA: As said in the LI-COR manual (all versions), "*You will do more harm than good, however, if* you *dutifully re-zero every day using chemicals…*". Another important thing to bear in mind is that "*If conditions (temperature, mostly) haven't changed a great deal since the last time you zeroed the IRGAs, it won't need adjusting*". So, we recommend to do the check of the *Zero* on a daily basis, but zeroing the IRGA only in laboratory conditions, with pure N_2 tanks. The best way is to connect the N_2 tank to a "Y" tube-connector that feed both *Reference* and *Sample* circuits to the LI-6400 head (avoiding the console), with a flow of about 0.5 to 1 L min^{-1}. Go to "Calib menu", then to IRGA, then IRGA *Zero*. Waiting for stabilization time is crucial at this moment, and more specifically for zeroing the water. The phenomenon of adsorption/desorption of water in many components (plastics, overall), induce a longer stabilization of the water zero. This procedure must also be done with a fully warmed-up machine. This means that a proper "*Zeroing*" procedure needs at least 30 min for warming and another 15–20 min for fully stable gases concentration.

Setting the Span: The span corresponds to the sensitivity of the sensor (here the IRGA) to the concentration measured. This corresponds to the slope of the relationship between measured concentration and real concentration. So, you will need a tank with a certified CO_2 concentration (for CO_2 span), or a dew-point generator that fixes a known concentration of water (for H_2O span). In principle, you need to use a known concentration that is within or slightly above the concentrations you are likely to measure. For example, the LI-COR factory uses tanks of 1500 μmol CO_2 mol^{-1}. The procedure consists on the same set-up as for "*Zeroing*" the IRGA: use a "Y" tube-connector, set a flow of about 0.5 to 1 L min^{-1}, and plug it directly to the LI-COR 6400XT head, then wait for stabilization. Then use the adjust button to match the reading concentration to the value of the tank (or the dew-point, in case of water span). The water span can be done with the LI-810 dew-point generator, that it was specially designed for this purpose.

2. Light calibration

As times passes, the PAR_{in} sensor or both LCF and LEDs chamber can drift, leading to an over or under- estimation of the real *PAR* reaching the leaf. Light sources can be sent back to LI-COR factory for calibration, but checks can be done manually by the user. This possible drift can take larger importance when the *PAR* is used for further calculations, for example for ETR calculation that is subsequently employed for several equations as photorespiration estimation or mesophyll conductance (Valentini et al. 1995; Flexas et al. 2012a). To do that, is it possible to fix a *PARout* sensor on the bottom part of the chamber (here it is really important than the sensor are centered and at the same exact position and distance to the light than leaves are placed in the chamber). Obviously, this sensor must be absolutely well calibrated, ideally a new fresh sensor from LI-COR factory or with less than 1-year-old calibration. Then, establish a calibration curve by changing the *PARin* value in let's say 4–5 steps, and record the reading value given by the *PARout* sensor. Then it is possible to calculate the slope and the intercept of this relationship, then use those values to set-up your next light intensity set-up.

3.2 *Making a Measurement*

Now, you are almost ready to perform a precise measurement of your leaves. The equipment is prepared but there are some important considerations that you need to care when you are measuring the leaf gas-exchange and fluorescence in plants.

3.2.1 Plants Need Time to Adapt to Your Measurements Conditions

The ideal case is to measure the plant without affecting its behaviour and physiological status, as leaves and plants are continuously interacting with the environment that means that leaves will need time to adapt to the new conditions in the measurement chamber. Thus, reach leaf steady-state becomes really important for the reproducibility of your data and as well for the comparison with another data from other researchers worldwide.

Also, take in mind that even the photosynthesis chamber is providing stable, uniform and regulated conditions between all your measurements, changing the condition of the rest of the plant will affect the behaviour of the portion of leaf that is inside the chamber. So, it could be an important source of variability in your data measure the photosynthesis inside the chamber meanwhile the rest of the plant is under dark conditions, or the contrary measure the respiration in the targeted leaf tissue inside the chamber meanwhile the rest of the plant is under light conditions.

3.2.2 Selecting Your Target Leaves

Selecting the same type of leaf can avoid a lot of variability between your biological replicates. Usually, the intra-plant variability (among all leaves within the same plant) is much larger than the inter-plant variability (among the same type of leaf among several plants grown in the same conditions). Conventionally, in the literature, the researcher classically uses the "youngest fully expanded leaf". This selection ensures to have a fully functional leaf that is not affected by ontology (leaf age). This leaf, in the vast majority of cases, must be a direct sun/light exposed leaf (not inside the canopy, not overlapped by other leaves). Of course, the leaf must be healthy and vigorous, not presenting any sign of herbivory, degradation, chlorosis, nitrogen deficiency, or any factor that can affect the physiology of the leaf.

3.2.3 The Leaf Inside the Chamber

Ensure Tight Closure Between Gaskets and the Leaf

When you have chosen the "good" leaf, then, how to clamp-it in the LI-6400 chamber? The aim is to tight the leaf enough to reduce leaks (CO_2 entry/exit between the chamber and the atmosphere) as much as possible, but not too much to do not damage the leaf. To do so, it is good to use the screw of the head handle to adjust how tight the leaf is: "*enough, but not too much*". To ensure that no or few leaks are present, after clamping the leaf and waited for ~30s to ensure a stabilization of the gas *Sample* circuit, you can gently blow around the chamber gaskets. Check for any variation within 1 μmol mol^{-1} of the CO2S ([CO_2] in the sample = leaf chamber).

Reading Correct Leaf Temperature Values

The reading value of *Tleaf* should be checked after placing the leaf inside the chamber. This value must be coherent with the *TBlk* and *Tair*, obviously depends on the light intensity selected (the higher radiation will tend to be higher the leaf temperature). Leaf temperature is a function of transpiration and stomatal conductance, so stressed plants with reduced stomatal conductance would tend to show higher *Tleaf* values. Keep in mind that measuring at field conditions with the open-top chamber (no light source, direct sun high radiations) and high air temperature will increase leaf temperature importantly over the selected *TBlk*. If some suspicious values occur or leaf temperature is unstable, check if the leaf perfectly contacts the thermocouple. The ultimate check consists of checking the reading of the Tleaf with a standardised thermocouple (external confirmation).

Area Correction

All the gas-exchange calculations are taking into account the area of the leaf. Photosynthesis chambers have a defined area, so if your leaf coverage the entire measurement area then calculations can be done automatically. Several species have leaves that do not allow you to cover the entire measurements area. Thus area correction is needed. Fortunately, the LI-COR output excel file provide all the formulae: so you just have to correct the area values and the rest of calculations change automatically. Area correction is typically done taking a picture of the exact piece of leaf inside your chamber, and then area calculation can be done with an image analysis software, for example, ImageJ (Carriquí et al. 2015; Tosens et al. 2016).

Light Saturation Measurements

In most of the cases, you want to measure your plant at saturating light to avoid changes in photosynthesis from intensities below saturating conditions. On the other hand, in some cases like shade species, too high saturating light radiation can induce photoinhibition. For this, it is recommended to determine the correct light intensity at which your plant saturates for light. To do so, perform a light response curve and take the minimum light intensity when the photosynthesis is saturated.

Air Vapor Pressure Deficit, Humidity and Stomata Interaction

Once the leaf is stabilized in the chamber and *Tleaf* reading is correct, others checks must be done concerning the water vapour inside the chamber. This parameter can be assessed when you read the value of *H2OS* (concentration of water vapor in the *Sample*) or *HR_S* (relative humidity in the *Sample*). Preferably, check the relative humidity and ensure that is comprised between 40% and 70%. Dry air will increase the Vapour Pressure Deficit (VPD_a) around the leaf provoking a stomatal closure (stomatal conductance, Cond) that can, in turn, induce a possible decrease of the photosynthetic rate (Photo) (Pérez-Martín et al. 2009). In the other hand, relative humidity higher than 80% in the LI-COR 6400XT can affect the stability of the CO_2 readings importantly, as well reduce the precision of estimation of the water-exchange.

It is also important to check as well that the difference in water vapor concentration between the surrounding atmosphere and that inside the chamber (*H2OS*). A too large gradient between chamber and atmosphere can provoke leaks of water vapor, which will affect the estimation of stomatal conductance and leaf transpiration rate (*Trmmol*). To check this, before clamping the leaf, do a check of the *H2OS* when the chamber is open, and let the surrounding air enter the chamber. This will give an idea of the water concentration and the relative humidity of the surrounding atmosphere during the measurements.

The Importance of Reaching the Leaf Steady-State

The leaf steady-state means the stabilization of all photosynthetic parameters before starting the measurements. Steady-state is a crucial point for good measurements. Once the leaf is inside the chamber, usually at saturating light, the photosynthesis is very likely to stabilize pretty rapidly. This comes from the fact that carbon fixation depends directly on the light available to feed the electrons transport rate. Those reactions are very fast, so any change in light is almost instantaneously reported on the photosynthetic rate.

The other factor limiting photosynthesis is the availability of CO_2 at the RubisCO site. This second factor is directly affected by the degree of aperture of the stomata, i.e., the stomatal conductance. This parameter, in turn, changes very slowly over time. The time needed to change from closed stomata to fully open can take 1 hr for some species. The steady-state is reached when both photosynthetic rate (Photo) and stomatal conductance (Cond) are fully stable. At that moment, the leaf is in steady-state, the measurements can begin. So once the leaf is clamped, wait minimum 15 min and then check the stability of Photo and Cond, over a time scale of 10 min in the LI-6400 graphs menu. If both are stable over a 10 min time lap, there is a good probability that the steady-state is reached. For very fine measurements or specific species, a steady-state of 1 h can be required. Typically, plants at field conditions reach the steady-state faster than plants from growing-chambers.

3.2.4 Ensuring the Precision of Your Measurement

The estimations of photosynthesis and transpiration rate are based on the difference of concentrations between the *Reference* and the *Sample*. Knowing that each IRGA has its own error of measurement (maximum deviation of ±5 μmol mol^{-1} from 0 to 1500 μmol mol^{-1}, and ± 10 μmol mol^{-1} from 1500 to 3100 μmol mol^{-1}), when the difference of concentration between *Reference* and *Sample* is very low, the precision of measure decreases. When the delta (of CO_2, for example) gets close to 0.5 μmol CO_2 mol^{-1}, then it is comprised within the measurement error. No reliable data can be obtained this way. One solution that can help to avoid or at least reduce this problem is to decrease the air flow through the chamber. This action decreases the air turn-over of the chamber and lets the leaf affecting more the [CO_2] and [H_2O] in the chamber. This provokes an increased delta, so we obtain a better precision but employing more time to make the measurements. Decreasing the air flow through the chamber is especially useful when measuring leaf respiration, plants with extreme low photosynthetic rates, or plants under a treatment (water stress, low light, low nitrogen, etc.). Since normal operating flows are 500 μmol s^{-1} (for 6 cm^2 chamber) or 300 μmol s^{-1} (for 2 cm^2 chamber), consider that the flow can be decreased up to 200–150 μmol s^{-1} depending on the chamber. Lower values would affect response time. In certain cases, e.g., for species with low exchange rate such as mosses, lower flow values could be attained. Another possible solution is to increase the area of measurement by choosing larger leaves.

Why do not work all the time at very low flows? Low flows also increase the influence of aside (chamber/surrounding atmosphere) CO_2 and H_2O exchanges (like leaks, typically). The second limit is the risk of condensation inside the chamber. At very low flow, the water coming from leaf transpiration can accumulate too much inside the chamber (rising air humidity). This accumulation increases the probability to reach the dew-point of the chamber (100% relative humidity), thus provoking water condensation inside the chamber. Condensation is a dramatic problem for the gas exchange user. Condensation will "trap" water inside the chamber inducing a wrong estimation of the transpiration rate. Moreover, once the water has condensed inside the air circuit of the LI-6400 (chamber, or in another part), it is very hard to fully re-evaporate this liquid water to come back to the proper conditions of measurement. If you have problems with condensation in your circuit, you must dry it. This can be done by connecting a dry air source to the system (console inlet or head) with a vent to avoid over or under-pressure. This can take several hours, overnight is recommended.

3.3 *Further Considerations and Useful Tips*

3.3.1 Leak Corrections

As said above, CO_2 (and H_2O) exchanges between inside the photosynthesis chamber and the surrounding atmosphere can be present. This is particularly the case during A/C_i curves since the chamber (*Sample*) [CO_2] changes dramatically (from 0 to 2000 μmol CO_2 mol^{-1}). Leakage can produce an artifactual photosynthesis rate that does not come from the leaf. For example, at low CO_2 the CO_2 is going from the atmosphere to inside the chamber, decreasing the estimate photosynthesis; at high CO_2, the CO_2 is entering from the inside of the chamber to atmosphere, increasing the estimate photosynthesis. The importance of the leaks flow will directly depend on the morphology of the measured leaf (thickness, regularity of the shape, size of leaf's vein).

In order to compensate this effect, "leaks curves" must be performed. The basic idea relies on employing the very same leaf to reproduce the leakage of its surface with the gaskets of the chamber, to check for the physical leakage. However, obviously, you do not want any biological gas-exchange from the leaf disturbing the physical leakage that you want to analyse. So, you can stop the biological gas-exchange using different manners: submerge the leaf in boiling water for 2–3 min, place in an oven at 110 °C for 1–2 min, or employ an oven for 2–3 min at 110 °C. Any case, you must ensure that the tissue is dead (thus no gas-exchange) with a Log with fluorescence measurement, and check the value of the ETR. If it is negative, then your leaf will not interact with the chamber atmosphere. Obviously, the method employed must preserve as much as possible the structure of the leaves to simulate the interaction between leaf surface and chamber gaskets that drives the leakage. Now place the leaf in the chamber and perform a classical A/C_i curve (ideally the same used to measure the functional leaf). The A/C_i curve performed with the dead leaf will produce a response curve of apparent photosynthesis (*Photo*) to the [CO_2] changes in the chamber (*CO2S*). This relationship is positive, relatively linear, with min and max values of apparent photosynthesis from −1 to 4 μmol m^{-2} s^{-1}. A trick: in theory, the *CO2S* at which *Photo* = 0 should correspond to the [CO_2] of the surrounding atmosphere at the moment of the measurement. The next step is to calculate the equation of the obtained relationship *Photo* = a**CO2S* + b where *a* and *b* are the slopes and the intercept of a linear function. This allows calculating the apparent photosynthesis (leaks) that occurred during the A/C_i curve performed with the functional leaf, using its own *CO2S*. The apparent photosynthesis – or leak, will be rested to the measured photosynthesis to obtain the leaf photosynthesis corrected for leaks. Keep in mind that any variable calculated from the photosynthesis rate, like C_i must be corrected as well. Fortunately, the excel data files (.xls) generated by the LI-COR 6400XT recalculate all those variables in consequence.

3.3.2 Correction of the ETR: Φ_{PSII} and Φ_{CO2} Under Non Photorespiratory Conditions

Some parameters, like ETR, need some specific parameterization procedure to be correctly estimated. As it was reported previously:

$$\mathrm{ETR} = PAR_i^{*}\Phi_{PSII}^{*}\alpha^{*}\beta$$

where PAR_i is the incident photosynthetically active radiation, Φ_{PSII} is the quantum yield of the PSII, α the leaf absorption (by default 0.87) and β the electrons portioning between PSI and PSII (by default 0.5). There are some methods which estimate them separately, but here we will see the main method used to estimate α * β product.

The aim is to generate a linear relationship between Φ_{PSII} and Φ_{CO2} under non-photorespiratory conditions (low, ~2% O_2 atmosphere). The source of variation can be light of CO_2 (knowing that the CO_2 method will need leaks correction). See Martins et al. (2013) for an extended description and test of the method. To do so,

A_N*/PAR* or A_N/C_i curves should be performed by feeding the LI-COR 6400XT with air without O_2 (typically N_2 air-compressed tank). This can be achieved by plugging the inlet of the LI-6400 console to a tank of pure N_2, with the caution a place vent (using a "Y") between them to do not damage the pump because of overpressure. To check that low O_2 air truly feeds the leaf, the steady-state photosynthesis at low O_2 should be around 20–30% higher than under 21% O_2 (because you are inhibiting the photorespiration activity of RubisCO). After checking the increase of photosynthesis (wait ~10 min after plugging to the N_2 source), you can start the *A/PAR* or A/C_i curve. Only the very linear part of the relationship will be employed in the calculations. Once the (positive) linear part of the Φ_{PSII} and Φ_{CO2} relationship is selected, extract the slope to have: $\alpha * \beta = 4/\text{slope}$.

3.3.3 Physiological "Tricks" to Keep in Mind

Once the machine is well calibrated and the leaf correctly placed in the chamber, there are some tricks that you are better to know to ensure that data provided are reliable.

- There is a "*general rule*" about the proportion of photosynthetic rate and the stomatal conductance, for the vast majority of the species. In general, when the photosynthesis is ca. 10 μmol m^{-2} s^{-1}, the g_s is about 0.1 mol H_2O m^{-2} s^{-1}. Species with high photosynthesis (>20 μmol m^{-2} s^{-1}) will systematically present high g_s values (0.2–0.3 mol m^{-2} s^{-1}). For example, a plant with very high photosynthesis cannot have very low g_s. The inverse can be more likely (low photosynthesis, high g_s) but this will be true only for specific species (typically from wet/flooded areas, or from shade conditions). Any case, you have plenty of data in the literature analyzing this relationship (for example Flexas et al. 2013; Gago et al. 2014) and, of course, the topic deserves from you a previous search in the literature to know reported photosynthetic data of your species.
- Combining gas-exchange with fluorescence data is very useful to check the ETR/A_N ratio. The theory says that photosynthesis needs at least 4 electrons to fix one molecule of CO_2 through the Calvin-Benson cycle; so, knowing that the photorespiration is also present (and also consuming electrons), the ETR/A_N ratio range from 8–10 for C3 species (Flexas et al. 2002). For C3 species, low ETR/A_N ratio indicates that it could be a problem in the estimation of the *ETR*. Alternatively, if you are using thick leaves, it could be an impairment between *ETR* (collected from the upper cell layer of the leaf) and net photosynthesis, that integrates all the layer of the leaf. Wrong estimation of $\alpha * \beta$ product can be the cause, wrong estimation of the *PAR*, or non-saturation of the PSII (too low F_m' values). Higher values of this ratio will indicate stress in your plants as typically CO_2 assimilation shows a steeper slope reduction under stress than the *ETR* (Flexas et al. 2002).

References

Atkin OK, Tjoelker MG (2003) Thermal acclimation and the dynamic response of plant respiration to temperature. Trends Plant Sci 8:343–351

Atkin OK, Bloomfield KJ, Reich PB, Tjoelker MG, Asner GP, Bonal D, Bönisch G, Bradford MG et al (2015) Global variability in leaf respiration in relation to climate, plant functional types and leaf traits. New Phytol 206:614–636

Bellasio C, Beerling DJ, Griffiths H (2016) An excel tool for deriving key photosynthetic parameters from combined gas exchange and chlorophyll fluorescence: theory and practice. Plant Cell Environ 39:1180–1197

Buckley TN, Díaz-Espejo A (2015) Partitioning changes in photosynthetic rate into contributions from different variables. Plant Cell Environ 38:1200–1211

Canadell JG, Le Quere C, Raupach MR, Field CB, Buitenhuis ET, Ciais P, Conway TJ, Gillett NP, Houghton RA, Marland G (2007) Contributions to accelerating atmospheric CO2 growth from economic activity, carbon intensity, and efficiency of natural sinks. Proc Natl Acad Sci U S A 104:18866–18870

Carriquí M, Cabrera HM, Conesa MÀ, Coopman RE, Douthe C, Gago J, Gallé A, Galmés J, Ribas-Carbo M, Tomás M, Flexas J (2015) Diffusional limitations explain the lower photosynthetic capacity of ferns as compared with angiosperms in a common garden study. Plant Cell Environ 38:448–460

Demmig-Adams B, Adams WW III, Barker DH, Logan BA, Bowling DR, Verhoeven AS (1996) Using chlorophyll fluorescence to assess the fraction of absorbed light allocated to thermal dissipation of excess excitation. Physiol Plant 98:253–264

Demmig-Adams B, Cohu CM, Muller O, Adams WW III (2012) Modulation of photosynthetic energy conversion efficiency in nature: from seconds to seasons. Photosynth Res 113:75–88

Epron D, Godard D, Cornic G, Genty B (1995) Limitation of net CO_2 assimilation rate by internal resistances to CO_2 transfer in the leaves of two tree species (*Fagus sylvatica* L and *Castanea sativa* mill). Plant Cell Environ 18:43–51

Ethier GJ, Livingston NJ (2004) On the need to incorporate sensitivity to CO_2 transfer conductance into the Farquhar-von Caemmerer-berry leaf photosynthesis model. Plant Cell Environ 27:137–153

Evans JR, Santiago LS (2014) PrometheusWiki gold leaf protocol: gas exchange using LI-COR 6400. Funct Plant Biol 41:223–226

Farquhar GD, von Caemmerer S, Berry JA (1980) A biochemical model of photosynthetic CO_2 assimilation in leaves of C3 species. Planta 149:78–90

Field CB, Ball JT, Berry JA (1989) Photosynthesis: principles and field techniques. In: Pearcy RW, Ehleringer JR, Mooney HA, Rundel PW (eds) Plant physiological ecology: field methods and instrumentation. Chapman & Hall, New York, pp 209–253

Field CB, Ball JT, Berry JA (2000) Photosynthesis: principles and field techniques. In: Pearcy RW, Ehleringer JR, Mooney HA, Rundel PW (eds) Plant physiological ecology. Springer, Dordrecht, pp 209–253

Flexas J, Bota J, Escalona JM, Sampol B, Medrano H (2002) Effects of drought on photosynthesis in grapevines under field conditions: an evaluation of stomatal and mesophyll limitations. Funct Plant Biol 29:461–471

Flexas J, Barbour MM, Brendel O, Cabrera HM, Carriquí M, Díaz-Espejo A, Douthe C, Dreyerc E, Ferrio JP, Gago J, Gallé A, Galmés J, Kodama N, Medrano H, Niinemets Ü, Peguero-Pina JJ, Pou A, Ribas-Carbó M, Tomás M, Tosens T, Warren CR (2012a) Mesophyll diffusion conductance to CO_2: an unappreciated central player in photosynthesis. Plant Sci 193:70–84

Flexas J, Loreto F, Medrano H (eds) (2012b) Terrestrial photosynthesis in a changing environment: a molecular, physiological, and ecological approach. Cambridge University Press, Cambridge ISBN: 9780521899413

Flexas J, Scoffoni C, Gago J, Sack L (2013) Leaf mesophyll conductance and leaf hydraulic conductance: an introduction to their measurement and coordination. J Exp Bot 64:3965–3981

Flexas J, Díaz-Espejo A, Conesa MA, Coopman RE, Douthe C, Gago J, Gallé A, Galmés J, Medrano H, Ribas-Carbó M, Tomàs M, Niinemets Ü (2016) Mesophyll conductance to CO_2 and Rubisco as targets for improving intrinsic water use efficiency in C3 plants. Plant Cell Environ 39:965–982

Gaastra P (1959) Photosynthesis of crop plants as influenced by light, carbon dioxide, temperature, and stomatal diffusion resistance. Doctoral dissertation, Wageningen University, Veenman

Gago J, Douthe C, Flórez-Sarasa I, Escalona JM, Galmés J, Fernie AR, Flexas J, Medrano H (2014) Opportunities for improving leaf water use efficiency under climate change conditions. Plant Sci 226:108–119

Gallé A, Flexas J (2010) Gas-exchange and chlorophyll fluorescence measurements in grapevine leaves in the field. In: Delrot S, Medrano H, Or E, Bavaresco L, Grando S (eds) Methodologies and results in grapevine research. Springer, Dordrecht, pp 107–121

Gallé A, Florez-Sarasa I, Tomás M, Pou A, Medrano H, Ribas-Carbó M, Flexas J (2009) The role of mesophyll conductance during water stress and recovery in tobacco (*Nicotiana sylvestris*): acclimation or limitation? J Exp Bot 60:2379–2390

Galmés J, Flexas J, Keys AJ, Cifre J, Mitchell RA, Madgwick PJ, Haslam RP, Medrano H, Parry MA (2005) RubisCO specificity factor tends to be larger in plant species from drier habitats and in species with persistent leaves. Plant Cell Environ 28:571–579

Galmés J, Molins A, Flexas J, Conesa MÀ (2017) Coordination between leaf CO_2 diffusion and Rubisco properties allows maximizing photosynthetic efficiency in *Limonium* species. Plant Cell Environ 40:2081–2094

Genty B, Harbinson J, Briantais JM, Baker NR (1990) The relationship between non-photochemical quenching of chlorophyll fluorescence and the rate of photosystem 2 photochemistry in leaves. Photosynth Res 25:249–257

Grassi G, Magnani F (2005) Stomatal, mesophyll conductance and biochemical limitations to photosynthesis as affected by drought and leaf ontogeny in ash and oak trees. Plant Cell Environ 28:834–849

Harley PC, Thomas RB, Reynolds JF, Strain BR (1992) Modelling photosynthesis of cotton grown in elevated CO_2. Plant Cell Environ 15:271–282

Hermida-Carrera C, Kapralov MV, Galmés J (2016) Rubisco catalytic properties and temperature response in crops. Plant Physiol 171:2549–2561

Jansson C, Wullschleger SD, Kalluri UC, Tuskan GA (2010) Phytosequestration: carbon biosequestration by plants and the prospects of genetic engineering. BioSci 60:685–696

Jones HG (1985) Partitioning stomatal and non-stomatal limitations to photosynthesis. Plant Cell Environ 8:95–104

Laisk A, Oja V (2018) Kinetics of photosystem II electron transport: a mathematical analysis based on chlorophyll fluorescence induction. Photosynth Res 136:63 https://doi.org/10.1007/s11120-017-0439-y

Long SP, Farage PK, Garcia RL (1996) Measurement of leaf and canopy photosynthetic CO_2 exchange in the field. J Exp Bot 47:1629–1642

Loriaux S, Avenson T, Welles J, McDermitt D, Eckles R, Riensche B, Genty B (2013) Closing in on maximum yield of chlorophyll fluorescence using a single multiphase flash of sub-saturating intensity. Plant Cell Environ 36:1755–1770

Martins SC, Galmés J, Molins A, DaMatta FM (2013) Improving the estimation of mesophyll conductance to CO_2: on the role of electron transport rate correction and respiration. J Exp Bot 64:3285–3298

Maxwell K, Johnson GN (2000) Chlorophyll fluorescence – a practical guide. J Exp Bot 51:659–668

Montero R, Ribas-Carbó M, Del Saz NF, El Aou-ouad H, Berry JA, Flexas J, Bota J (2016) Improving respiration measurements with gas exchange analyzers. J Plant Physiol 207:73–77

Murchie EH, Niyogi KK (2011) Manipulation of photoprotection to improve plant photosynthesis. Plant Physiol 155:86–92

Niinemets Ü (2016) Within-canopy variations in functional leaf traits: structural, chemical and ecological controls and diversity of responses. In: Hikosaka K, Niinemets Ü, Anten N (eds) Canopy photosynthesis: from basics to applications. Springer, Berlin, pp 101–141

Niinemets Ü, Cescatti A, Rodeghiero M, Tosens T (2005) Leaf internal diffusion conductance limits photosynthesis more strongly in older leaves of Mediterranean evergreen broad-leaved species. Plant Cell Environ 28:1552–1566

Niinemets Ü, Díaz-Espejo A, Flexas J, Galmes J, Warren CR (2009) Importance of mesophyll diffusion conductance in estimation of plant photosynthesis in the field. J Exp Bot 60:2271–2282

Osmond B, Förster B (2006) Photoinhibition: then and now. In: Demmig-Adams B, Adams WW III, Mattoo AK (eds) Photoprotection, Photoinhibition, gene regulation, and environment: advances in photosynthesis and respiration. Kluwer, Dordrecht, pp 11–22

Osmond CB, Ludlow MM, Davis R, Cowan IR, Powles SB, Winter K (1979) Stomatal responses to humidity in *Opuntia inermis* in relation to control of CO_2 and H_2O exchange patterns. Oecologia 41:65–76

Pérez-Martín A, Flexas J, Ribas-Carbó M, Bota J, Tomás M, Infante JM, Díaz-Espejo A (2009) Interactive effects of soil water deficit and air vapour pressure deficit on mesophyll conductance to CO_2 in *Vitis vinifera* and *Olea europaea*. J Exp Bot 60:2391–2405

Pons TL, Flexas J, von Caemmerer S, Evans JR, Genty B, Ribas-Carbó M, Brugnoli E (2009) Estimating mesophyll conductance to CO_2: methodology, potential errors, and recommendations. J Exp Bot 60:2217–2234

Sharkey TD (2016) What gas exchange data can tell us about photosynthesis. Plant Cell Environ 39:1161–1163

Sharkey TD, Bernacchi CJ, Farquhar GD, Singsaas EL (2007) Fitting photosynthetic carbon dioxide response curves for C3 leaves. Plant Cell Environ 30:1035–1040

Tosens T, Nishida K, Gago J, Coopman RE, Cabrera HM, Carriquí M, Laanisto L, Morales L, Nadal M, Rojas R, Talts E, Tomas M, Hanba Y, Niinemets Ü, Flexas J (2016) The photosynthetic capacity in 35 ferns and fern allies: mesophyll CO_2 diffusion as a key trait. New Phytol 209:1576–1590

Valentini R, Epron D, Deangelis P, Matteucci G, Dreyer E (1995) In-situ estimation of net CO_2 assimilation, photosynthetic electron flow and photorespiration in Turkey oak (*Q. cerris* L) leaves – diurnal cycles under different levels of water-supply. Plant Cell Environ 18:631–640

Valentini R, Matteucci G, Dolman AJ, Schulze ED, Rebmann C, Moors EJ, Granier A, Gross P, Jensen NO, Pilegaard K, Lindroth A, Grelle A, Bernhofer C, Grünwald T, Aubinet M, Ceulemans R, Kowalski AS, Vesala T, Rannik Ü, Berbigier P, Loustau D, Guðmundsson J, Thorgeirsson H, Ibrom A, Morgenstern K, Clement R, Moncrieff J, Montagnani L, Minerbi S, Jarvis PG (2000) Respiration as the main determinant of carbon balance in European forests. Nature 404(6780):861–865

von Caemmerer S, Farquhar GD (1981) Some relationships between the biochemistry of photosynthesis and the gas-exchange of leaves. Planta 153:376–387

Walker BJ, Skabelund DC, Busch FA, Ort DR (2016) An improved approach for measuring the impact of multiple CO_2 conductances on the apparent photorespiratory CO_2 compensation point through slope–intercept regression. Plant Cell Environ 39:1198–1203

Walker BJ, Orr DJ, Carmo-Silva E, Parry MA, Bernacchi CJ, Ort DR (2017) Uncertainty in measurements of the photorespiratory CO_2 compensation point and its impact on models of leaf photosynthesis. Photosynth Res 132:245–255

Yin X, Struik PC, Romero P, Harbinson J, Evers JB, Van der Putten, Vos J (2009) Using combined measurements of gas exchange and chlorophyll fluorescence to estimate parameters of a biochemical C3 photosynthesis model: a critical appraisal and a new integrated approach applied to leaves in a wheat (*Triticum aestivum*) canopy. Plant Cell Environ 32:448–464

Yin X, Sun Z, Struik PC, Gu J (2011) Evaluating a new method to estimate the rate of leaf respiration in the light by analysis of combined gas exchange and chlorophyll fluorescence measurements. J Exp Bot 62:3489–3499

For Further Details

LI-COR (2012) Using the LI-6400/LI-6400XT portable photosynthesis system – Version 6. LI-COR Biosciences Inc., Lincoln

Chapter 5
Chlorophyll Fluorescence: A Practical Approach to Study Ecophysiology of Green Plants

Amarendra Narayan Mishra

1 Introduction

Photosynthesis is the primary solar harvesting system on earth. The photosynthetic process absorbs solar energy and transduces into organic chemical bond energy. However, a small fraction of solar insolation on earth is absorbed by photosynthetic machinery in plants, cyanobacteria and algae (Falkowski and Raven 2007). A major part of incident light is emitted back with a time lag. The process of absorption or molecular excitement and emission or de-excitation of photo-excitable molecules is a photo- physical process in physical, chemical and biological entities.

The principle of photo-absorption or excitation of molecules follows Planks law of $\Delta E = hv$, with ΔE, energy difference between ground and excited state; h, Planck quantum; and v, frequency of radiation (Rabinowitch and Govindjee 1969). The absorption of a photon by a molecule excites it from the ground state to higher electronic excited states <10–15 s^{-1}. The excite molecule relaxes back to the ground de-excited state by emitting photon. This luminescence from the molecule excited to the singlet state to its ground state without any change in the electron spin is known as fluorescence. This de-excitation phenomenon takes place in a very short time span of less than 10^{-8} s (Noomnarm and Clegg 2009). The internal vibrational loss of energy makes this fluorescence emission energetically lower than the exciting photon (Rabinowitch and Govindjee 1969).

Photosynthetic pigments chlorophylls and carotenoids present in the photosynthetic antenna molecules of the thylakoid membranes of green plants primarily absorb light energy (Maxwell and Johnson 2000; Strasser et al. 2000, 2004; Govindjee 2004). Ultimately, the exciton in the pigment bed of photosystems (PSII and PSI) is transferred to the reaction center pigments and elicits the photochemistry

A. N. Mishra (✉)
Khallikote Cluster University, Berhampur, Odisha, India

Centre for Life Science, Central University of Jharkhand, Ranchi, Jharkhand, India

© Springer International Publishing AG, part of Springer Nature 2018
A. M. Sánchez-Moreiras, M. J. Reigosa (eds.), *Advances in Plant Ecophysiology Techniques*, https://doi.org/10.1007/978-3-319-93233-0_5

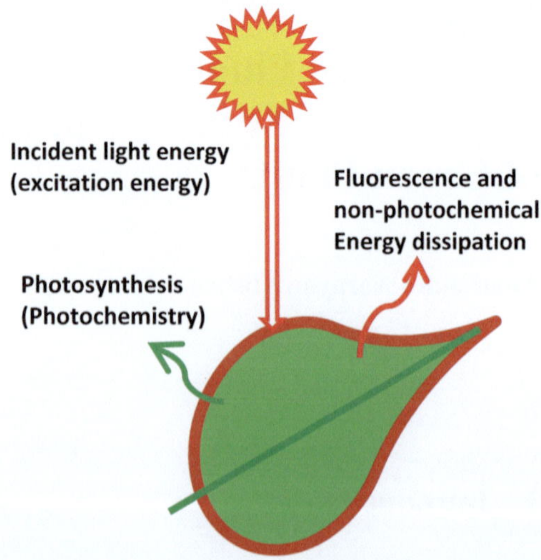

Fig. 5.1 Light energy absorption in green leaf, which induces photochemistry of photosynthesis (green color) and dissipated of heat or emitted as fluorescence emission (in red color). The quantum efficiency of photosynthesis can be measured by measuring these light absorption and emission phenomena

or photochemical electron transport in photosystems (Misra et al. 2001a, b, 2012). The de-excitation of these pigments in a heterogeneous molecular system in green plants follows the above principle of internal decay of excited molecules by fluorescence emission, along with additional processes of external pathways like the transfer of excitation energy to another molecule with a similar energy gap and/or through the electron transfer from or to another molecule. Such external energy transfer processes in photosynthetic light-harvesting antennae and charge separation in the reaction centers of both PSII and PSI contribute to the decay of the excited chlorophyll molecules, which usually low the fluorescence yield in plants (Misra et al. 2012). This is how the 30% quantum yield of chlorophyll *a* fluorescence in organic solvent is reduced to 0.6–3% in the complex photosynthetic materials (Latimer et al. 1956; Trissl et al. 1993). Light energy that is absorbed by photosynthetic pigments in chloroplasts elicits several competing processes out of which the process of photochemistry, and dissipated of heat or emitted as fluorescence emission are important for the measurement of quantum efficiency of photosynthesis (Fig. 5.1). So, measuring any of these three will give a relative quantitative picture about the photosynthetic function of plants under a given environment.

The photosynthetic process in green plants is an end result of two co-operative light reactions operating in photosystem PSII and PSI reaction centers (RC) (Fig. 5.2) (Misra et al. 2012). The photosynthetic reactions occurring at room temperature show a characteristic Chl *a* fluorescence from the antenna of PSII with an emission peak at 685 nm (F685) (Govindjee 2004). The fast photophysical formation of excited Chl (Chl*) formation in PSII antenna within few femtoseconds leads to a sequence of energy transfer reactions from the antenna (LHC) to the photosynthetic RC. A part of the absorbed photon energy is emitted as fluorescence or is lost as heat (Fig. 5.2). The room temperature fluorescence emission by Chl molecules is determined by the absorbing Chl molecules, the energy transfer intermediates, and the orientation of the fluorescing Chl molecules in the lipo-protein complexes of the

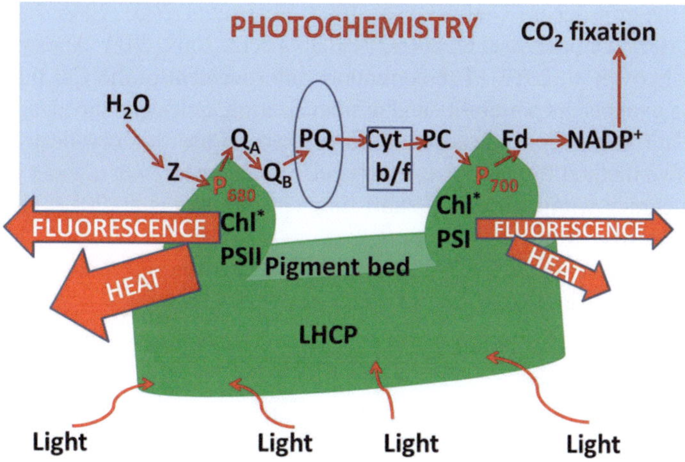

Fig. 5.2 Cooperativity between photosystem II (PSII) and photosystem I (PSI). (Adapted from Schreiber et al. 1994)

thylakoid membrane, redox state of the donors and acceptors of photosystems, and thylakoid stacking, etc. (Strasser et al. 2005). In green plants, Chl fluorescence is the most widely used parameter for monitoring the modulation and alterations in the photosynthetic process/systems under stress (Misra et al. 2012).

There are different types of fluorescence measurements used under different type of study and the suitability of the photosynthetic system. The analysis of Chl fluorescence intensity, peak, bandwidth, etc., gives an insight to the structure and function of the photosynthetic system, energy transducing and the pigment protein architecture in the green plants under stress or non-stress conditions. Despite of the fact that there are several Chl fluorescence measurement methods (for details see Misra et al. 2012; Kalaji et al. 2014, 2017a, b), in the following sections only the most used common methods are described for the ease of understanding and routine use by the researchers of environmental physiology. Mostly, the non-destructive method of measurement of Chl fluorescence is described. This method is used for measurement of aquatic, forest, terrestrial, including desert, productivity (Misra et al. 2012; Kalaji et al. 2014, 2017a, b).

In the last decade, the exploitation of this non-destructive technique in field experiments have paved way to an ease of working and interpreting the Chl fluorescence data for stress studies in eco-physiology (Misra et al. 2012; Kalaji et al. 2014, 2017a, b). The literature that cites the maximum number of publications is in the field of 'photo-inhibition', which is defined as the decrease in the quantum yield of primary photochemical reaction (photochemistry) of dark-adapted samples or the open reaction centers of PSII, measured by the Fv/Fm. The variable fluorescence to maximum fluorescence measure is a widely used parameter and invariably in most of the stress factors this is found to be the most sensitive parameter (Misra et al. 2001a, b, 2007, 2011, 2012). Besides these uses, Chl fluorescence can be used as a sensitive parameter for biosensors in photosynthetic system using thylakoid membranes as the transducer (Koblizek et al. 1998; Misra et al. 2003, 2006, 2011; Giardi and Pace 2005; Dobrikova et al. 2009; Vladkova et al. 2009, 2011; Apostolova et al. 2011; Rashkov et al. 2012). The continuous improvement of the Chl fluorescence measuring gadgets, its portability and miniaturization, quick and thousand of simultaneous data acquisition helped the field workers to use this non-destructive and non-invasive method for various uses in land, air and water with a wider utility and greater accuracy within a defined short time span (Misra et al. 2012; Kalaji et al. 2014, 2017a, b).

The portable methods of Chl fluorescence so far used by field researchers in plant biology and agriculture, forestry and aquatic science are:

1. Room temperature fluorescence.
2. Fast Chl fluorescence.
3. Pulse amplitude modulation, PAM, fluorometry.

2 Room Temperature Fluorescence

The Chl fluorescence emission at room temperature is a net result of dark reduction of plastoquinone, heat dissipation, linear and cyclic electron flow, and leakage of electrons from the thylakoid. The Chl fluorescence intensity at room temperature is also taken as quantitative measure of the Chl content in leaves (Buschmann 2007). The room temperature Chl fluorescence at 685 nm is by PSII and at 720–740 nm is by PSI. Usually, the room temperature Chl fluorescence is at 685 nm whereas the low temperature (77 K) fluorescence is at 735 nm (Fig. 5.3).

The fluorescence emission is usually measured at right angle (90°) to the photo-excitating light at the blue or red band of the spectrum or with a saturating pulse of

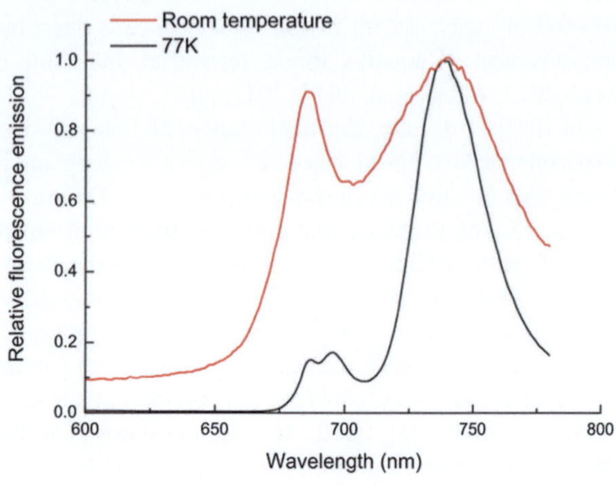

Fig. 5.3 Chlorophyll fluorescence emission spectrum at room temperature and 77 K

visible (white) light. The fluorescence yield is lowered by photochemical charge separation and/or dissipation of excess energy as heat.

3 Fast Chlorophyll Fluorescence Kinetics

In a dark adapted photosynthetic material, in which the duration of darkness is sufficient to bring down all the RC Chl molecules to their ground state, otherwise called 'open RC, Chl *a* fluorescence is emitted in a transient manner giving rise to an 'induction curve' named as Kautsky curve (Kautsky and Hirsh 1931). This induction curve under continuous light has a fast exponential phase lasting less than a second and a slow decay lasting for few minutes. The fast phase gives rise to the OJIP' fluorescence transients (Fig. 5.4) and a detailed computational analysis taking into consideration several theoretical assumptions gives rise to the derived photosynthetic parameters as described by Strasser et al. (2004) and Stirbet and Govindjee (2011) (for details see Misra et al. 2012). The fast phase is one of most widely studied in stress physiology. However, the slow phase or the SMT'decay is also studied for specific purposes such as energy transduction, or State transition etc. (Stirbet and Govindjee 2011; Bernet et al. 2018).

Instruments for the measurement of OJIP or SMT inductions are manufactured by Hansatech (UK), PSI (Czech) and few other manufacturers.

Nomenclature and definitions for 'OJIP'curve: O is for origin or F_0 level of Chl fluorescence at 20–50 μs, J at 2 ms, I at 30 ms, and P at 200 ms is the peak or maximal fluorescence otherwise known as F_M. There are several intermediate peaks reported in foraminifers, zooxanthellae and lichens and are designated as G and H

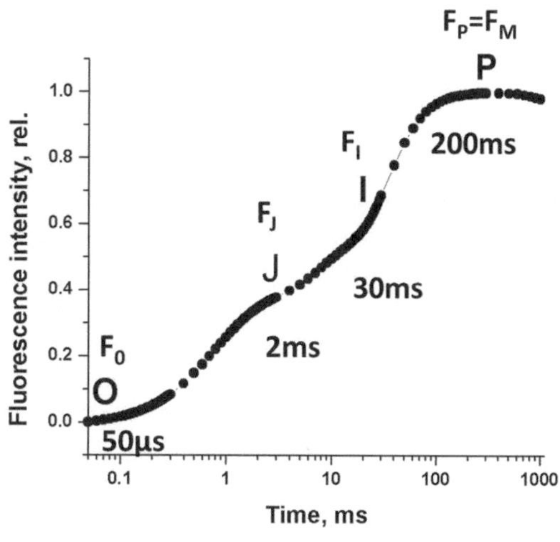

Fig. 5.4 A typical Kautsky curve of a green leaf and the exponential 'OJIP' curve. The timeline for O-J-I-P fluorescence peaks are shown in below the curve and corresponding fluorescence nomenclature is given above OJIP nomenclature. These fluorescence intensity at different time scales of OJIP are taken for calculations and deriving the photosynthetic parameters as shown in Table 5.1

or P peak (Tsimilli-Michael et al. 1998; Ilik et al. 2006). A K peak at around 300 μs is reported as a result of heat-stress to photosynthetic materials (Guisse et al. 1995; Srivastava et al. 1995, 1997; Strasser 1997; Misra et al. 2001b, 2007). The G peak is proposed to be originating as a result of an early activation of the FNR or Fd-NADP reductase (Ilik et al. 2006).

The OJIP transient is used routinely by authors for the measurement of photochemical quantum yield of PSII, and the electron transport properties within PSII. It is routinely used to monitor the modulation of structure, function and architecture of photosynthetic apparatus under normal and climatic (environmental) stress conditions as well for titrating photosynthetic inhibitors like pesticides and herbicides or agrochemicals (Misra et al. 2012).

The open RC of the photosynthetic materials kept in darkness does not elicit any electron transport in dark and so the acceptor side of PSII, i.e. Q_A remains in the oxidized state and the Chl fluorescence emission is zero ('O') or called F_0. When the open RC is excited with a strong intensity of light that theoretically excite almost all the Chl molecules in the thylakoid membranes elicits a fast electron flow within 2 ms giving rise to a transient rise of fluorescence from 'O' to 'J'. Subsequently relatively slow phases of J–I and I–P transients arise. The P level of fluorescence (F_M) occurs within 1 s and represents the closed PSII centers with complete reduction of Q_A molecules, which drive a saturating electron flow on the acceptor side of PSII (Schansker et al. 2005; Misra et al. 2012; Kalaji et al. 2014, 2017a, b). The measurement of parameters derived from the Chl fluorescence induction curve commonly uses F_0 or F_M values. The difference these two, known as F_V is used to measure the most widely used photosynthetic measured parameter F_V/F_M ratio or the maximum quantum yield of primary photochemistry of PSII (Butler and Kitajima 1975). In a healthy leaf F_V/F_M value is between 0.78–0.84 (Bjorkman and Demmig 1987). Strasser and his group extensively analyzed the OJIP induction curves and derived an exhaustive list of Chl *a* fluorescence parameters, which are summarized in Table 5.1.

4 Pulse Amplitude Modulated (PAM) Fluorescence

Another widely used Chl fluorescence technique is the pulse modulated fluorescence measurement method described by Schrieber et al. (1986). This technique uses photochemical quenching parameters using saturation pulse illumination of photosynthetic material, gradually reducing the acceptor side of PSII and inducing the electron transport in the thylakoid membrane (Schreiber et al. 1986; Schreiber 2004; Misra et al. 2012; Kalaji et al. 2014, 2017a, b). The ground fluorescence (F_o) of dark-adapted leaf is measured by a weak modulating light (ML) and a saturating pulse (SP) (about 8000 μmol m^{-2} s^{-1} for 0.6–1 s) light raises the fluorescence to the maximum level, Fm. As the fluorescence measurements are done in saturating pulse

Table 5.1 JIP parameters proposed by Strasser et al. (2004, 2010) and modified by Stirbet and Govindjee (2011)

Information selected from the fast OJIP fluorescence induction (data necessary for the calculation of the so-called JIP parameters)	
$Fo = F_{20\mu s}$ or $F_{50\mu s}$	Initial fluorescence after the onset of actinic illumination
$F_{300\mu s}$	Fluorescence value at 300 μs
$F_J \equiv F_{2ms}$	Fluorescence value at 2 ms (J-level)
$F_I \equiv F_{30ms}$	Fluorescence value at 30 ms (I-level)
$F_P(\equiv F_M)$	Fluorescence maxima under saturating illumination
t_{Fmax}	Time to reach F_M
Area	Area under F_o and F_M
$V_v = F_t - F_o$	Variable Chl fluorescence
$F_V = F_M - F_o$	Maximum variable Chl fluorescence
$V_t = (F_t - F_o)/(F_M - F_o)$	Relative variable Chl fluorescence
$M_o = (d_V/d_t)_o = 4\ ms^{-1}. (F_{300\mu s} - F_o)/(F_V)$	Value of the initial slope of curve under V_t
$S_m = Area/F_V$	Normalized area (proportional to the number of reduction and oxidation of one Q_A-molecules or the number of electron carriers per electron transport chain)
Quantum yields and efficiencies	
$\varphi_{Po} \equiv J_o^{TR}/J^{ABS} = 1 - F_o/F_M$	Maximum quantum yield of primary PSII photochemistry
$\varphi_{Pt} \equiv J^{TR}/J^{ABS} = 1 - F_t/F_M = \varphi_{Po} - (1 - V_t)$	Quantum yield of primary PSII photochemistry
$\varphi_{ETo} \equiv J_o^{ET2}/J^{ABS} = 1 - F_J - F_M = \varphi_{Po}. (1 - V_J)$	Quantum yield of the electron transport from Q_A to Q_B
$\varphi_{RE1o} \equiv J_o^{RE1}/J^{ABS} = 1 - F_I/F_M\ \varphi_{Po}. (1 - V_I)$	Quantum yield of the electron transport upto the PSI electron acceptors
$\psi_{ET2o} \equiv J_o^{ET2}/J_o^{TR} = 1 - V_J$	Efficiency of trapped electron transfer from Q_A to Q_B
$\psi_{ET1o} \equiv J_o^{RE1}/J_o^{TR} = 1 - V_I$	Efficiency of electron transfer from PSII to PSI acceptors
$\delta_{RE1o} \equiv J_o^{RE1}/J_o^{ET2} = (1 - V_I)/(1-V_j)$	Efficiency of electron transfer from Q_B to PSI acceptors
Specific energy fluxes (per active PSII RC)	
$J^{ABS}/RC = (M_o/V_J). (1/\varphi_{Po})$	Absorbed photon flux per PSII RC (apparent antenna size of active PSII)
$\gamma_{RC2} = Chl_{RC}/Chl_{tot}$	Probability that a PSII Chl functions as RC
$RC/J^{ABS} = \varphi_{Po}. V_J/M_o = \gamma_{RC2}/(1 - \gamma_{RC2})$	Number of Q_A reducing RCs per PSII antenna Chl
$J_o^{TR}/RC = M_o/V_J$	Maximum trapped exciton flux per PSII
$J_o^{ET2}/RC = (M_o/V_J). (1 - V_J)$	Electron transport from Q_A to Q_B per PSII RC
$J_o^{RE1}/RC = (M_o/V_J). (1 - V_I)$	Electron transport to PSI acceptors per PSII RC
Phenomenological energy fluxes (per CS_o)	
$J^{ABS}/CS_o = F_o$ or $J^{ABS}/CS_M = F_M$	Absorbed photon flux per cross section (apparent PSII antenna size)
$RC/CS = (RC/J^{ABS}). (J^{ABS}/CS)$	The number of active PSII RCs per cross section

(continued)

Table 5.1 (continued)

Information selected from the fast OJIP fluorescence induction (data necessary for the calculation of the so-called JIP parameters)	
$J_o^{TR}/CS = (J_o^{TR}/J^{ABS}). (J^{ABS}/CS)$	Maximum trapped exciton flux per cross section
$J_o^{ET2}/CS = (J_o^{ET2}/J^{ABS}). (J^{ABS}/CS)$	Electron transport from Q_A to Q_B per cross section
$J_o^{RE1}/CS = (J_o^{RE1}/J^{ABS}). (J^{ABS}/CS)$	Electron transport flux until PSI acceptors per cross section
Performance index	
$PI_{ABS} = [\gamma_{RC2}/(1 - \gamma_{RC2})].[\varphi_{Po}/(1 - \varphi_{Po})].[\psi_{ET2o}/(1 - \psi_{ET2o})]$	Performance index for energy conservation from photons absorbed by PSII antenna, to the reduction of Q_B
$PI_{ABS}^{tota} = PI_{ABS}. [\delta_{RE1o}/(1 - \delta_{RE1o})]$	Performance index for energy conservation n from photons absorbed by PSII antenna, until the reduction of PSI acceptors
$PI_{CSo}^{total} = F_o. PI_{ABS}^{total}; PI_{CSM}^{total} = F_M. PI_{ABS}^{total}$	Performance index on cross section basis
Driving forces (for photochemical activity)	
$DF_{ABS}^{total} = \log(PI_{ABS}^{total})$	Driving force on absorption basis
$DF_{CS}^{total} = \log(PI_{CS}^{total})$	Driving force on cross section basis

of light, it allows one to measure the maximum quantum efficiency of photosystem II (PSII) primary photochemistry (F_v/F_m) or the 'intrinsic quantum yield' (Kitajima and Butler 1975). After the first light pulse the actinic light (AL) is switched on and the SP is turned on repeatedly. This induced F_m' (F_m at light adapted state). After the initial increase of F_m' with few SP, it starts declining. The initial phase of rise in fluorescence in light adapted state is called 'photochemical quenching' which helps in generating reductants to ultimately reduce CO_2 in leaves (van Kooten and Snell 1990; Edwards and Baker 1993). Sequential SP interrupted with dark period gradually reduces the intensity of fluorescence emission and is commonly known as 'non-photochemical quenching' or NPQ (Walter and Horton 1991; Johnson et al. 1993; Niyogi et al. 1997; Oxbrough and Baker 1997). A typical PAM fluorescence measurement is shown in Fig. 5.5.

The PAM fluorescence measurement systems are manufactured and marketed by Walz (Germany). There are several models of PAM instruments and can be used as per the user preferences. The calculation of quenching parameters and the terminology suggested by van Kooten and Snell (1990) and then modified by Maxwell and Johnson (2000) and Baker (2008) is given below. The details of the quenching parameters calculated from the PAM fluorescence measurements are given in Table 5.2.

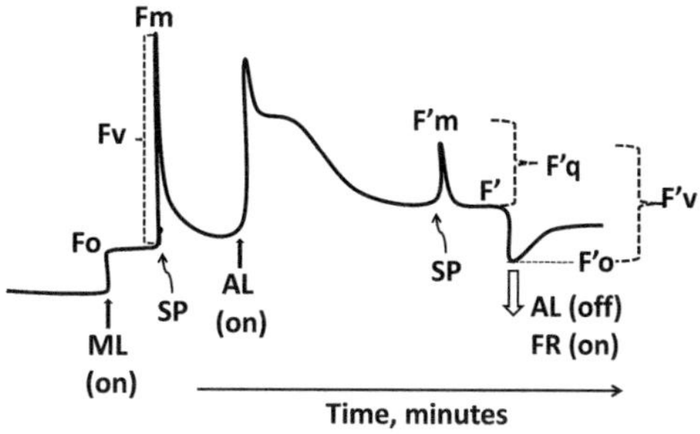

Fig. 5.5 A typical PAM fluorescence signal of a leaf disc. The fluorescence in dark-adapted leaves are denoted by F and in the light adapted state F′ are recorded and different quenching parameters are measured. (Adapted from Misra et al. 2012)

Table 5.2 The commonly used photochemical and non-photochemical quenching parameters derived from PAM fluorescence

Parameter	Definition
Photochemical quenching parameters	
$F_v/F_m = (F_m - F_o)/F_m$	Maximum quantum efficiency of PSII photochemistry
$F_q'/F_m' = \Delta F/F_m' = (F_m' - F')/F_m'$	Photochemical quenching or Genty parameter or the quantum efficiency of PSII photochemistry in the light adapted state
$qP = F_q'/F_v' = (F_m - F)/(F_m - F_o')$	Coefficient of photochemical quenching
$qL = \mathrm{qP}(F_o'/F')$	Photochemical quenching parameter or the fraction of open PSII RC
1 − qP	Proportion of closed centers or the “excitation pressure”
ETR = F_q'/F_m'·PFD·αL·(PSII/PSI)	Electron transport rate (ETR) in PSII
Non-photochemical quenching parameters	
qN = $F_m - F_m'/F_m - F_o$	Coefficient of nonphotochemical fluorescence quenching
NPQ = $(F_m - F_m')/(F_m') = (F_m/F_m') - 1$	Rate constant for excitation quenching by regulated thermal dissipation (k’N). It is the sum total for the photo-protective mechanisms (qE), state transition quenching (qT) and
	Photo-inhibition (qI) i.e. NPQ = qE + qT + qI

For details refer Misra et al. (2012) and Kalaji et al. (2014, 2017a, b)

4.1 Method

The F_o measurement of the dark adapted (10–30 min) sample is measured with a low intensity modulating light (ML) and then a saturation pulse (SP) of light, which is high enough to theoretically completely close all the primary electron acceptors (Q_A) in PSII by completely reducing PSII RC, resulting in F_m. After the first SP, an actinic light (AL) is turned on and the fluorescent signal declines slowly. This is ascribed to the onset of CO_2 fixation and the fluorescence emission reaches a steady state. During this period, photochemical quenching occurs, which is a measure of open PSII centers, photo-protective non-photochemical quenching and heat dissipation. Saturation pulses during steady state photosynthesis gives rise to maximal fluorescence in light adapted state F_m'.

5 Changes of Fluorescence Parameters Under Stress Conditions

A decrease in the maximum quantum efficiency of PSII calculated by the ratio of F_v/F_m is commonly known as 'photo-inhibition'. It decreases with lowering of F_m and/or increase in F_o. The dissociation of LHCII from the PSII core complex usually results in an increase in F_o (Misra et al. 2001a, b; Misra and Terashima 2003; Misra et al. 2007). Under stressful conditions, F_v/F_m is used commonly to measure the extent of photoinhibition of photosynthesis. However, an increase in NPQ under stress conditions also induces a decrease in F_v/F_m. So this quantification under the conditions of increasing NPQ can be erroneous. Interestingly, during photoinhibitory conditions, NPQ value can decreases due to low F_m (Misra et al. 2006, 2011). *qP* is defined as the coefficient of photochemical fluorescence quenching and is a measure of the fraction of open PSII reaction centers (van Kooten and Snel 1990). 1 – qP, reflects the "excitation pressure" on PSII (Maxwell et al. 1994; Misra et al. 2006, 2011). The excitation pressure also increases under adverse environmental conditions. ETR of PSII is the total sum of carbon fixation, photorespiration, nitrate assimilation, and Mehler reaction. A perturbation in any of these parameters affects ETR and so under stress conditions the ETR decreases drastically (Misra et al. 2012). NPQ is a measure of heat dissipation and is taken as a mechanism of photo-protection, state transition, and photo-inhibition (Krause and Weis 1991; Muller et al. 2001; Finazzi et al. 2006). NPQ occurs at low light intensity also. But under stress such as high light intensity or photoinhibition, salinity, heavy metal toxicity, drought or chilling (low temperature) NPQ value increases. So, a decrease in F_v/F_m or an increase in NPQ is taken as a stress index. The increase in NPQ under photo-inhibitory conditions is partly regulated by the 'Xanthophyll cycle' (Bilger and Björkman 1990; Misra et al. 2006, 2011). The details of the stress-induced changes in fluorescence parameters are summarized in Tables 5.3 and 5.4 respectively for fast Chl fluorescence (OJIP) and PAM fluorescence parameters.

Table 5.3 Effect of abiotic stress on the changes in fast Chl fluorescence (OJIP) parameters

Environmental stress	Changes in fluorescence parameters	References
Photoinhibition	Decrease in quantum efficiency of photochemistry F_v/F_m, qP, F_V/F_M is the most widely characterized parameter. This is also taken as an indicator for stress induced photoinhibition in chloroplasts.	Srivastava et al. (1995) and Misra et al. (2001a, b)
Heat/high temperature	Significant increase in F_o, ET efficiency decreased.	Sharkey and Schrader (2006) and Mathur et al. (2014)
	Reduced Q_A/RC and Q_B^- to Q_A^-	Kalaji et al. (2016)
	F_V/F_M decreased, F_m decreased, F_o increased	Misra et al. (2001b), Chen et al. (2009), Mathur et al. (2011a, b) and Brestic et al. (2013)
	K peak	Srivastava and Strasser (1995), Strasser et al. (2000) and Laza'r (2006)
	Performance index (PI) decreased	Misra et al. (2001b), Stefanov et al. (2011), Brestic et al. (2012) and Brestic and Zivcak (2013)
Chilling/low temperature	Short duration protects PSII photochemistry and so the fluorescence parameters	Misra et al. (2001b)
Drought/water stress	PI decreases	Oukarroum et al. (2007), Zivcak et al. (2008), Boureima et al. (2012), Guha et al. (2013), Jedmowski et al. (2013) and Zivcak et al. (2014a)
	Increase of ABS/RC due to inactivation of PSII RCs and/or an increase in antenna size.	Van Heerden et al. (2007) and Gomes et al. (2012)
	Decrease in I-P amplitude	Oukarroum et al. (2009) and Ceppi et al. (2012)
Salinity	Decreased F_v/F_m, increase in DI. Reduced PSII efficiency in light, electron transport chain efficiency, and the efficiency of PSII open reaction centers in light, more damage to donor side than acceptor side	Misra et al. (2001a,b) and Yang et al. (2008)
		He et al. (2009) and Zhang and Sharkey (2009)
		Mehta et al. (2010) and Singh-Tomar et al. (2012)

(continued)

Table 5.3 (continued)

Environmental stress	Changes in fluorescence parameters	References
Nutrient deficiency	N deficiency: decreased photochemical efficiency and RC.	Redillas et al. (2011) and Li et al. (2012)
		Dudeja and Chaudhary (2005)
	N supplement: increased PI	Van Heerden et al. (2004), Li et al. (2012) and Zivcak et al. (2014b)
	P deficiency: decreased photochemical efficiency F_v/F_m	Kruger et al. (1997), Tsimilli-Michael and Strasser (2008) and Schweiger et al. (1996)
	Calcium deficiency: decreased photochemical efficiency F_v/F_m	Misra et al. (2001a, b), Lauriano et al. (2006) and Liu et al. (2009)
	Magnesium deficiency	Smethurst et al. (2005)
	Iron deficiency	Molassiotis et al. (2006)
	Nutrient [N, P, K, Ca, S, Mn, Cu, Fe, etc.] deficiency: Decrease in F_v/F_m, φPo, φEo, ψEo	Baker and Rosenquist (2004) and Kalaji et al. (2014a, 2017a, b)
Heavy metal (excess/toxicity)	Cd: decrease in F_v/F_m, RC/CS, ET_O/CS, and in the activity of OEC	Janeczko et al. (2005) and Han et al. (2009)
	Pb: decrease in I-P intensity, appearance of K peak. High abs and DI, low TR and ET	Kalaji and Łoboda (2007)
		Laza'r and Jablonsky (2009)
	Al: decrease in F_v/F_m, RC/CS, ET_O/CS, and in the activity of OEC, decrease in I-P intensity, appearance of K peak. High abs and DI, low TR and ET	Jiang et al. (2009)
	Bo	Oz et al. (2014)
UV radiation	UV-C: increase in effective antenna size of active RC, DI flux and decrease in F_v/F_m, ET and TR, PI	Beneragama et al. (2014)
	UV-B: decrease in F_v/F_m, ET0/TR0, TR0/ABS. Increase in DIo/RC, increased F_o level	Aguilera et al. (1999), Wobeser et al. (2000) and Albert et al. (2005)
		Force et al. (2003)
	Decrease in φPo	Manes et al. (2001), Nussbaum et al. (2001) and Paoletti et al. (2004)

Table 5.4 Effect of abiotic stress on the changes in PAM Chl fluorescence parameters

Environmental stress	Changes in fluorescence parameters	References
Photoinhibition	Decrease in F_v/F_m, photochemical quenching (qP, Φ_P, Φ_{II}) and increase in non-photochemical quenching (qN, NPQ).	Bilger and Björkman (1990) and Frank et al. (1994); Horton et al. (1994), Demmig-Adams et al. (1996), Fracheboud and Leipner (2003) and Misra et al. (2003, 2006, 2011)
Heat/high temperature	Increase in F_o, decrease in the quantum efficiency of PSII, increase in NPQ	Schreiber and Berry (1977), Bilger and Björkman (1990), Krause and Weis (1991), Yamane et al. (1997), Muller et al. (2001), Fracheboud and Leipner (2003), Finazzi et al. (2006), Suzuki et al. (2011), Schreiber et al. (2012) and Dankov et al. (2014)
Chilling/low temperature,	Short duration protects PSII photochemistry and so the fluorescence parameters. But long duration causes damage to PSII.	Fryer et al. (1998), Fracheboud and Leipner (2003) and Yang et al. (2008).
Drought/water stress	Decrease in the quantum efficiency of PSII, increase in NPQ	Fracheboud and Leipner (2003)
Salinity	Decrease in F_v/F_m, increase in NPQ reduced PSII efficiency in light, ETR, and the efficiency of PSII open reaction centers in light	Misra et al. (2003, 2006, 2011)
	More damage to donor side than acceptor side	
Heavy metals	e.g. Cu decreases F_v/F_m, qP, increase NPQ	Dobrikova et al. (2009)
UV radiation	UV-B: Decrease in F_v/F_m, qP and increase in NPQ of mesophilic green algae and cyanobacteria	Apostolova et al. (2014)
Nitric oxide	SNP enhances PSII photochemistry	Vladkova et al. (2011) and Misra et al. (2014)

6 Conclusion

Fluorimetric techniques are extensively used for estimating plant productivity and photosynthetic efficiency in agriculture, horticulture, forestry and marine/aquatic environment under stress or non-stress conditions (for recent reviews refer Misra et al. 2012; Kalaji et al. 2014, 2017a, b). The use of fast Chl fluorescence and PAM fluorimetry in plant stress studies is exhaustive. The most widely studied stress and the common stress effect in plants is the 'photoinhibition of photosynthesis'. Our group used chlorophyll fluorescence as sensitive parameter for developing biosensors (Koblizek et al. 1998; Misra et al. 2003, 2006, 2011; Giardi and Pace 2005; Dobrikova et al. 2009; Vladkova et al. 2009, 2011; Apostolova et al. 2011; Rashkov et al. 2012).

The improvement on the size and efficacy of data acquisition is one of the primary needs of the day for field workers to use Chl fluorescence studies, along with other tools for global productivity studies in the changing environment (Misra et al. 2012).

References

Aguilera J, Jimenez C, Figueroa FL, Lebert M, Hader DP (1999) Effect of ultraviolet radiation on thallus absorption and photosynthetic pigments in the red alga *Porphyra umbilicalis*. J Photochem Photobiol B Biol 48:75–82

Albert KR, Mikkelsen TN, Ro-Poulsen H (2005) Effects of ambient versus reduced UV-B radiation on high arctic *Salix arctica* assessed by measurements and calculations of chlorophyll-a fluorescence parameters from fluorescence transients. Physiol Plant 124:208–226

Apostolova E, Dobrikova AG, Rashkov GD, Dankov KG, Vladkova RS, Misra AN (2011) Prolonged sensitivity of immobilized thylakoid membranes in cross-linked matrix to atrazine. Sensors Actuators B 156:140–146

Apostolova EL, Rashkov G, Misra AN, Pouneva I, Dankov K (2014) Effect of UV-B radiation on photosystem II functions in Antarctic and mesophilic strains of a green alga *Chlorella vulgaris* and a cyanobacterium *Synechocystis salina*. Indian J Plant Physiol 19:111–118

Baker NR (2008) Chlorophyll fluorescence: a probe of photosynthesis in vivo. Annu Rev Plant Biol 59:659–668

Baker NR, Rosenqvist E (2004) Applications of chlorophyll fluorescence can improve crop production strategies: an examination of future possibilities. J Exp Bot 55:1607–1621

Beneragama CK, Balasooriya BLHN, Perera TMRS (2014) Use of O-J-I-P chlorophyll fluorescence transients to probe multiple effects of UV-C radiation on the photosynthetic apparatus of *Euglena*. Int J Appl Sci Biotechnol 2:553–558

Bernat G, Steinbach G, Kaňa R, Govindjee MAN, Prašil O (2018) On the origin of the slow M–T chlorophyll *a* fluorescence decline in cyanobacteria: interplay of short-term light-responses. Photosynth Res 136:183. https://doi.org/10.1007/s11120-017-0458-8

Bilger W, Björkman O (1990) Role of the xanthophyll cycle in photoprotection elucidated by measurements of light-induced absorbance changes, fluorescence and photosynthesis in leaves of *Hedera canariensis*. Photosynth Res 25:173–185

Bjorkman O, Demmig B (1987) Photon yield of O_2 evolution and chlorophyll fluorescence characteristics at 77 K among vascular plants of diverse origins. Planta 170:489–504

Boureima S, Oukarroum A, Diouf M, Cisse N, Van Damme P (2012) Screening for drought tolerance in mutant germplasm of sesame (*Sesamum indicum*) probing by chlorophyll a fluorescence. Environ Exp Bot 81:37–43

Brestic M, Zivcak M (2013) PSII Fluorescence techniques for measurement of drought and high temperature stress signal in crop plants: protocols and applications. In: Molecular stress physiology of plants. Springer, Berlin, pp 87–131

Brestic M, Zivcak M, Kalaji MH, Carpentier R, Allakhverdiev SI (2012) Photosystem II thermostability in situ: environmentally induced acclimation and genotype-specific reactions in *Triticum aestivum* L. Plant Physiol Biochem 57:93–105

Brestic M, Zivcak M, Olsovska K, Repkova J (2013) Involvement of chlorophyll a fluorescence analyses for identification of sensitiveness of the photosynthetic apparatus to high temperature in selected wheat genotypes. In: Photosynthesis research for food, fuel and the future. Springer, Berlin, pp 510–513

Buschmann C (2007) Variability and application of the chlorophyll fluorescence emission ratio red/far-red of leaves. Photosynth Res 92:261–271

Butler WL, Kitajima M (1975) Fluorescence quenching in photosystem II of chloroplasts. Biochim Biophys Acta 376:116–125

Ceppi MG, Oukarroum A, Cicek N, Strasser RJ, Schansker G (2012) The IP amplitude of the fluorescence rise OJIP is sensitive to changes in the photosystem I content of leaves: a study on plants exposed to magnesium and sulfate deficiencies, drought stress and salt stress. Physiol Plant 144:277–288

Chen LS, Li P, Cheng L (2009) Comparison of thermotolerance of sun-exposed peel and shaded peel of 'Fuji' apple. Environ Exp Bot 66:110–116

Dankov K, Rashkov G, Misra AN, Apostolova EL (2014) Temperature sensitivity of photosystem II in isolated thylakoid membranes from fluridone treated pea leaves. Turk J Bot 39:420–428. https://doi.org/10.3906/bot-1407-46

Demmig-Adams B, Gilmore AM, Adams WW III (1996) In vivo functions of carotenoids in higher plants. FASEB J 10:403–412

Dobrikova A, Vladkova R, Rashkov G, Busheva M, Taneva SG, Misra AN, Apostolova E (2009) Assessment of sensitivity of photosynthetic oxygen evolution and chlorophyll florescent parameters to copper for use in biosensors. C R Acad Bulg Sci 62:723–728

Dudeja SS, Chaudhary P (2005) Fast chlorophyll fluorescence transient and nitrogen fixing ability of chickpea nodulation variants. Photosynthetica 43:253–259

Edwards GE, Baker NR (1993) Can CO_2 assimilation in maize leaves be predicted accurately from chlorophyll fluorescence analysis? Photosynth Res 37:89–102

Falkowski PG, Raven JA (2007) Aquatic photosynthesis, 2nd edn. Princeton University Press, Princeton, 484 pp

Finazzi G, Johnson GN, Dall'Osto L, Zito F, Bonente G, Bassi R, Wollman FA (2006) Non-photochemical quenching of chlorophyll fluorescence in *Chlamydomonas reinhardtii*. Biochemistry 45:1490–1498

Force L, Critchly C, Van Rensen JJS (2003) New fluorescence parameters for monitoring photosynthesis in plants. 1. The effect of illumination on the fluorescence parameters of the JIP-test. Photosynth Res 78:17–33

Fracheboud Y, Leipner J (2003) The application of chlorophyll fluorescence to study light, temperature, and drought stress. In: JR DE, PMA T (eds) Practical applications of chlorophyll fluorescence in plant biology. Kluwer Academic Publishers, Dordrecht, pp 125–150

Frank HA, Cua A, Chynwat V, Young A, Gosztola D, Wasielewski MR (1994) Photophysics of the carotenoids associated with the xanthophyll cycle in photosynthesis. Photosynth Res 41:389–395

Fryer MJ, Andrews JR, Oxborough K, Blowers DA, Baker NR (1998) Relationship between CO_2 assimilation, photosynthetic electron transport, and active O_2 metabolism in leaves of maize in the field during periods of low temperature. Plant Physiol 116:571–580

Giardi MT, Pace E (2005) Photosynthetic proteins for technological applications. Trends Biotech 23:257–263

Gomes MTG, da Luz AC, dos Santos MR, Batitucci MDCP, Silva DM, Falqueto AR (2012) Drought tolerance of passion fruit plants assessed by the OJIP chlorophyll a fluorescence transient. Sci Hortic 142:49–56

Govindjee (2004) Chlorophyll a fluorescence: a bit of basics and history. In: Papageorgiou GC, Govindjee (eds) Chlorophyll a fluorescence: a signature of photosynthesis. Advances in photosynthesis and respiration, vol 19. Springer, Dordrecht, pp 1–41

Guha A, Sengupta D, Reddy AR (2013) Polyphasic chlorophyll a fluorescence kinetics and leaf protein analyses to track dynamics of photosynthetic performance in mulberry during progressive drought. J Photochem Photobiol B 119:71–83

Guisse B, Srivastava A, Strasser RJ (1995) The polyphasic rise of the chlorophyll a fluorescence (OKJIP) in heat stressed leaves. Arch Sci Geneve 48:147–160

Han S, Tang N, Jiang H-X, Yang L-T, Li Y, Chen L-S (2009) CO_2 assimilation, photosystem II photochemistry, carbohydrate metabolism and antioxidant system of citrus leaves in response to boron stress. Plant Sci 176:143–153

He Y, Zhu Z, Yang J, Ni X, Zhu B (2009) Grafting increases the salt tolerance of tomato by improvement of photosynthesis and enhancement of antioxidant enzymes activity. Environ Exp Bot 66:270–278

Horton P, Ruba AV, Walters RG (1994) Regulation of light harvesting in green plants. Plant Physiol 106:415–420

Ilik P, Schansker G, Kotabova E, Vaczi P, Strasser RJ, Bartak M (2006) A dip in the chlorophyll fluorescence induction at 0.2–2 s in Trebouxia-possesing lichens reflects a fast reoxidation of photosystem I. A comparison with higher plants. Biochim Biophys Acta 1757:12–20

Janeczko A, Koscielniak J, Pilipowicz M, Szarek-Lukaszewska G, Skoczowski A (2005) Protection of winter rape photosystem 2 by 24-epibrassinolide under cadmium stress. Photosynthetica 43:293–298

Jedmowski C, Ashoub A, Brüggemann W (2013) Reactions of Egyptian landraces of *Hordeum vulgare* and *Sorghum bicolor* to drought stress, evaluated by the OJIP fluorescence transient analysis. Acta Physiol Plant 35:345–354

Jiang H–X, Tang N, Zheng J–G, Chen L-S (2009) Antagonistic actions of boron against inhibitory effects of aluminum toxicity on growth, CO_2 assimilation, ribulose-1,5-bisphosphate carboxylase/oxygenase, and photosynthetic electron transport probed by the JIP-test, of *Citrus grandis* seedlings. BMC Plant Biol 9:102

Johnson GN, Yong AJ, Scholes JD, Horton P (1993) The dissipation of excess excitation energy in British plant species. Plant Cell Environ 16:673–679

Kalaji HM, Łoboda T (2007) Photosystem II of barley seedlings under cadmium and lead stress. Plant Soil Environ 53:511–516

Kalaji HM, Schansker G, Ladle RJ, Goltsev V, Bosa K, Allakhverdiev SI, Brestic M, Bussotti F, Calatayud A, Dąbrowski P, Elsheery NI, Ferroni L, Guidi L, Hogewoning SW, Jajoo A, Misra AN, Nebauer SG, Pancaldi S, Penella C, Poli DB, Pollastrini M, Romanowska-Duda ZB, Rutkowska B, Serôdio J, Suresh K, Szulc W, Tambussi E, Yanniccari M, Zivcak M (2014) Frequently asked questions about in vivo chlorophyll fluorescence: practical issues. Photosynth Res 122:121–158. https://doi.org/10.1007/s11120-014-0024-6

Kalaji HM, Oukarroum A, Alexandrov V et al (2014a) Identification of nutrient deficiency in maize and tomato plants by in vivo chlorophyll a fluorescence measurements. Plant Physiol Biochem 81:16–25

Kalaji HM, Jajoo A, Oukarroum A, Brestic M, Zivcak M, Samborska IA, Cetner MD, Łukasik I, Goltsev V, Ladle RJ (2016) Chlorophyll a fluorescence as a tool to monitor physiological status of plants under abiotic stress conditions. Acta Physiol Plant 38:102

Kalaji HM, Schansker Brestic M, Bussotti F, Calatayud A, Ferroni L, Goltsev V, Guidi L, Jajoo A, Li P, Losciale P, Mishra VK, Misra AN, Nebauer SG, Pancaldi S, Pancaldi S, Penella C, Pollastrini M, Suresh K, Tambussi E, Yanniccari M, Zivcak M, Cetner MD, Samborska IA, Stirbet A, Olsovska K, Kunderlikova K, Shelonzek H, Rusinowski S, Baba W (2017a) Frequently asked questions about in vivo chlorophyll fluorescence, the sequel. Photosynth Res 132:13–66

Kalaji HM, Bąba W, Gediga K, Goltsev V, Samborska IA, Cetner MD, Dimitrova S, Piszcz U, Bielecki K, Karmowska K, Dankov K, Kompała-Bąba A (2017b) Chlorophyll fluorescence as a tool for nutrient status identification in rapeseed plants. Photosynth Res 136:329. https://doi.org/10.1007/s11120-017-0467-7

Kautsky H, Hirsch A (1931) Neue Versuche zur Kohlensaure assimilation. Naturwissenschaften 19:964

Kitajima M, Butler WL (1975) Quenching of chlorophyll fluorescence and primary photochemistry in chloroplasts by dibromo-thymoquinone. Biochim Biophys Acta 376:105–115

Koblizek M, Masojidek J, Komenda J, Kusera T, Pilloton R, Mattoo AK, Giardi MT (1998) A sensitive photosystem II based biosensor for detection of a class of herbicides. Biotech Bioenerg 60:664–669

Krause GH, Weis E (1991) Chlorophyll fluorescence and photosynthesis: the basics. Annu Rev Plant Physiol Plant Mol Biol 42:313–349

Kruger GHT, Tsmilli-Michael M, Strasser RJ (1997) Light stress provokes plastic and elastic modifications in structure and function of photosystem II in camelia leaves. Physiol Plant 101:265–277

Latimer P, Bannister TT, Rabinowitch E (1956) Quantum yields of fluorescence of plant pigments. Science 124:585–586

Lauriano JA, Ramalho JC, Lidon FC, Ce'umatos M (2006) Mechanisms of energy dissipation in peanut under water stress. Photosynthetica 44:404–410

Laza'r D (2006) The polyphasic chlorophyll a fluorescence rise measured under high intensity of exciting light. Funct Plant Biol 33:9–30

Laza'r D, Jablonsky J (2009) On the approaches applied in formulation of a kinetic model of photosystem II: different approaches lead to different simulations of the chlorophyll a fluorescence transients. J Theor Biol 257:260–269

Li G, Zhang ZS, Gao HY, Liu P, Dong ST, Zhang JW, Zhao B (2012) Effects of nitrogen on photosynthetic characteristics of leaves from two different stay-green corn (*Zea mays* L.) varieties at the grain-filling stage. Can J Plant Sci 92:671–680

Liu WJ, Chen YE, Tian WJ, Du JB, Zhang ZW, Xu F, Zhang F, Yuan S, Lin HH (2009) Dephosphorylation of photosystem II proteins and phosphorylation of CP29 in barley photosynthetic membranes as a response to water stress. Biochim Biophys Acta 1787:1238–1245

Manes F, Donato E, Vitale M (2001) Physiological response of *Pinus halepensis* needles under ozone and water stress conditions. Physiol Plant 113:249–257

Mathur S, Mehta P, Jajoo A, Bharti S (2011a) Analysis of elevated temperature induced inhibition of Photosystem II using Chl a fluorescence induction kinetics. Plant Biol 13:1–6

Mathur S, Allakhverdiev SI, Jajoo A (2011b) Analysis of high temperature stress on the dynamics of antenna size and reducing side heterogeneity of photosystem II in wheat leaves (*Triticum aestivum*). Biochim Biophys Acta 1807:22–29

Mathur S, Agrawal D, Jajoo A (2014) Photosynthesis: limitations in response to high temperature stress. J Photochem Photobiol B Biol 137:116. https://doi.org/10.1016/j.jphotobiol.2014.01.010

Maxwell K, Johnson GN (2000) Chlorophyll fluorescence-a practical guide. J Exp Bot 51:659–668

Maxwell DP, Falk S, Trick CG, Huner N (1994) Growth at low temperature mimics high-light acclimation in *Chlorella vulgaris*. Plant Physiol 105:535–543

Mehta P, Jajoo A, Mathur S, Bharti S (2010) Chlorophyll a fluorescence study revealing effects of high salt stress on Photosystem II in wheat leaves. Plant Physiol Biochem 48:16–20

Misra AN, Terashima I (2003) Changes in photosystem activities during adaptation of *Vicia faba* seedlings to low, moderate and high temperatures. Plant cell physiology. In: Abstract, annual symposium JSPP, Nara, Japan 27–29 March 2003

Misra AN, Srivastava A, Strasser RJ (2001a) Utilisation of fast chlorophyll *a* fluorescence technique in assessing the salt/ion sensitivity of mung bean and brassica seedlings. J Plant Physiol 158:1173–1181

Misra AN, Srivastava A, Strasser RJ (2001b) Fast chlorophyll a fluorescence kinetic analysis for the assessment of temperature and light effects: a dynamic model for stress recovery phenomena. Photosynthsis: PS2001. CSIRO Publishers, Melbourne S3–007

Misra AN, Latowski D, Strzalka K (2003) De-epoxidation state of lutein and violaxanthin in the seedlings of salt sensitive and salt tolerant plants grown under NaCl salt stress. Plant Biology, Honolulu, Hawaii, USA, 25–30 July 2003

Misra AN, Latowski D, Strzalka K (2006) The xanthophylls cycle activity in kidney bean and cabbage leaves under salinity stress. Russ J Plant Physiol 53:102–109

Misra AN, Srivastava A and Strasser RJ (2007) Elastic and plastic responses of Vicia faba leaves to high temperature and high light stress. Gordon Conference on "Temperature stress in plants", Ventura, USA 25–30 Jan 2007

Misra AN, Latowski D, Strzalka K (2011) Violaxanthin de-epoxidation in aging cabbage (*Brassica oleracea* L.) leaves play as a sensor for photosynthetic excitation pressure. J Life Sci 5:182–191

Misra AN, Misra M, Singh R (2012) Chlorophyll fluorescence in plant biology. In: Misra AN (ed) Biophysics. Intech Open, pp 171–192. http://www.intechopen.com

Misra AN, Vladkova R, Singh R, Misra M, Dobrikova AG, Apostolova EL (2014) Action and target sites of nitric oxide in chloroplasts. Nitric Oxide 39:35–45

Molassiotis A, Tanou G, Diamantidis G, Patakas A, Therios I (2006) Effects of 4-month Fe deficiency exposure on Fe reduction mechanism, photosynthetic gas exchange, chlorophyll fluorescence and antioxidant defense in two peach rootstocks differing in Fe deficiency tolerance. J Plant Physiol 163:176–185

Muller P, Li XP, Niyogi KK (2001) Non-photochemical quenching. A response to excess light energy. Plant Physiol 125:1558–1566

Niyogi KK, Bjorkman O, Grossman AR (1997) *Chlamydomonas* xanthophyll cycle mutants identified by video imaging of chlorophyll fluorescence quenching. Plant Cell 9:1369–1380

Noomnarm U, Clegg R (2009) Fluorescence lifetimes: fundamentals and interpretations. Photosynth Res 101:181–194

Nussbaum S, Geissmann M, Eggenberg P, Strasser RJ, Fuhrer J (2001) Ozone sensitivity in herbaceous species as assessed by direct and modulated chlorophyll fluorescence techniques. J Plant Physiol 158:757–766

Oukarroum A, Madidi SE, Schansker G, Strasser RJ (2007) Probing the responses of barley cultivars (*Hordeum vulgare* L.) by chlorophyll a fluorescence OLKJIP under drought stress and rewatering. Environ Exp Bot 60:438–446

Oukarroum A, Schansker G, Strasser RJ (2009) Drought stress effects on photosystem I content and photosystem II thermotolerance analyzed using Chl a fluorescence kinetics in barley varieties differing in their drought tolerance. Physiol Plant 137:188–199

Oxborough K, Baker NR (1997) Resolving chlorophyll *a* fluorescence images of photosynthetic efficiency into photochemical and non-photochemical components: calculation of qP and Fv'/Fm' without measuring Fo. Photosynth Res 54:135–142

Öz MT, Turan Ö, Kayihan C, Eyidoğan F, Ekmekçi Y, Yücel M, Öktem HA (2014) Evaluation of photosynthetic performance of wheat cultivars exposed to boron toxicity by the JIP fluorescence test. Photosynthetica 52:555–563

Paoletti E, Bussotti F, Della Rocca G, Lorenzini G, Nali C, Strasser RJ (2004) Fluorescence transient in ozonated Mediterranean shrubs. Phyton Annales Rei Botanicae 44:121–131

Rabinowitch E, Govindjee (1969) Photosynthesis. Wiley, New York, 273 pp

Rashkov GD, Dobrikova AG, Pouneva ID, Misra AN, Apostolova E (2012) Sensitivity of *Chlorella vulgaris* to herbicides. Possibility of using it as a biological receptor in biosensors. Sensors Actuators B 161:151–155

Redillas MCFR, Jeong JS, Strasser RJ, Kim YS, Kim JK (2011) JIP analysis on rice (*Oryza sativa* cv Nipponbare) grown under limited nitrogen conditions. J Korean Soc Appl Biol Chem 54:827–832

Schansker G, Toth SZ, Strasser RJ (2005) Methylviologen and dibromothymoquinone treatments of pea leaves reveal the role of photosystem I in the Chl a fluorescence rise OJIP. Biochim Biophys Acta 1706:250–261

Schreiber U (2004) Pulse-amplitude-modulation (PAM) fluorometry and saturation pulse method: an overview. In: Papageorgiou GC, Govindjee (eds) Chlorophyll a Fluorescence: a signature of photosynthesis, advances in photosynthesis and respiration, vol 19. Springer, Dordrecht, pp 279–319

Schreiber U, Berry JA (1977) Heat-induced changes of chlorophyll fluorescence in intact leaves correlated with damage of the photosynthetic apparatus. Planta 136:233–238

Schreiber U, Schliwa U, Bilger W (1986) Continuous recording of photochemical and non-photochemical chlorophyll fluorescence quenching with a new type of modulation fluorometer. Photosynth Res 10:51–62

Schreiber U, Bilger W, Neubauer C (1994) Chlorophyll fluorescence photosynthesis. Ecophysiology of photosynthesis. Springer, Berlin, pp 49–70

Schreiber U, Klughammer C, Kolbowski J (2012) Assessment of wavelength-dependent parameters of photosynthetic electron transport with a new type of multi-color PAM chlorophyll fluorometer. Photosynth Res 113:127–144

Schweiger J, Lang M, Lichtenthaler HK (1996) Differences in fluorescence excitation spectra of leaves between stressed and non-stressed plants. J Plant Physiol 148:536–547

Sharkey TD, Schrader SM (2006) High temperature stress. Physiology and molecular biology of stress tolerance in plants. Springer, Berlin, pp 101–129

Singh-Tomar R, Mathur S, Allakhverdiev SI, Jajoo A (2012) Changes in PSII heterogeneity in response to osmotic and ionic stress in wheat leaves (*Triticum aestivum*). J Bioenerg Biomembr 44:411–419

Smethurst CF, Garnett T, Shabala S (2005) Nutritional and chlorophyll fluorescence responses of lucerne (*Medicago sativa*) to waterlogging and subsequent recovery. Plant Soil 270:31–45

Srivastava A, Strasser RJ (1995) How do land plants respond to stress temperature and stress light? Arch Sci Geneve 48:135–146

Srivastava A, Greppin H, Strasser RJ (1995) Acclimation of land plants to diurnal changes in temperature and light. In: Mathis P (ed) Photosynthesis: from light to biosphere, vol 4. Kluwer Academic Publishers, Dordrecht, pp 909–912

Srivastava A, Guisse B, Greppin H, Strasser RJ (1997) Regulation of antenna structure and electron transport in PSII of *Pisum sativum* under elevated temperature probed by the fast polyphasic chlorophyll a fluorescence transient OKJIP. Biochim Biophys Acta 1320:95–106

Stefanov D, Petkova V, Denev ID (2011) Screening for heat tolerance in common bean (*Phaseolus vulgaris* L.) lines and cultivars using. JIP-test Sci Hortic 128:1–6

Stirbet A, Govindjee (2011) On the relation between the Kautsky effect (chlorophyll a fluorescence induction) and Photosystem II: basics and applications of the OJIP fluorescence transient. J Photochem Photobiol B Biol 104:236. https://doi.org/10.1016/j.jphotobiol.2010.12.010

Strasser BJ (1997) Donor side capacity of photosystem II probed by chlorophyll a fluorescence transients. Photosynth Res 52:147–155

Strasser RJ, Srivastava A, Tsimilli-Michael M (2000) The fluorescence transient as a tool to characterize and screen photosynthetic samples. In: Yunus M, Pathre U, Mohanty P (eds) Probing photosynthesis: mechanism, regulation and adaptation. Taylor and Francis, London, pp 443–480

Strasser RJ, Tsimilli-Michael M, Srivastava A (2004) Analysis of the chlorophyll fluorescence transient. In: Papageorgiou GC, Govindjee (eds) Chlorophyll fluorescence: a signature of photosynthesis, advances in photosynthesis and respiration, vol 19. Springer, Dordrecht, pp 321–362

Strasser RJ, Tsimilli-Michael M, Srivastava A, Srivastava A (2005) Analysis of the chlorophyll a fluorescence transient. In: Papageorgiou GC, Govindjee (eds) Advances in photosynthesis and respiration chlorophyll a Fluorescence: a signature of photosynthesis. Kluwer Academic Publisher, Dordrecht, pp 321–362

Strasser RJ, Tsimilli-Michael M, Qiang S, Goltsev V (2010) Simultaneous in vivorecording of prompt and delayed fluorescence and 820-nm reflection changesduring drying and after rehydration of the resurrection plant Haberlea rhodopen-sis. Biochim Biophys Acta 1797:1313–1326

Suzuki K, Ohmori Y, Ratel E (2011) High root temperature blocks both linear and cyclic electron transport in the dark during chilling of the leaves of rice seedlings. Plant Cell Physiol 52:1697–1707

Trissl HW, Gao Y, Wulf K (1993) Theoretical fluorescence induction curves derived from coupled differential equations describing the primary photochemistry of photósystem II by an exciton-radical pair equilibrium. Biophys J 64:974–988

Tsimilli-Michael M, Strasser RJ (2008) In vivo assessment of plants' vitality: applications in detecting and evaluating the impact of mycorrhization on host plants. In: Varma A (ed) Mycorrhiza: state of the art, genetics and molecular biology, eco-function, biotechnology, eco-physiology, structure and systematics, 3rd edn. Springer, Dordrecht, pp 679–703

Tsimilli-Michael M, Pecheux M, Strasser RJ (1998) Vitality and stress adaptation of the symbionts of coral reef and temperate foraminifers probed in hospite by the fluorescence kinetics OJIP. Archs Sci Geneve 51:205–240

Van Heerden PD, Strasser RJ, Krüger GH (2004) Reduction of dark chilling stress in N_2-fixing soybean by nitrate as indicated by chlorophyll a fluorescence kinetics. Physiol Plant 121:239–249

Van Heerden PDR, Swanepoel JW, Krüger GHJ (2007) Modulation of photosynthesis by drought in two desert scrub species exhibiting C3-mode CO_2 assimilation. Environ Exp Bot 61:124–136

van Kooten O, Snel JFH (1990) The use of chlorophyll fluorescence nomenclature in plant stress physiology. Photosynth Res 25:147–150

Vladkova R, Ivanova PI, Krastera V, Misra AN, Apostolova E (2009) Assessment of chlorophyll florescent and photosynthetic oxygen evolution parameters in pea thylakoid membranes for use in biosensors against Q_B binding herbicide- atrazine. C R Acad Bulg Sci 62:355–360

Vladkova R, Dobrikova AG, Singh R, Misra AN, Apostolova E (2011) Photoelectron transport ability of chloroplast thylakoid membranes treated with NO donor SNP: changes in flash oxygen evolution and chlorophyll fluorescence. Nitric Oxide Biol Chem 24:84–90

Walters RG, Horton P (1991) Resolution of non-photochemical chlorophyll fluorescence quenching in barley leaves. Photosynth Res 27:121–133

Wobeser EAV, Figueroa FL, Cabello-Pasini A (2000) Effect of UV radiation on photoinhibition of marine macrophytes in culture systems. J Appl Phycol 12:159–168

Yamane Y, Kashino Y, Koike H, Satoh K (1997) Increases in the fluorescence Fo level and reversible inhibition of photosystem II reaction center by high-temperature treatments in higher plants. Photosynth Res 52:57–64

Yang X, Liang Z, Wen X, Lu C (2008) Genetic engineering of the biosynthesis of glycinebetaine leads to increased tolerance of photosynthesis to salt stress in transgenic tobacco plants. Plant Mol Biol 66:73–86

Zhang R, Sharkey TD (2009) Photosynthetic electron transport and proton flux under moderate heat stress. Photosynth Res 100:29–43

Zivcak M, Brestic M, Olsovska K, Slamka P (2008) Performance index as a sensitive indicator of water stress in *Triticum aestivum*. Plant Soil Environ 54:133–139

Zivcak M, Kalaji MH, Shao HB, Olsovska K, Brestic M (2014a) Photosynthetic proton and electron transport in wheat leaves under prolonged moderate drought stress. J Photochem Photobiol B Biol 137:107–115. https://doi.org/10.1016/j.jphotobiol.2014.01.007

Zivcak M, Olsovska K, Slamka P, Galambosova J, Rataj V, Shao HB, Brestic M (2014b) Application of chlorophyll fluorescence performance indices to assess the wheat photosynthetic functions influenced by nitrogen deficiency. Plant Soil Environ 60:210–215

Chapter 6
Thermoluminescence: A Tool to Study Ecophysiology of Green Plants

Amarendra Narayan Mishra

1 Introduction

Thermally induced light emission in physical chemical or biological systems is known as thermoluminescence or TL (Demeter and Govindjee 1989; Misra and Ramaswamy 2001; Misra et al. 2001a, b, 2012; Ducruet 2003; Maslenkova 2010; Sane et al. 2012; Misra 2013). This phenomena is the characteristic of a solid state or semi-conductor, in which thermally activated recombination of electrons with positive holes is generated by particle or electromagnetic radiation at room or low temperature prior to their heating in dark (Randall and Wilkins 1945; Demeter and Govindjee 1989). Luminescence occurs in materials absorbing light. Light energy absorbed by a system induces photochemical reactions and transduces light/photon energy to kinetic and/or chemical energy. Excess light energy that is not utilized by photochemical processes are emitted back or dissipated in various forms of luminescence *viz.* fluorescence, phosphorescence, delayed luminescence, chemiluminescence and thermoluminescence (Misra et al. 2001a, b, 2012). The time course of the emission lifetime of this luminescence is given in Table 6.1. These are the phenomena of de-excitation of any photo-excited materials. The quantum yield of the de-excited system is less than the excited state, due to internal conversion of energy and/or heat dissipation. Thermoluminescence (TL) is the characteristic of a system that emits light at a characteristic temperature due to the chemiluminescence properties, radical pair states, or electron hole pairs (Misra et al. 2001a, b, 2012; Ducruet 2003). The biophysical analysis of the charge recombination shows that the phenomenon in darkness is the reversal of the primary photochemical processes in PS II (Misra et al. 2001a, b; Sane 2004). In the present chapter, the practical use of

---

A. N. Mishra (✉)
Khallikote Cluster University, Berhampur, Odisha, India

Centre for Life Science, Central University of Jharkhand, Ranchi, Jharkhand, India

© Springer International Publishing AG, part of Springer Nature 2018
A. M. Sánchez-Moreiras, M. J. Reigosa (eds.), *Advances in Plant Ecophysiology Techniques*, https://doi.org/10.1007/978-3-319-93233-0_6

Table 6.1 The lifetime of different luminescence from an excited material

Luminescence	Temperature dependent or independent	Half life of emission after excitation (τ_c)
Fluorescence	Independent	$<10^{-8}$ s
Phosphorescence	Dependent	$>10^{-8}$ s
Delayed luminescence	Independent	Minutes $< \tau_c <$ years
Thermoluminescence	Dependent	Minutes $< \tau_c < 4.6 \times 10^9$ years

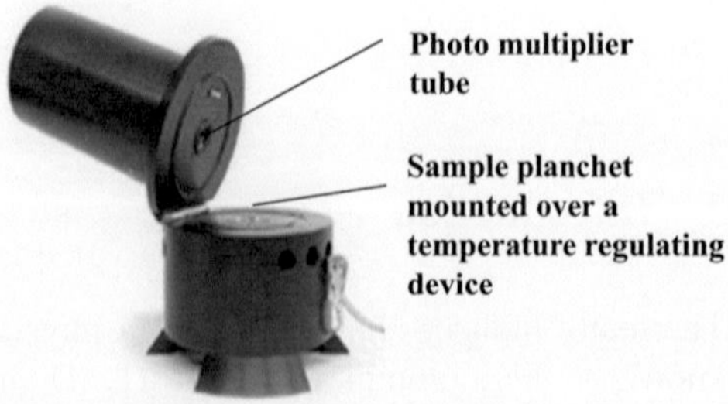

Fig. 6.1 A thermoluminiscence apparatus set up

TL for the study of the assessment of environmental impact on the changes in the primary photochemical processes of PSII is explained.

2 Instrumentation

TL measurement is done usually with an assembly of dark chamber, a copper planchet with a temperature sensor, a manual or peltier cooling/heating device, red-sensitive photo multiplier tube, signal amplifier, and a X-Y recorder or data acquisition instrument/computer (Tatake et al. 1971; Ducruet and Miranda 1992; Zeinalov and Maslenkova 1996; Bhatnagar et al. 2002; Ducruet 2003; Gilbert et al. 2004b). The samples of photosynthetic materials are photo-excited by several (8–10 nos.) flashes of short (5 ms) duration and cooled either to liquid nitrogen temperature in order to keep the charge particles in a physically separated state. Depending on the experimental requirement, one can also cool the samples to sub Zero temperatures. Then the samples are heated in a gradual, slow and linear heating mode to induce charge recombination, giving rise to a set of different TL emission bands as a result of recombination of different charge pairs at a particular temperature (Misra et al. 2001a, b, 2012). A picture of the TL set-up is shown in Fig. 6.1. These characteristic TL bands are used in the study of various biotic and abiotic stress factors in green plants (Misra et al. 2012). Photosynthetic materials are directly placed on the sample holder and excited by (i) continuous light during freezing or (ii) excited by flash(s) of saturating pulse, series, or (iii) excited by combining

(i) with (ii) at a particular temperature prior to flash freezing the sample. (Ducruet and Vass 2009; Sane et al. 2012). There are several TL apparatus commercially available by Photon Systems Instruments (Brno, Czech Republic), which provides spectral deconvolution programs (Ducruet and Miranda 1992).

3 Thermoluminescence Glow Peaks

Photosynthetic materials, such as isolated photosystem II (PS II) pigment-protein complexes, thylakoid membranes, chloroplasts, cyanobacteria, algae and green leaves illuminated with a saturating flash of light, induce charge separation in PS II. The TL measurement is done by photoexcitation of the leaf sample, then cooling it at liquid nitrogen or to a low temperature, followed by heating in dark and recording the photo-emittance during heating (Tatake et al. 1971; Zeinalov and Maslenkova 1996; Misra et al. 2001a, b, 2012; Ducruet 2003; Gilbert et al. 2004a, b; Bhagwat and Bhattacharjee 2005; Ducruet and Vass 2009). But recent commercial instruments use a peltier cooling and heating system. The TL emission is then measured with a sensitive photomultiplier. The emission around 730 nm vs. temperature is plotted in a graph sheet. Arnold and Azzi (1968) showed the occurrence TL glow peaks between −40 °C and +50 °C in photosynthetic materials. In pre-irradiated photosynthetic materials (pigment protein complexes, thylakoid membranes, intact chloroplasts or green leaves) TL glow peaks arise in darkness (Misra and Ramaswamy 2001; Misra et al. 2001a, b, 2012; Misra 2013). The separated charge pairs recombine and emit photon. Saturating and sequential short pulses (in ms scale) light generates S0, S1, S2, and S3 states in the water oxidizing Mn cluster of PS II. The S0 and S1 states remain stable during darkness. Upon illumination, these states are photo-converted and the S^o and S^1 states are distributed in 25% and 75% approximately in the photosynthetic materials. In leaves, approximately 40% of Q_B is reduced (Rutherford et al. 1984a) and the $Q_B{}^-/Q_B$ ratio oscillates with a periodicity of 2 flashes. When leaf photosynthesis is inhibited or the electron transfer from Q_A to Q_B in PS II is blocked, only $Q_A{}^-$ charge accumulates. The recombination of charges and holes at a particular temperature and emission of photon is designated by specific nomenclatures as shown in Table 6.2. This charge recombination of $Q_A{}^-$ and $Q_B{}^-$ with S2/S3 results in Q band and B band, respectively, at around 5 °C and at 20–35 °C. The B- band is the major TL band observed in any photosynthetic material studied so far (Fig. 6.2). The charge recombination of S1/S2 states with $Q_A{}^-$ is less stable than that of B band and recombines quicker than $Q_B{}^-$. These recombinations are very sensitive to redox changes in the charge pairs (Misra and Ramaswamy 2001; Misra et al. 2001a, b, 2012; Misra 2013). Thus, any change in the stable environment of PS II can be measured by the changes in TL glow peaks. As it also oscillates with each flash number, the redox state of Mn cluster can be titrated with TL measurements (Misra et al. 2001a, b, 2012; Misra 2013).

Table 6.2 Characteristic thermoluminescence (TL) bands from photosynthetic materials. These bands are reported in PSII particles, thylakoid membranes, chloroplasts, cyanobacteria, algae and green leaves

TL band	Temperature	Charge recombination	References
Very low temperature TL peaks (LTL)	−200 to 250 °C	Chlorophyll aggregates	Sane et al. (2012)
Z	−160 °C	$Chl^+ Chl^-$	Misra et al. (2001a, b, 2012)
Zv (variable)	−80 to 30 °C	$P680^+ Q_A^-$	Sane et al. (2012)
A	−15 °C	Tyr $Z^+ QB^-$	Misra et al. (2001a, b, 2012)
AT	−10 °C	$S_3Q_A^-$	Tatake et al. (1971), Inoue et al. (1977), Rosza and Demeter (1982), Demeter et al. (1985) and Homann (1999)
Q	+5 °C	$S_2Q_A^-$	Misra et al. (2001a, b, 2012)
B1	+20 °C	$S_3Q_B^-$	Inoue (1976), Joliot and Joliot (1980), Vass et al. (1981), Rutherford et al. (1982, 1984b), Demeter and Sallai (1986) and Miranda and Ducruet (1995b)
B2	+30 °C	$S_2Q_B^-$	
C	+50 °C	$TyrD^+Q_A^-$	Misra et al. (2001a, b, 2012)
AG	+40 to 50 °C	$S_2/S_3Q_B^-$	Bertsch and Azzi (1965), Bjorn (1971), Inoue (1996), Nakamoto et al. (1988), Sundblad et al. (1988), Hideg et al. (1991), Johnson et al. (1994) and Miranda and Ducruet (1995a, b)
High temperature TL peaks (HTL)	50 to 160 °C	Oxidative chemi-luminescence	Venediktov et al. (1989), Vavilin et al. (1991), Merzlyak et al. (1992), Hideg and Vass (1993), Stallaert et al. (1995), Misra et al. (1997), Marder et al. (1998), Vavilin and Ducruet (1998), Ducruet and Vavilin (1999), Havaux and Niyogi (1999), Skotnica et al. (1999), Havaux and Niyogi (1999) and Ducruet and Vavilin (1999)

For more details see Sane and Rutherford (1986), Inoue (1996), Misra and Ramaswamy (2001), Misra et al. (2001a, b, 2012), Sane et al. (2012) and Misra (2013)

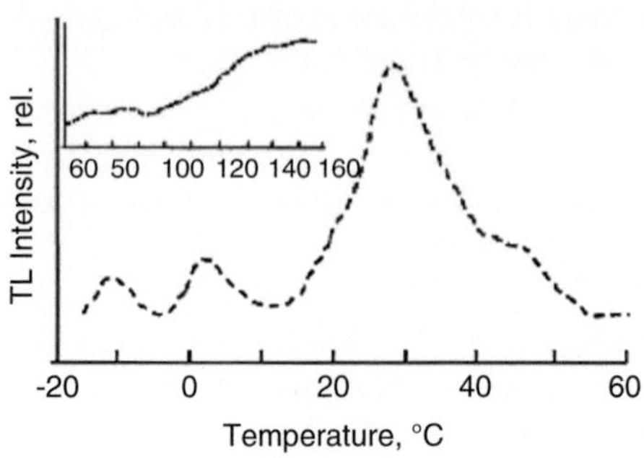

Fig. 6.2 Typical thermoluminescence (TL) peaks observed in a green leaf with a linear heating rate of ≤20 °C/s in darkness. (From Misra 2013)

4 Stress Induced Changes in TL Glow Peaks

The TL glow peaks as depicted in Table 6.2 clearly show that TL is a useful tool for the study of PS II electron transfer, both at the donor and acceptor sides (Misra and Ramaswamy 2001; Misra et al. 2001a, b, 2012; Misra 2013). Extensive reports are available to suggest that biotic and abiotic stress bring about a qualitative and quantitative change in the TL peak temperature and intensity. The changes in the TL characteristics and the environmental factors affecting it are summarized in Table 6.3.

Table 6.3 Changes in TL glow peaks of photosynthetic materials induced by developmental and stress (biotic and abiotic) responses (Misra and Ramaswamy 2001; Misra et al. 2001a; b, 2012; Misra 2013)

Environmental stress and plastid development	Changes in TL glow peaks	References
Plastid development		
Etiolated leaf	Major TL peaks missing due to a lack of functional pigment-protein complexes associated with PS II and do not develop Mn cluster	Inoue (1996), Sane et al. (1977), Misra et al. (1998a, b, c) and Dilnawaz et al. (2000)
Greening leaf	Q-band and B band intensity increases gradually from base to apex of that wheat leaves greening under continuous illumination	Misra et al. (1998b)
	Leaf greening under intermittent illumination leaves does not show the TL bands, as these plastids do not develop Mn cluster properly	Inoue (1996) and Sane et al. (1977)
Aging and senescence of leaf	Decrease in Q and B-band.	Biswal et al. (2001)
	The titre shows a gradual decrease in quinone pool and a block in electron flow between QA to QB.	
Genetic modification	Origin of Tl glow peaks	Farineau (1993) and Homann (1999)
Biotic stress		
Pathogen (viral) infection	Decreased B-band intensity and higher peak temperature.	Stallaert et al. (1995) and Rahoutei et al. (1999)
	A new TL peak at 70 °C	
Abiotic stress		
Salinity	Affects Q-band and B-band in a dose and duration dependent manner. B-band comparatively more affected.	Misra et al. (1998c), Sahu et al. (1998, 1999), Biswal et al. (2001) and Zurita et al. (2005)
	Back flow of electrons in PS II.	
Water/drought	Temperature shift in the TL glow peaks due to redox shift in the charge pairs	Ducruet and Vavilin (1999), Janda et al. (1999) and Misra et al. (2002)

(continued)

Table 6.3 (continued)

Environmental stress and plastid development	Changes in TL glow peaks	References
Anoxia	Reduces the B and C band intensities	Soltnev et al. (1999)
Mineral supplement (N, P, K)	A band shift from nearly −13 °C to 8 °C	Soltnev et al. (1999)
Heat/high temperature	Decrease in Q-band and B-band intensity	Misra et al. (1997, 1998a)
Photoinhibition	B-band affected with less effect on C-band	Misra et al. (1997, 1998a)
State transition	B-band affected	Bernat et al. (2018)
Heavy metal (Cu, Ni or Zc)	B-band affected	Mohanty et al. (1989)
Inert gas (N, He, Ar or Xe)	Reduces the B and C band intensities	Soltnev et al. (1999)

5 Future Perspective

Environmental and developmental changes affect the photosynthetic machinery (Joshi et al. 2013). Thermoluminescence is a non-invasive method and can give insight into the qualitative and quantitative changes in the Q_A and Q_B environment of PS II and thus give an insight to the donor/acceptor side structure and function, and also the oxidative state of thylakoid membranes (Misra and Ramaswamy 2001; Misra et al. 2001a, b, 2012; Misra 2013). Recently, TL signals have been used as 'sensors' for the study of photosynthetic materials (Zhang et al. 2007). TL technique gives a wide array of information about the redox state of electron donors, acceptors and charge accumulation in PS II of green leaves, and TL studies have an extensive and wide use in eco-physiological and stress studies in photosynthesis, as well as in agriculture.

References

Arnold WA, Azzi JR (1968) Chlorophyll energy levels and electron flow in photosynthesis. Proc Natl Acad Sci U S A 61:29–35

Bernat G, Steinbach G, Kaňa R, Govindjee, Misra AN, Prašil O (2018) On the origin of the slow M–T chlorophyll *a* fluorescence decline in cyanobacteria: interplay of short-term light-responses. Photosynth Res (in press) https://doi.org/10.1007/s11120-017-0458-8

Bertsch WF, Azzi JR (1965) A relative maximum in the decay of long-term delayed light emission from the photosynthetic apparatus. Biochim Biophys Acta 94:1526

Bhagwat AS, Bhattacharjee SK (2005) Thermoluminescence as a tool in the study of photosynthesis. Taylor and Francis Group, LLC, Boca Raton

Bhatnagar R, Saxena P, Vora HS, Dubey VK, Sarangpani KK, Shirke ND, Bhattacharjee SK (2002) Design and performance of a versatile, computer controlled instrument for studying low temperature thermoluminescence from biological samples. Meas Sci Technol 13:2017–2026

Biswal AK, Dilnawaz F, David KAV, Ramaswamy NK, Misra AN (2001) Increase in the intensity of thermoluminescence Q-band during leaf ageing is due to a block in the electron transfer from Q_A to Q_B. Luminescence 16:309–313

Bjorn LO (1971) Far-red induced, long-live after glow from photosynthetic cells. Size of after glow unit and path of energy accumulation and dissipation. Photochem Photobiol 13:5–20

Demeter S, Govindjee (1989) Thermoluminescence in plants. Physiol Plantarum 75:121–130

Demeter S, Sallai A (1986) Effect of pH on the thermoluminescence of spinach chloroplasts in the presence and absence of photosystem II inhibitors. Biochim Biophys 851:267–275

Demeter S, Rosza Z, Vass I, Hideg E (1985) Thermoluminescence study of charge recombination in photosystem II at low temperature II: oscillatory properties of the Z and a thermoluminescence bands in chloroplasts dark adapted for various time periods. Biochim Biophys Acta 809:379–387

Dilnawaz F, Vass I, Misra AN (2000) Thermoluminescence glow peaks of chloroplasts along the axis of wheat leaf lamina. Photosynthesis: PS2001. CSIRO Publishing. Melbourne, Australia S2-001

Ducruet JM (2003) Chlorophyll thermoluminescence of leaf discs: simple instruments and progress in signal interpretation open the way to new ecophysiological indicators. J Exp Bot 54:2419–2430

Ducruet JM, Miranda T (1992) Graphical and numerical analysis of thermoluminescence and fluorescence F0 emission in photosynthetic material. Photosynth Res 33:15–27

Ducruet JM, Vass I (2009) Thermoluminescence: experimental. Photosynth Res 101:195–204

Ducruet JM, Vavilin D (1999) Chlorophyll high-temperature thermoluminescence emission as an indicator of oxidative stress. Perturbating effects of oxygen and leaf water content. Free Rad Res 31S:187–192

Farineau J (1993) Compared thermoluminescence characteristics of pea thylakoids studied in vitro and in situ (leaves). The effect of photoinhibitory treatments. Photosynth Res 36:25–34

Gilbert M, Skotnica J, Weingart I, Wilhelm C (2004a) Effects of UV irradiation on barley and tomato leaves: thermoluminescence as a method to screen the impact of UV radiation on crop plants. Funct Plant Biol 31:825–845

Gilbert M, Wagner H, Weingart I, Skotnica J, Nieber K, Tauer G, Bergmann F, Fischer H, Wilhelm C (2004b) A new type of thermoluminometer: a highly sensitive tool in applied photosynthesis research and plant stress physiology. J Plant Physiol 161:641–651

Havaux M, Niyogi KK (1999) The violaxanthin cycle protects plants from photooxidative damage by more than one mechanism. Proc Natl Acad Sci U S A 96:8762–8767

Hideg E, Vass I (1993) The 75°C thermoluminescence band of green tissues: chemiluminescence from membrane-chlorophyll interaction. Photochem Photobiol 58:280–283

Hideg E, Kobayashi M, Inaba H (1991) The far-red induced slow component of delayed light from chloroplasts is emitted from photosystem II. Evidence from emission spectroscopy. Photosynth Res 29:107–112

Homann PH (1999) Reliability of photosystem II thermoluminescence measurements after sample freezing: few artifacts with photosystem II membranes but gross distortions with certain leaves. Photosynth Res 62:219–229

Inoue Y (1976) Manganese catalyst as a possible cation carrier in thermoluminescence. FEBS Lett 72:279–282

Inoue Y (1996) Photosynthetic thermoluminescence as a simple probe of photosystem II electron transport. In: Amesz J, Hoff AJ (eds) Biophysical techniques in photosynthesis, Advances in photosynthesis, vol 3. Kluwer Academic Publishers, Dordrecht, pp 93–107

Inoue Y, Yamasita T, Kobayashi Y, Shibata K (1977) Thermoluminescence changes during inactivation and reactivation of the oxygen-evolving system of isolated chloroplasts. FEBS Lett 82:303–306

Janda T, Szalai G, Giauffret C, Paldi E, Ducruet JM (1999) The thermoluminescence 'After glow' band as a sensitive indicator of abiotic stresses in plants. Z Naturforsch 54c:629–633

Johnson GN, Boussac A, Rutherford AW (1994) The origin of 40–50°C thermoluminescence bands in photosystem II. Biochim Biophys Acta 1184:85–92

Joliot P, Joliot A (1980) Dependence of delayed luminescence upon adenosine triphosphatase activity in *Chlorella*. Plant Physiol 65:691–696

Joshi P, Misra AN, Nayak L, Biswal B (2013) Chapter 28: Response of mature, developing and senescing chloroplast to environmental stress. In: Biswal B, Krupinska K, Biswal UC (eds) Plastid development in leaves during growth and senescence, Advances in photosynthesis and respiration, vol 36. Springer, Dordrecht, pp 641–668

Marder JB, Droppa M, Caspi V, Raskin VI, Horvath G (1998) Light-independent thermoluminescence from greening barley leaves: evidence for involvement of oxygen radicals and free chlorophylls. Physiol Plantarum 104:713–719

Maslenkova L (2010) Thermoluminescence from photosynthesizing systems as a method for detection of early plant stress symptoms. Effect of desiccation on thermoluminescence emission parameters in mesophytic and poikilohydric plants. Gener Appl Plant Physiol 36:87–99

Merzlyak MN, Pavlov VK, Zhigalova TV (1992) Effect of desiccation on chlorophyll high temperature chemiluminescence in *Acer platanoides* L. and *Aesculushippo castanum* L. leaves. J Plant Physiol 139:629–631

Miranda T, Ducruet JM (1995a) Characterization of the chlorophyll thermoluminescence after glow in dark-adapted or far-red-illuminated plant leaves. Plant Physiol Biochem 33:689–699

Miranda T, Ducruet JM (1995b) Effects of dark- and light induced proton gradients in thylakoids on the Q and B thermoluminescence bands. Photosynth Res 43:251–262

Misra AN (2013) Thermoluminescence in plants: concept and application. Int J LifeSci Biotechnol Pharm Res 2:10–17

Misra AN, Ramaswamy NK (2001) Thermoluminescence of green plants (review, new trends in photosciences). Indian Photobiol Soc News Lett 40:45–50

Misra AN, Ramaswamy NK, Desai TS (1997) Thermoluminescence studies on photoinhibition of pothos leaf discs at chilling, room and high temperature. J Photochem Photobiol 38:164–168

Misra AN, Ramaswamy NK, Desai TS (1998a) Thermoluminescence properties and changes in D1 polypeptide of spinach chloroplasts during photoinhibition of spinach leaf discs at chilling, room and high temperatures. In: Garab G (ed) Photosynthesis: mechanism and effects, vol III. Kluwer Academic Publisher, Dordrecht, pp 2213–2216

Misra AN, Sahu SM, Dilnawaz F, Mahapatra P, Misra M, Ramaswamy NK, Desai TS (1998b) Photosynthetic pigment-protein content, electron transport activity and thermoluminescence properties of chloroplasts along the developmental gradient in greening wheat (*Triticum aestivum* L.) leaves. In: Garab G (ed) Photosynthesis: mechanism and effects, vol IV. Kluwer Academic Publisher, Dordrecht, pp 3179–3182

Misra AN, Sahu SM, Misra M, Ramaswamy NK, Desai TS (1998c) Sodium chloride salt stress induced changes in thylakoid pigment-protein complexes, PS II activity and TL glow peaks of chloroplasts from mungbean (*Vigna radiata* L.) and Indian mustard (*Brassica juncea* Coss.) seedlings. Z Naturforsch C 54:640–644

Misra AN, Dilnawaz F, Misra M, Biswal AK (2001a) Thermoluminescence in chloroplasts as an indicator of alterations in photosystem II reaction center by biotic and abiotic stress. Photosynthetica 39:1–9

Misra AN, Dilnawaz F, Misra M, Biswal AK (2001b) Thermoluminescence in chloroplasts. In: Padha Saradhi B (ed) Biophysical processes in living systems. Oxford/IBH, Science Publisher, New Delhi/Enfield/Plymouth, pp 303–311

Misra AN, Biswal AK, Misra M (2002) Physiological, biochemical and molecular aspects of water stress responses in plants, and the biotechnological applications. Proc Natl Acad Sci (India) 72B(II):115–134

Misra AN, Misra M, Singh R (2012) Thermoluminescence in chloroplast thylakoid. In: Misra AN (ed) Biophysics. Intech Open, London, pp 155–170

Mohanty N, Vass I, Demeter S (1989) Copper toxicity affects photosystem II electron transport at the secondary quinone acceptor, Q_B. Plant Physiol 90:175–179

Nakamoto I, Sundblad LG, Garderstro ÈP, Sundbom E (1988) Far-red stimulated long-lived luminescence from barley protoplasts. Plant Sci 55:1–7

Rahoutei J, Baron M, Garcia-Luque I, Droppa M, Nemenyi A, Horvath G (1999) Effect of Tobamo virus infection on thermoluminescence characteristics of chloroplasts from infected plants. Z Naturforsch 54c:634–639

Randall JT, Wilkins MHF (1945) Phosphorescence and electron traps. Proc R Soc (London) Ser A 184:366–408

Rosza Z, Demeter S (1982) Effect of inactivation of the oxygen evolving system on the thermoluminescence of isolated chloroplasts. Photochem Photobiol 36:705–708

Rutherford AW, Crofts AR, Inoue Y (1982) Thermoluminescence as a probe of photosystem II photochemistry. The origin of the flash-induced glow peaks. Biochim Biophys Acta 682:457–465

Rutherford AW, Govindjee, Inoue Y (1984a) Charge accumulation and photochemistry in leaves studied by thermoluminescence and delayed light emission. Proc Natl Acad Sci U S A 81:1107–1111

Rutherford AW, Renger G, Koike H, Inoue Y (1984b) Thermoluminescence as a probe of photosystem II. The redox and protonation states of the secondary acceptor quinone and O_2-evolving enzyme. Biochim Biophys Acta 767:548–556

Sahu SM, Misra AN, Misra M, Ramaswamy NK, Desai TS (1998) Sodium chloride salt stress induced changes in thylakoid pigment-protein complexes, PS II activity of mungbean (*Vigna radiata* L.) seedlings. In: Garab G (ed) Photosynthesis: mechanism and effects, vol IV. Kluwer Academic Publisher, Dordrecht, pp 2625–2628

Sahu SM, Dilnawaz F, Meera I, Misra M, Ramaswamy NK, Desai TS, Misra AN (1999) Photosynthetic efficiency of mung bean (*Vigna radiata* L. Wilczek) and Indian mustard (*Brassica juncea* Coss.) during seedling establishment under NaCl salinity. In: Srivastava GC, Singh K, Pal M (eds) Plant physiology for sustainable agriculture. Pointer Publisher, Jaipur, pp 388–391

Sane PV (2004) Thermoluminescence. A technique for probing photosystem II. In: Carpentier R (ed) Photosynthesis research protocols, Methods in molecular biology. Humana Press Inc, Totowa, pp 229–248

Sane P, Rutherford AW (1986) Thermoluminescence from photosynthetic membranes. In: Govindjee, Amesz J, Fork DC (eds) Light emission by plants and bacteria. Academic, Orlando, pp 291–329

Sane PV, Desai TS, Tatake VG, Govindjee (1977) On the origin of glow peaks in *Euglena* cells, spinach chloroplasts and subchloroplast fragments enriched in system I or II. Photochem Photobiol 26:33–39

Sane PV, Ivanov A, Oquist G, Huner N (2012) Thermoluminescence. In: Eaton-Rye JJ, Tripathy BC, Sharkey TD (eds) Photosynthesis: plastid biology, energy conversion and carbon assimilation, Advances in photosynthesis and respiration, vol 34. Springer, Dordrecht/New York, pp 445–474

Soltnev MK, Tashish V, Karavayev V, Khomutov A (1999) Specific action of iodine ions on the regulatory processes in photosynthetic membranes. Biophysics 40:1273–1275

Skotnica J, Fiala J, Ilik P, Dvorak L (1999) Thermally induced chemiluminescence of barley leaves. Photochem Photobiol 69:211–217

Stallaert VM, Ducruet JM, Tavernier E, Blein JP (1995) Lipid peroxidation in tobacco leaves treated with the elicitor cryptogein: evaluation by high-temperature thermoluminescence emission and chlorophyll fluorescence. Biochim Biophys Acta 1229:290–295

Sundblad LG, Schroder WP, Akerlung HE (1988) S-state distribution and redox state of QA in barley in relation to luminescence decay kinetics. Biochim Biophys Acta 973:47–52

Tatake VG, Desai TS, Bhattacharjee SK (1971) A variable temperature cryostat for thermoluminescence studies. J Physics E: Sci Instrum 4:755–762

Vass I, Horvath G, Herczeg T, Demeter S (1981) Photosynthetic energy conservation investigated by thermoluminescence. Activation energy and half life of thermoluminescence bands of chloroplasts determined by mathematical resolution of glow curves. Biochim Biophys Acta 634:140–152

Vavilin DV, Ducruet JM (1998) The origin of 120–130°C thermoluminescence bands in chlorophyll-containing material. Photochem Photobiol 68:191–198

Vavilin DV, Matorin DN, Kafarov RS, Bautina AL, Venediktov PS (1991) High temperature thermoluminescence of chlorophyll in lipid peroxidation. Biologich Membr 8:89–98

Venediktov PS, Matorin DN, Kafarov RS (1989) Chemiluminescence of chlorophyll upon lipid photoperoxidation in thylakoid membranes. Biofizika (Moscow) 34:241–245

Zeinalov Y, Maslenkova L (1996) A computerized equipment for thermoluminescence investigations. Bulg J Plant Physiol 22:88–94

Zhang L, Xing D, Wang J (2007) A non-invasive and real-time monitoring of the regulation of photosynthetic metabolism biosensor based on measurement of delayed fluorescence in vivo. Sensors 7:52–66

Zurita JL, Roncel M, Aguilar M, Ortega JM (2005) A thermoluminescence study of photosystem II back electron transfer reactions in rice leaves – effects of salt stress. Photosynth Res 84:131–137

Chapter 7
Determining Plant Water Relations

Gorka Erice, María Luisa Pérez-Bueno, Mónica Pineda, Matilde Barón, Ricardo Aroca, and Mónica Calvo-Polanco

1 Introduction

As for all living organisms, water is fundamental for plants, not only because water is the origin of life (Daniel et al. 2006), but because plants use water loss by transpiration as a mechanisms of heat dissipation and cooling of leaves (Curtis 1936; Pallas et al. 1967), and furthermore, water is essential to maintain plant cell turgor and favor plant growth. Thus, the percentage of water in plant tissues (the term plant in this chapter will be referred only to higher plants) may range from 30 to 50% in woody trunks (Borchert 1994) to around 90% in leaves and roots of several plant species. However, since plants are sessile organisms they have need to develop mechanisms to get water homeostasis mostly when water availability is scare. The main mechanisms that plants have to regulate their water content are the regulation of stomatal aperture and of root water uptake properties (Vaadia et al. 1961; Aroca et al. 2012).

Based on what is written above and before to continue with this chapter will be useful to define the term *plant water relations*. In the literature there is not a clear definition. Historically, *plant water relations* have been mostly linked to the response of plants to drought. In a database search including the terms *plant water relations*, 14% of the outputs are related to drought and 33% to stresses in general.

G. Erice · R. Aroca (✉)
Department of Soil Microbiology and Symbiotic Systems, Estación Experimental del Zaidín (CSIC), Granada, Spain
e-mail: raroca@eez.csic.es

M. L. Pérez-Bueno · M. Pineda · M. Barón
Department of Plant Biochemistry, Cellular and Molecular Biology, Estación Experimental del Zaidín (CSIC), Granada, Spain

M. Calvo-Polanco
Department of Plant Biochemistry and Molecular Physiology, SupAgro/INRA UMR 5004, Montpellier Cedex 2, France

© Springer International Publishing AG, part of Springer Nature 2018
A. M. Sánchez-Moreiras, M. J. Reigosa (eds.), *Advances in Plant Ecophysiology Techniques*, https://doi.org/10.1007/978-3-319-93233-0_7

Hence, in an early review about *plant water relations*, it addressed mostly how plants control its water content under conditions of soil water deficit (Vaadia et al. 1961). Nevertheless, since plants grow in a constantly changing environment and each change of temperature, light intensity or humidity may change stomatal aperture and water uptake capacity (Hirasawa et al. 1992; Ache et al. 2010), the term plant water relations cannot be only linked to stressful conditions. Therefore, we could define *plant water relations* as the study of water in plants, from its entry to the plant by roots until its exit from leaves, including also its transport along the stems and how it is retained in plant tissues. Thus, this chapter will start describing the methods for measuring root water uptake capacity, followed by methods determining xylem sap flow, tissue water status and stomatal aperture.

2 Root Water Uptake Capacity

2.1 General Concepts

There are in the literature several reviews addressing how plants regulate root water uptake capacity (Aroca et al. 2012; Kudoyarova et al. 2013). Two main forces determine the water uptake capacity of roots, namely hydrostatic and osmotic forces. Hydrostatic force is caused by transpiration demand, and the water goes mainly by the apoplastic path, since less resistance is found. However under conditions where transpiration is restricted (at night or under some abiotic stresses like drought or salinity) water goes mainly by the cell-to-cell path, which includes symplastic and transcellular paths. Symplastic path compromises the water flowing through plasmodesmata, while transcellular path compromises water flowing across cellular membranes. But most important is that these paths are interconnected and water exchange among them may occur depending on the environmental conditions at each time. Prof. Steudle named the above described water routes as the composite water transport model, since root water uptake is a dynamic process finely regulated, even at the molecular level thanks to aquaporins (Steudle 1997, 2000; Steudle and Peterson 1998).

2.2 Root Hydraulic Conductivity

Root hydraulic conductivity (L) is the parameter to be determined in order to estimate root water uptake capacity (Javot and Maurel 2002). As any conductivity, L is determined by the whole flow and by the gradient, which causes such flow. In the case of L it is calculated as the ratio between the root water flow (expressed in a root dry weight or area basis) and the hydrostatic or osmotic gradient expressed in pressure units, mostly in MPa. There are mainly four methods to determine L namely free exudation method, pressure chamber, high-pressure flow meter (HPFM) and root pressure probe. Each of the methods has its own advantages and disadvantages and they will be described in the following sections.

2.2.1 Free Exudation Method

The free exudation method is the simplest way to calculate L. The method consists in cutting the stem the closest to root as possible, attach a silicon tube or another kind of tubing taking special care to avoid sap licks, discard the first minutes of sap exuded to avoid phloem contaminations, and record the volume of sap exuded in a fixed period of time (from few minutes to hours, depending on the rate of exudation). Next, the osmotic potential of the exuded sap and the solution surrounding roots is determined and L can be calculated as follow:

$$L = \frac{Jv}{\Delta\Psi s}$$

Where Jv is the xylem sap flux in a root units basis as dry weight, surface or volume, and $\Delta\Psi$s is the osmotic potential difference between the exuded sap and the soil solution. By this method, since there is no transpiration, L corresponds only to the cell-to-cell path (Steudle and Peterson 1998). Under atmospheric pressure conditions (no external pressure is applied) the water fluxes through the root thanks to the osmotic gradient generated by the uptake of solutes from the soil solution by the roots. However the above formula assumed the roots behave as a perfect osmometer, which is a perfect semi-permeable membrane avoiding totally the pass of solutes. However root membranes are not perfect osmometers and the calculation of osmotic reflection coefficient (σ) is needed to calculate L as:

$$L = \left(\frac{Jv}{\Delta\Psi s \times \sigma} \right)$$

However, the calculation of σ is not easy and it is mostly estimated (Fiscus 1977), and it is usually close to 1 (Fiscus 1977, 1986). Thus several authors omitted σ from the calculation of L (Sánchez-Romera et al. 2014; Mahdieh et al. 2016), and others trait to introduce the term composite root hydraulic conductivity ($\sigma \times$ L; Bigot and Boucaud 2000; Aroca et al. 2001), but this term has not been used that much. Then, σ can be calculated by the pressure chamber (Fiscus 1977) or by the root pressure probe (Steudle et al. 1987; Miyamoto et al. 2001). Readers are referred to these articles to see how σ could be calculated.

As summary, the free exudation method is ideal for a first screening on how any environmental trait modifies L, but its significance is limited to the cell-to-cell path and should be taken into account the value of σ, since in some species as rice its value can be as low as 0.4 (Miyamoto et al. 2001). However it could be used in comparative studies.

2.2.2 Pressure Chamber and High-Pressure Flow Meter (HPFM)

The use of the pressure chamber to calculate L was initiated in the 70s (Fiscus 1977; Markhart et al. 1979). In this method decapitated roots are immersed in a chamber filled with nutrient solution and the roots are pressurized in small steps from 0.1 MPa to around 0.7 MPa (depending on the plant species). At each step, after some minutes of stabilization, the xylem sap exuded is collected as in the free exudation method, but only during few minutes (from 5 to 10 min). Thereafter the sap volume at each pressure (normalized by a root dry weight, surface or volume basis) is plotted against the corresponding pressure. The slope of the resulted trend line is equal to L (Markhart et al. 1979). To be considered to calculate L, the R value of the trend line should be at least greater than 0.90. By this method, since the roots are pressurized, both water pathways are considered (cell-to-cell and apoplastic ones). Commonly, the L values determined by the pressure chamber are higher than the values determined by the free exudation method (Miyamoto et al. 2001; Ranathunge et al. 2003), since the later only consider flow caused by osmotic forces. However, in some plants or under certain environmental conditions, were cell-to-cell path dominates, L determined by the free exudation method could be higher than one determined by the pressure chamber (Steudle and Jeschke 1983; Aroca et al. 2001). At the same time, when measuring L by the pressure chamber, it is also possible to determine the percentage of water that is flowing by the apoplastic path. By dissolving different apoplastic tracers (compounds unable to cross living membranes) in the nutrient solution inside the pressure chamber, recording its absorbance in the exuded sap and by applying the following formula:

$$\%AP = \left(\frac{AXyl}{ASol}\right) \times 100$$

Where, % AP is the percentage of water circulating by the apoplastic path, AXyl the absorbance of the exuded sap, and ASol the absorbance of the nutrient solution (Bárzana et al. 2012). The tracers mostly used among others are light green SF yellowish and trisodium 3-hydroxy-5,8,10-pyrene-trisulfonate (Zimmermann and Steudle 1998; Bárzana et al. 2012).

While in the pressure chamber the water circulates in the same sense as it does under growing conditions (from roots to shoots), under the high-pressure flow meter (HPFM) the water goes from the stems to the roots. HPFM was first used by the group of Prof. Tyree (Tyree et al. 1995). In this method, the roots remain in the soil substrate or in the hydroponics, without suffering any potential disturbance or damage as could happen in the pressure chamber method. When the stem is cut under water to avoid lost of xylem conductivity, water is perfused into the root system and the pressure is increasing in constant rate of about 3–7 kPa s^{-1}. The system is measuring at the same time water flow and pressure, and as in the pressure chamber method, the slope of the plot of flow versus applied pressure is L normalized by a root units (Tyree et al. 1995). Although HPFM method was first employed in trees (Tyree et al. 1995; Nardini and Tyree 1999), it can be used also in herbaceous plants

(Tsuda and Tyree 2000; Calvo-Polanco et al. 2014). The main problem of using HPFM to determine L is its technical complexity. In fact there is only one company located in USA that commercializes this apparatus. In contrast, the acquisition of a pressure chamber is more reasonable. Obviously, under this method both water pathways are determined at the same time (cell-to-cell and apoplastic).

2.2.3 Root Pressure Probe

In 1987 the group of Prof. Steudle applied the use of the pressure probe to an intact maize roots (Steudle et al. 1987). In this method, a single intact root is connected to the pressure probe by a capillary with is filled with water in root contact zone and thereafter with silicon oil. Since both fluids are not miscible, a meniscus is formed, which can be monitored by a microscope. Changes in root pressure can be obtained by the use of a movable rod (hydrostatic) or adding different solutes to the root medium (osmotic). These changes can be recorder by a pressure transducer and the displacement of the meniscus can be monitored. After changing root pressure (hydrostatic or osmotic), if the root has a high L, it recovers quickly the original pressure, but if the root has lower L, the time to recover the initial pressure is larger. By this method, both osmotic and hydrostatic L can be calculated as well σ. The main problem of this technique it is a very complex handling, and that it has to be home made. There are no commercial companies distributing this apparatus. Readers are advising to see Steudle et al. (1987) and Tomos and Leigh (1999) to get more information about this technique.

3 Xylem Root Hydraulics

3.1 Introduction

The transport of water through the xylem is a critical piece of information to define how water loss is replaced during transpiration, preventing desiccation and maintaining photosynthetic rates. Direct or indirect measurements of the stem hydraulic properties have provided fundamental information of a plant's capacity to supply water to photosynthetic and growing tissues and information on plants sensitivity to adverse environmental stresses (Tyree and Zimmermann 2002; Holbrook and Zwieniecki 2005; Brodribb et al. 2010; Melcher et al. 2012). The water transport process is affected by increasing negative pressures in the liquid water inside the xylem conduits (Pockman et al. 1995), with the risk that the column of water collapses and vapor nucleation occurs (risk of cavitation). These pressures can lead to cavitation, and creating an embolus that impairs water transport (Zimmermann 1983). One of the most useful and extended ways to quantify the relationship between the hydraulic conductivity of the xylem and the xylem pressure is to generate a 'vulnerability curve' (Tyree and Sperry 1989a).

3.2 Measurements of Xylem Hydraulic Conductivity

The xylem hydraulic conductivity is defined as the flux for a given driving force (Q/ΔP), normalized by the length of the segment and referenced either to the cross-section of the area of the xylem (K_s: xylem specific conductivity) or to the leaf area supported by the xylem (K_L: leaf specific conductivity):

- $K_s = QL/(\Delta P\, A_{SW})$
- $K_L = QL/(\Delta P\, A_L)$

Where Q is the recorded flux (gravimetric or volumetric flow rate), L is the length of the measured segment, ΔP is the pressure drop across the segment, A_{sw} is the cross-sectional area of the conducting sapwood, A_L is the total leaf area supplied by the measured segment. The flow rate measurement of fluid flows can be determined gravimetrically or using conductivity apparatus. In the gravimetrically determinations, the segment of interest will be connected to a water reservoir on one side, and to a pipette or an analytical balance with a computer interface to the other (Tyree et al. 1983; Sperry et al. 1988; Nardini et al. 2001). With the pipette, we will record the change in meniscus position as a function of time gives the volume flow rate, while the height difference between the water leave in the reservoir and the meniscus in the horizontal pipette gives the driving gradient. With the balance, we can register the change in mass with time with a computer interface. An assumption of these measurements is that the system is at constant flow rate for a stable driving gradient.

An alternative for the previous methods are the hydraulic flow meters, where pressures across a calibrated resistance tube can be measured using pressure transducers and data logging systems to provide continuous flows. This technique was developed by Sperry et al. (1988) with the low pressure flow meter (LPFM), and by Cochard et al. (2005) with the Xyl'EM system (Bronkhorst, France SAS; http://www.bronkhorst.fr/fr/produits/xylem_embolie-metre/). It is also possible the construction of these devices in an inexpensive way as it is detailed in Sack et al. (2011) (http://prometheuswiki.publish.csiro.au/tiki-index.php?page=Constructing+and+operating+a+hydraulics+flow+meter).

3.3 Measuring the Vulnerability of the Xylem

3.3.1 Methods to Induce Xylem Cavitation and Embolism

- Dehydration method (Sperry 1986; Tyree and Sperry 1989a)
  The method consists in the progressively desiccation of the sample as to simulate various xylem pressures. The hydraulic conductivity is measured gravimetrically, or using the hydraulic flow meters explained before. The measurements will be conducted before and after any air in the conduit system has been

dissolved which will give the decrease in hydraulic conductivity associated with the minimum xylem pressure during the dehydration. The biggest disadvantage is the time consuming of the procedure and different branch systems are dried to different negative pressures and the curve assembled from conductivity measurements made on multiple segments. This method has been recently shown to be prone to artifacts caused by cutting xylem under tension (Wheeler et al. 2013; Torres-Ruiz et al. 2014).

- The air injection method (Cochard 1992; Sperry and Saliendra 1994)
  The air injection method is based on the observation that the pressure required to push air into the functional xylem and decrease its conductivity, and is approximately equal, and opposite to the negative pressure causing cavitation. The sample (stem segment) is inserted through a double-ended pressure bomb. The hydraulic conductivity of the xylem is measured progressively at higher air pressures inside the chamber, and once the air pressure has displaced the water from the xylem, the hydraulic conductivity decreases. The major improvement of this method is that allows having the vulnerability curve in a single stem. On the contrary, the procedure is also time consuming and only provides an estimate of the vulnerability curve.
- Centrifugal methods
  These methods are based in the fact that centrifugal force can be used to create negative pressure in the xylem of woody vascular plants (Pockman et al. 1995). Alder et al. (1997) formalized the first centrifugal-technique, known as the 'standard centrifuge method'. The measurements of the hydraulic conductivity of xylem segments are done before and after spinning them in a centrifuge rotor to generate progressively negative xylem pressure. To determine the hydraulic conductivity, the segments need to be removed from the rotor between intervals of spinning (Pockman et al. 1995; Alder et al. 1997). This method was modified to allow measurements of conductivity while stem segments are spinning under negative pressure (Cochard et al. 2005; Li et al. 2008). In this 'spin' method, the stem ends are immersed in water as they are in the gravity method, but the water level in the upstream reservoir is higher that the downstream position (Cochard et al. 2005).

3.3.2 Construction of Vulnerability Curves

Vulnerability curves (VC) represent percentage of cavitation or embolism. They are plotted as percentage loss of hydraulic conductivity (PCL) vs xylem pressure. Any of the methods explained to measure cavitation can be combined with the ones to induce water stress to determine VC. With the method of modified centrifugal techniques, we can obtain the entire curve relatively quickly on a single xylem segment, and the segment can be treated ahead of time to remove any native embolism. The shape of the resulting vulnerability curve (VC) varies from sigmoidal (s-shaped) to logarithmic (r-shaped). The r-shaped curves imply a high amount of embolism formation or high native PCL levels, and will imply that refilling of embolized conduits occurs on a daily basis. The s-shaped curves predict that embolism formation occurs probably only during periods of severe drought, and embolism refilling is either minimal or occurring on a seasonal basis.

In the recent years, it has been a big debate on the accuracy of the measurements to construct VCs due to the different methodologies applied in different experiments. In conifers, the standard centrifuge and the centrifuge modifications seems not to alter the results of VC (Li et al. 2008; Lopez et al. 2016; Torres-Ruiz et al. 2016). However, in angiosperms, different experimentations have shown different results that are thought to be directly related on how the samples are measured and the VCs constructed (Cochard et al. 2010, 2013; Wheeler et al. 2013; Torres-Ruiz et al. 2014, 2015; Hacke et al. 2015; Venturas et al. 2015; Charrier et al. 2016). Up-to-recent years, there were not truly reference methods that could validate the results from the centrifuges methodologies. With the development of non-destructive methodologies as magnetic resonance imaging MRI and X-ray micro-tomography, new tools to advance in the knowledge of water transport and embolism in plants are available (Torres-Ruiz et al. 2016; Nardini et al. 2017).

3.4 Methods to Measure Xylem Cavitation and Embolism

3.4.1 Light Microscope, Cryo-Scanning and Dye Coloration

Direct observations under eye or light microscope gave the first evidence of embolized conduits in plants (Richard 1838; Dixon 1914; Sperry et al. 1988), although this technique gives difficulties as for the preparation of the samples (Lewis et al. 1994) and the limited field of observation.

More sophisticated technologies have been also used for direct observation of the samples as in Canny (1997), where a scanning electron microscope equipped with a cryogenic stage was used. This technique is also laborious, but it has high spatial resolution, and has been used to determine different aspects of xylem water transport, as well as the water content on small tracheids and wood fibers (e.g. Utsumi et al. 1998; Cochard et al. 2000, 2004; Sano et al. 2011; Aoyama et al. 2014).

Dye coloration of the xylem is a simple and effective way to visualize embolism in the xylem (Cai and Tyree 2010). The dye is induced on branches by leaf transpiration or by infiltration with a small gravimetric pressure (Sperry 1986). As it is quite a laborious process, other techniques are preferred.

3.4.2 Acoustic Measurements (AEs)

This method provided the first demonstration that plants could cavitate. It is a non-destructive technique, which its principle is based on the detection of AEs (Milburn and Johnson 1966) and ultra-AEs (Tyree and Sperry 1989b) due to the energy relaxed in the xylem when cavitation occurs. The main advantage is that can be used easily under field conditions and that have a very high temporal resolution that is able to detect the time of occurrence of a cavitation with great accuracy. The main disadvantage is that ultra-AEs is not produced only in xylem conduits (Kikuta 2003), and that the method is more qualitative than quantitative. With the commercialization of a new generation of acoustic systems, it has becoming possible to filter out all the irrelevant ultra-AEs, and the technique is having a renewed interest (Wolkerstorfer et al. 2012; Vergeynst et al. 2016).

3.4.3 Magnetic Resonance Imaging (MRI)

MRI is an imaging method based on nuclear magnetic resonance (NMR). It is a non-destructive and non-invasive technique that allows for direct in vivo observation of water content and water movement in plants under regular or environmental stress conditions (Van As et al. 2009). MRI experiments are based on the interaction between the magnetic moment of certain nuclei (especially ^{1}H) and an external magnetic field. The magnetization of the sample is manipulated by radio frequency (RF) pulses, which are detected by a coil placed around the sample (Köckenberger et al. 1997, 2004). Although MRI suffers from poor spatial resolution (20–30 μm), it delivers an abundant amount of information at a reasonable spatial and temporal resolution (Van As et al. 2009). This methodology has become a growing discipline in the study of the dynamics of water (Windt et al. 2006; Homan et al. 2007; Van As and Windt 2008) and also in the occurrence of cavitation and its repair (Zwieniecki et al. 2013; Wang et al. 2013; Ogasa et al. 2016).

3.4.4 Synchrotron and X-Ray Microtomography (Micro-CT)

This technology presents the same advantages than the MRI but with higher spatial resolution (around 1 μm). The analyses of samples start with a 3D reconstruction of the selected area, and a transverse two-dimensional (2D) cross section from the middle of the volume, where the determination of embolism is done. The access to synchrotron facilities and the X-ray microtomography systems have giving the opportunity to test indirect hydraulic conductivity measurements in plants with quite positive results (Choat et al. 2016). It is also giving interesting results on how embolism initiates and spreads among plant species (Brodersen et al. 2013; Dalla-Salda et al. 2014; Torres-Ruiz et al. 2016). The emergence of powerful and affordable desktop-based micro-CT systems offers the opportunity to propose this technology as a reference for xylem embolism studies (Charra-Vaskou et al. 2012; Suuronen et al. 2013; Torres-Ruiz et al. 2014, 2016; Knipfer et al. 2016; Nardini et al. 2017) that will allow to confirm previous experimentation and to give reference information on the study of xylem embolism and refilling.

4 Measurement of the Leaf Water Status

4.1 *Introduction*

Life in our planet is intimately linked to water and the knowledge of drought stress physiology is of increasing interest as water is becoming limiting in a climate change scenario. The International Panel on Climate Change predicts a worldwide increase in aridity (IPCC 2013) and subsequent water stress for crops. As a consequence, water deficit is becoming the most important abiotic stress. Therefore, strategies for sustainable use of water (Kang et al. 2003) and to improve drought resistance are urgent and should integrate conventional breeding and biotechnological approaches (Chaves et al. 2003). A key factor determining plant productivity under limited water supply is water use efficiency (WUE) (Saranga et al. 1999) and it is mentioned as a strategy to improve crop performance under water limited conditions (Araus et al. 2002). Plant physiology can increase the knowledge about how to improve plants water use efficiency either by using varieties that are more drought and salt resistant or by using strategic agricultural practices.

Therefore it is essential that experiments conducted to understand the mechanisms involved in drought tolerance and/or resistance to drought characterize plant water status. Incorrect determination of plant water content can easily lead the researchers to wrong interpretation of the results. In this sense the selection of the appropriate technique to measure plant water status is elementary to achieve the objectives of the research. Extensive research has been done measuring plant water status trough determining soil or even stem (trunk) water content but leaf water status is thought to be more sensitive with regard to reflecting water deficit (Zheng et al. 2015).

In this line it is assumed that transpiring parts of the plant, especially leaves, usually develop greater and longer water deficits (Kramer and Boyer 1995; Erice et al. 2006). Consequently, authors working on plant responses to drought have focused on measuring leaf water status as a method to highlight the differences among water regimes. In this section the methods to measure plant water status will be compiled and the applications will be discussed in terms of the objective of the research.

4.2 *Volumetric Techniques*

Leaf water status can be quantified in terms of **water content (WC)**, which is the amount of water contained in a piece of leaf tissue. This content is dependent of the size of the considered biomass and needs to be normalized and expressed as fraction of the leaf mass. If expressed on fresh mass units it can be considered as the proportion of water that is contained in the leaf. WC also may be expressed on dry weight basis. The interpretation of this numeric value normalized by dry mass is more difficult but it has the advantage of being more stable than the WC on fresh mass basis.

$$WC = \frac{(FW - DW)}{FW}$$

or

$$WC = \frac{(FW - DW)}{DW}$$

where FW is the leaf fresh weight and DW is the leaf dry weight. DW is obtained after drying at 80 °C the leaf tissue until stable weight (normally 48 h) (Erice et al. 2006).

Nevertheless the WC is submitted to dairy and seasonal changes and normalization by fresh or dry weight also makes difficult the comparison between species with different grade of tissue hydration potential (i.e. succulent plants). This problem is solved by including in the equation the normalization by turgid weight (TW); the resulted parameter is denominated **relative water content (RWC)**. The RWC was originally defined as "relative turgidity" by Weatherley (1950) and defined as:

$$RWC = \left(\frac{FW - DW}{TW - DW}\right) * 100$$

A RWC of 100 means full turgor, whereas smaller value is indicative of leaf water deficit. In the same way it is defined the complementary parameter of the RWC, the **water saturation deficit** (WSD):

$$WSD = \left(\frac{TW - FW}{TW - DW}\right) * 100$$

Addition of RWC and WSD is 100. The determination of RWC or WSD requires the measurement of the TW of the leaf. Leaf tissues use to be fully hydrated after 24 h in a saturated humidity environment (typically a closed vial with wet cotton or floating on the surface of water) at 4 °C.

4.3 Thermodynamic Techniques

Above described parameters (WC, RWC and WSD) are refer to the proportion of water contained in the leaf and do not allow to determine the direction of the water being translocated between the different organs of the plant or between the soil and the plant. The movement of the water from the soil and across the different parts of the plant, roots, stems and leaves, follows the magnitude of the chemical potential, the partial molar Gibss free energy. That means that water flows following the differences in the thermodynamic potential between the different parts of the so-called soil-plant-atmosphere continuum system. Plant physiologists use the term **water potential, Ψ**, to refer to this concept. It integrates the different potentials that bind water to plant tissues. Pure water has the reference Ψ of zero and all the Ψ associated to water have negative values. Ψ was expressed in pressure units by Slatyer (1967):

$$\Psi = \left(\frac{RT}{V_w}\right)\log_e\left(\frac{e}{e_0}\right)$$

where R is the universal gas constant (8.31 J K^{-1} mol^{-1}), T is temperature (K), V_w is the partial molal volume of water, e is the vapour pressure in equilibrium with the atmosphere and e_0 is the saturation vapour pressure.

In the soil-plant atmosphere system it is considered that the total water potential is the sum of various components following the equation:

$$\Psi = \Psi_p + \Psi_\pi + \Psi_m + \Psi_g$$

where Ψ_p is the pressure potential, the hydrostatic pressure opposite to the action of vacuolar pressure on the cell wall resulting from cell turgor. It increases the water energy status in the cell and it is positive as long as the vacuole presses the cell wall (the cell is not plasmolyzed).

Ψ_π is the osmotic potential, consequence of the dissolved solutes. It decreases the water free energy. It is negative and its value decreases with the concentration of solutes. A useful approach to the relationship between solutes and the Ψ_π is the Van't Hoff equation:

$$\Psi_\pi = -RTc_s$$

where c_s is the solute concentration (mol m^{-3}).

Ψ_m is the matrix (or matric) potential and as the Ψ_π it decreases the water free energy. Ψ_m is due mainly to the binding of water by capillarity and adsorption to the cellular wall fibers and cytoplasmic macromolecules and colloids. In adult plant cell most of the volume is occupied by the vacuole so it is assumed that the water exchange is controlled by the tonoplast. In this situation the Ψ_m could be disregarded.

Ψ_g is the gravitational potential, the consequence of the potential energy change between locations with different height from the reference point. It is often disregarded but it definitively has to be taken into account when water potential is considered in trees. That is why except in the cases of studies involving tall trees plant water potential (Ψ) is generally supposed to be dependent exclusively of the osmotic potential (Ψ_π), matrix potential (Ψ_m) and pressure potential (Ψ_p).

$$\Psi = \Psi_p + \Psi_\pi + \Psi_m$$

Moreover in parenchymatous adult cells where the vacuole occupies most of the cell volume and considering both tonoplast and cell membranes semipermeables matrix potential may be disregarded:

$$\Psi = \Psi_p + \Psi_\pi$$

Measurement of the plant water potential is done using dry thermocouple psychrometry, plant pressure probe and the pressure chamber.

Psychrometric method consists of the encapsulation of the plant tissue in a hermetically closed chamber until the equilibration with the atmosphere and the thermocouple in contact with a solution of known potential. The change in vapour pressure could be measured using fresh and frozen-thaw tissue to break the membranes and eliminate the pressure potential component. Thus from the result of the Slatyer equation of Ψ for fresh (Ψ_π + Ψ_p) and thaw (Ψ_π) values it is possible to calculate Ψ_p. Any plant tissue Ψ can be measured by this method regardless its size so psychrometry has been used as a very versatile technique.

Pressure probe is a method focused on cell level water potential. In this method a capillary probe is inserted inside the cell. The probe is filled with silicone oil and connected to a pressure sensor. Ψ_p makes oil go backwards and it is registered by the sensor. Pressure probe is considered an accurate method and it is possible to use it on individual cells. Another advantage of the probe is that Ψ_p can be combined with the Ψ_π from the extracted cell content.

Nevertheless, the most extensively used method for plant Ψ measure is the pressure chamber initially described by Scholander et al. (1965). As above described methods (psychrometry and cell probe) are adaptable to any plant organ, pressure chamber is specifically developed for leaves. Leaves are introduced in a sealed chamber while the petiole protrudes the lid. Pressure inside the chamber is increased gradually due to compressed nitrogen gas. As the pressure reaches the value to restore the xylem water column (balance pressure) the water flows outside the leaf. This pressure value is the matrix apoplastic potential and assuming that the resistance to flux between symplast and apoplast is low and apoplastic water osmotic pressure close to zero it is considered similar to water potential (Ψ).

This technique is relatively fast and the equipment is suitable to use in field conditions. Moreover the pressure chamber can be also used to measure the pressure-volume parameters. Applying pressure values above the balance point the volume of xylem sap can be applied to determine the pressure-volume curve (Nilsen and Orcutt 1996). From the analysis of the curve osmotic potential and leaf tissue elasticity can be calculated.

4.4 Measuring Considerations of the Water Status Parameters

Water status measurement in plant leaves is normally used to establish the experimental conditions, to impose a water stress treatment of a defined severity, and then to relate the plant tolerance mechanisms to the water stress quantification. It is essential to select the best parameter to accomplish with the objectives of the research. For this purpose it is crucial to choose not only the right parameter but also to measure it when it can reflect the process we want to study (i.e. adaptive plant responses to stress or the movement of water in the plant).

Furthermore, studies focused on water relations, have to take into account the wide range of time scales in which plant processes altering water status may occur. There is a diurnal dynamic due to the loss of water by transpiration in exchange of CO_2. This loss of water may be also altered in a shorter scale of minutes due to environmental changes (i.e. light or temperature). In the same way there are long-term changes that potentially may alter plant tolerance responses to drought. Short-term responses comprise stomatal response, gene expression, leaf movement, wilting and osmotic adjustment whereas long-term strategies include canopy development, leaf senescence and root system modification (Passioura 1996). As longer time is available to develop long-term strategies to face drought, plants present acclimation to water deficit. In this sense, volumetric parameters, as RWC, have been reported to show differences due to drought treatment particularly when the combination between time scale and stress harshness did not allow plants to acclimate. However as time scale increases, as typically occurs in sustained water limitation studies, plants can adjust their leaf osmotic potential, enhance root to shoot ratio and limit leaf area development enable plants to maintain RWC at steady value (Erice et al. 2012).

Thermodynamic measurements, as Ψ, are also modified by the time scale. Leaf Ψ has been describe as extremely sensitive to environmental conditions and drought treatment differences are often smaller than its short-term fluctuations (Jones et al. 1983). When measured pre-dawn the plant Ψ is considered equilibrated with the soil potential so the daily short-term dynamic may be smoothed as it may be used as a tentative index of soil water availability. Nevertheless, after dawn the plant opens stomata to start photosynthesize and transpire. This water movement requires the decrease of leaf Ψ below the soil water potential. At midday hours the leaf Ψ reaches the minimum value as the physiological drought experienced by the plant tissue and its measurement is more related to water transport across the soil-plant-atmosphere system (Jones 2007) than to the actual hydration of the leaf tissue.

4.5 High-Throughput Methods to Assess Leaf Water Status

In this chapter classic methodology to measure leaf water status has been discussed. The presented techniques have been proven to be accurate but they are also laborious. The RWC requires collecting leaves and weighting them to obtain fresh, turgid and dry mass whereas water potential involves the use of the pressure chamber and to slowly reach the potential of the tissue. These methods could be also considered time-consuming for efficient screening of large population of plants for breeding purposes (Winterhalter et al. 2011). When measuring extensive number of plants the need to develop new rapid techniques is becoming a problem for the scientists in order to cope with their objectives. Indeed, it has been pointed that high-throughput physiology and phenotyping have become a new bottleneck in plant biology and crop breeding (Furbank 2009). Moreover, recording of real-time leaf water content is an important feature to implement precision irrigation, which is the optimal water management both spatially and temporally.

In order to assess leaf water status by high-throughput methods different approaches are being developed. In this line two main different strategies are being developed: one is based on electrical properties of the leaves and the other focuses on leaf optical reflectance.

Leaf capacitance and impedance were used by Mizukami et al. (2006) to estimate the moisture content of tea leaves. The accuracy of the method was influenced by the electrical condition of the cell wall so leaves needed to be separated by maturity. Regression equations obtained resulted on high correlation coefficients with satisfactory levels of standard error. More recently Zheng et al. (2015) developed a four-electrode method to evaluate leaf water content by the voltage drop between electrodes. Leaf electrical properties were shown to be correlated better to RWC rather than WC. Even if this method is revealed as effective to evaluate the water stress interestingly the correlation coefficient had a downward trend as the growth period advanced.

Other methods involve the leaf reflectance properties to estimate plant water status rapidly using model building and spectral indices or ratios between reflectance values at specific wavelengths. Multispectral imagery was used to determine water status by water potential by Kriston-Vizi et al. (2008). These authors found that Ψ is correlated with red or even green leaf reflectance. Using a partial least square regression (PLSR) method Gillon et al. (2004) developed a model based on near infra-red (NIR) spectra from 400 to 2500 nm. They concluded that it was possible to estimate the initial WC of the fresh leaves from its spectral characteristics when dried. They also presented evidences that some biochemical properties of the leaves may be associated to leaf water status causing changes in spectral reflectance even in dried samples. Zhang et al. (2012) explored the same wavelength spectrum reflectance to determine the water content in the leaf. This study featured that PLSR combined with spectrum preprocessing methods could construct accurate model and present a satisfactory prediction precision.

Despite of the efforts to establish and normalize the spectral reflectance to estimate leaf water status it remains unclear how to compare spectral characteristics across plant developmental stages. As Elsayed et al. (2011) pointed maybe changes in Ψ are difficult to detect spectrometrically because of autocorrelation with leaf WC. Thus, these authors still highlight the leaf reflectance as an accurate method for screening large numbers of plants even if a reduced calibration dataset from pressure chamber is needed.

5 Stomatal Aperture and Transpiration

5.1 *Measurements of Transpiration Using Porometers or Infra-Red Gas Analyzers*

Transpiration is the process of water evaporation from aerial parts of the plant. It is dynamically regulated by stomata, pores in the epidermis of leaves formed by one pair of guard cells, which control the gas exchange. The stomatal aperture can be measured indirectly as stomatal conductance to water vapour using a porometer or an infra-red gas analyzer (IRGA). Most of these are open systems with a sample cuvette for the leaf to be measured through which a constant air stream flows. The system determines the water vapour pressure in the air fluxes going in and out of the sample cuvette. Their difference allows the calculation of the transpiration rate, from which stomatal conductance is estimated. Both are expressed as a water flow rate in mol H_2O m^{-2} s^{-1}.

Several factors need to be taken in account when estimating stomatal conductance. Relative humidity of the air is the driving force for transpiration, and this is greatly influenced by the temperature. That is because temperature influences the water holding capacity of the air. For example, warmer air can hold more water, so and increase in air temperature decreases its relative humidity, making it drier and increasing the evaporation rate from the leaf surface. Generally, the temperature on the incoming air flow is fixed constant for more stability of the system. Light is another important factor, since it triggers stomata opening to allow the entry of CO_2 to the leaf and make it available for photosynthesis. Stomata are most sensitive to blue light, the predominant at sunrise. In the dark, stomata are closed in most plants. Hence, light intensity and quality before and after the measurement are very important and may affect the results greatly.

On the other hand, the wind has a strong effect on transpiration. The boundary layer around leaves is a layer of still air that offers resistance to water evaporation and the wind helps removing water molecules away from the surface of the leaves (Long and Hällgren 1993). This boundary layer can be modified by plants as a long term response to the water availability and relative humidity in the air, by altering a variety of structural features such as trichomes or leaf size. The boundary layer resistance is important when estimating stomatal conductance, since it limits transpiration, causing underestimation of stomatal conductance. To prevent this, the boundary layer needs to be broken using an air mixer inside the sample cuvette. Moreover, the boundary layer resistance, as a parameter in the equation of stomatal conductance needs to be corrected according to the distribution of stomata in adaxial and abaxial sides of the leaves. This stomatal ratio depends upon the plant species but not on growth conditions (Redmann 1985). For an extensive review on sources of error in the determination of stomatal conductance, please see (McDermitt 1990). An adequate calibration of the system before starting the measurements is essential. Firstly, it is very important to achieve an accurate zeroing by using dry air.

Secondly, the water vapour span must be checked, and corrected if necessary. For that purpose, a dew point generator should be used to safely provide a constant stream of water saturated air into the system. Before addressing the calibration, it is important to make sure there are no leaks in the system.

Stomatal aperture and closure are controlled by many intrinsic and extrinsic signals (Assmann 1993), and the study of these processes can be tackled by the determination of stomatal conductance under constant conditions of temperature, light, relative humidity and CO_2 concentration, which can be equal or different to the growth conditions. Also the response of stomata to these parameters can be evaluated by performing response curves. When carrying out CO_2 response curves, the rest of variables should be held constant, and the light intensity should be high enough not to limit photosynthesis. On the other hand, the response to light intensity and quality can be evaluated by light response curves, for which is important to ensure a steady temperature throughout the measurement. It can be advisable to hold the leaf temperature constant.

5.2 *Thermal Imaging*

Although it is feasible to measure stomatal conductance by means of leaf porometers or IRGA, these determinations present some disadvantages. First, leaf gasometry involves contact with leaves, which often interferes with their function. Furthermore, during the monitoring of point measurements on randomly selected leaf areas, the information about gradients over the whole leaf surface is missed unless one makes time-consuming, repeated point measurements. In contrast, the thermal imaging systems, originated from the use of thermography point sensors, which have been applied in the field, offer rapid analysis and high spatial resolution from the leaf to the remote sensing scale (Chaerle and Van der Straeten 2001; Jones et al. 2002; Jones 2004a; Costa et al. 2013; Fiorani and Schurr 2013).

It is well known from the 1960s (Fuchs and Tanner 1966) that leaf temperature correlates with transpiration and stomatal conductance. Plants are cooled by transpiration and when the stomata are closed due to different environmental conditions, the decreased transpiration induces a temperature increase.

Thermal cameras measure emitted radiation in the thermal range (3–14 μm) and allows imaging the differences in surface temperature of plant leaves and canopies. These sensors convert patterns of radiation to visual pseudocolour images representing temperature levels, where each image pixel contains the temperature value. With proper calibration, thermal maps can be converted directly into maps of stomatal conductance (Jones 2004a). Although thermal imaging does not directly measure stomatal conductance, in any given environment stomatal variation is the dominant cause of changes in canopy temperature (Jones 1999).

The energy balance of a plant, and thus its temperature, is dependent on environmental factors such as light intensity, temperature, relative humidity, wind speed, etc. From this reason, imaging sensor calibration and atmospheric correction are usually required. Since the closure of stomata also affects photosynthesis, imaging of chlorophyll fluorescence parameters could be a complementary technique to thermography.

Thermography can visualize either dynamic or heterogeneous patterns of stomatal closure called patchiness (Jones 2004a). It has also been used to screen Arabidopsis mutants affected in stomatal regulation (Chaerle and Van der Straeten 2001; Merlot et al. 2002; de Marcos et al. 2015). Early uses of thermal imaging are reviewed in Jones et al. (2009) and some studies demonstrated stomatal closure as a result of exposure to pollutants such as NO_2, SO_2, and O_3. In other studies it has been shown that nitrogen fertilization also can affect the canopy temperature of paddy rice (Wakiyama 2016). Leaf or canopy temperature as an indicator of water stress and the estimation of stomatal conductance from infrared thermography is widely used (Jones 1999, 2004a, b; Zhang and Kovacs 2012; Li et al. 2014). Indices of crop water stress were developed for irrigation scheduling purposes but also for estimation of evaporation rates from plant canopies. These indices are largely assumed to reflect changes in stomatal opening and evaporation rate as water becomes limiting (Jones et al. 2009; Costa et al. 2013).

The regulation of the stomatal closure plays a key role on plant defense to prevent microbial invasion (Sawinski et al. 2013). Very often pathogen infection triggers stomatal closure in plants, resulting on decrease of transpiration (Barón et al. 2016; Mahlein 2016). Some pathogens are able to prevent stomatal closure upon detection of pathogens or are even able to reopen them (Zeng et al. 2010). Thermal imaging has been used to monitor plant stress caused by viruses (Chaerle et al. 1999, 2006), bacteria (Pérez-Bueno et al. 2015, 2016) and fungi (Chaerle et al. 2004; Aldea et al. 2006; Granum et al. 2015). Furthermore, thermal imaging combined with chlorophyll fluorescence has allowed the analysis of spatial and temporal heterogeneity on leaf transpiration and to correlate it with photosynthetic activity (Barón et al. 2012, 2016). This combination of techniques can reveal presymptomatic responses to pathogens. Thermography also facilitates phenotyping for disease-resistant plant genotypes, – one of the main challenge in plant breeding (Walter et al. 2015).

In addition to laboratory and proximal level of thermal sensing, thermal imaging sensors have been integrated into phenotyping platforms under controlled-environment (http://www.lemnatec.com/). Applications of high throughput phenotyping in the search for drought tolerant varieties are reviewed in (Humplik et al. 2015); being connected the physiological responses to drought and high temperature stresses, similar approaches can be used to study the tolerance of plants to both stress factors. Mahlein (2016) have showed the applications of thermal imaging in precision agriculture and plant phenotyping for plant disease detection, screening of plant resistance and assessment of plant stress reactions. A number of field experiments not only of water stress have been carried out with thermal sensors using ground-base (http://www.lemnatec.com/) or aerial-based methods (Sobrino et al. 2009; Jones 2011; Liebisch et al. 2015).

However, thermal cameras as single sensors provide information on changes in stomatal opening. For stress diagnosis when stomata can be affected by many different primary stresses a multi-sensor approach is required, where thermal sensing is combined with reflectance, fluorescence and other sensing techniques (Fiorani and Schurr 2013).

5.3 Stomata Visualization

Changes in measured leaf temperature or gas exchange could be due to alterations in stomatal density or aperture (Smith 1941; Nobel 1991). When discussing plant hydraulics under stress conditions, it should be determine the number of stomata per leaf surface and the degree of ostiole aperture, in order to provide an explanation to the observed phenomena. If stomatal density is not significantly different between treatments, then the stress factor does not modify the normal development of stomata in the plant. Thus, observed changes should be attributed to stomatal regulation upon stress (Daszkowska-Golec and Szarejko 2013; McLachlan et al. 2014). On the other hand, if ostiole aperture does not change, results obtained by means of porometers or thermal cameras can be used to look for plants affected in stomatal development in plant phenotyping programs (Fiorani and Schurr 2013; Mahlein 2016).

When no knowledge is available about if studied plant is epistomatic, hypostomatic or amphistomatic, it is a good advice to examine both adaxial and abaxial sides of the leaves. Stomata can be observed directly under the microscope or indirectly by means of epidermal impressions. In the first case, the positive image can be achieved using different microscopic techniques. Sometimes it is possible grasping the epidermis with fine forceps, then slowly peeling the epidermis away from the leaf and mounts it in a microscope slide with a drop of water in order to observe it under light microscopy. To enhance the contrast in unstained, transparent samples such as peelings or whole leaves, it is possible to use differential interference contrast microscopy or Nomarski microscopy. Optical topometry by means of confocal microscopy is a powerful, non-destructive method for quantitative characterization of the plant epidermis requiring no sample preparation (Haus et al. 2015). Scanning electron microscopy is another method to visualize the plant epidermis, but it requires fixation, desiccation and metallization of the sample.

Indirect visualization of stomatal implies the realization of epidermal impressions using some clear-to-light material, such us nail polish, liquid plaster or water-soluble glue. The organic solvents in the material used for impressions may harm the leaves; thus, it is advisable to try different materials to find the best one, moreover if the leaf will continue attached to the plant. With the help of a brush or spatula, the substance should be carefully extended in order not to capture air bubbles. The film has to be as thin as possible and cover at least 1 cm. Once completely dry, the negative impression should be gently removed with the help of a forceps or clear cellophane tape, and mounted in a microscope slide for its visualization under light microscope in the form of a dry mount. These leaves impressions require closing the condenser aperture diaphragm due to the low contrast of the samples, or alternatively, using oblique illumination, resulting in a characteristic 3-dimensional appearance. Alternatively, epidermal negative imprints could be done using dental resin. This negative impression can be archived and used to make a positive cast (with clear nail polish for example) whenever desired. Despite the technique used to visualize stomata, the use of grid reticules is a good option to establish the stomatal density. Several imaging handling software (such as the free available ImageJ or CellProfiler) are useful to determine the ostiole length.

References

Ache P, Bauer H, Kollist H, Al-Rasheid KAS, Lautner S, Hartung W, Hedrich R (2010) Stomatal action directly feeds back on leaf turgor: new insights into the regulation of the plant water status from non-invasive pressure probe measurements. Plant J 62:1072–1082

Aldea M, Frank TD, DeLucia EH (2006) A method for quantitative analysis of spatially variable physiological processes across leaf surfaces. Photosynth Res 90(2):161–172

Alder NN, Pockman WT, Sperry JS, Nuismer S (1997) Use of centrifugal force in the study of xylem cavitation. J Exp Bot 48:665–674

Aoyama Y, Suzuki K, Tabe Y, Chikahisa T (2014) Observation of water transport in the microporous of a PEFC with freezing method and cryo-SEM. Electrochem Commun 41:72–75

Araus J, Slafer G, Reynolds M, Royo C (2002) Plant breeding and drought in C-3 cereals: what should we breed for? Ann Bot 89:925–940

Aroca R, Tognoni F, Irigoyen JJ, Sánchez-Díaz M, Pardossi A (2001) Different root low temperature response of two maize genotypes differing in chilling sensitivity. Plant Physiol Biochem 39:1067–1073

Aroca R, Porcel R, Ruiz-Lozano JM (2012) Regulation of root water uptake under abiotic stress conditions. J Exp Bot 63:43–57

Assmann SM (1993) Signal transduction in guard cells. Annu Rev Cell Biol 9:345–375

Barón M, Flexas J, Delucia EH (2012) Photosynthetic responses to biotic stress. In: Flexas J, Loreto F, Medrano H (eds) Terrestrial photosynthesis in a changing environment: a molecular, physiological, and ecological approach, vol 1. Cambridge University Press, Cambridge, pp 331–350

Barón M, Pineda M, Pérez-Bueno ML (2016) Picturing pathogen infection in plants. Z Naturforsch C Bio Sci 71:355–368. https://doi.org/10.1515/znc-2016-0134

Bárzana G, Aroca R, Paz JA, Chaumont F, Martínez-Ballesta MC, Carvajal M, Ruiz-Lozano JM (2012) Arbuscular mycorrhizal symbiosis increases relative apoplastic water flow in roots of the host plant under both well-watered and drought stress conditions. Ann Bot 109:1009–1017

Bigot J, Boucaud J (2000) Effects of Ca-signalling inhibitors on short-term cold-acclimation of hydraulic conductivity in roots of *Brassica rapa* plants. J Plant Physiol 157:7–12

Borchert R (1994) Electric resistance as a measure of tree water status during seasonal drought in a tropical dry forest in Costa Rica. Tree Physiol 14:299–312

Brodersen CR, McElrone AJ, Choat B, Lee EF, Shackel KA, Matthews MA (2013) In vivo visualizations of drought-induced embolism spread in *Vitis vinifera*. Plant Physiol 161:1820–1829

Brodribb TJ, Bowman DMJS, Nichols S, Delzon S, Burlett R (2010) Xylem function and growth rate interact to determine recovery rates after exposure to extreme water deficit. New Phytol 188:533–542

Cai J, Tyree MT (2010) The impact of vessel size on vulnerability curves: data and models for within-species variability in saplings of aspen, *Populus tremuloides* Michx. Plant Cell Environ 33:1059–1069

Calvo-Polanco M, Molina S, Zamarreño AM, García-Mina JM, Aroca R (2014) The symbiosis with the arbuscular mycorrhizal fungus *Rhizophagus irregularis* drives root water transport in flooded tomato plants. Plant Cell Physiol 55:1017–1029

Canny MJ (1997) Vessel contents during transpiration: embolisms and refilling. Am J Bot 84:1223–1230

Chaerle L, Van der Straeten D (2001) Seeing is believing: imaging techniques to monitor plant health. Biochim Biophys Acta 1519(3):153–166

Chaerle L, Van Caeneghem W, Messens E, Lambers H, Van Montagu M, Van der Straeten D (1999) Presymptomatic visualization of plant-virus interactions by thermography. Nat Biotechnol 17(8):813–816

Chaerle L, Hagenbeek D, De Bruyne E, Valcke R, Van der Straeten D (2004) Thermal and chlorophyll-fluorescence imaging distinguish plant-pathogen interactions at an early stage. Plant Cell Physiol 45(7):887–896

Chaerle L, Pineda M, Romero-Aranda R, Van der Straeten D, Barón M (2006) Robotized thermal and chlorophyll fluorescence imaging of pepper mild mottle virus infection in *Nicotiana benthamiana*. Plant Cell Physiol 47(9):1323–1336

Charra-Vaskou K, Badel E, Burlett R, Cochard H, Delzon S, Mayr S (2012) Hydraulic efficiency and safety of vascular and non-vascular components in *Pinus pinaster* leaves. Tree Physiol 32:1161–1170

Charrier G, Torres-Ruiz JM, Badel E, Burlett R, Choat B, Cochard H, Delmas CE, Domec JC, Jansen S, King A, Lenoir N, Martin-StPaul N, Gambetta GA, Delzon S (2016) Evidence for hydraulic vulnerability segmentation and lack of xylem refilling under tension. Plant Physiol 172(3):1657–1668

Chaves M, Maroco J, Pereira J (2003) Understanding plant responses to drought – from genes to the whole plant. Funct Plant Biol 30:239–264

Choat B, Badel E, Burlett R, Delzon S, Cochard H, Jansen S (2016) Non-invasive measurement of vulnerability to drought induced embolism by X-ray microtomography. Plant Physiol 170:273–282

Cochard H (1992) Vulnerability of several conifers to air embolism. Tree Physiol 11:73–83

Cochard H, Bodet C, Ameglio T, Cruiziat P (2000) Cryoscanning electron microscopy observations of vessel content during transpiration in walnut petioles. Facts or artifacts? Plant Physiol 124:1191–1202

Cochard H, Froux F, Mayr S, Coutand C (2004) Xylem wall collapse in water-stressed pine needles. Plant Physiol 134:401–408

Cochard H, Damour G, Bodet C, Tharwat I, Poirier M, Améglio T (2005) Evaluation of a new centrifuge technique for rapid generation of xylem vulnerability curves. Physiol Plant 124:410–418

Cochard H, Herbette S, Barigah T, Badel E, Ennajeh M, Vilagrosa A (2010) Does sample length influence the shape of xylem embolism vulnerability curves? a test with the Cavitron spinning technique. Plant Cell Environ 33:1543–1552

Cochard H, Badel E, Herbette S, Delzon S, Choat B, Jansen S (2013) Methods for measuring plant vulnerability to cavitation: a critical review. J Exp Bot 64:4779–4791

Costa JM, Grant OM, Chaves MM (2013) Thermography to explore plant-environment interactions. J Exp Bot 64(13):3937–3949

Curtis OF (1936) Transpiration and the cooling of leaves. Am J Bot 23:7–10

Dalla-Salda G, Fernández ME, Sergent AS, Rozenberg P, Badel E, Martinez-Meier A (2014) Dynamics of cavitation in a Douglas-fir tree-ring: transition-wood, the lord of the ring? J Plant Hydraul 1:e0005

Daniel I, Oger P, Winter R (2006) Origins of life and biochemistry under high-pressure conditions. Chem Soc Rev 35:858–875

Daszkowska-Golec A, Szarejko I (2013) Open or close the gate – stomata action under the control of phytohormones in drought stress conditions. Front Plant Sci 4:138. https://doi.org/10.3389/fpls.2013.00138

de Marcos A, Triviño M, Pérez-Bueno ML, Ballesteros I, Barón M, Mena M, Fenoll C (2015) Transcriptional profiles of Arabidopsis stomataless mutants reveal developmental and physiological features of life in the absence of stomata. Front Plant Sci 6:456. https://doi.org/10.3389/fpls.2015.00456

Dixon HH (1914) Transpiration and the ascent of sap in plants. MacMillan, London

Elsayed S, Mistele B, Schmidhalter U (2011) Can changes in leaf water potential be assessed spectrally? Funct Plant Biol 38:523–533

Erice G, Irigoyen J, Perez P, Martinez-Carrasco R, Sanchez-Diaz M (2006) Effect of elevated CO_2, temperature and drought on dry matter partitioning and photosynthesis before and after cutting of nodulated alfalfa. Plant Sci 170:1059–1067

Erice G, Sanz-Sáez A, Aranjuelo I, Irigoyen JJ, Sánchez-Díaz M (2012) Future environmental conditions will limit yield in N_2 fixing alfalfa. In: Aroca R (ed) Plant responses to drought stress: from morphological to molecular features. Springer, Berlin/Heidelberg, pp 363–382

Fiorani F, Schurr U (2013) Future scenarios for plant phenotyping. Annu Rev Plant Biol 64:267–291

Fiscus EL (1977) Determination of hydraulic and osmotic properties of soybean root systems. Plant Physiol 59:1013–1020

Fiscus EL (1986) Diurnal changes in volume and solute transport coefficients of *Phaseolus* roots. Plant Physiol 80:752–759

Fuchs M, Tanner CB (1966) Infrared thermometry of vegetation. Agron J 58:597–601

Furbank R (2009) Plant phenomics: from gene to form and function. Funct Plant Biol 36:V–VI

Gillon D, Dauriac F, Deshayes M, Valette JC, Moro C (2004) Estimation of foliage moisture content using near infrared reflectance spectroscopy. Agric Forest Meteorol 124:51–62

Granum E, Pérez-Bueno ML, Calderón CE, Ramos C, de Vicente A, Cazorla FM, Barón M (2015) Metabolic responses of avocado plants to stress induced by *Rosellinia necatrix* analysed by fluorescence and thermal imaging. Eur J Plant Pathol 142(3):625–632

Hacke UG, Venturas MD, MacKinnon ED, Jacobsen AL, Sperry JS, Pratt RB (2015) The standard centrifuge method accurately measures vulnerability curves of long-vesselled olive stems. New Phytol 205:116–127

Haus MJ, Kelsch RD, Jacobs TW (2015) Application of optical topometry to analysis of the plant epidermis. Plant Physiol 169(2):946–959

Hirasawa T, Tsuchida M, Ishihara K (1992) Relationship between resistance to water transport and exudation rate and the effect of the resistance on the midday depression of stomatal aperture in rica plants. Jpn J Crop Sci 61:145–152

Holbrook NM, Zwieniecki MA (2005) Vascular transport in plants. Elsevier/Academic, Oxford

Homan N, Windt CW, Vergeldt FJ, Gerkema E, Van As H (2007) 0.7 and 3 T MRI and sap flow in intact trees: xylem and phloem in action. Appl Magn Reson 32:157–170

Humplik JF, Lazar D, Husickova A, Spichal L (2015) Automated phenotyping of plant shoots using imaging methods for analysis of plant stress responses – a review. Plant Methods 11:29. https://doi.org/10.1186/s13007-015-0072-8

IPCC (2013) Climate Change 2013: The physical science basis. Cambridge University Press, New York

Javot H, Maurel C (2002) The role of aquaporins in root water uptake. Ann Bot 90:301–313

Jones HG (1999) Use of thermography for quantitative studies of spatial and temporal variation of stomatal conductance over leaf surfaces. Plant Cell Environ 22(9):1043–1055

Jones HG (2004a) Application of thermal imaging and infrared sensing in plant physiology and ecophysiology. Adv Bot Res 41:107–163

Jones HG (2004b) Irrigation scheduling: advantages and pitfalls of plant-based methods. J Exp Bot 55(407):2427–2436

Jones H (2007) Monitoring plant and soil water status: established and novel methods revisited and their relevance to studies of drought tolerance. J Exp Bot 58:119–130

Jones HG (2011) Remote detection of crop water "stress" and distinguishing it from other stresses. In: Fernandez JE, Ferreira MI (eds) Xxviii international horticultural congress on science and horticulture for people, Acta Horticulturae, vol 922. International Society for Horticultural Science, Leuven, pp 23–34

Jones H, Luton M, Higgs K, Hamer P (1983) Experimental control of water status in an apple orchard. J Hortic Sci 58:301–316

Jones HG, Stoll M, Santos T, de Sousa C, Chaves MM, Grant OM (2002) Use of infrared thermography for monitoring stomatal closure in the field: application to grapevine. J Exp Bot 53(378):2249–2260

Jones HG, Serraj R, Loveys BR, Xiong L, Wheaton A, Price AH (2009) Thermal infrared imaging of crop canopies for the remote diagnosis and quantification of plant responses to water stress in the field. Funct Plant Biol 36:978–989

Kang S, Shan L, Davies B, Cai H (2003) International Conference on Water-Saving Agriculture and Sustainable Use of Water and Land Resources (ICWSAWLR) – organized by Northwest Sci-Tech University of Agriculture and Forestry (NWSUAF), China and the Lancaster Environment Centre (LEC), Lancaster University, UK – Yangling, Shaanxi, PR China – 26–29 October 2003 – Preface. J Exp Bot 54:2–2

Kikuta SB (2003) Ultrasound acoustic emissions from bark samples differing in anatomical characteristics. Phyton 43:161–178

Knipfer T, Cuneo IF, Brodersen CR, McElrone AJ (2016) In situ visualization of the dynamics in xylem embolism formation and removal in the absence of root pressure: a study on excised grapevine stems. Plant Physiol 171:1024–1036

Köckenberger W, Pope JM, Xia Y, Komor E, Jeffrey KR, Callaghan PT (1997) A non-invasive measurement of phloem and xylem water flow in castor bean seedlings by nuclear magnetic resonance microimaging. Planta 201:53–63

Köckenberger W, De Panfilis C, Santoro D, Dahiya P, Rawsthorne S (2004) High resolution NMR microscopy of plants and fungi. J Microsc 214:182–189

Kramer PJ, Boyer JS (1995) Water relations of plants and soils. Academic, New York

Kriston-Vizi J, Umeda M, Miyamoto K (2008) Assessment of the water status of mandarin and peach canopies using visible multispectral imagery. Biosyst Eng 100:338–345

Kudoyarova GR, Kholodova VP, Veselov DS (2013) Current state of the problem of water relations in plants under water deficit. Russ J Plant Physiol 60:165–175

Lewis M, Harnden VD, Tyree MT (1994) Collapse of water-stress emboli in the tracheids of *Thuja occidentalis* L. Plant Physiol 106:1639–1646

Li Y, Sperry JS, Taneda H, Bush SE, Hacke UG (2008) Evaluation of centrifugal methods for measuring xylem cavitation in conifers, diffuse- and ring-porous angiosperms. New Phytol 177:558–568

Li L, Zhang Q, Huang D (2014) A review of imaging techniques for plant phenotyping. Sensors 14(11):20078–20111. https://doi.org/10.3390/s141120078

Liebisch F, Kirchgessner N, Schneider D, Walter A, Hund A (2015) Remote, aerial phenotyping of maize traits with a mobile multi-sensor approach. Plant Methods 11:9. https://doi.org/10.1186/s13007-015-0048-8

Long SP, Hällgren J-E (1993) Measurement of CO2 assimilation by plants in the field and the laboratory. In: Hall DO, JMO S, Bolhàr-Nordenkampf HR, Leegood RC, Long SP (eds) Photosynthesis and production in a changing environment: a field and laboratory manual. Chapman and Hall, London, pp 129–219

Lopez R, Cano FJ, Choat B, Cochard H, Gil L (2016) Plasticity in vulnerability to cavitation of *Pinus canariensis* occurs only at the driest end of an aridity gradient. Front Plant Sci 7:769

Mahdieh M, Mostajeran A, Katsuhara M (2016) Phosphorous deprivation effects on water relations of *Nicotiana tabacum* plant via reducing plasma membrane permeability. Russ J Plant Physiol 63:54–61

Mahlein A-K (2016) Plant disease detection by imaging sensors – parallels and specific demands for precision agriculture and plant phenotyping. Plant Dis 100(2):241–251. https://doi.org/10.1094/pdis-03-15-0340-fe

Markhart AH, Fiscus EL, Naylor AW, Kramer PJ (1979) Effect of abscisic acid on root hydraulic conductivity. Plant Physiol 64:611–614

McDermitt DK (1990) Sources of error in the estimation of stomatal conductance and transpiration from porometer data. Hortscience 25:1538–1548

McLachlan DH, Kopischke M, Robatzek S (2014) Gate control: guard cell regulation by microbial stress. New Phytol 203(4):1049–1063

Melcher PJ, Holbrook NM, Burns MJ, Zwieniecki MA, Cobb AR, Brodribb TJ, Sack L (2012) Measurements of stem xylem hydraulic conductivity in the laboratory and field. Meth Ecol Evol 3:685–694

Merlot S, Mustilli AC, Genty B, North H, Lefebvre V, Sotta B, Vavasseur A, Giraudat J (2002) Use of infrared thermal imaging to isolate Arabidopsis mutants defective in stomatal regulation. Plant J 30(5):601–609

Milburn JA, Johnson RPC (1966) The conduction of sap. II. Detection of vibrations produced by sap cavitation in *Ricinus* xylem. Planta 69:43–52

Miyamoto N, Steudle E, Hirasawa T, Laffite R (2001) Hydraulic conductivity of rice roots. J Exp Bot 52:1835–1846

Mizukami Y, Sawai Y, Yamaguchi Y (2006) Moisture content measurement of tea leaves by electrical impedance and capacitance. Biosyst Eng 93:293–299

Nardini A, Tyree MT (1999) Root and shoot hydraulic conductance of seven *Quercus* species. Ann For Sci 56:371–377

Nardini A, Tyree MT, Salleo S (2001) Xylem cavitation in the leaf of *Prunus laurocerasus* and its impact on leaf hydraulics. Plant Physiol 125:1700–1709

Nardini A, Savi T, Losso A, Petit G, Pacilè S, Tromba G, Mayr S, Trifilò P, Lo Gullo MA, Salleo S (2017) X-ray microtomography observations of xylem embolism in stems of *Laurus nobilis* are consistent with hydraulic measurements of percentage loss of conductance. New Phytol 213(3):1068–1075

Nilsen ET, Orcutt DM (1996) The physiology of plants under stress. Wiley, New York

Nobel PS (1991) Leaves and fluxes. In: Nobel PS (ed) Physicochemical and environmental plant physiology. Academic, San Diego, pp 393–472

Ogasa MY, Utsumi Y, Miki NH, Yazaki K, Fukuda K (2016) Cutting stems before relaxing xylem tension induces artefacts in *Vitis coignetiae,* as evidenced by magnetic resonance imaging. Plant Cell Environ 39:329–337

Pallas JE, Michel BE, Harris DG (1967) Photosynthesis, transpiration, leaf temperature and stomatal activity of cotton plants under varying water potentials. Plant Physiol 42:76–88

Passioura J (1996) Drought and drought tolerance. Plant Growth Regul 20:79–83

Pérez-Bueno ML, Pineda M, Díaz-Casado ME, Barón M (2015) Spatial and temporal dynamics of primary and secondary metabolism in *Phaseolus vulgaris* challenged by *Pseudomonas syringae*. Physiol Plant 153(1):161–174

Pérez-Bueno ML, Granum E, Pineda M, Flors V, Rodríguez-Palenzuela P, López-Solanilla E, Barón M (2016) Temporal and spatial resolution of activated plant defense responses in leaves of *Nicotiana benthamiana* infected with *Dickeya dadantii*. Front Plant Sci 6:1209. https://doi.org/10.3389/fpls.2015.01209

Pockman WT, Sperry JS, O'Leary JW (1995) Sustained significant negative water pressure in xylem. Nature 378:715–716

Ranathunge K, Steudle E, Latiffe R (2003) Control of water uptake by rice (*Oryza sativa* L.): role of the outer part of the root. Planta 217:193–205

Redmann RE (1985) Adaptation of grasses to water stress-leaf rolling and stomata distribution. Ann Missouri Bot Gard 72(4):833–842

Richard A (1838) Nouveaux Élémens de Botanique et de Physiologie Végétale, 6th edn. Béchet Jeune, Paris

Sack L, Bartlett M, Creese C, Guyot G, Scoffoni C (2011) Constructing and operating a hydraulics flow meter (Prometheus Wiki, 2011). http://prometheuswiki.publish.csiro.au/tiki-index.php?page=Constructing+and+operatinga+hydraulics+flow+meter

Sánchez-Romera B, Ruiz-Lozano JM, Li G, Luu DT, Martínez-Ballesta MC, Carvajal M, Zamarreño AM, García-Mina JM, Maurel C, Aroca R (2014) Enhancement of root hydraulic conductivity by methyl jasmonate and the role of calcium and abscisic acid in this process. Plant Cell Environ 37:995–1008

Sano Y, Morris H, Shimada H, Ronse De Craene LP, Jansen S (2011) Anatomical features associated with water transport in imperforate tracheary elements of vessel-bearing angiosperms. Ann Bot 107:953–967

Saranga Y, Flash I, Paterson A, Yakir D (1999) Carbon isotope ratio in cotton varies with growth stage and plant organ. Plant Sci 142:47–56

Sawinski K, Mersmann S, Robatzek S, Bohmer M (2013) Guarding the green: pathways to stomatal immunity. Mol Plant-Microbe Interact 26(6):626–632. https://doi.org/10.1094/MPMI-12-12-0288-CR

Scholander P, Hammel H, Bradstreet E, Hemmingsen E (1965) Sap pressure in vascular plants – negative hydrostatic pressure can be measured in plants. Science 148:339

Slatyer RO (1967) Plant-water relationships. Academic, London

Smith HB (1941) Variation and correlation of stomatal frequency and transpiration rate in *Phaseolus vulgaris*. Am J Bot 28:722–725

Sobrino JA, Jimenez-Munoz JC, Zarco-Tejada PJ, Sepulcre-Canto G, de Miguel E, Soria G, Romaguera M, Julien Y, Cuenca J, Hidalgo V, Franch B, Mattar C, Morales L, Gillespie A, Sabol D, Balick L, Su Z, Jia L, Gieske A, Timmermans W, Olioso A, Nerry F, Guanter L, Moreno J, Shen Q (2009) Thermal remote sensing from airborne hyperspectral scanner data in the framework of the SPARC and SEN2FLEX projects: an overview. Hydrol Earth Syst Sci 13(11):2031–2037

Sperry JS (1986) Relationship of xylem embolism to xylem pressure potential, stomatal closure, and shoot morphology in the palm *Rhapis excelsa*. Plant Physiol 80:110–116

Sperry JS, Saliendra NZ (1994) Intra- and inter-plant variation in xylem cavitation in *Betula occidentalis*. Plant Cell Environ 17:1233–1241

Sperry JS, Donnelly JR, Tyree MT (1988) A method for measuring hydraulic conductivity and embolism in xylem. Plant Cell Environ 11:35–40

Steudle E (1997) Water transport across plant tissue: role of water channels. Biol Cell 89:259–273

Steudle E (2000) Water uptake by plant roots: an integration of views. Plant Soil 226:45–56

Steudle E, Jeschke WD (1983) Water transport in barley roots. Planta 158:237–248

Steudle E, Peterson CA (1998) How does water get through roots? J Exp Bot 49:775–788

Steudle E, Oren R, Schulze ED (1987) Water transport in maize roots. Measurement of hydraulic conductivity, solute permeability, and of reflection coefficients of excised roots using the root pressure probe. Plant Physiol 84:1220–1232

Suuronen JP, Peura M, Fagerstedt K, Serimaa R (2013) Visualizing water-filled versus embolized status of xylem conduits by desktop x-ray microtomography. Plant Methods 9:11

Tomos AD, Leigh RA (1999) The pressure probe: a versatile tool in plant cell physiology. Annu Rev Plant Physiol Plant Mol Biol 50:447–472

Torres-Ruiz JM, Cochard H, Mayr S, Beikircher B, Diaz-Espejo A, Rodriguez-Dominguez CM, Badel E, Fernandez JE (2014) Vulnerability to cavitation in *Olea europaea* current-year shoots: further evidence of an open-vessel artifact associated with centrifuge and air-injection techniques. Physiol Plant 153:465–474

Torres-Ruiz JM, Jansen S, Choat B, McElrone A, Cochard H, Brodribb TJ, Badel E, Burlett R, Bouche PS, Brodersen C, Li S, Morris H, Delzon S (2015) Direct X-ray microtomography observation confirms the induction of embolism upon xylem cutting under tension. Plant Physiol 167:40–43

Torres-Ruiz JM, Cochard H, Mencuccini M, Delzon S, Badel E (2016) Direct observation and modelling of embolism spread between xylem conduits: a case study in Scots pine. Plant Cell Environ 39(12):2774–2785

Tsuda M, Tyree MT (2000) Plant hydraulic conductance measured by high pressure flow meter in crop plants. J Exp Bot 51:823–828

Tyree MT, Sperry JS (1989a) Vulnerability of xylem to cavitation and embolism. Ann Rev Plant Physiol Molec Biol 40:19–38

Tyree MT, Sperry JS (1989b) Characterization and propagation of acoustic emission signals in woody plants: towards an improved acoustic emission counter. Plant Cell Environ 12:371–382

Tyree MT, Zimmermann MH (2002) Xylem structure and the ascent of sap, 2nd edn. Springer, Berlin, 283 p

Tyree MT, Graham MED, Cooper KE, Bazos LJ (1983) The hydraulic architecture of *Thuja occidentalis*. Can J Bot 61:2105–2111

Tyree MT, Patiño S, Bennink J, Alexander J (1995) Dynamic measurements of root hydraulic conductance using a high pressure flow meter in the laboratory and field. J Exp Bot 46:83–94

Utsumi Y, Sano Y, Fujjikawa S, Funada S, Ohtani J (1998) Visualization of cavitated vessels in winter and refilled vessels in spring in diffuse-porous trees by cryo-scanning electron microscopy. Plant Physiol 117:1463–1471

Vaadia Y, Raney FC, Hagan RM (1961) Plant water deficits and physiological processes. Annu Rev Plant Physiol 12:265–292

Van As H, Windt CW (2008) Magnetic resonance imaging of plants: water balance and water transport in relation to photosynthetic activity. In: Aartsma TJ, Matysik J (eds) Biophysical techniques in photosynthesis II. Springer, Berlin, pp 55–75

Van As H, Scheenen T, Vergeldt FJ (2009) MRI of intact plants. Photosynth Res 102:213–222

Venturas MD, Mackinnon ED, Jacobsen AL, Pratt RB (2015) Excising stem samples underwater at native tension does not induce xylem cavitation. Plant Cell Environ 38:1060–1068

Vergeynst LL, Sause MGR, Steppe K (2016) Clustering reveals cavitation-related acoustic emission signals from dehydrating branches. Tree Physiol 36:786–796

Wakiyama Y (2016) The relationship between SPAD values and leaf blade chlorophyll content through the rice development cycle. Jarq – Jpn Agric Res Q 50:329–334

Walter A, Liebisch F, Hund A (2015) Plant phenotyping: from bean weighing to image analysis. Plant Methods 11(1):1–11. https://doi.org/10.1186/s13007-015-0056-8

Wang MT, Tyree MT, Wasylishen RE (2013) Magnetic resonance imaging of water ascent in embolized xylem vessels of grapevine stem segments. Can J Plant Sci 93:879–893

Weatherley P (1950) A convenient volumenometer for biological work. J Exp Bot 1:244–248

Wheeler JK, Huggett BA, Tofte AN, Rockwell FE, Holbrook NM (2013) Cutting xylem under tension or supersaturated with gas can generate PLC and the appearance of rapid recovery from embolism. Plant Cell Environ 36:1938–1949

Windt CW, Vergeldt FJ, de Jager PA, Van As H (2006) MRI of long distance water transport: a comparison of the phloem and xylem flow characteristics and dynamics in poplar, castor bean, tomato and tobacco. Plant Cell Environ 29:1715–1729

Winterhalter L, Mistele B, Jampatong S, Schmidhalter U (2011) High-throughput sensing of aerial biomass and above-ground nitrogen uptake in the vegetative stage of well-watered and drought stressed tropical maize hybrids. Crop Sci 51:479–489

Wolkerstorfer SV, Rosner S, Hietz P (2012) An improved method and data analysis for ultrasound acoustic emissions and xylem vulnerability in conifer wood. Physiol Plant 146:184–191

Zeng W, Melotto M, He SY (2010) Plant stomata: a checkpoint of host immunity and pathogen virulence. Curr Opin Biotechnol 21(5):599–603

Zhang C, Kovacs JM (2012) The application of small unmanned aerial systems for precision agriculture: a review. Precis Agric 13:693–712

Zhang Q, Li Q, Zhang G (2012) Rapid determination of leaf water content using VIS/NIR spectroscopy analysis with wavelength selection. Spectroscopy 27:13.7

Zheng L, Wang Z, Sun H, Zhang M, Li M (2015) Real-time evaluation of corn leaf water content based on the electrical property of leaf. Comput Electron Agric 112:102–109

Zimmermann MH (1983) Xylem structure and the ascent of sap. Springer, Berlin

Zimmermann HM, Steudle E (1998) Apoplastic transport across young maize roots: effects of the exodermis. Planta 206:7–19

Zwieniecki MA, Melcher PJ, Ahrens ET (2013) Analysis of spatial and temporal dynamics of xylem refilling in *Acer rubrum* L. using magnetic resonance imaging. Front Plant Sci 4:265

Chapter 8
Thermal Imaging and Infrared Sensing in Plant Ecophysiology

Hamlyn G. Jones

1 Introduction

1.1 Why Thermal Sensing?

Leaf temperature is important for plants both because of its effects on the rates of critical physiological processes and because of the damaging effects of temperature extremes. Notwithstanding these phenomena, remote estimation of leaf temperature also provides a powerful tool for the study of plant water relations and some other phenomena such as freezing. Most ecological and physiological applications of thermal remote sensing are based on the fact that the latent heat of vaporization required to evaporate from plant leaves is a major part of the leaf or canopy energy balance such that increasing evaporation or transpiration through the stomata leads to a cooling of the surface. The degree of cooling can therefore be used as an indicator of transpiration rate and stomatal opening, and hence as a measure of plant response to environmental stresses such as drought. A major practical problem in the application of this result in agriculture and in natural ecosystems is that, in addition to stomatal closure, many plant factors such as leaf orientation or reflectance and environmental factors such as radiation, wind speed and humidity also impact on leaf or canopy temperature over a rapid timescale. Practical application of this result for ecophysiological studies therefore requires approaches to allow for the various physiological and environmental determinants of canopy temperature (Jones 1999a, b, 2004), as will be discussed in detail below.

Thermal imaging also has applications in fields other than plant water relations. For example the high latent heat of fusion of water means that a large amount of heat

H. G. Jones (✉)
University of Dundee, Dundee, Scotland, UK

University of Western Australia, Crawley, WA, Australia
e-mail: h.g.jones@dundee.ac.uk

© Springer International Publishing AG, part of Springer Nature 2018

A. M. Sánchez-Moreiras, M. J. Reigosa (eds.), *Advances in Plant Ecophysiology Techniques*, https://doi.org/10.1007/978-3-319-93233-0_8

is released when plant tissue freezes and various workers have used thermal imaging to follow the (rapid) progress of freezing (Fuller and Wisniewski 1998; Hamed et al. 2000; Carter et al. 2001; Stier et al. 2003). Similarly, thermal dynamics can provide useful information at scales from the functioning of ecosystems (Bendoricchio and Jørgensen 1997; Aerts et al. 2004) through soil moisture content estimation (Verstraeten et al. 2006) to the aerodynamic properties of single leaves (Jones 2014).

The principles underlying the control of leaf temperature were formalised by the early 1960s (Raschke 1956, 1960) and, with the development of infrared thermometers, were first used for the study of plant water relations by Tanner and Fuchs (Tanner 1963; Fuchs and Tanner 1966). Extensive subsequent studies by Jackson, Idso and colleagues (1977, 1981; Idso 1982) promoted thermal sensing as a tool for irrigation scheduling.

The greatest uptake of thermal sensing, however, came with the widespread availability of thermal imaging and especially, from about the year 2000, uncooled microbolometer sensors that operate in the long-wave (c.9–14 μm) thermal infrared (Jones 2004). For thermal sensing in the field it is essential to use only imagers that operate in the long-wave thermal region rather than shorter wavelengths as the latter are much more subject to interference by reflected sunlight (Leigh et al. 2006). Imagery has many advantages over the use of point sensors, especially the possibility of multiple replication within an image, and the ability to ensure that the temperature is that of the material of interest (e.g. a leaf) without any contamination by background.

Thermal imagery can be used in several ways for ecophysiological purposes. The simplest is in a purely qualitative mode where visual comparison of temperatures of different canopies can be used to identify plants, or areas of vegetation, under particular stress (where those areas having particularly high temperatures indicate stomatal closure). The second mode is semi-quantitative, and is widely used in plant phenotyping and involves ranking plots or genotypes for temperature (and hence stomatal closure). The most powerful approach, however, involves quantitative application to estimation of stomatal conductance, evaporation rate or aerodynamic properties based on application of energy balance and measurement of the necessary meteorological variables.

1.2 Basic Energy Balance Calculations

Application of thermal sensing as an ecophysiological tool is largely based on an understanding of leaf or canopy energy balance. Useful summaries of the energy balance equations and their application to thermal imagery may be found in a number of texts (Monteith and Unsworth 2008; Jones and Vaughan 2010; Jones 2014) and reviews (Jones 2004; Maes and Steppe 2012).

For a plant leaf (or equivalently for a canopy) at any instant, the net energy fluxes into and out of the system must equal the rate of energy storage

$$\mathbf{R}_{\mathrm{n}} - \mathbf{C} - \lambda\mathbf{E} = \mathbf{M} + \mathbf{S} \tag{8.1}$$

where $\mathbf{R}_n$, is the net heat gain from radiation, **C** is the net 'sensible' heat loss, **λE** is the net latent heat loss, **M** is the net heat stored in biochemical reactions (usually negligible) and **S** is the net physical storage. In the steady state net storage is zero, so the net radiation input is balanced by evaporative and sensible heat losses. All fluxes are expressed per unit leaf area as flux densities in W m^{-2}.

Leaf temperature affects all these fluxes, with sensible heat loss given by

$$\mathbf{C} = g_H \left(\rho c_p\right)\left(T_s - T_a\right) \tag{8.2}$$

where g_H is the leaf conductance to heat transfer, ρ is the density of air, c_p is the specific heat of air, T_s is the leaf temperature and T_a is the air temperature (Jones 2014).

Similarly one can write the evaporative heat loss as

$$\lambda\mathbf{E} = g_W \left(\rho c_p / \gamma\right)\left(D + s\left(T_s - T_a\right)\right) \tag{8.3}$$

where g_W is the total conductance to water vapour, γ is the psychrometer constant, D is water vapour pressure difference between the inside of the leaf and the free air and s is the slope of the saturation vapour pressure-temperature curve.

Substituting these equations into the steady state energy balance and rearranging gives the leaf to air temperature difference as the following function of environmental and physiological variables:

$$T_s - T_a = \left(\gamma \mathbf{R}_n - g_W \rho c_p D\right) / \left(\rho c_p \left(g_H \gamma + g_W s\right)\right) \tag{8.4}$$

It is convenient to eliminate the need for a direct measurement of the net radiation actually received by the leaf by using the concept of net isothermal radiation ($\mathbf{R}_{ni}$), defined as the net radiation that would be received if the leaf were at air temperature (Jones 2014), and rewriting Eq. (8.4) as

$$T_s - T_a = \left(\gamma \mathbf{R}_{ni} - g_W \rho c_p D\right) / \left(\rho c_p \left(g_{HR} \gamma + g_W s\right)\right) \tag{8.5}$$

where g_{HR} is the parallel resistance to sensible heat and radiation.

2 Applications

2.1 *Stress Indices and Normalisation of Leaf Temperature*

A number of approaches have been used over the years to develop semi-quantitative approaches to normalisation of leaf or canopy temperature to account for variation in environmental conditions. One of the main drivers has been the wish to develop indicators of plant water-deficit stress as indicated by stomatal closure. The simplest

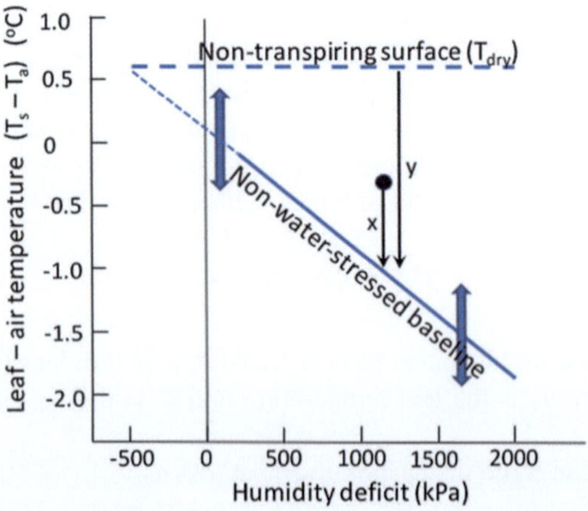

Fig. 8.1 Illustration of the calculation of CWSI. The non-water stressed baseline temperatures are obtained empirically for a given crop/site combination. The upper limit represents the temperature of a non-transpiring surface in the same environment. For any measured canopy temperature, CWSI is calculated as x/y. The two broad arrows indicate potential errors in temperature measurement and show that a given measurement error has a much greater relative impact on errors in CWSI in humid environments than in more arid situations

approach is to normalise data with reference to canopy air temperature, with early studies defining stress as the canopy-air temperature difference around mid-day (Jackson et al. 1977) or as the temperature difference between the canopy of interest and a comparable well-irrigated canopy (Gardner et al. 1981). More powerful was the introduction by Idso and colleagues (Jackson et al. 1981; Idso 1982) of a crop water stress index (CWSI; see Fig. 8.1) that takes account of variation in atmospheric humidity and makes use of the temperature of a notional non-transpiring canopy as well as the temperature of a well watered canopy transpiring at its potential rate (T_{nws}) using

$$\mathrm{CWSI} = \left(T_s - T_{nws}\right) / \left(T_{dry} - T_{nws}\right) \tag{8.6}$$

This approach has been very widely used since the 1980s for irrigation scheduling, but it suffers from a number of disadvantages that limit its application in ecological systems and in more humid or temperate climates. In particular it requires a standard empirical non-water stressed base line derived for a comparable crop in a similar environment, but does not take account of any potential variation in irradiance or wind speed, though extensions that overcome some of these limitations have been suggested (Keener and Kircher 1983). Another problem with this formulation is that it is very non-linearly related to evaporation rate or stomatal conductance, which are usually of most interest in plant water relations studies.

The use of actual physical references measured simultaneously by thermography is a particularly powerful approach to improvement of this approach that reduces the effects of short-term environmental fluctuations. The use of reference surfaces to mimic dry or wet leaves (Brough et al. 1986; Qiu et al. 1996; Jones et al. 1997) allows one to eliminate the need for one or more of these environmental variables. Using such references having aerodynamic and radiative properties as similar as possible to real leaves, Jones (1999b) defined a stress index (SI_{CWSI}) analogous to Idso's CWSI, but using a wet surface as reference (with infinite surface conductance), rather than a well-watered vegetation surface. Importantly, this paper also proposed a 'conductance index' (SI_{gl}) that is proportional to stomatal conductance and can be expressed as

$$SI_{gl} = (T_{dry} - T_s)/(T_s - T_{wet}) \tag{8.7}$$

where T_{wet} is the temperature of a wet freely evaporating mimic leaf.

2.2 *Use of Temperature Variance As a Stress Index*

An alternative approach to the use of thermal data for the study of differences in stomatal conductance is based on the variability of temperature between different leaves in the canopy. This approach was proposed by Fuchs (1990) and is based on the idea that temperature variance increases as stomata close, because the importance of the radiative component of the energy balance increases as stomata close. This idea was further developed by Bryant and Moran (1999), who derived a histogram-derived crop water stress index (HCWSI) based on the skewness or kurtosis of the temperature frequency distribution. In principle the variance-based approaches will be most sensitive for sunlit leaves which maximise the temperature variation as compared with shaded leaves (Jones 2004). Unfortunately attempts to use these approaches have thus far had only variable success (Grant et al. 2007, 2016).

2.3 *Stomatal Conductance and Evaporation*

More quantitative data can be obtained by use of the full energy balance equation; for example Eq. (8.5) can be rearranged (Leinonen et al. 2006; Guilioni et al. 2008) to allow one to estimate the total conductance to water vapour according to

$$g_W = \gamma\left(\left(R_{ni} / \rho c_p\right) - g_{HR}\left(T_s - T_a\right)\right) / \left(s\left(T_s - T_a\right) + D\right) \tag{8.8}$$

Unfortunately, in addition to the requirement for an accurate estimate of leaf temperature use of this equation for estimation of conductance, requires information on air temperature, net radiation, humidity and boundary layer conductance which are not all easy to estimate at the level of the leaf. Useful reductions in environmental data requirements can be obtained by using temperatures of the 'mimic' reference surfaces discussed above. For example, measurement of the temperature of a dry reference surface that mimics the radiative and aerodynamic properties of the transpiring leaves (T_{dry}) allows one to eliminate R_{ni} from Eq. (8.8), giving (Leinonen et al. 2006),

$$g_W = g_{HR}\gamma\left(T_{dry} - T_s\right)/\left(s\left(T_s - T_a\right) + D\right) \tag{8.9}$$

If one adds the temperature (T_{wet}) of a wet 'mimic' surface with no surface resistance to water loss, one can simplify this further (Guilioni et al. 2008; Leinonen et al. 2006). The precise formulation depends on whether the reference is wetted on one or both sides and whether the stomatal conductance is the same or different for the two leaf surfaces (Guilioni et al. 2008), however, for an amphistomatous (isolateral) leaf with a wet reference wetted on both sides this reduces to an estimate of the stomatal conductance (g_s) as

$$g_s = \left(\left(T_{dry} - T_s\right)/\left(T_s - T_{wet}\right)\right)\left[2\gamma g_{HR} / \left(\left(\gamma g_{HR} / g_{aW}\right) + 2s\right)\right] \tag{8.10}$$

where g_{aW} is the boundary layer conductance to water vapour for one side of the leaf. This equation is of the same form as Eq. 8.7, where the term in square brackets is a multiplier that depends primarily on the boundary layer conductance. Advantages of the use of reference surfaces include the fact that where all temperatures are measured by the same infrared thermometer/camera, errors in absolute temperature calibration can be eliminated.

In practice it is difficult to design a wet reference surface that remains wet continuously in hot, dry environments, though various designs have been proposed (Maes et al. 2016). Therefore most systems are based on the use of only dry references combined with humidity measurements (Eq. 8.9) with only a small reduction in accuracy (Leinonen et al. 2006).

2.4 *Evaporation*

Although it is the most difficult component of the surface energy balance to obtain from thermal sensing, rates of evapotranspiration (**ET**) can be conveniently estimated from remotely-sensed canopy temperature if appropriate assumptions are made (for a detailed discussion see Jones and Vaughan 2010). The best approach depends on the type of data available. The usual approach for satellite thermal imagery is to estimate regional **ET** as the residual term in the canopy energy balance using one-source energy balance algorithms such as SEBAL (Bastiaanssen et al. 1998a, b) and METRIC (Allen et al. 2007) or the more sophisticated two-source algorithms that treat vegetated and non-vegetated areas separately (Kustas 1990; Kustas and Anderson 2009).

Other more approximate approaches can be valuable in many situations. One of these is to make use of the fact that the upper limit of **ET** is set by the available energy (R_n) and that for large well vegetated areas **ET** is often well approximated by the 'equilibrium evaporation rate' (McNaughton and Jarvis 1983) given by

$$\mathbf{ET} = 1.26s\left(\mathbf{R}_n - \mathbf{G}\right)/\left(\lambda\left(s+\gamma\right)\right) \tag{8.11}$$

in which λ is the latent heat of evaporation, **G** is heat storage in the soil and the constant 1.26 is known as the Priestley-Taylor constant. Where ground cover is not continuous it is necessary to modify the calculation by multiplying the result by a 'evaporative fraction' that takes account of the fraction of the surface that is evaporating at the potential rate (Sobrino et al. 2007). This can be achieved by making use of the linear relationship between **ET** and surface temperature and combining this with remotely sensed information on canopy cover.

Temperature gives a direct estimate of evaporation rate, while combination with information on the fractional canopy cover (f_{veg}, often estimated from the Normalised Difference Vegetation Index (NDVI) as obtained from multispectral sensing (Jones and Vaughan 2010)) allows one to further derive information on the degree of stomatal closure (Fig. 8.2).

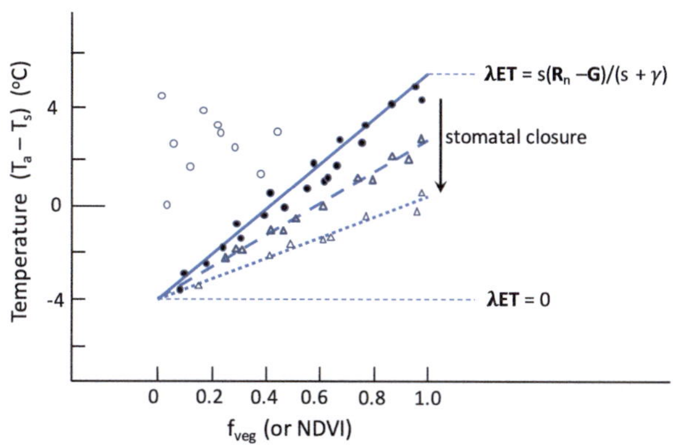

Fig. 8.2 Illustration of the general relationship between surface temperature and f_{veg} showing the trend to increasing Evaporation as canopy temperature decreases ($T_a - T_s$) increases. The solid circles show temperatures decrease as f_{veg} increases for dry soil and well watered plants, while the open triangles show the corresponding temperatures expected for water stressed vegetation showing increasing stomatal closure, increasing evaporation above that expected for transpiring vegetation alone. The open circles indicate samples with areas of wet soil or areas of open water. The line described by λ**ET** = s (**R**$_n$ – **G**)/($s + \gamma$) indicates the temperatures that correspond to the potential rate of **ET** as determined by incoming energy

2.5 Dynamics

The dynamics of temperature changes can provide important ecophysiological information, whether at the scale of single leaves or plants, or at the plant community or even regional scales. At the single leaf scale it can be used as an alternative method for estimation of aerodynamic resistances.

The rate of change of temperature (dT/dt) depends on the tissue heat capacity ($\rho^* c_p^* \ell^*$) according to (Jones 2014)

$$\mathrm{dT/dt} = \mathbf{S}/\left(\rho^* c_p^* \ell^*\right) \tag{8.12}$$

where the ρ^* and c_p^* are the density and heat capacity of the tissue material, and ℓ^* is the thickness. **S** is the rate of heat storage ignoring any **M** (=$\mathbf{R}_n$ – **C** – λ**E**). Following a step change in the environment that affects the rate of heat storage (such as a change in radiation), leaf temperature changes according to the following

$$\mathrm{T} = \mathrm{T_2} - \left(\mathrm{T_2} - \mathrm{T_1}\right)\exp\left(-\mathrm{t}/\tau\right) \tag{8.13}$$

where T_1 is the initial temperature and T_2 is the final equilibrium temperature and $\boldsymbol{\tau}$ is the time for c.63.2% of the total change to occur (known as the time constant). The time constant is given by

$$\tau = r^* c_p^* \ell^* / \left(rc_p\left(\left(1/\mathrm{r_{HR}}\right) + \left(s/\left(\gamma\left(\mathrm{r_{aW}} + \mathrm{r_{\ell W}}\right)\right)\right)\right)\right) \tag{8.14}$$

which for a non-transpiring surface where $r_{\ell W}$ is infinite, can be simplified and rearranged to give the aerodynamic resistance as

$$\mathrm{r_{aW}} = \tau\rho c_p / \left(\left(\rho^* c_p^* \ell^*\right) - \tau\rho c_p / \mathrm{r_g}\right) \tag{8.15}$$

Any suitable curve-fitting method can be used to estimate $\boldsymbol{\tau}$ (and hence r_{aW}) from the time course of temperature change after an environmental perturbation. For non-transpiring leaves (for example covered in petroleum jelly) this provides a potentially useful method for estimation of aerodynamic resistances, though it does not appear to have been used previously, probably because it requires accurate estimates of the heat capacity per unit area of tissue.

For more massive bodies such as fruits, and at larger scales for soils or regions, where temperature does not rapidly equilibrate throughout the body, the time course of surface temperature change is more complex. In this case the rate of change of surface temperature depends on the thermal conductance from the surface down into the depths and on the heat capacity of the material. A high thermal conductivity leads to surface temperature changing relatively slowly for a given energy input giving a material with a high thermal inertia (a strong resistance to temperature change).

The thermal dynamics of fruit surfaces can provide a useful tool for the detection of internal bruising or other damage. For example, Baranowski and colleagues (2008) have demonstrated that the thermal inertial was substantially greater for apple fruits with water core than for healthy fruits and that bruising can be readily detected (Baranowski et al. 2009). This difference results from the higher tissue water content and hence higher thermal conductivity in the damaged fruits.

The high conductivity of water lies behind the widespread application of the use of thermal inertia, as estimated from diurnal changes in surface temperature obtained from satellite imagery such as from MODIS which has two overpasses per day, as an indicator of soil moisture content (Verstraeten et al. 2006; Veroustraete et al. 2012; Zhang and Zhou 2016). Soils with higher water content have significantly greater thermal inertia together with a delay in the phase maximum.

Thermal inertia, otherwise known as thermal buffer capacity, has been proposed as an overall indicator of ecosystem integrity, summarising biotic and abiotic controls over energy flows in ecosystems (Aerts et al. 2004). In practice such a measure is, however, dominated by differences in vegetation, with forests and high soil moisture favouring high thermal inertia, while dry desert or grassland will have the lowest thermal inertia (most rapid temperature changes).

2.6 *Boundary Layer Conductance*

A number of approaches are available for estimation of the boundary layer conductance required for application of the methods described above. The value of g_H can be estimated at a leaf scale from aerodynamic theory using the relationships between wind speed and leaf size from

$$g_H = 6.62\left(u / l\right)^{0.5} \tag{8.16}$$

where u (m s^{-1}) is the wind speed and l (m) is the characteristic dimension of the leaf (Jones 2014). Alternatively, at a canopy scale, one can use wind profile theory (Monteith and Unsworth 2008; Jones 2014) to estimate g_{aH} as

$$g_H = \kappa^2 u_z / \left[\ln\left(\left(z - d\right) / z_o\right)\right]^2 \tag{8.17}$$

where κ is von Karmann's constant (=0.41), u_z is the wind speed at canopy height z, z_o is the roughness length (often assumed equal to 0.64*z), and d is the zero plane displacement (often assumed equal to 0.13*z).

A third approach that is well adapted to the leaf scale and for continuous monitoring in the field would be to calculate g_H from the temperatures of heated and unheated replica leaves mounted in the canopy (Brenner and Jarvis 1995).

A number of other, thermally-based approaches are at least theoretically possible. One of these is to use the cooling dynamics of heated leaves (as described above), while another would be to use thermal imagery with leaf 'mimic' reference surfaces wetted on 0, 1 or 2 surfaces.

2.7 Some Other Qualitative Applications

There has been a recent increase in the range of applications of thermal imaging in other eco-physiological studies. For example, thermal sensing can be a powerful tool for the study of plant disease, as many diseases lead to alterations in leaf temperature that can be detected by thermography, with the precise temperature responses and their sequence depending on the species/disease combination (Chaerle et al. 1999; Chaerle and Van Der Straeten 2001; Chaerle et al. 2004). Although mostly applied in a proximal or near-field situation for early detection of infections, there is evidence that useful information can be obtained from satellite or airborne thermal imagery (Lindenthal et al. 2005; Stoll et al. 2008).

Another area of interest has been in the study of thermogenic respiration in flowers (Skubatz et al. 1991; Bermadinger-Stebentheiner and Stabentheiner 1995; Lamprecht et al. 2002) as well as the role of solar heating in floral physiology, especially in arctic and alpine plants (Lamprecht et al. 2002, 2006; Dietrich and Korner 2014). Thermal imaging has also been used in forestry to investigate tree species diversity, with typical canopy temperatures being shown to vary between coniferous and broad-leaved forests (Leuzinger and Körner 2007).

3 Some Practical Aspects of Thermal Imaging

3.1 Absolute Accuracy

The absolute accuracy of thermal measurements is only critical for applications where absolute temperatures are required (e.g. for estimation of ET or stomatal conductance) and is not particularly critical for relative studies as in crop phenotyping. Most imagers currently available have thermal resolution fully adequate for plant water relations studies (i.e. <0.1 K), but their absolute accuracy is often no better than ±1 or 2 °C, so care is needed when absolute rather than relative values are of interest. This poor absolute accuracy is one of the strongest justifications for the use of reference surfaces in calculations because in such situations only temperature differences (which are generally available at higher precision) are required.

The accuracy of surface temperature measurement using thermography is critically dependent on the surface emissivity chosen and on settings for the

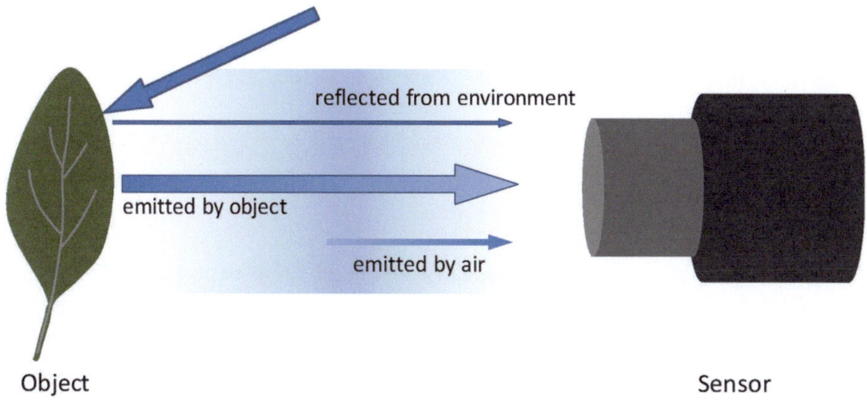

Fig. 8.3 Illustration of the radiation streams impinging on a thermal sensor, all of which need to be allowed for to obtain accurate temperature estimates. These corrections are commonly available in the thermal imaging software provided with the thermal cameras. The required corrections are greater, the more remote the sensor is from the object

'environmental temperature' and for the air temperature, humidity and distance between object and sensor. The temperature of a surface (T_s) is normally estimated by inversion of the following equation

$$R \cong \tau \varepsilon \sigma T_s^4 \tag{8.18}$$

where R is the total radiant flux density (W m^{-2}) received at unit area of sensor surface, τ is the longwave transmission by the atmosphere between the object and the sensor, ε is the surface emissivity and σ is the Stefan-Boltzmann constant (5.6697×10^{-8} W m^{-2} K^{-4}). Software in the sensor usually converts the received radiance to the radiance expected for the long wave band (correcting for the spectral sensitivity of the sensor) to produce an estimate of T_s. However, as shown in Fig. 8.3, the total longwave radiance arriving at the sensor is the sum of the thermal radiation emitted by the object (given by Eq. 8.18), the longwave radiation emitted by the atmosphere intervening between the object and the sensor, and the environmental longwave radiation reflected by the object towards the sensor, together with any of the emitted radiation absorbed by the intervening atmosphere.

For close-range sensing (less than about 5–10 m) the atmosphere is normally assumed to have a negligible effect on the at sensor radiance, though at high humidities it can be necessary to make some correction. The object emissivity also affects the temperature estimate, through its effect on both the emitted and reflected environmental radiation streams. The environmental radiation impinging on the object can have quite substantial effects on the apparent object temperature, especially for surfaces with emissivities substantially different from 1; for example, dry soils may have emissivities substantially below 0.9. The importance of emissivity to the correct estimation of temperature is illustrated by the fact that a 1% error in emissivity can give rise to an error in the estimated temperature of more than 0.5 K (Jones and Vaughan 2010), suggesting that particular care is needed when comparing different types of surface in a single image.

3.2 Illumination of Object

The temperature of any surface is critically dependent on its exposure to the incoming solar radiation (Eq. 8.4). As pointed out by Jones et al. (2009) it follows therefore that leaves on the shaded side of trees will be cooler than those on the sunlit side. Similarly, even within a homogeneous canopy, the temperature of different leaves will vary over quite a wide range (Fuchs 1990) dependent on their incident radiation (which is a function of leaf orientation and any shading). A continuing discussion point is whether it is better to orient a thermal sensor with the sun behind the observer (to maximise the proportion of sunlit leaves), or whether better discrimination between plants is obtained when orienting the sensor towards the sun so that more of the observed leaves are shaded. The former has the advantages that temperatures are higher so that there is a greater temperature response for given differences in conductance, while conductances themselves tend also to be higher than for shaded leaves. The latter approach of studying shaded leaves has the advantage that temperature is much less sensitive to leaf orientation, but the lower temperatures and conductance can reduce sensitivity (Jones et al. 2002). Other studies, however, have suggested that thermal measurements on the top of canopies rather than the sides can produce more reliable estimates of conductance indices than viewing from the side (Grant et al. 2016), supporting the value of imagery from UAVs.

3.3 Design of Reference Surfaces

In order to give useful estimates of dry or wet reference temperatures for use in the equations outlined above, it is critical that the reference surface has similar radiative and aerodynamic properties to the surface of interest (e.g. a plant canopy). The choice of reference surfaces varies from the use of reference crops grown under similar conditions (Idso 1982) through simple temperature approximations (Irmak et al. 2000) to a range of physical references (Jones et al. 2017). For the dry reference it has frequently been proposed (Irmak et al. 2000; Ben-Gal et al. 2009; Meron et al. 2010) that approximating the dry reference as T_a + 5 °C is adequate in Mediterranean climates. Unfortunately as pointed out by Jones et al. (2017) there can in practice be substantial deviations from this simple approximation, especially in low radiation or high wind speed environments, leading to large errors in the derived estimates of conductance or **ET**. Similarly, large errors can arise if the spectral absorptance differs from the actual leaves (Jones et al. 2009) or where the size or thermal properties of the reference do not mimic the real leaves.

Fig. 8.4 Illustration of the ArduCrop dry reference sensor, showing the field of view of the downward-looking thermal infrared sensor and the dry reference hemisphere whose temperature is detected by an upward pointing infrared sensor. Power to drive the sensor and its connection to a local Wireless Sensor Network is provided by the solar panel shown. (see Jones et al. 2017, for further details)

Most reference surfaces have been constructed as flat leaf replicas made of green filter paper or similar material (Maes and Steppe 2012), but such surfaces cannot effectively replicate the range of illumination experienced by typical leaves in a plant canopy. In order to better represent the temperature of the range of leaves in a canopy, especially as solar elevation changes during the day, Jones et al. (2017) proposed use of the average temperature of a hemispherical surface (constructed of a half table-tennis ball; Fig. 8.4) on the basis that for a typical canopy with leaves oriented according to a spherical distribution (Monteith and Unsworth 2008) this would provide a closer approximation of the effective mean leaf temperature than would a flat surface.

Because the aerodynamic properties of the hemispherical reference may not exactly mimic those of leaves in a canopy, and also because of potential differences in solar absorptance from actual leaves, calculated estimates of conductance or **ET** may require modification by the use of a constant correction factor as outlined by Jones et al. (2017). A particular advantage of this type of reference is that it allows continuous monitoring of stomatal conductance and **ET**, allowing long-term and diurnal studies of water relations.

4 Conclusions

Thermal infrared sensing is becoming an increasingly powerful tool for the study of plant water relations and a range of other ecophysiological processes in both natural and agricultural ecosystems. Thermal imaging, whether from in-field sensors or from airborne sensors mounted on drones, aircraft or even satellites, has particular advantages over the use of simple infrared thermometers with their single field of view because of the ability to make use of information about the variance of temperature of any object. A further advantage of imagery is that it can also be readily combined with multispectral or hyperspectral imagery to enhance the information about the object, for example through separating plant material from background and allowing the extraction of temperature of the plant material without confounding by background (Leinonen and Jones 2004). Nevertheless, arrays of simple cheap infrared thermometers, especially when combined with dry reference sensors (Jones et al. 2017), can be valuable for long term remote monitoring of evaporation and stomatal conductance. The recent advances in automated image analysis have the potential to greatly increase the power of thermal imagery for the study of plant water relations as it is becoming possible to avoid or reduce what is often a labour-intensive step in the application of thermography (Fuentes et al. 2012).

References

Aerts R, November E, Behailu M, Deckers J, Muys B (2004) Ecosystem thermal buffer capacity as an indicator of the restoration status of protected areas in the northern Ethiopian highlands. Restor Ecol 12:586–596

Allen RG, Tasumi M, Morse A, Trezza R (2007) Satellite-based energy balance for mapping evapotranspiration with internalized calibration (METRIC) – model. J Irrig Drain Eng 133:380–394

Baranowski P, Lipecki J, Mazurek W, Walczak RT (2008) Detection of watercore in 'Gloster' apples using thermography. Postharvest Biol Technol 47:358–366

Baranowski P, Mazurek W, Walczak W, Slawinski C (2009) Detection of early apple bruises using pulsed-phase thermography. Postharvest Biol Technol 53:91–100

Bastiaanssen WGM, Menentia M, Feddes RA, Holtslag AAM (1998a) A remote sensing surface energy balance algorithm for land (SEBAL). 1. Formulation. J Hydrol 213:198–212

Bastiaanssen WGM, Menentia M, Feddes RA, Holtslag AAM (1998b) A remote sensing surface energy balance algorithm for land (SEBAL) – 2. Validation. J Hydrol 213:213–229

Bendoricchio G, Jørgensen SE (1997) Exergy as goal function of ecosystems dynamic. Ecol Model 102:5–15

Ben-Gal A, Agam N, Alchanatis V, Cohen Y, Yermiyahu U, Zipori I, Presnov E, Sprintsin M, Dag A (2009) Evaluating water stress in irrigated olives: correlation of soil water status, tree water status, and thermal imagery. Irrig Sci 27:367–376

Bermadinger-Stebentheiner E, Stabentheiner A (1995) Dynamics of thermogenesis and structure of epidermal tiussues in inflorescences of *Arum maculatum*. New Phytol 131:41–50

Brenner AJ, Jarvis PG (1995) A heated leaf replica technique for determination of leaf boundary layer conductance in the field. Agric For Meteorol 72:261–275

Brough DW, Jones HG, Grace J (1986) Diurnal changes in water content of the stems of apple trees, as influenced by irrigation. Plant Cell Environ 9:1–7

Bryant RB, Moran MS (1999) Determining crop water stress from crop canopy temperature variability. ERIM International, Ann Arbor

Carter J, Brennan R, Wisniewski M (2001) Patterns of ice formation and movement in blackcurrant. HortSci 36:1027–1032

Chaerle L, Van Der Straeten D (2001) Seeing is believing: imaging techniques to monitor plant health. Biochim Biophys Acta-Gene Struct Express 1519:153–166

Chaerle L, Van Caeneghem W, Messens E, Lambers H, Van Montagu M, Van Der Straeten D (1999) Presymptomatic visulaization of plant-virus interactions by thermography. Nat Biotechnol 17:813–816

Chaerle L, Hagenbeek D, De Bruyne E, Valcke R, Van Der Straeten D (2004) Thermal and chlorophyll-fluorescence imaging distinguish plant-pathogen interactions at an early stage. Plant Cell Physiol 45:887–896

Dietrich L, Korner C (2014) Thermal imaging reveals massive heat accumulation in flowers across a broad spectrum of alpine taxa. Alpine Bot 124:27–35

Fuchs M (1990) Infrared measurement of canopy temperature and detection of plant water stress. Theor Appl Climatol 42:253–261

Fuchs M, Tanner CB (1966) Infrared thermometry of vegetation. Agron J 58:597–601

Fuentes S, De Bei R, Pech J, Tyerman S (2012) Computational water stress indices obtained from thermal image analysis of grapevine canopies. Irrig Sci 30:523–536

Fuller MP, Wisniewski M (1998) The use of infrared thermal imaging in the study of ice nucleation and freezing of plants. J Thermal Biol 23:81–89

Gardner BR, Blad BL, Watts DG (1981) Plant and air temperature in differentially irrigated corn. Agric Meteorol 25:201–207

Grant OM, Tronina L, Jones HG, Chaves MM (2007) Exploring thermal imaging variables for the detection of stress responses in grapevine under different irrigation regimes. J Exp Bot 58:815–825

Grant OM, Ochagavía H, Baluja J, Diago MP, Tardáguila J (2016) Thermal imaging to detect spatial and temporal variation in the water status of grapevine (*Vitis vinifera* L.). J Hort Sci Biotech 91:44–55

Guilioni L, Jones HG, Leinonen I, Lhomme JP (2008) On the relationships between stomatal resistance and leaf temperatures in thermography. Agric For Meteorol 148:1908–1912

Hamed F, Fuller MP, Telli G (2000) The pattern of freezing of grapevine shoots during early bud growth. Cryo-Lett 21:255–260

Idso SB (1982) Non-water-stressed baselines - a key to measuring and interpreting plant water-stress. Agric Meteorol 27:59–70

Irmak S, Dorota ZH, Bastug R (2000) Determination of crop water stress index for irrigation timing and yield estimation of corn. Agron J 92:1221–1227

Jackson RD, Reginato RJ, Idso SB (1977) Wheat canopy temperature: a practical tool for evaluating water requirements. Water Resour Res 13:651–656

Jackson RD, Idso SB, Reginato RJ, Pinter PJ Jr (1981) Canopy temperature as a crop water-stress indicator. Water Resour Res 17:1133–1138

Jones HG (1999a) Use of infrared thermometry for estimation of stomatal conductance as a possible aid to irrigation scheduling. Agric Forest Meteorol 95:139–149

Jones HG (1999b) Use of thermography for quantitative studies of spatial and temporal variation of stomatal conductance over leaf surfaces. Plant Cell Environ 22:1043–1055

Jones HG (2004) Application of thermal imaging and infrared sensing in plant physiology and ecophysiology. Adv Bot Res 41:107–163

Jones HG (2014) Plants and microclimate: a quantitative approach to environmental plant physiology, 3rd edn. Cambridge University Press, Cambridge

Jones HG, Vaughan RA (2010) Remote sensing of vegetation: principles, techniques, and applications. Oxford University Press, Oxford

Jones HG, Aikman D, McBurney TA (1997) Improvements to infra-red thermometry for irrigation scheduling. Acta Hort 449:259–266

Jones HG, Stoll M, Santos T, de Sousa C, Chaves MM, Grant OM (2002) Use of infrared thermography for monitoring stomatal closure in the field: application to grapevine. J Exp Bot 53:2249–2260

Jones HG, Serraj R, Loveys BR, Xiong LH, Wheaton A, Price AH (2009) Thermal infrared imaging of crop canopies for the remote diagnosis and quantification of plant responses to water stress in the field. Funct Plant Biol 36:978–989

Jones HG, Hutchinson PA, May T, Jamali H, Deery DM (2017) A practical method using a network of fixed infrared sensors for estimating crop canopy conductance and evaporation rate. Biosyst Eng 165:59–69

Keener ME, Kircher PL (1983) The use of canopy temperature as an indicator of drought stress in humid regions. Agric Meteorol 28:339–349

Kustas WP (1990) Estimates of evapotranspiration with one- and two-layer model of heat transfer over partial land cover. J Appl Meteorol 29:704–715

Kustas WP, Anderson M (2009) Advances in thermal infrared remote sensing for land surface modeling. Agric For Meteorol 149:2071–2081

Lamprecht I, Schmolz E, Blanco L, Romero CM (2002) Flower ovens: thermal investigations on heat producing plants. Thermochim Acta 391:107–118

Lamprecht I, Maierhofer C, Röllig M (2006) A thermographic promenade through the Berlin Botanic Garden. Thermochim Acta 446:4–10

Leigh A, Close JD, Ball MC, Siebke K, Nicotra AB (2006) Light cooling curves: measuring leaf temperature in sunlight. Funct Plant Biol 33:515–519

Leinonen I, Jones HG (2004) Combining thermal and visible imagery for estimating canopy temperature and identifying plant stress. J Exp Bot 55:1423–1431

Leinonen I, Grant OM, Tagliavia CP, Chaves MM, Jones HG (2006) Estimating stomatal conductance with thermal imagery. Plant Cell Environ 29:1508–1518

Leuzinger S, Körner C (2007) Tree species diversity affects canopy leaf temperatures in a mature temperate forest. Agric For Meteorol 146:29–37

Lindenthal M, Steiner U, Dehne H-W, Oerke E-C (2005) Effect of downy mildew development on transpiration of cucumber leaves visualised by digital infrared thermography. Phytopathology 95:233–240

Maes WH, Steppe K (2012) Estimating evapotranspiration and drought stress with ground-based thermal remote sensing in agriculture: a review. J Exp Bot 63:4671–4712

Maes WH, Baert A, Steppe K, Huete AR, Minchin PEH, Snelgar WP (2016) A new wet reference target method for continuous infraredthermography of vegetations. Agric For Meteorol 226:119–131

McNaughton KG, Jarvis PG (1983) Predicting the effects of vegetation changes on transpiration and evaporation. In: Kozlowski TT (ed) Water deficits and plant growth. Academic, New York, pp 1–47

Meron M, Alchanatis V, Cohen Y, Tsipris J, Orlov V (2010) Crop water stress mapping for site-specific irrigation by thermal imagery and artificial reference surfaces. Precis Agric 11:148–162

Monteith JL, Unsworth MH (2008) Principles of environmental physics, 3rd edn. Academic, Burlington

Qiu G-Y, Yano T, Momii K (1996) Estimation of plant transpiration by imitation leaf temperature – application of imitation leaf temperature for detection of crop water stress (II). Trans JSIDRE 185:43–49

Raschke K (1956) Über die physikalischen Beziehungen zwischen Wärmeübergangszahl, Strahlungsaustausch, Temperatur und transpiration eines Blattes [The physical relationships between heat-transfer coefficients, radiation exchange, temperature and transpiration of a leaf.]. Planta 48:200–238

Raschke K (1960) Heat transfer between the plant and the environment. Annu Rev Plant Physiol 11:111–126

Skubatz H, Nelson TA, Meeuse BJ, Bendich AJ (1991) Heat production in the Voodoo lily (*Sauromatum guttatum*) as monitored by infrared thermography. Plant Physiol 95:1084–1088

Sobrino JA, Gómez M, Jiménez-Muñoz JC, Olioso A (2007) Application of a simple algorithm to estimate daily evapotranspiration from NOAA–AVHRR images for the Iberian Peninsula. Remote Sens Environ 110:139–148

Stier JC, Filiault DL, Wisniewski M, Palta JP (2003) Visualization of freezing progression in turfgrasses using infrared video thermography. Crop Sci 43:415–420

Stoll M, Schultz HR, Baecker G, Berkelmann-Loehnertz B (2008) Early pathogen detection under different water status and the assessment of spray application in vineyards through the use of thermal imagery. Precis Agric 9:407–417

Tanner CB (1963) Plant temperatures. Agron J 55:210–211

Veroustraete F, Li Q, Verstraeten WW, Chen X, Bao A, Dong Q, Liu T, Willems P (2012) Soil moisture content retrieval based on apparent thermal inertia for Xinjiang province in China. Int J Remote Sens 33:3870–3885

Verstraeten WW, Veroustraete F, van der Sande CJ, Grootaers I, Feyen J (2006) Soil moisture retrieval using thermal inertia, determined with visible and thermal spaceborne data, validated for European forests. Remote Sens Environ 101:299–314

Zhang D, Zhou G (2016) Estimation of soil moisture from optical and thermal remote sensing: a review. Sensors 16:1308

Chapter 9
Photoprotection and Photo-Oxidative Stress Markers As Useful Tools to Unravel Plant Invasion Success

Erola Fenollosa and Sergi Munné-Bosch

1 Introduction

Light is essential for life, but also potentially dangerous, particularly for plants. As sessile and photosynthetic organisms, plants benefit from solar irradiation but must also cope with it when too much light is received. The meaning of 'excess light' strictly refers to the amount of energy not used for photosynthesis in chloroplasts of plant leaves. A number of factors determine excess energy in chloroplasts, including not only the amount of solar radiation but also its quality and duration, the plant physiological status (including the development stage), plant stress tolerance, and the availability of other resources for plant growth (Demmig-Adams et al. 2017). As the name of "photosynthesis" itself reveals, light is the main resource for photosynthesis, this is, the conversion of light into chemical energy stored in carbohydrate molecules, synthesized from carbon dioxide and water, releasing oxygen. Under optimal conditions, light is captured by the light harvesting complexes (LHC) at the photosystems (PSI and PSII), which are found at the thylakoid membrane inside the chloroplast of the photosynthetic tissues (Croce and Van Amerongen 2011). Photosynthetic pigments, such as chlorophylls and carotenoids, are responsible of light capture and transference into reaction centres which ultimately allow redox reactions through the electron transport chain (ETC) leading to the ultimate reduction of NADP to NADPH. In addition, this creates a proton gradient across the chloroplast membrane, which is used by ATP synthase in the synthesis of ATP. The NADPH and ATP generated after the ETC are essential for carbon assimilation through their use in the Calvin cycle.

E. Fenollosa · S. Munné-Bosch (✉)
Faculty of Biology, Department of Evolutionary Biology,
Ecology and Environmental Sciences, University of Barcelona, Barcelona, Spain
e-mail: smunne@ub.edu

© Springer International Publishing AG, part of Springer Nature 2018
A. M. Sánchez-Moreiras, M. J. Reigosa (eds.), *Advances in Plant Ecophysiology Techniques*, https://doi.org/10.1007/978-3-319-93233-0_9

There are different factors that can lead to suboptimal conditions for photosynthesis. For instance, low concentrations of the substrate for the Calvin cycle, i.e. CO_2, may lead to an accumulation of $NADPH^+$ at the ETC. A common plant response to stress is stomatal closure, which reduces water losses through transpiration but at the same time slows down the photosynthetic machinery. Besides low internal CO_2 concentration, high light itself may collapse the photosynthetic apparatus by an energy excess that cannot be used due to saturation on the ETC components. In PSII, a bound quinone (Q_A) receives the electron transferred from water splitting via the initial acceptor pheophytin. However, Q_A is not able to accept another electron from PSII until it has passed its electron to the next carrier, Q_B (Kalaji et al. 2014). In this state, the reaction centers are considered to be 'closed', leading to an accumulation of molecules of excited chlorophyll ($^3Chl^*$). This, in turn, will inevitably cause a decline in quantum efficiency of PSII and damage on it due to the consequent generation of reactive oxygen species (ROS) (Apel and Hirt 2004). After damage on the PSII reaction centre by light excess, it must be dissembled and repaired. The D1 protein is the only compound that, when damaged, needs to be synthesized de novo (Goh et al. 2012). When the oxidation of D1 overcomes its regeneration capacity, photoinhibition occurs, thus leading to a light-induced reduction of the photosynthetic capacity (Takahashi and Badger 2011).

Not only at the PSII, but also at the PSI, the high energy received and the high tensions of oxygen found inside the chloroplast may lead to the formation of ROS, such as singlet oxygen ($^1O_2$), superoxide ion ($^{\cdot}O_2^-$), hydrogen peroxide (H_2O_2) and hydroxyl radical ($^{\cdot}OH$) (Asada 2006). Singlet oxygen is formed at the PSII due to an accumulation of excited chlorophylls ($^3Chl^*$) (Havaux and Triantaphylides 2009). Singlet oxygen seems to be the major ROS involved in photo-oxidative stress-induced cell death, and is therefore a very interesting ROS to quantify, despite its high reactivity. The superoxide ion is formed at the PSI rapidly leading to hydrogen peroxide by the action of superoxide dismutase, potentially leading thereafter to the formation of hydroxyl radical, a very reactive ROS (Asada 2006). Here, we will use the term "photo-oxidative stress" as the imbalance between pro-oxidants (such as ROS) and antioxidant defences caused by excess energy in chloroplasts.

If photo-oxidative stress is not properly counterbalanced by antioxidant defences, oxidative damage occurs over different biomolecules, causing peroxidation of lipids, oxidation of proteins, and/or damaging nucleic acids. Photo-oxidative damage is therefore characterized by alterations of the membrane properties (changes in fluidity, ion transport), a loss of enzymatic activity, protein cross-linking, inhibition of protein synthesis, DNA damage and at the end, the death of the cells (Sharma et al. 2012). When this occurs irreversible by lack of sufficient regeneration capacity within cells and organs, this photo-oxidative damage in chloroplasts leads to photo-oxidative damage at the cellular, organ and eventually organism levels.

However, there are multiple photo-protective mechanisms that plants have developed to protect the chloroplast from photoinhibition and photo-oxidative stress (Takahashi and Badger 2011). These include from structural changes that reduce light collection, to an increase in the amounts of antioxidants that quench and/or scavenge ROS ("quenching" is considered here as the physical process eliminating ROS, while "scavenging" involves a chemical reaction for ROS elimination). All these responses reflect the plant's physiological status and correlate with different stresses intensity, being therefore highly informative to understand stress responses, compare genotypes and give insight into new alternatives to improve environmental management. Likewise, stress markers based on photo-oxidative stress may be helpful on some global ecological problems such as invasive plant species that constitute the second main threat to biodiversity. The utility of photo-oxidative stress markers in invasive plants studies lies in the fact that invasive vigour is determined by their physiological capacity overcoming the native coexistent species.

In this chapter, we aim at compiling existing information on the photoprotective and photo-oxidative stress markers used in plant invasion biology studies: from the study of the light harvesting complexes composition or the photosynthetic efficiency, to ROS formation and the accumulation of antioxidants and its oxidation. Much emphasis will be put on providing the essential information that each marker offers, but also their limitations and the actual and potential use in plant invasion studies.

2 Photoprotection and Photo-Oxidative Stress Markers: How to Measure Them

A **photo-oxidative stress marker** could be considered any molecule, ratio, index or general descriptor that responds to excess light and is related to oxidative stress. The different approaches to quantify photoprotection and photo-oxidative stress comprise the different defense levels that the plants trigger to respond to it and its measurement may include both in situ and ex situ measurements (Fig. 9.1). At the first level, the composition of light harvesting complexes regulates the light capture process at the thylakoid membrane (Walters 2005). Hence, plant pigments play a crucial role on the capacity to transfer light energy into the ETC that will ultimately lead into the production of ATP and storing the reducing power as NADPH. The efficiency by which the electrons are transferred can constitute also a stress marker, providing information on the actual degree of photoinhibition of the photosynthetic apparatus (Kalaji et al. 2014). If the energy exceeds the photosynthetic capacity, ROS are generated and, thus, an estimation of ROS production and/or the accumulation of antioxidants (preferably including their redox state) is another way to get a proxy of the extent of photo-oxidative stress. Finally, the accumulation of oxidation products is also a measure of the degree of oxidative stress. We are going to get through the different approaches to measure photoprotection and photo-oxidative stress and present the most used techniques (resumed in Table 9.1), taking into account what information we are really getting from them, including their limitations and requirements.

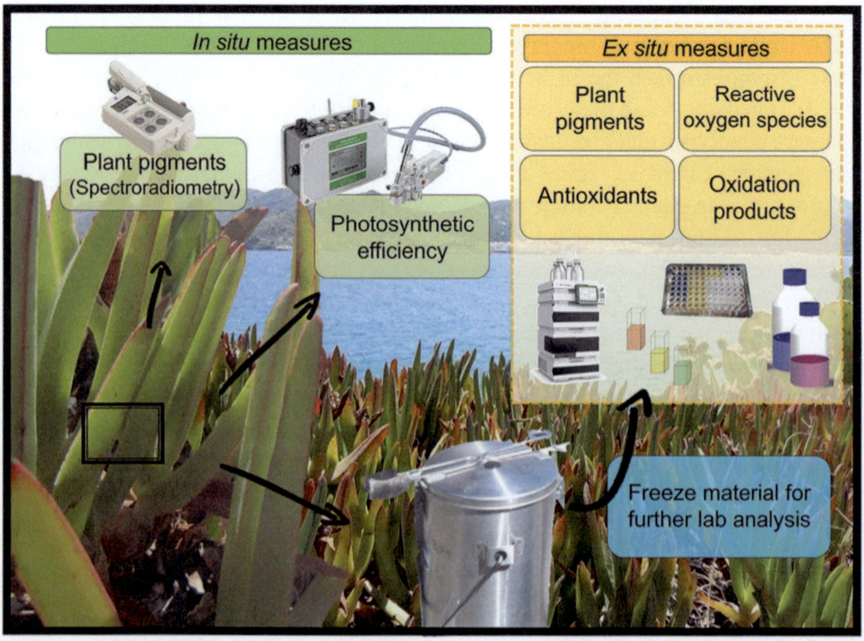

Fig. 9.1 Form field to lab in an invasion biology study, here exemplified with the aggressive invasive plant species *Carpobrotus edulis*, using photo-oxidative stress markers, measured in situ and ex situ. As examples, SPAD (MCL502, Minolta SPAD 502, GIS Ibérica, Spain), MiniPam II (Heinz WalzGmbH, Germany) and an Agilent HPLC are included here

2.1 Sampling Design

Before going through the different approaches and techniques used to measure photoprotection and photo-oxidative stress markers, it is important to point out some common requirements related to the sampling design.

At first, **representativeness** must be seriously considered when designing our experimental set, taking into account the high biological diversity at multiple levels. It is well known that biological diversity is wide, not only at the species level but also among different individuals from the same species, and even at the intraindividual level. For instance, there is an incredible variability considering the different organs within an individual taking into account the cellular structure and its biochemistry.

Choosing an appropriate number of replicates is essential to capture the biological variation, but it may depend on the study scale (growth chamber, common garden, field, ecosystem, international, etc.). In general, pseudo-replications are not recommended if we are after a real representation on the plant response to its environment. The number of replicates must increase after the variability on the environmental conditions and the differences among individuals (age, size, number and

Table 9.1 Summary of techniques and stress markers used to assess photoprotection and photo-oxidative stress in invasion biology and other ecophysiological studies, including a qualitative evaluation of their difficulty, accuracy, costs and dependency on other markers

	Photo-oxidative stress markers	Technique	Difficulty	Accuracy	Costs	Dependency on other markers
Plant pigments	Chl T	Spectrophotometry or HPLC	L H	M H	L H	L
	Chl a/b	Spectrophotometry or HPLC	L H	M H	L H	L
	Car/Chl	Spectrophotometry or HPLC	L H	M H	L H	L
	VAZ	HPLC	H	H	H	L
	DPS	HPLC	H	H	H	L
	Lut	HPLC	H	H	H	H
	β-Car	HPLC	H	H	H	L
	Anthocyanins	Spectrophotometry	L	M	L	M
	SPAD	Spectroradiometry	L	L	L	H
	NDVI	Spectroradiometry	L	L	L	H
	PRI	Spectroradiometry	L	L	L	M
Photosynthetic efficiency	F_v/F_m	Fluorescence	L	H	L	L
	Φ_{PSII} or ETR	Fluorescence	L	H	L	L
	NPQ	Fluorescence	L	H	L	L
ROS	H_2O_2	Spectrophotometry	M	M	H	M
Antioxidants	AsA	Spectrophotometry	M	M	H	L
	AsA/(AsA+DHA)	Spectrophotometry	M	M	H	L
	α-, β-, γ-, δ-Toc	HPLC	H	H	H	L
Oxidation products	LOOH	Spectrophotometry	M	M	H	M
	MDA	Spectrophotometry or HPLC	M H	M H	M H	M
	Protein carbonylation	Spectrophotometry or HPLC	M H	M H	M H	H
	β-CC	GC/MS	H	H	H	H

This last parameter refers to the possibility of understanding photo-oxidative stress with the stress marker alone. The color code refers to the goodness of the qualification, from green (adequate) to orange (not adequate)
L low, *M* medium, *H* high

position of leaves, phenology, reproductive effort, etc.). For measures of photoprotection and photo-oxidative stress markers on ecophysiological studies performed under natural conditions, a number of replicates between 10 and 20 is in general recommended per sampling point, treatment and genotype, with a minimum of at least 8 individuals. The standard deviation can indeed be used as a measure of variation in each particular case.

To guarantee representativeness, some considerations of what and when we should sample are recommended, such as limit our sampling material to **leaves at the same developmental stage** and sample always under **similar environmental conditions**. Attention should be paid to avoid other factors influencing potential differences such as light incidence or biotic stress. We recommend sampling fully-expanded young leaves that receive direct solar radiation to minimize heterogeneity.

As photoprotection and photo-oxidative stress markers are strongly light-dependent, it is crucial to choose similar sampling environmental conditions and time of the day for measurements. We recommend performing **samplings during midday** (when the sun is at its zenith) on clear, sunny days.

Most of the techniques here presented require laboratory analysis (Fig. 9.1) and they need special **considerations to prevent sample degradation** and **additional measures** to estimate the final concentration. During sampling, samples must be immediately frozen in liquid nitrogen and stored at −80 °C until analysis to prevent sample degradation and changes on the cellular redox state. Moreover, as some of the molecules are highly reactive, thermo- or light sensitive, it is recommended to perform all analysis under cold conditions (4 °C) and protecting the samples against direct light. It is necessary to calculate the fresh weight/dry weight ratio of each sample and if possible the leaf mass area (LMA), to present the quantified molecules by fresh weight, dry weight and leaf area.

2.2 *Plant Pigments*

Among the Viridiplantae subkingdom (vascular plants, mosses and green algae), pigment composition has been shown to be remarkably constant, with chlorophylls *a* (Chl *a*) and *b* (Chl *b*) and six carotenoids: lutein (Lut), β-carotene (β-Car), neoxanthin (Neo), violaxanthin (Vio), antheraxanthin (Ant) and zeaxanthin (Zea), being found in all species (Young et al. 1997). Each of these pigments plays a specific role and is distinctively located within the photosynthetic apparatus (Croce and Van Amerongen 2011; Takahashi and Badger 2011). All these pigments play a dual role by collecting light through the light harvesting complexes (LHC) and offering photoprotection at the photosystem II, where light is initially collected.

Chlorophylls are the main photosynthetic pigments responsible for light capture, constituting therefore good photo-oxidative stress markers. Chlorophylls are found in cyanobacteria, algae and plants and are composed by a large heterocyclic aromatic ring with a magnesium ion at the centre of it. Chl *a* is present in the reaction centres and the antennae of PSI and PSII, whereas the presence of Chl *b* is restricted to light-harvesting systems (Croce and Van Amerongen 2011). Therefore, the ratio Chl *a/b* could be an indicator of the degree of sun/shade acclimation or the intensity of a stress (Esteban et al. 2015). Also, the content of total Chl itself respond to the widest variety of stressors (Esteban et al. 2015). Chlorophyll loss is a process associated with both intense stress and senescence processes (Zimmermann and Zentgraf 2005).

On the other hand, **carotenoids** belong to the category of tetraterpenoids, which take the form of a polyene hydrocarbon chain, which is sometimes terminated by rings. This group of isoprenoids play a dual role in the photosynthetic machinery, as light-harvesting pigments (Bontempo e Silva et al. 2012), but also they protect against photooxidative damage (Lambrev et al. 2012). This group is subdivided into carotenes, of which β-carotene (β-Car) is the most abundant, and xanthophylls, which contain oxygen in is chemical structure and include lutein, violaxanthin, zeaxanthin, antheraxanthin and neoxanthin. β-Car is especially efficient at eliminating the singlet oxygen ($^1O_2$) generated in photosystem II (PSII) from excited triplet chlorophyll ($^3Chl^*$) (Ramel et al. 2012). Lutein is the most abundant xanthophyll species in plants and is essential for protein folding and ^{3}Chl* quenching (Dall'Osto et al. 2006). Moreover, xanthophylls are crucial as physical quenchers that promote thermal dissipation or non-photochemical quenching (NPQ), an efficient energy-dissipation mechanism in plants (Demmig-Adams and Adams 1996). The de-epoxidation of Vx to Ax and Zx (components of the VAZ cycle) responds to different environmental stresses (Demmig-Adams et al. 2012).

There are other plant pigments with an important role on photoprotection and widely distributed among the plant kingdom: **anthocyanins**, a class of flavonoids. These water-soluble pigments consist of an aromatic ring bound to a heterocyclic ring that contains oxygen, which is linked through a carbon-carbon bond to a third aromatic ring (forming the anthocyanidins), in some cases all bound to a sugar moiety (forming the corresponding anthocyanins) (Castañeda-Ovando et al. 2009). Anthocyanins are responsible of screening ultraviolet (UV) light and therefore constitute an important photoprotective defense for plants, as UV comprises 7–9% of the total solar radiation energy (Jansena et al. 1998), protecting plants from PSII damage (Takahashi et al. 2010). Anthocyanins are responsible for some of the colors on leaves, flowers, fruits and seeds, and are not localized on the chloroplast but accumulated in vacuoles, especially in the leaf epidermis cells, together with other phenolic compounds that accomplish the same screening photoprotective function (Takahashi and Badger 2011). The synthesis of phenolic compounds (including anthocyanins) is enhanced under strong light, particularly UV and blue light conditions (Winkel-Shirley 2002).

There are different techniques to measure plant pigments, all based on its specific light absorption spectrum (Table 9.1). With a liquid solvent (methanol, ethanol or acetone with different purity) we can easily extract all plant pigments. Calibration curves have been defined for determination of Chl *a*, Chl *b* and total carotenoids (Car) through **spectrophotometry** using different solvents (Lichtenthaler 1987). It was not until the late 1980s when good protocols for an easy and precise separation of the different carotenoids through **high-performance liquid chromatography** (HPLC), usually employing acetonitrile as the mobile phase, were developed (Thayer and Björkman 1990; Munné-Bosch and Alegre 2000). This is a relatively expensive approach, but it allows quantifying all the carotenes from one extract, offering the possibility to have a deeper understanding on the plant physiological status. Through this methodology, one can quantify how much energy the plant is dissipating through the xanthophyll cycle, by calculating the proportion of de-epoxidated xanthophylls, i.e. the de-epoxidation state ((DPS = (Zx + Ax)/Vx). Not only the DPS but the total amount of Vx, Ax and Zx (so called VAZ) increases in response to stress (Demmig-Adams and Adams 1996).

Anthocyanins can also be measured both by spectrophotometry and HPLC, taking advantage of the absorption range of the spectrum among 500–530 nm of these reddish pigments. There are several methods that show different specificity. At first, the most used method, due to its simplicity, is to estimate total anthocyanins by acidifying the methanol extract with 1% HCl and reading absorbance at 535 nm (Siegelman and Hendricks 1958; Fuleki and Francis 1968), always subtracting unspecific absorbance at 700 nm. Despite the simplicity of this method, it shows low specificity as all reddish pigments are quantified as anthocyanins, as phlobaphenes (Winkel-Shirley 2002). Another method is the pH differential method or the total monomeric anthocyanin method, designed to measure only single anthocyanin units (Giusti and Wrolstad 2001). Monomeric anthocyanins can change their colour under different acidic conditions, and the lectures at pH 1 and 4.5 comparison allows the removal of the interference of other reddish pigments, being an interesting method for several species (Lee et al. 2005; Dandena et al. 2011). Spectrophotometric methods usually use cyanidin-3-glucoside chloride as a standard, taking its extinction coefficient, as the main anthocyanin found in plants. HPLC methods can be used not only for a more precise quantification but also for identification of the precise anthocyanin composition. Different procedures have been proposed, including an acid hydrolysis that breaks the glyosidic bond of monomeric anthocyanins, releasing anthocyanidins (Lao and Giusti 2016).

Different indices and techniques based on **spectroradiometry** have been described also for chloroplastic pigments determination. Based on leaf transmittance, SPAD (MCL502, Minolta SPAD 502, GIS Ibérica, Spain) is a simple and portable apparatus that determines a relative quantity of chlorophylls by a simple non-destructive leaf measurement (Richardson et al. 2002) that can be measured in situ (Fig. 9.1). The relative measures show a high correlation with the total chlorophyll content and therefore it is a simple and fast alternative to laboratory analysis. However, this measure shows a high variability and needs calibration depending on the species and the environmental conditions. Based on leaf reflectance there are a whole plethora of different defined indexes with different applications. The broadest index used is the normalized difference vegetation index (NDVI), which is known for its good correlation with this chlorophyll content (Richardson et al. 2002). Another commonly used spectral reflectance index is the photochemical reflectance index (PRI) that often strongly correlates with total carotenoids or chlorophyll *a/b*, but also with radiation use efficiency, chlorophyll fluorescence parameters, DPS, net CO_2 uptake, Jmax or water content (Garbulsky et al. 2011). NDVI and PRI can be calculated at different scales, using different platforms where we use the spectroradiometer such as a drone, a balloon, planes or satellites. Spectroradiometric indexes are a promising tool for high-throughput phenotyping, but this technique requires calibration depending on the species, the season, the environmental conditions, etc., and normally there is a huge variability associated.

2.3 *Photosynthetic Efficiency*

Once a chlorophyll molecule gets excited, it helps the transference of an electron through the different proteinic complexes that form the electron transport chain (ETC). At the end, through the generation of an H^+ gradient, this process will generate ATP and accumulate reducing power as NADPH, necessary for Calvin cycle. An easy, in situ, and non-destructive way to measure the efficiency by which electrons pass through the ETC and detect photoinhibition is the measurement of **chlorophyll**

***a* fluorescence** (Fig. 9.1). The illumination of the photosynthetic tissue with photosynthetic active radiation leads to the emission of fluorescence (680–760 nm), mainly associated with chlorophyll *a* on the PSII. This fluorescence is one of the three ways where chlorophylls excitation energy is distributed, apart from the photochemical reactions on the ETC and the thermal dissipation (explained above). As the three processes are competitive, it is possible to estimate them from the chlorophyll fluorescence measurements. A range of instruments has been developed focusing on different aspects of photosynthesis and on different properties of Chl *a* fluorescence, but most authors are using only a limited set of experimental protocols based on methods that have been developed over time (Kalaji et al. 2014). One of the favorite techniques involves the use of a modulated measuring system, which allows the quantification of the contribution of the photochemical and non-photochemical quenching. In darkness, all the PSII reaction centers are open (all the quinone pool is reduced) and when a leaf is transferred from the darkness into light, PSII reaction centers close progressively. The comparison between the fluorescence emitted after a short duration saturation flash light (that immediately reduces the whole quinone pool) under natural light and after darkness adaptation allows the differentiation between the energy derived to photochemical and non-photochemical processes. In darkness, with all the reaction centers open, the increase on the fluorescence emission ($Fv = Fm - F_0$, variance, maximum and basal fluorescence) due to the saturation flash light indicates the maximum capacity of the PSII to transport electrons. One of the most widely used photo-oxidative stress markers is the maximum efficiency of the PSII (F_v/F_m), calculated from the parameters presented above, and measured by all the modulated fluorimeters. For unstressed leaves, the value of F_v/F_m is highly consistent, with values of ~0.83, and correlates to the maximum quantum yield of photosynthesis (Demmig and Björkman 1987). F_v/F_m below 0.75 reflect damage on the PSII, photoinhibition, and therefore it is an extremely informative stress marker.

More information can be obtained from chlorophyll *a* fluorescence analysis. The same calculation from the basal and maximum fluorescence after a saturating flash pulse under natural light gives the relative efficiency of the PSII (φ_{PSII}) and the electron transport rate (ETR). The latter requires the use of the average ratio of light absorbed by the leaf (around 0.84) and the average ratio of PSII reaction centers to PSI reaction centers (0.50) for calculation. Another parameter obtained from chlorophyll fluorescence analysis is NPQ. This parameter estimates the non-photochemical quenching and its calculation involves both light and dark-adapted measures. NPQ is calculated as $(Fm - Fm')/Fm'$, with the prima parameters corresponding to those taken under light conditions. NPQ strongly correlates with DPS (Demmig-Adams et al. 2012; Jahns and Holzwarth 2012), being a cheaper and non-destructive alternative to the measurement of the xanthophyll cycle by HPLC. The most attractive feature of Chl *a* fluorescence is its non-invasive character, but it is common to commit some pitfalls with the measures. Several reviews have elegantly compiled common pitfalls, questions and conflictive points of view of Chl *a* fluorescence techniques (Maxwell and Johnson 2000; Logan et al. 2007; Murchie and Lawson 2013; Kalaji et al. 2014).

2.4 *Reactive Oxygen Species*

We define oxidative stress as the imbalance between prooxidants and antioxidants. Therefore, the amount of reactive oxygen species (ROS), among other prooxidants, gives us information about the status of the imbalance during a stress response. Chloroplasts are quantitatively and qualitatively one of the most important sources of ROS in illuminated plant cells (Foyer and Noctor 2003). Thus, the measurement of singlet oxygen ($^1O_2$), superoxide ion ($^{\cdot}O_2^-$), hydrogen peroxide (H_2O_2), and hydroxyl radical ($^{\cdot}OH$) are good markers to evaluate the status of the photosynthetic apparatus.

There are three approaches for measuring ROS in plant tissues: (1) monitoring ROS released into a medium where the cell culture grows, (2) in vivo ROS visualization and (3) quantification of ROS production (Noctor et al. 2016). The third group is indeed the best suited for ecophysiology experiments. Here we will present the measures of hydrogen peroxide as it is the most stable of the group of the four primary ROS (H_2O_2, superoxide ion, hydroxyl radical and singlet oxygen), and therefore it is quantifiable after direct extraction (third approach).

Hydrogen peroxide (H_2O_2) can be quantified through spectral changes of different substances when they are oxidized by this molecule. For instance, the ferrous xylenol orange (FOX) assay is based on the oxidation of ferrous to ferric ions by H_2O_2 producing a chromophore complex which absorbs strongly at 540–600 nm (Cheeseman 2006); however, there are some matrix effects that may be taken into consideration (Queval et al. 2008). Another method is the use of Amplex Red (10-acetyl-3,7-dihydroxyphenoxazine) which is converted to the fluorescent resorufin, easily quantified with a fluorescence spectrophotometer (Zhou et al. 1997). For estimation of the extent of photo-oxidative stress, chloroplasts can be isolated from leaves under reducing conditions and the amount of ROS measured thereafter. This is essential for ROS that can be produced in various cellular compartments, as it occurs with hydrogen peroxide (Munné-Bosch et al. 2013).

2.5 *Antioxidants*

An antioxidant is a molecule that prevents the oxidation of other molecules. One of the most common responses to stress is the activation of antioxidant defences. We can classify the antioxidants into enzymatic and non-enzymatic, and the latter, depending on their affinity to water can be classified into hydrophilic and lipophilic antioxidants. The enzymatic antioxidants, (such as superoxide dismutase (SOD), ascorbate peroxidase (APX) and glutathione reductase (GR)), among others) are found in chloroplasts, but also in other cellular compartments reducing oxidative stress. Therefore, they are not necessarily only related to photo-oxidative stress and specific chloroplastic isoforms should therefore be investigated to relate them to excess light energy. In contrast, a clear and strong relationship has been established between photo-oxidative stress and the accumulation and tocopherols (or vitamin E) and carotenoids (Car), since both are exclusively located in chloroplasts.

The group of **tocopherols (Toc)** include the α-, β-, γ-, and δ-tocopherols, which are differentiated by the number and position of methyl groups on the chromanol ring. It is specifically this chromanol head that provides the molecule its antioxidant scavenging properties as it can donate electrons to various acceptors such as ˙OH or $^1O_2$. Tocopherols also deactivate singlet oxygen by (physical) quenching, being this latter function the most important quantitatively, protecting PSII from photo-oxidative damage. Tocopherols, which are located on the thylakoid membrane, but accumulate as well in the plastoglobuli (where they are stored), have also an essential role in preventing the propagation of lipid peroxidation (scavenging lipid peroxyl radicals, Munné-Bosch and Alegre 2002a). The contents of tocopherols, in agreement with their antioxidant function, increase in plants adapted to drought and other abiotic stresses (Munné-Bosch 2005). The four tocopherol homologues can be measured after an extraction with methanol by HPLC with a mixture of n-hexane and p-dioxane as a mobile phase, using a fluorescence detector, emitting at 330 nm and with detection at 295 nm (Amaral et al. 2005). The major homologue found in leaves is the α-tocopherol, followed by its immediate precursor, γ-tocopherol. β-, and δ-tocopherols are usually present at very low concentrations in leaves.

Ascorbic acid (AsA) is the most abundant hydrophilic antioxidant in plant leaves, and it is mainly accumulated in the chloroplast (Queval and Noctor 2007). Ascorbate can be oxidized to monodehydroascorbate radical (MDHA) or dehydroascorbate (DHA). Not only the total amount of AsA, but also the redox state of the ascorbic acid pool (AsA/(AsA + DHA)), particularly when measured in isolated chloroplasts, constitute excellent photo-oxidative stress markers and have been described to be very sensitive to several stresses. The most popular techniques for measuring AsA are based on the molecule's absorbance at 256 nm. To determine the amount of reduced and oxidised AsA is common to use reducing agents such as dithiothreitol (DTT) and ascorbate oxidase (AO), that reduce/oxide the whole sample extract in an acid medium and compare the maximum and the minimum absorbance with the initial one (Queval and Noctor 2007). AsA has an intimal relationship with tocopherols as it mediates their regeneration. At the same time, AsA is regenerated from DHA to AsA by **glutathione (GSH)**, another hydrophilic antioxidant found in most organelles.

2.6 Oxidation Products

As a result of photo-oxidative stress, if ROS are not counterbalanced by antioxidant defences, oxidative damage occurs over different biomolecules. The derived products of this process: oxidized compounds, such as primary or secondary lipid peroxidation products and modified proteins constitute excellent photo-oxidative stress markers focusing on the consequences after the damage. The enhanced production of ROS during environmental stresses can pose a threat to cells by causing peroxidation of lipids, oxidation of proteins, and/or damage to nucleic acids, thus causing enzyme inhibition, alterations of the membrane properties (changes in fluidity, ion transport), protein cross-linking, inhibition of protein synthesis, DNA damage and at the end, the death of the cells (Sharma et al. 2012).

Over lipids, free radicals or ROS can inflict direct damage, leading to lipid peroxidation that at the same time can inflict damage over DNA or the protein complexes of the PSII (Pospísil and Yamamoto 2017). This is the process under which free radicals attack polyunsaturated fatty acids (PUFAs) of the phospholipidic membrane from the cell or its organelles, essential for cell survival (Ayala and Muñoz 2014). Hydroxyl radical ($HO^{\cdot}$) and hydroperoxyl ($HO^{\cdot}_2$) are the most dangerous ROS for lipids, and a single molecule of ROS can result in multiple peroxided PUFAs as they trigger a cyclic chain reaction that propagates itself very fast (Sharma et al. 2012). The overall process of lipid peroxidation consists of three steps: initiation, propagation and termination (Schneider 2009). During initiation, ROS react with methylene groups of PUFA forming lipid peroxy radicals and hydroperoxides (LOOH). These lipidic products formed are highly reactive and attack other lipids propagating the chain reaction at the propagation phase. After that reactions several reactive species including lipid alkoxyl radicals, aldehydes (malonyldialdehyde, among others), alkanes, lipid epoxides and alcohols are formed by the decomposition of lipid hydroperoxides (Davies 2000). In the termination phase, antioxidants such as vitamin E donate a hydrogen atom to the lipid peroxyl radical ($LOO^{\cdot}$) species forming tocopheroxyl radical that reacts with another $LOO^{\cdot}$ forming nonradical products (Ayala and Muñoz 2014).

Proteins can be affected directly or indirectly by ROS. Direct modifications include modification of its activity trough nitrosylation, carbonylation, disulphide bond formation and glutathionylation, while indirect effects include protein conjugation with lipid peroxidation products (Sharma et al. 2012). Protein carbonylation is defined as an irreversible post-transcriptional modification that yields a reactive carbonyl moiety in a protein, such as an aldehyde or ketone (Fedorova et al. 2014). This is the most common protein modification derived from the oxidation by a ROS, and an enhanced modification of proteins has been reported in plants under various stresses, therefore considering it a major hallmark of oxidative stress (Dalle-Donne et al. 2006). The accumulation of carbonylated proteins results in biomolecule malfunctions that can lead to cell death (Curtis et al. 2013).

Reactive oxygen species, specially $^{\cdot}OH$ and $^1O_2$ constitute the main source of DNA damage resulting in deoxyribose oxidation, strand breakage, removal of nucleotides and a variety of modifications in the organic bases of the nucleotides (Sharma et al. 2012). Despite the fact that ROS can inflict damage to nuclear, mitochondrial and chloroplast DNA, the two last are more susceptible to oxidative damage than nuclear DNA, due to the lack of protective protein, histones, and because they are very close to locations where ROS is produced (Manova and Gruszka 2015).

Among the different biomolecules damaged by ROS, some are better suited than others to become good stress markers. For lipid damage, the accumulation of lipid peroxides or the secondary product malondialdehyde constitute good markers of lipid peroxidation. Protein carbonylation is being also used as a good marker of oxidative stress (Levine et al. 1994). However, neither lipid peroxidation nor protein carbonylation are exclusively formed in chloroplasts and their use as markers of photo-oxidative stress should be interpreted carefully. Protein carbonyls are in turn more stable (in a scale of hours/days) than lipid peroxidation products, which are removed within minutes (Weber et al. 2015).

The accumulation of **lipid peroxides** (LOOH) are key indicators of the degree of lipid peroxidation, and constitute a good stress marker (Niki 2014). There are multiple approaches to measure the accumulation of LOOH. For plant samples, an easy method can be performed after a methanol extraction through **spectrophotometry** using again the FOX method, which measures the oxidation from ferrous to ferric ions by LOOH, in comparison with an extract where all LOOH are reduced by adding triphenylphospine (TPP). The ferric ions form a chromophore complex with the xylenol orange that absorbs at 540–600 nm (Bou et al. 2008).

Malondialdehyde (MDA) is one of the oxidation products derived from lipid peroxidation and usually measured by several studies assessing the degree of oxidative stress (see some examples in Table 9.2). The assay of thiobarbituric acid-reactive substances (TBARS) is the most used method to assess the breakdown products from lipid peroxidation, including MDA. The TBARS assay includes a liquid extraction with 80% ethanol and measure at 440, 532 and 600 nm with the **spectrophotometer** (Du and Bramlage 1992; Hodges et al. 1999) after an incubation with thiobarbituric acid. Higher precision can be obtained by **HPLC**, using a similar procedure (Iturbe-Ormaetxe et al. 1998; Munné-Bosch and Alegre 2002a).

3 Photoprotection and Photo-Oxidative Stress Markers in Invasion Biology Studies

The economical, demographic and technological development has allowed us to access to almost every biome causing some impacts, altering ecosystems functions. Moreover, globalization has led to the possibility that some species move along with humans, jumping off the geographical barriers that define the realized niche of each species, impacting on native ecosystems by changes on function and composition. Indeed, invasive species are considered the second major threat for the global biodiversity, after habitat loss (Simberloff et al. 2013). Invaders are supposed to have an increased vigor and/or an increased phenotypic plasticity underlying its ability to displace native species (Higgins and Richardson 2014). Therefore, invasive species may have increased physiological performance responding better to the environmental local conditions. In that way, the use of photoprotection and photo-oxidative stress markers may be helpful on invasive studies allowing a better comprehension of the boundaries of the physiological niche by understanding their stress tolerance and adaptation.

Photoprotection and photo-oxidative stress markers may be useful in invasion studies to understand the differences that may lead invasive species to outcompete natives through the description of their capacities under different environmental conditions and the affectation over native species. The comparison of invasive

Table 9.2 Compilation of the plant invasion studies using photoprotection and photo-oxidative stress markers during the last decade (since 2007)

Measurement	Methodology or type	References	Number of studies
Plant pigments	Spectro-photometry	Kim et al. (2008), Li et al. (2008), Liu et al. (2008), Mateos-Naranjo et al. (2008, 2010), Qaderi and Reid (2008), Qaderi et al. (2008), Zhang and Wen (2008), Feng (2008), Feng and Fu (2008), Funk (2008), Andrews et al. (2009), Hussner and Meyer (2009), Küpper et al. (2009), Yang et al. (2009), Feng et al. (2009), Zheng et al. (2012), Kaur et al. (2013), Funk et al. (2013), Castillo et al. (2014), Oliveira et al. (2014), Díaz-Barradas et al. (2015), Al Hassan et al. (2016), Huangfu et al. (2016), Lechuga-Lago et al. (2016), Lyu et al. (2016), Zhang et al. (2016), González-Teuber et al. (2017), Rotini et al. (2017), Souza-Alonso and González (2017), Varone et al. (2017), and Choi et al. (2017)	32
	HPLC	Cela et al. (2009), Song et al. (2010), Cela and Munné-Bosch (2012), Molina-Montenegro et al. (2012), Fleta-Soriano et al. (2015), Lassouane et al. (2016), Fenollosa et al. (2017), and Pintó-Marijuan et al. (2017)	8
	Spectro-radiometry	Ge et al. (2008), Spencer et al. (2008), Hestir et al. (2008), Funk and Zachary (2010), Naumann et al. (2010), Godoy et al. (2011), Roiloa et al. (2013, 2014, 2016), Wang et al. (2016), Yu et al. (2016), Heberling and Fridley (2016), and Roiloa and Retuerto (2016)	13
Photosynthetic efficiency	F_v/F_m only	Wang et al. (2008), Li et al. (2008), Liu et al. (2008), Bihmidine et al. (2009); Naumann et al. (2010); Funk and Zachary (2010), Redondo-Gómez et al. (2011), Immel et al. (2011), Waring and Maricle (2012), Roiloa et al. (2014, 2016), Madawala et al. (2014), Díaz-Barradas et al. (2015), Fleta-Soriano et al. (2015), Lechuga-Lago et al. (2016), Lyu et al. (2016), and Souza-Alonso and González (2017)	17
	F_v/F_m, NPQ, ΦPSII	Qaderi and Reid (2008), Richards et al. (2008), Zhang and Wen (2008), Funk (2008), Mateos-Naranjo et al. (2008. 2010), Cela et al. (2009), Wu et al. (2009), Yang et al. (2009), Song et al. (2010), Cela and Munné-Bosch (2012), Molina-Montenegro et al. (2012, 2016), Roiloa et al. (2013), Funk et al. (2013), Quinet et al. (2015), Li et al. (2015), Roiloa and Retuerto (2016), Lassouane et al. (2016), Pintó-Marijuan et al. (2017), Varone et al. (2017), Fenollosa et al. (2017), and Lukatkin et al. (2017)	23
ROS	H_2O_2	Kaur et al. (2013), Oliveira et al. (2014), and Mamik and Sharma (2017)	3

(continued)

Table 9.2 (continued)

Measurement	Methodology or type	References	Number of studies
Antioxidants	Enzymatic	Lu et al. (2007); Zhang and Wen (2008); Li et al. (2008); Immel et al. (2011), Redondo-Gómez et al. (2011), Morais et al. (2012), Huang et al. (2013), Kaur et al. (2013), Oliveira et al. (2014), Al Hassan et al. (2016), Zhang et al. (2016), and Mamik and Sharma (2017)	12
	Non-enzymatic	Cela et al. (2009), Cela and Munné-Bosch (2012), Huang et al. (2013), Fleta-Soriano et al. (2015), Al Hassan et al. (2016), and Pintó-Marijuan et al. (2017)	6
Oxidation products	MDA	Lu et al. (2007) ,(2008), Zhang and Wen 2008, Li et al. (2008), Immel et al. (2011), Falleh et al. (2012), Huang et al. (2013), Kaur et al. (2013), Oliveira et al. (2014), Quinet et al. (2015), Fleta-Soriano et al. (2015), Al Hassan et al. (2016), Molina-Montenegro et al. (2016), Zhang et al. (2016), Lassouane et al. (2016), and Mamik and Sharma (2017)	16

Measures of carotenoids are included on "Plant pigments" despite some of them have a role as antioxidants. Also, due to the high number of studies measuring the total amount of phenolic compounds, they have not been considered here inside the antioxidants group, although they are non-enzymatic antioxidants

species with coexistent natives had led to the conclusions that invaders have higher capacities to respond to stress or that they have broader physiological niches. Moreover, photo-oxidative stress markers can be helpful to predict plant responses to new environmental conditions, such as climate change. The direction of the community changes due to a new climatic framework can only be predicted with a complete ecophysiological approach. Finally, it is important to describe the extend of the differences between the genotypes from the invasive and the native ranges of one species. An in-depth understanding of these differences with the use of physiological descriptors (such as photo-oxidative stress markers) may undoubtedly help predict new invasions.

Although the interest to study the invasion process **using a complete ecophysiological approach** has increased recently, studies considering in-depth physiological processes are still limited (Pintó-Marijuan and Munné-Bosch 2013). However, photo-oxidative stress markers are being used more and more, and constitute indeed a promising tool for a better understanding of the invasion process. As we can see in Table 9.2, the most common photo-oxidative stress markers measured in plant invasion studies are photosynthetic pigments and chlorophyll *a* fluorescence parameters.

Considering the methodologies used, the most common measurements are photosynthetic pigments through spectrophotometry and the measure only of F_v/F_m. Indeed, only a few studies on the last decade include different measurements of photo-oxidative stress markers, which guarantee a complete understanding of the plant response. The most common combination is the measurement of photosynthetic pigments through spectrophotometry, the F_v/F_m, and the extend of lipid peroxidation through MDA analysis.

4 Some Limitations and Perspectives

The techniques to measure photo-oxidative stress markers present some common limitations. First, as Pintó-Marijuan and Munné-Bosch (2014) pointed out, it is very difficult to differentiate photo-oxidative damage caused by stress from that caused by **leaf senescence**. Senescence is the physiological deterioration with aging, and some of the hallmarks of senescence are chlorophyll loss and an increase of oxidative stress (Munné-Bosch and Alegre 2002b). Sampling fully-expanded young leaves throughout the experiment is the only way to separate stress- vs. senescence-related effects.

Another point to consider is the **localization** of the measured compound. For example, the measurement of ascorbic acid in leaves. A significant percentage of the ascorbic acid is normally found outside the chloroplast and an increase of this antioxidant can be a consequence of other processes rather than photo-oxidative stress. An easy (but time-consuming) way to ensure that we are measuring a photo-oxidative stress marker is to isolate chloroplasts.

The **matrix effect** must be cheeked every time we work with a new species or a known species under different conditions. Some of the protocols and the authors describing them indeed propose some alternatives to reduce matrix effects. One must keep in mind that some protocols are pH-dependent and the sample pH will depend on the species and its conditions.

It is essential to understand what **information** do we get from each photo-oxidative stress marker, and be aware of the fact that depending on the stress intensity we will see changes on different markers. Sometimes a **combined approach** with different photo-oxidative stress markers would be the most appropriate solution and the selection must follow the question we are trying to answer with our experiment.

As said before, there is a need to use combined stress markers. Photo-oxidative stress is a final consequence of the imbalance of different processes. It is not until antioxidant systems have been taken down that we can measure an accumulation of reactive oxygen species or oxidation products. Therefore, combined markers provide complementary information about the stress response. Here we propose some tips to perform a multiple approach to understand the global plant response.

If we focus on the obtained information, it would be ideal to take one photo-oxidative stress marker from the following groups: plant pigments, photosynthetic efficiency, reactive oxygen species, antioxidants and oxidation products. However, that represents multiple assays and a large amount of samples. If a faster and efficient protocol is needed, it is possible to connect different protocols. For instance, the extraction of plant pigments (chlorophylls, carotenoids and anthocyanins), tocopherols and lipid hydroperoxides have a common start, and all molecules can be extracted with methanol. Thus, it is possible to save time by performing a common extraction. If the limitation is the economy we can use the cheapest techniques, such as chlorophyll *a* fluorescence and spectroradiometric indexes, such as NDVI or PRI, which with the appropriate models can estimate some photo-oxidative stress markers (Table 9.1). The same techniques are useful if we have a high-scale experimental design (e.g. for large-scale phenotyping).

5 General Conclusions

Photoprotection and photo-oxidative stress are central elements of plant responses to a variety of stresses. Markers based on photoprotection and photo-oxidative stress may be extremely useful for understanding plant acclimation, and constitute a promising tool for the study of invasion success. Working with photoprotection and photo-oxidative stress markers requires the understanding of the meaning of every specific marker within the whole framework of the photoprotective mechanisms. As discussed in this chapter, a combined approach is required to better understand the ecophysiology of invasive vs. native species, using several markers providing complementary information. Here, we have provided some essential tools for a correct choosing of the fittest photoprotection and photo-oxidative stress markers, encouraging its use on invasion studies to help unravelling invaders success.

References

Al Hassan M, Chaura J, López-Gresa MP, Borsai O, Daniso E, Donat-Torres MP, Mayoral O, Vicente O, Boscaiu M (2016) Native-invasive plants vs. halophytes in mediterranean salt marshes: stress tolerance mechanisms in two related species. Front Plant Sci 7:1–21

Amaral JS, Casal S, Torres D, Seabra RM, Oliveira BPP (2005) Simultaneous determination of tocopherols and tocotrienols in hazelnuts by a normal phase liquid chromatographic method. Anal Sci 21:1545–1548

Andrews M, Hamish GM, Hodge S, Cherrill A, Raven JA (2009) Seed dormancy, nitrogen nutrition and shade acclimation of *Impatiens glandulifera*: implications for successful invasion of deciduous woodland. Plant Ecol Divers 2:145–153

Apel K, Hirt H (2004) Reactive oxygen species: metabolism, oxidative stress, and signal transduction. Annu Rev Plant Biol 55:373–399

Asada K (2006) Production and scavenging of reactive oxygen species in chloroplasts and their functions. Plant Physiol 141:391–396

Ayala A, Muñoz MF (2014) Lipid peroxidation: production, metabolism, and signaling mechanisms of malondialdehyde and 4-hydroxy-2-nonenal. Oxidative Med Cell Longev 2014:360438

Bihmidine S, Bryan NM, Payne KR, Parde MR, Okalebo JA, Cooperstein SE, Awada T (2009) Photosynthetic performance of invasive *Pinus ponderosa* and *Juniperus virginiana* seedlings under gradual soil water depletion. Plant Biol 12:668–675

Bontempo e Silva EA, Hasegawa SF, Ono K, Sumida A, Uemura S, Hara T (2012) Differential photosynthetic characteristics between seedlings and saplings of *Abies sachalinensis* and *Picea glehnii*, in the field. Ecol Res 27:933–943

Bou R, Codony R, Tres A, Decker EA, Guardiola F (2008) Determination of hydroperoxides in foods and biological samples by the ferrous oxidation-xylenol orange method: a review of the factors that influence the method's performance. Anal Biochem 377:1–15

Castañeda-Ovando A, Pacheco-Hernández MDL, Páez-Hernández ME, Rodríguez JA, Galán-Vidal CA (2009) Chemical studies of anthocyanins: a review. Food Chem 113:859–871

Castillo JM, Grewell BJ, Pickart A, Bortolus A, Peña C, Figueroa E, Sytsma M (2014) Phenotypic plasticity of invasive *Spartina densiflora* (Poaceae) along a broad latitudinal gradient on the pacific coast of North America. Am J Bot 101:448–458

Cela J, Munné-Bosch S (2012) Acclimation to high salinity in the invasive CAM plant *Aptenia cordifolia*. Plant Ecol Divers 53:403–410

Cela J, Arrom L, Munné-Bosch S (2009) Diurnal changes in photosystem II photochemistry, photoprotective compounds and stress-related phytohormones in the CAM plant, *Aptenia cordifolia*. Plant Sci 177:404–410

Cheeseman JM (2006) Hydrogen peroxide concentrations in leaves under natural conditions. J Exp Bot 57:2435–2444

Choi D, Watanabe Y, Guy RD, Sugai T, Toda H, Koike T (2017) Photosynthetic characteristics and nitrogen allocation in the black locust (*Robinia pseudoacacia* L.) grown in a FACE system. Acta Physiol Plant 39:1–12

Croce R, Van Amerongen H (2011) Light-harvesting and structural organization of Photosystem II: from individual complexes to thylakoid membrane. J Photoch Photobio B 104:142–153

Curtis JM, Hahn WS, Long EK, Burrill JS, Arriaga EA, Bernlohr DA (2013) Protein carbonylation and metabolic control systems. Trends Endocrin Met 23:399–406

Dall'Osto L, Lico C, Alric J, Giuliano G, Havaux M, Bassi R (2006) Lutein is needed for efficient chlorophyll triplet quenching in the major LHCII antenna complex of higher plants and effective photoprotection in vivo under strong light. BMC Plant Biol 6:32

Dalle-Donne I, Aldini G, Carini M, Colombo R, Rossi R, Milzani A (2006) Protein carbonylation, cellular dysfunction, and disease progression. J Cell Mol Med 10:389–406

Dandena A, Leimane I, Kletnieks U (2011) Validation of monomeric anthocianin determination method for bilberry juice and marc extracts. J Life Sci 6:1378–1382

Davies KJ (2000) Oxidative stress, antioxidant defenses, and damage removal, repair, and replacement systems. IUBMB Life 50:279–289

Demmig B, Björkman O (1987) Comparison of the effect of excessive light on chlorophyll fluorescence (77K) and photon yield of O_2 evolution in leaves of higher plants. Planta 171:171–184

Demmig-Adams B, Adams W (1996) The role of xanthophyll cycle carotenoids in the protection of photosynthesis. Trends Plant Sci 1:21–26

Demmig-Adams B, Cohu CM, Muller O, Adams WW (2012) Modulation of photosynthetic energy conversion efficiency in nature: from seconds to seasons. Photosynth Res 113:75–88

Demmig-Adams B, Stewart JJ, Iii WWA (2017) Environmental regulation of intrinsic photosynthetic capacity: an integrated view. Curr Opin Plant Biol 37:34–41

Díaz-Barradas MC, Zunzunegui M, Álvarez-Cansino L, Esquivias MP, Collantes MB, Cipriotti PA (2015) Species-specific effects of the invasive *Hieracium pilosella* in Magellanic steppe grasslands are driven by nitrogen cycle changes. Plant Soil 397:175–187

Du Z, Bramlage WJ (1992) Modified thiobarbituric acid assay for measuring lipid oxidation in sugar-rich plant tissue extracts. J Agric Food Chem 40:1566–1570

Esteban R, Barrutia O, Artetxe U, Fernández-Marín B, Hernández A, García-Plazaola JI (2015) Internal and external factors affecting photosynthetic pigment composition in plants: a meta-analytical approach. New Phytol 206:268–280

Falleh H, Jalleli I, Ksouri R, Boulaaba M, Guyot S, Magné C (2012) Effect of salt treatment on phenolic compounds and antioxidant activity of two *Mesembryanthemum edule* provenances. Plant Physiol Biochem 52:1–8

Fedorova M, Bollineni RC, Hoffmann R (2014) Protein carbonylation as a major hallmark of oxidative damage: update of analytical strategies. Mass Spectrom Rev 33:79–97

Feng YL (2008) Photosynthesis, nitrogen allocation and specific leaf area in invasive *Eupatorium adenophorum* and native *Eupatorium japonicum* grown at different irradiances. Physiol Plantarum 133:318–326

Feng YL, Fu GL (2008) Nitrogen allocation, partitioning and use efficiency in three invasive plant species in comparison with their native congeners. Biol Invasions 10:891–902

Feng Y-L, Lei Y-B, Wang R-F, Callaway RM, Valiente-Banuet A, Inderjit LY-P, Zheng Y-L (2009) Evolutionary tradeoffs for nitrogen allocation to photosynthesis versus cell walls in an invasive plant. Proc Natl Acad Sci U S A 106:1853–1856

Fenollosa E, Munné-Bosch S, Pintó-Marijuan M (2017) Contrasting phenotypic plasticity in the photoprotective strategies of the invasive species *Carpobrotus edulis* and the coexisting native species *Crithmum maritimum*. Physiol Plantarum 160:185–200

Fleta-Soriano E, Pintó-Marijuan M, Munné-Bosch S (2015) Evidence of drought stress memory in the facultative cam, *Aptenia cordifolia*: possible role of phytohormones. PLoS One 10:e0135391

Foyer CH, Noctor G (2003) Redox sensing and signalling associated with reactive oxygen in chloroplasts, peroxisomes and mitochondria. Physiol Plantarum 119:355–364

Fuleki T, Francis FJ (1968) Quantitative methods for anthocyanins. J Food Sci 33:72–77

Funk JL (2008) Differences in plasticity between invasive and native plants from a low resource environment. J Ecol 96:1162–1173

Funk JL, Zachary ÆVA (2010) Physiological responses to short-term water and light stress in native and invasive plant species in southern California. Biol Invasions 12:1685–1694

Funk JL, Glenwinkel LA, Sack L (2013) Differential allocation to photosynthetic and non-photosynthetic nitrogen fractions among native and invasive species. PLoS One 8:e64502

Garbulsky MF, Peñuelas J, Gamon J, Inoue Y, Filella I (2011) The photochemical reflectance index (PRI) and the remote sensing of leaf, canopy and ecosystem radiation use efficiencies. A review and meta-analysis. Remote Sens Environ 115:281–297

Ge S, Carruthers RI, Spencer DF, Yu Q (2008) Canopy assessment of biochemical features by ground-based hyperspectral data for an invasive species, giant reed (*Arundo donax*). Environ Monit Assess 147:271–278

Giusti MM, Wrolstad RE (2001) Characterization and measurement of anthocyanins by UV visible spectroscopy. CPFAC F1.2.1-F1.2.13

Godoy O, Saldaña A, Fuentes N, Valladares F, Gianoli E (2011) Forests are not immune to plant invasions: phenotypic plasticity and local adaptation allow *Prunella vulgaris* to colonize a temperate evergreen rainforest. Biol Invasions 13:1615–1625

Goh CH, Ko SM, Koh S, Kim YJ, Bae HJ (2012) Photosynthesis and environments: photoinhibition and repair mechanisms in plants. J Plant Biol 55:93–101

González-Teuber M, Quiroz CL, Concha-Bloomfield I, Cavieres LA (2017) Enhanced fitness and greater herbivore resistance: implications for dandelion invasion in an alpine habitat. Biol Invasions 19:647–653

Havaux M, Triantaphylides C (2009) Singlet oxygen in plants: production, detoxification and signaling. Trends Plant Sci 14:219–228

Heberling JM, Fridley JD (2016) Invaders do not require high resource levels to maintain physiological advantages in a temperate deciduous forest. Ecology 97:874–884

Hestir EL, Khanna S, Andrew ME, Santos MJ, Viers JH, Greenberg JA, Rajapakse SS, Ustin SL (2008) Identification of invasive vegetation using hyperspectral remote sensing in the California Delta ecosystem. Remote Sens Environ 112:4034–4047

Higgins SI, Richardson DM (2014) Invasive plants have broader physiological niches. Proc Natl Acad Sci U S A 111:10610–10614

Hodges DM, DeLong JM, Forney CF, Prange RK (1999) Improving the thiobarbituric acid-reactive-substances assay for estimating lipid peroxidation in plant tissues containing anthocyanin and other interfering compounds. Planta 207:604–611

Huang Y, Bai Y, Wang Y, Kong H (2013) Allelopathic effects of the extracts from an invasive species *Solidago canadensis* L. on *Microcystis aeruginosa*. Lett Appl Microbiol 57:451–458

Huangfu C, Li H, Chen X, Liu H, Wang H, Yang D (2016) Response of an invasive plant, *Flaveria bidentis*, to nitrogen addition: a test of form-preference uptake. Biol Invasions 18:3365–3380

Hussner A, Meyer C (2009) The influence of water level on the growth and photosynthesis of *Hydrocotyle ranunculoides* L.fil. Flora 204:755–761

Immel F, Renaut J, Masfaraud J (2011) Physiological response and differential leaf proteome pattern in the European invasive Asteraceae *Solidago canadensis* colonizing a former cokery soil. J Proteome 75:1129–1143

Iturbe-Ormaetxe I, Escuredo PR, Arrese-Igor C, Becana M (1998) Oxidative damage in pea plants exposed to water deficit or paraquat. Plant Physiol 116:173–181

Jahns P, Holzwarth AR (2012) The role of the xanthophyll cycle and of lutein in photoprotection of photosystem II. BBA-Bioenergetics 1817:182–193

Jansena MAK, Gaba V, Greenberg BM (1998) Higher plants and UV-B radiation: balancing damage, repair and acclimation. Trends Plant Sci 3:131–135

Kalaji HM, Schansker G, Ladle RJ, Goltsev V, Bosa K, Allakhverdiev SI, Brestic M et al (2014) Frequently asked questions about in vivo chlorophyll fluorescence: practical issues. Photosynth Res 122:121–158

Kaur T, Bhat HA, Raina A, Koul S, Vyas D (2013) Glutathione regulates enzymatic antioxidant defence with differential thiol content in perennial pepperweed and helps adapting to extreme environment. Acta Physiol Plant 35:2501–2511

Kim YO, Rodriguez RJ, Lee EJ, Redman RS (2008) *Phytolacca americana* from contaminated and noncontaminated soils of South Korea: effects of elevated temperature, CO_2 and simulated acid rain on plant growth response. J Chem Ecol 34:1501–1509

Küpper H, Götz B, Mijovilovich A, Küpper FC, Meyer-Klaucke W (2009) Complexation and toxicity of copper in higher plants: characterization of copper accumulation, speciation, and toxicity in *Crassula helmsii* as a new copper accumulator. Plant Physiol 151:702–714

Lambrev PH, Miloslavina Y, Jahns P, Holzwarth AR (2012) On the relationship between non-photochemical quenching and photoprotection of Photosystem II. Biochim Biophys Acta 1817:760–769

Lao F, Giusti MM (2016) quantification of purple corn (*Zea mays* L.) anthocyanins using spectrophotometric and HPLC approaches: method comparison and correlation. Food Anal Methods 9:1367–1380

Lassouane N, Aïd F, Lutts S (2016) Drought inhibits early seedling establishment of *Parkinsonia aculeata* L. under low light intensity: a physiological approach. Plant Growth Regul 80:115–126

Lechuga-Lago Y, Sixto-Ruiz M, Roiloa SR, González L (2016) Clonal integration facilitates the colonization of drought environments by plant invaders. AoB Plants 8:plw023

Lee J, Durst RW, Wrolstad RE (2005) Determination of total monomeric anthocyanin pigment content of fruit juices, beverages, natural colorants, and wines by the pH differential method: collaborative study. J AOAC Int 88:1269–1278

Levine RL, Williams JA, Stadtman EP, Shacter E (1994) Carbonyl assays for determination of oxidatively modified proteins. Method Enzymol 233:346–357

Li H, Qiang S, Qian Y (2008) Physiological response of different croftonweed (*Eupatorium adenophorum*) populations to low temperature. Weed Sci 56:196–202

Li W, Luo J, Tian X, Soon Chow W, Sun Z, Zhang T, Peng S, Peng C (2015) A new strategy for controlling invasive weeds: selecting valuable native plants to defeat them. Sci Rep 5:11004

Lichtenthaler HK (1987) Chlorophyll fluorescence signatures of leaves during the autumnal chlorophyll breakdown. Plant Phys 131:101–110

Liu J, He WM, Zhang SM, Liu FH, Dong M, Wang RQ (2008) Effects of clonal integration on photosynthesis of the invasive clonal plant *Alternanthera philoxeroides*. Photosynthetica 46:299–302

Logan BA, Adams WW, Demmig-Adams B (2007) Avoiding common pitfalls of chlorophyll fluorescence analysis under field conditions. Funct Plant Biol 34:853–859

Lu P, Sang WG, Ma KP (2007) Activity of stress-related antioxidative enzymes in the invasive plant crofton weed (*Eupatorium adenophorum*). J Integr Plant Biol 49:1555–1564

Lu P, Sang WG, Ma KP (2008) Differential responses of the activities of antioxidant enzymes to thermal stresses between two invasive *Eupatorium* species in China. J Integr Plant Biol 50:393–401

Lukatkin AS, Tyutyaev EV, Sharkaeva ES, Lukatkin AA, Teixeira da Silva JA (2017) Mild abiotic stresses have different effects on chlorophyll fluorescence parameters in leaves of young woody and herbaceous invasive plants. Acta Physiol Plant 39:1–7

Lyu XQ, Zhang YL, You WH (2016) Growth and physiological responses of *Eichhornia crassipes* to clonal integration under experimental defoliation. Aquat Ecol 50:153–162

Madawala S, Hartley S, Gould KS (2014) Comparative growth and photosynthetic responses of native and adventive iceplant taxa to salinity stress. New Zeal J Bot 52:37–41

Mamik S, Sharma AD (2017) Analysis of boiling stable antioxidant enzymes in invasive alien species of *Lantana* under abiotic stress-like conditions. Rev Bras Bot 37:129–141

Manova V, Gruszka D (2015) DNA damage and repair in plants – from models to crops. Front Plant Sci 6:1–26

Mateos-Naranjo E, Redondo-Gómez S, Cambrollé J, Luque T, Figueroa ME (2008) Growth and photosynthetic responses to zinc stress of an invasive cordgrass, *Spartina densiflora*. Plant Biol 10:754–762

Mateos-Naranjo E, Redondo-Gómez S, Álvarez R, Cambrollé J, Gandullo J, Figueroa ME (2010) Synergic effect of salinity and CO_2 enrichment on growth and photosynthetic responses of the invasive cordgrass *Spartina densiflora*. J Exp Bot 61:1643–1654

Maxwell K, Johnson GN (2000) Chlorophyll fluorescence – a practical guide. J Exp Bot 51:659–668

Molina-Montenegro MA, Peñuelas J, Munné-Bosch S, Sardans J (2012) Higher plasticity in ecophysiological traits enhances the performance and invasion success of *Taraxacum officinale* (dandelion) in alpine environments. Biol Invasions 14:21–33

Molina-Montenegro MA, Galleguillos C, Oses R, Acuña-Rodríguez IS, Lavín P, Gallardo-Cerda J, Torres-Díaz C, Diez B, Pizarro GE, Atala C (2016) Adaptive phenotypic plasticity and competitive ability deployed under a climate change scenario may promote the invasion of *Poa annua* in Antarctica. Biol Invasions 18:603–618

Morais MC, Panuccio MR, Muscolo A, Freitas H (2012) Salt tolerance traits increase the invasive success of *Acacia longifolia* in Portuguese coastal dunes. Plant Physiol Biochem 55:60–65

Munné-Bosch S (2005) The role of alpha-tocopherol in plant stress tolerance. Plant Physiol 162:743–748

Munné-Bosch S, Alegre L (2000) Changes in carotenoids, tocopherols and diterpenes during drought and recovery, and the biological significance of chlorophyll loss in *Rosmarinus officinalis* plants. Planta 210:925–931

Munné-Bosch S, Alegre L (2002a) Interplay between ascorbic acid and lipophilic antioxidant defences in chloroplasts of water-stressed Arabidopsis plants. FEBS Lett 524:145–148

Munné-Bosch S, Alegre L (2002b) Plant aging increases oxidative stress in chloroplasts. Planta 214:608–615

Munné-Bosch S, Queval G, Foyer CH (2013) The impact of global change factors on redox signalling underpinning stress tolerance. Plant Physiol 161:5–19

Murchie EH, Lawson T (2013) Chlorophyll fluorescence analysis: a guide to good practice and understanding some new applications. J Exp Bot 64:3983–3998

Naumann JC, Bissett SN, Anderson JE (2010) Diurnal patterns of photosynthesis, chlorophyll fluorescence, and PRI to evaluate water stress in the invasive species, *Elaeagnus umbellata* Thunb. Trees 24:237–245

Niki E (2014) Biomarkers of lipid peroxidation in clinical material. Biochim Biophys Acta 1840:809–817

Noctor G, Mhamdi A, Foyer CH (2016) Oxidative stress and antioxidative systems: recipes for successful data collection and interpretation. Plant Cell Environ 39:1140–1160

Oliveira MT, Matzek V, Medeiros CD, Rivas R, Falcao HM, Santos MG (2014) Stress tolerance and ecophysiological ability of an invader and a native species in a seasonally dry tropical forest. PLoS One 9:e105514

Pintó-Marijuan M, Munné-Bosch S (2013) Ecophysiology of invasive plants: osmotic adjustment and antioxidants. Trends Plant Sci 18:660–666

Pintó-Marijuan M, Munné-Bosch S (2014) Photo-oxidative stress markers as a measure of abiotic stress-induced leaf senescence: advantages and limitations. J Exp Bot 65:3845–3857

Pintó-Marijuan M, Cotado A, Fleta-Soriano E, Munné-Bosch S (2017) Drought stress memory in the photosynthetic mechanisms of an invasive CAM species, *Aptenia cordifolia*. Photosynth Res 131:241–253

Pospísil P, Yamamoto Y (2017) Damage to photosystem II by lipid peroxidation products. Biochim Biophys Acta 1861:457–466

Qaderi M, Reid D (2008) Combined effects of temperature and carbon dioxide on plant growth and subsequent seed germinability of *Silene noctiflora*. Int J Plant Sci 169:1200–1209

Qaderi MM, Yeung EC, Reid DM (2008) Growth and physiological responses of an invasive alien species, *Silene noctiflora*, during two developmental stages to four levels of ultraviolet-B radiation. Ecoscience 15:150–159

Queval G, Noctor G (2007) A plate reader method for the measurement of NAD, NADP, glutathione, and ascorbate in tissue extracts: application to redox profiling during Arabidopsis rosette development. Anal Biochem 363:58–69

Queval G, Hager J, Gakière B, Noctor G (2008) Why are literature data for H_2O_2 contents so variable? A discussion of potential difficulties in the quantitative assay of leaf extracts. J Exp Bot 59:135–146

Quinet M, Descamps C, Coster Q, Lutts S, Jacquemart A-L (2015) Tolerance to water stress and shade in the invasive *Impatiens parviflora*. Int J Plant Sci 176:848–858

Ramel F, Birtic S, Cuiné S, Triantaphylidès C, Ravanat J-L, Havaux M (2012) Chemical quenching of singlet oxygen by carotenoids in plants. Plant Physiol 158:1267–1278

Redondo-Gómez S, Andrades-Moreno L, Mateos-Naranjo E, Parra R, Valera-Burgos J, Aroca R (2011) Synergic effect of salinity and zinc stress on growth and photosynthetic responses of the cordgrass, *Spartina densiflora*. J Exp Bot 62:5521–5530

Richards CL, Walls RL, Bailey JP, Parameswaran R, George T, Pigliucci M (2008) Plasticity in salt tolerance traits allows for invasion of novel habitat by Japanese knotweed s l (*Fallopian japonica* and *F. bohemica*, Polygonaceae). Am J Bot 95:931–942

Richardson AD, Duigan SP, Berlyn GP, Richardson AD (2002) An evaluation of noninvasive methods to estimate foliar chlorophyll content. New Phytol 1:185–194

Roiloa SR, Retuerto R (2016) Effects of fragmentation and seawater submergence on photochemical efficiency and growth in the clonal invader *Carpobrotus edulis*. Flora 225:45–51

Roiloa SR, Rodrı S, Rube HF (2013) Developmentally-programmed division of labour in the clonal invader *Carpobrotus edulis*. Biol Invasions 9:1895–1905

Roiloa S, Rodiguez-Echeverria S, Lopez-Otero A, Retuerto R, Freitas H (2014) Adaptative plasticity to heterogeneous environments increases capacity for division of labor in the clonal invader *Carpobrotus edulis* (Aizoaceae). Am J Bot 101:1301–1308

Roiloa SR, Retuerto R, Campoy JG, Novoa A, Barreiro R (2016) division of labor brings greater benefits to clones of *Carpobrotus edulis* in the non-native range: evidence for rapid adaptive evolution. Front Plant Sci 7:1–13

Rotini A, Mejia AY, Costa R, Migliore L, Winters G (2017) Ecophysiological plasticity and bacteriome shift in the seagrass *Halophila stipulacea* along a depth gradient in the northern red sea. Front Plant Sci 7:1–12

Schneider C (2009) An update on products and mechanisms of lipid peroxidation. Mol Nutr Food Res 53:315–321

Sharma P, Jha AB, Dubey RS, Pessarakli M (2012) Reactive oxygen species, oxidative damage, and antioxidative defense mechanism in plants under stressful conditions. J Bot 2012:1–26

Siegelman HW, Hendricks SB (1958) Photocontrol of alcohol, aldehyde, and anthocyanin production in apple skin. Plant Physiol 6:409–413

Simberloff D, Martin J-L, Genovesi P, Maris V, Wardle DA, Aronson J, Courchamp F, Galil B, García-Berthou E, Pascal M, Pyšek P, Sousa R, Tabacchi E, Vilà M (2013) Impacts of biological invasions: what's what and the way forward. Trends Ecol Evol 28:58–66

Song L, Chow WS, Sun L, Li C, Peng C (2010) Acclimation of photosystem II to high temperature in two *Wedelia* species from different geographical origins: implications for biological invasions upon global warming. J Exp Bot 61:4087–4096

Souza-Alonso P, González L (2017) Don't leave me behind: viability of vegetative propagules of the clonal invasive *Carpobrotus edulis* and implications for plant management. Biol Invasions 19:2171–2183

Spencer DF, Tan W, Liow P-S, Ksander GG, Whitehand LC (2008) Evaluation of a late summer imazapyr treatment for managing giant reed (*Arundo donax*). J Aquat Plant Manag 47:40–43

Takahashi S, Badger MR (2011) Photoprotection in plants: a new light on photosystem II damage. Trends Plant Sci 16:53–60

Takahashi S, Milward SE, Yamori W, Evans JR, Hillier W, Badger MR (2010) The solar action spectrum of photosystem. Plant Physiol 153:988–993

Thayer SS, Björkman O (1990) Leaf xanthophyll content and composition in sun and shade determined by HPLC. Photosynth Res 23:331–343

Varone L, Catoni R, Bonito A, Gini E, Gratani L (2017) Photochemical performance of *Carpobrotus edulis* in response to various substrate salt concentrations. S Afr J Bot 111:258–266

Walters RG (2005) Towards an understanding of photosynthetic acclimation. J Exp Bot 56:435–447

Wang N, Yu FH, Li PX, He WM, Liu FH, Liu JM, Dong M (2008) Clonal integration affects growth, photosynthetic efficiency and biomass allocation, but not the competitive ability, of the alien invasive *Alternanthera philoxeroides* under severe stress. Ann Bot 101:671–678

Wang C, Liu J, Xiao H, Zhou J (2016) Differences in leaf functional traits between *Rhus typhina* and native species. Clean Air 44:1591–1597

Waring EF, Maricle BR (2012) Photosynthetic variation and carbon isotope discrimination in invasive wetland grasses in response to flooding. Environ Exp Bot 77:77–86

Weber D, Davies MJ, Grune T (2015) Determination of protein carbonyls in plasma, cell extracts, tissue homogenates, isolated proteins: focus on sample preparation and derivatization conditions. Redox Biol 5:367–380

Winkel-Shirley B (2002) Biosynthesis of flavonoids and effects of stress. Curr Opin Plant Biol 5:218–223

Wu Y, Liu C, Li P, Wang J, Xing D, Wang B (2009) Photosynthetic characteristics involved in adaptability to Karst soil and alien invasion of paper mulberry (*Broussonetia papyrifera* (L.) Vent.) in comparison with mulberry (*Morus alba* L.). Photosynthetica 47:155–160

Yang L, Liu N, Ren H, Wang J (2009) Facilitation by two exotic Acacia: *Acacia auriculiformis* and *Acacia mangium* as nurse plants in South China. For Ecol Manag 257:1786–1793

Young AJ, Phillip D, Savill J (1997) Carotenoids in higer plant photosynthesis. In: Pessarakli M (ed) Handbook of photosynthesis, 2nd edn. Marcel Dekker, Tucson, p 1027

Yu HW, Yang JX, Gao Y, He WM (2016) Soil organic nitrogen endows invasive *Solidago canadensis* with greater advantages in low-phosphorus conditions. Ecosphere 7:1–10

Zhang LL, Wen DZ (2008) Photosynthesis, chlorophyll fluorescence, and antioxidant enzyme responses of invasive weed *Mikania micrantha* to *Bemisia tabaci* infestation. Photosynthetica 46:457–462

Zhang KM, Shen Y, Fang YM, Liu Y (2016) Changes in gametophyte physiology of *Pteris multifida* induced by the leaf leachate treatment of the invasive *Bidens pilosa*. Environ Sci Pollut R 23:3578–3585

Zheng YL, Feng YL, Lei YB, Liao ZY (2012) Comparisons of plastic responses to irradiance and physiological traits by invasive *Eupatorium adenophorum* and its native congeners. Plant Physiol 169:884–891

Zhou M, Diwu Z, Panchuk-Voloshina N, Haugland RP (1997) A stable nonfluorescent derivative of resorufin for the fluorometric determination of trace hydrogen peroxide: applications in detecting the activity of phagocyte NADPH oxidase and other oxidases. Anal Biochem 253:162–168

Zimmermann P, Zentgraf U (2005) The correlation between oxidative stress and senescence during plant development. Cell Mol Biol 10:515–534

Chapter 10
Reactive Oxygen Species and Antioxidant Enzymatic Systems in Plants: Role and Methods

Teresa Papalia, Maria Rosaria Panuccio, Maria Sidari, and Adele Muscolo

1 Introduction

Reactive oxygen species (ROS) were initially recognized as toxic by-products of aerobic metabolism. In recent years, it has become apparent that ROS play an important signaling role in plants, controlling processes such as growth, development and especially response to biotic and abiotic environmental stimuli. The major members of the ROS family include free radicals like $O_2^{\bullet-}$, $OH^{\bullet}$ and non-radicals like H_2O_2 and O_2. Organelles with a highly oxidizing metabolic activity or with an intense rate of electron flow, such as chloroplast, mitochondrion and peroxisome, are the major source of ROS production in plants (Foyer and Shigeoka 2011). Along with these organelles, peroxidases present in cell walls and NADPH oxidase located in the plasma membrane are also enzymatic source of ROS. Plant NADPH oxidases, also known as respiratory burst oxidase homologs (RBOHs), have cytosolic FAD- and NADPH-binding domains in the C-terminal region, and transmembrane domains that correspond to those in mammalian NADPH oxidases (Suzuki et al. 2011). In addition, plant RBOHs have a cytosolic N-terminal extension, contain regulatory regions such as calcium-binding EF-hands and phosphorylation target sites that are important for the function and regulation of the plant NADPH oxidases (Oda et al. 2010; Suzuki et al. 2011). Increasing evidence demonstrated NADPH oxidases as key signaling nodes in the ROS regulation network of plants, integrating numerous signal transduction pathways with ROS signaling and mediating multiple important biological processes, including cell growth and plant development, abiotic stress response and adaptation, plant–microbe pathogenic and symbiotic interactions (Torres and Dangl 2005; Suzuki et al. 2011; Marino et al. 2012). The increased production of ROS during stress conditions acts as signal for the activation of stress response in terms of efficient enzymatic and non-enzymatic

T. Papalia · M. R. Panuccio · M. Sidari · A. Muscolo (✉)
Dipartimento di Agraria, Università Mediterranea Feo di Vito, Reggio Calabria, Italy
e-mail: amuscolo@unirc.it

© Springer International Publishing AG, part of Springer Nature 2018

A. M. Sánchez-Moreiras, M. J. Reigosa (eds.), *Advances in Plant Ecophysiology Techniques*, https://doi.org/10.1007/978-3-319-93233-0_10

antioxidant pathways (Baxter et al. 2014). As option to this antioxidant system, plants produce alternative oxidases (AOX) that are able to prevent the excess generation of ROS in the electron transport chains of mitochondria (Maxwell et al. 1999). By diverting electrons flowing through electron-transport chains, AOX can decrease the possibility of electron leaking to O_2 to generate $O_2^{\bullet-}$. Other mechanisms, such as leaf movement and curling, or photosynthetic apparatus rearranging, may also represent an attempt to avoid the over-reduction of ROS by balancing the amount of energy absorbed by the plant with the availability of CO_2 (Mittler 2002). Our attention is mainly focused on the different ROS scavenging mechanism mediated by enzymes, highlighting the role of superoxide dismutase (SOD), ascorbate peroxidase (APX), catalase (CAT), glutathione peroxidase (GPX), monodehydroascorbate reductase (MDHAR), dehydroascorbate reductase (DHAR), glutathione reductase (GR), glutathione S-transferase (GST), guaiacol peroxidase (POX) and peroxiredoxin (PRX) in stress tolerance. These antioxidant enzymes are located in different sites of plant cells where ROS are generally produced under both normal and stressful conditions (chloroplasts, mitochondria, peroxisomes, plasma membranes, ER and the cell wall), and work together to detoxify ROS (Fig. 10.1).

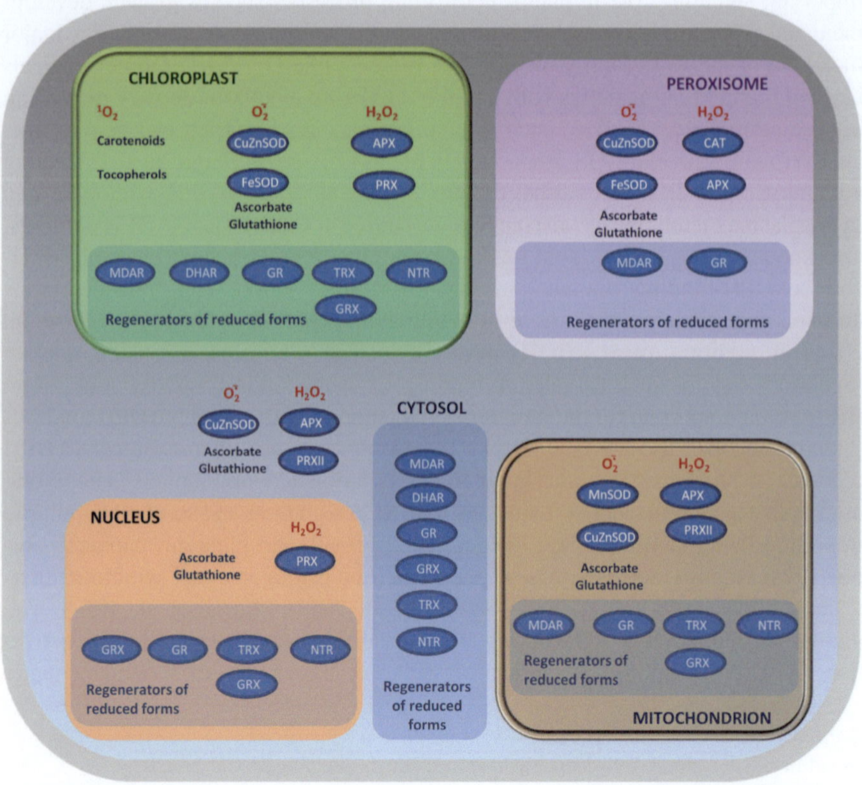

Fig. 10.1 Distribution of the main antioxidant enzymes in plant cells (Source: Noctor et al. 2017)

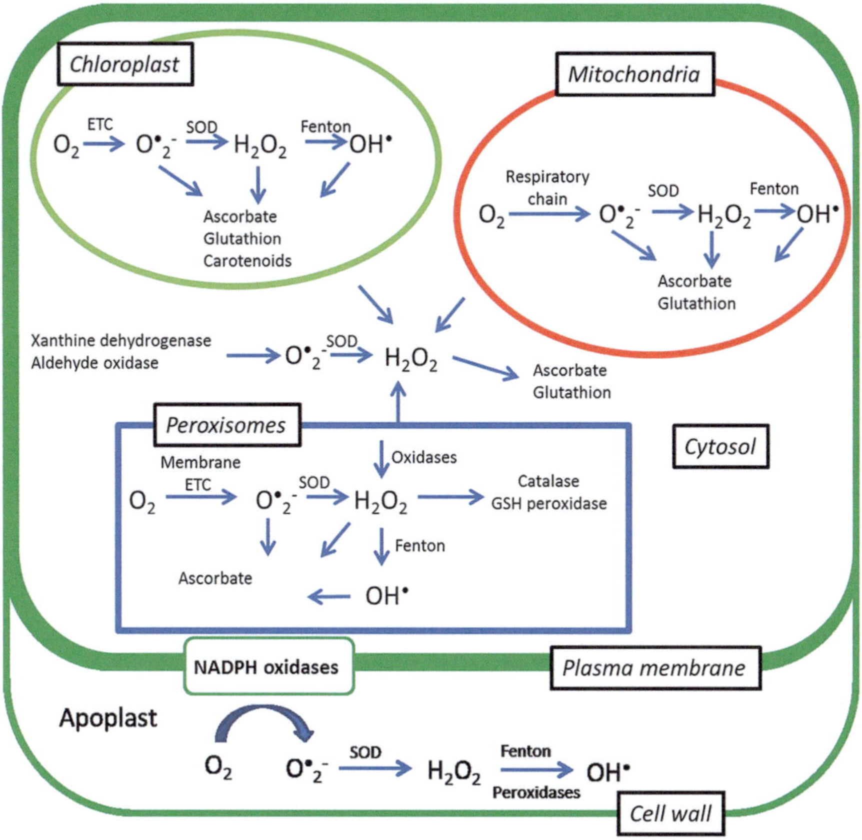

Fig. 10.2 Superoxide dismutase (SOD) role (Source: Jajic et al. 2015)

SOD is not only the first line of defense but also the only enzyme capable of converting $O_2^{\bullet-}$ into H_2O_2 and decreasing the risk of the formation of the hydroxyl radical ($HO^{\bullet}$), which is among the most reactive oxygen species (Kehrer 2000). Therefore SOD holds a key position within the antioxidant network (Fig. 10.2).

CAT, APX, and GPX detoxify H_2O_2 in different cellular sites. H_2O_2, being moderately reactive, does not cause extensive damage and at low concentrations acts as regulatory signal for essential physiological processes, cell cycle, growth and development (Das and Roychoudhury 2014).

Catalase is the unique antioxidant enzyme not requiring reducing equivalent, in contrast APX requires an ascorbic acid (AsA) and/or a glutathione (GSH) regenerating cycle involved MDHAR, DHAR, and GR. GPX, GST reduce H_2O_2 and organic hydroperoxides through ascorbate-independent thiol-mediated pathways using GSH, thioredoxin (TRX) or glutaredoxin (GRX) as nucleophile (Dietz et al. 2006; Meyer et al. 2012; Noctor et al. 2014).

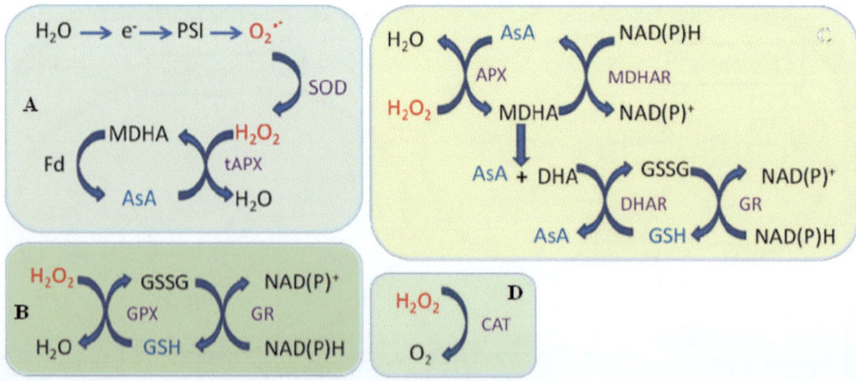

Fig. 10.3 Different pathways for reactive oxygen species (ROS) scavenging in plants. Modified from Mittler et al. (2004). (**a**) Water–water cycle (Mehler reaction); (**b**) Glutathione peroxidase (GPX) cycle; (**c**) Ascorbate–glutathione cycle; (**d**) Catalase (CAT)

A variety of in vitro spectrophotometric methods are used to determine the antioxidant enzymatic activities in plants, the below reported assays, differ in the concentration of same reagents in reaction mixture, and have been selected from the most cited in the updated literature (Fig. 10.3).

2 Extraction of Antioxidant Enzymes

1. Crude extract is prepared by homogenization of frozen plant sample in buffer medium; 10 g of the sample is cut quickly into thin slices and homogenized in 50 mL of 100 mM sodium phosphate buffer (pH 7.0) containing 1 mM ascorbic acid (ASA) and 0.5% (w/v) polyvinyl pyrrolidone (PVPP) for 5 min at 4 °C. The homogenate is filtered through three layers of cheesecloth and then the filtrate is centrifuged at 12000 × *g* for 15 min, and the supernatant is collected.
2. Crude extract is prepared by homogenization of frozen plant sample in four volumes of 50 mM potassium phosphate buffer, pH 7.5, containing 1 mM ethilendyaminotetracetic acid (EDTA), 1 mM phenylmethylsulfonyl fluoride (PMSF), 5 mM sodium ascorbate and 5% (w/v) PVPP. The homogenate is strained through two layers of Miracloth and centrifuged at 17000 *g* for 10 min. All above operations are carried out at 0 ± 4 °C. (Garciâ-Limones et al. 2002).

2.1. Superoxide dismutases (**SOD**, EC 1.15.1.1) are metalloproteins catalyzing the dismutation of the superoxide free radical to molecular oxygen and hydrogen peroxide. There are three forms of SOD identified in plants. One is a copper/zinc combination, the second is formed from manganese, and the third is formed from iron.

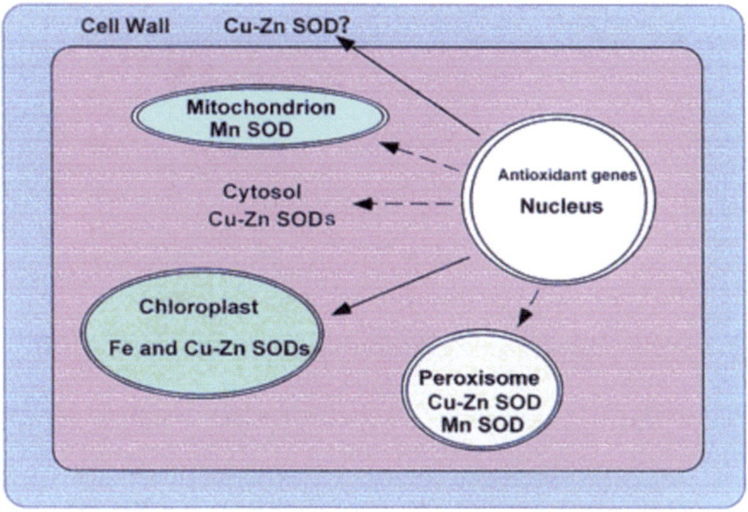

Fig. 10.4 Locations of SODs throughout the plant cell (Source: Alscher et al. 2002)

1. Analyses of SOD activity by the method of Dhindsa et al. (1981), by measuring its ability to inhibit the photochemical reduction of nitroblue tetrazolium (NBT). The reaction mixture (2.725 mL) contains 50 mM phosphate buffer, pH 7.8, 26 mM methionine, 20 μM riboflavin, 750 μM NBT and 1 μM EDTA. After adding enzyme solution (25 μL) and distilled water (250 μL), the reaction is allowed to run 15 min under 4000 lx light. The absorbance by the reaction mixture is read at 560 nm.
2. Determination of SOD activity by measuring the inhibition in photoreduction of NBT by SOD enzyme. The reaction mixture contains 50 mM sodium phosphate buffer (pH 7.6), 0.1 mM EDTA, 50 mM sodium carbonate, 12 mM L-methionine, 50 μM NBT, 10 μM riboflavin and 100 μL of crude extract in a final volume of 3.0 mL. A control reaction is performed without crude extract. SOD reaction is carried out by exposing the reaction mixture to white light for 15 min at room temperature. After 15 min incubation, absorbance is recorded at 560 nm using a spectrophotometer. One unit (U) of SOD activity is defined as the amount of enzyme causing 50% inhibition of photochemical reduction of NBT (Kumar et al. 2012) (Fig. 10.4).

2.2. Peroxidase (POX, EC 1.11.1.7) is a heme-containing enzyme composed of 40–50 kDa monomers, which eliminates excess H_2O_2 both during normal metabolism as well as during stress. It is well known its pivotal role in the biosynthesis of lignin as well as defense against biotic stress by degrading indole acetic acid (IAA) and utilizing H_2O_2 in the process.

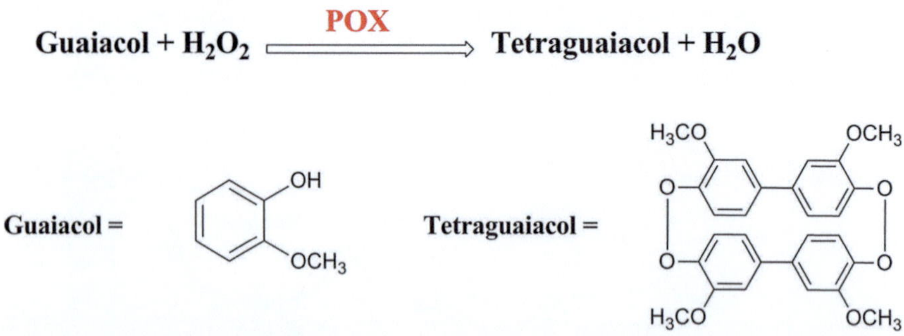

Fig. 10.5 POX assay (Source: Corban et al. 2011)

1. Measurement of POX activity using a reaction medium containing 50 mM phosphate buffer (pH 7.0), 9 mM guaiacol, and 19 mM H_2O_2 (Lin and Kao 1999). The kinetic evolution of absorbance at 470 nm is measured during 1 min. Peroxidase activity is calculated using the extinction coefficient (ε = 26.6 mM^{-1} cm^{-1} at 470 nm). One unit of peroxidase is defined as the amount of enzyme that causes the formation of 1 mM of tetraguaiacol per minute.
2. Determination of POX activity using 4-methylcatechol as substrate. Increase in the absorption caused by oxidation of 4-methylcatechol by H_2O_2 is measured spectrophotometrically at 420 nm. The reaction mixture contains 100 mM sodium phosphate buffer (pH 7.0), 5 mM 4-methylcatechol, 5 mM H_2O_2 and 500 μL of crude extract in a total volume of 3.0 mL at room temperature. One unit of enzyme activity is defined as 0.001 change in absorbance per min, under assay conditions (Onsa et al. 2004) (Fig. 10.5).

2.3. Catalases (CAT, EC 1.11.1.6) are tetrameric heme containing enzymes with the potential to directly dismutate H_2O_2 into H_2O and O_2. CAT is important in the removal of H_2O_2 generated in peroxisomes by oxidases involved in β-oxidation of fatty acids, photorespiration and purine catabolism. CAT activity is generally low under normal growth conditions and it increases only at relatively high H_2O_2 concentration or under stress condition to support APX, SOD and other peroxidases, primarily involved in ROS homeostasis (Garg and Manchanda 2009). The CAT isozymes have been studied extensively in higher plants (Polidoros and Scandalios 1999).

1. Determination of CAT activity by monitoring the disappearance of H_2O_2 at 240 nm (ε = 40 mM^{-1} cm^{-1}) according to the method of Aebi (1984). The reaction mixture contains 50 mM K-phosphate buffer (pH 7.0), 33 mM H_2O_2 and enzyme extract. Decrease in absorbance is recorded by the addition of H_2O_2 at 240 nm for 5 min.
2. Assay of CAT activity according to Garciâ-Limones et al. (2002). The reaction medium consists of 50 mM potassium phosphate buffer, pH 7.0, 20 mM H_2O_2 and between 10 and 30 mL of enzyme extract. The reaction is started by adding H_2O_2, and the decrease in A240 (e = 36 mM^{-1} cm^{-1}), due to H_2O_2 breakdown, is

Table 10.1 Role of APX in plant abiotic stress tolerance

Cellular location	Function	References
Cytosolic	Cellular response to oxidative stress, ROS, salinity stress tolerance	Wang et al. (2005)
Peroxisomal	Salinity and drought stress	Teixeira et al. (2004)
Stomatal	Salinity stress	Hong et al. (2007)
Tylakoid	Water stress and water cycle	Zhu et al. (2013)

recorded. One CAT unit is defined as the amount of enzyme necessary to decompose 1 mmol min^{-1} H_2O_2 under the above assay conditions.

2.4. Ascorbate Peroxidase (APX, EC 1.11.1.11) has been reported to increase on exposure to drought, salt, cold, heat, pathogen infection, wound stress, and other biotic or abiotic stresses.

However, the quantitative expression varies in different sub-cellular compartments and is also dependent on the developmental stages of the plant and the imposed stress conditions (Table 10.1). Increase in APX activity is many times supplemented with the activity of other antioxidant enzymes that work in tandem with APX (Teixeira et al. 2006; Lee et al. 2007).

1. Determination of APX activity from the decrease in absorbance (ABS) at 290 nm (ε = 2.8 mM^{-1} cm^{-1}) due to the H_2O_2 – dependent oxidation of ascorbate. An amount of 2 mL reaction mixture contains 50 mM potassium phosphate (pH 7.5), 0.5 mM ascorbic acid, 1 mM H_2O_2, 0.1 mM EDTA and enzyme extract (30 μL) at 25 °C (Barka 2001). The activity is expressed as nmol min^{-1} mg^{-1} protein.
2. Assay of APX activity according to Temizgul et al. (2016). An aliquot of 50 μL of crude extract into a solution containing 50 mM potassium phosphate ($KHPO_4$, pH 7.0), 0.15 mM ascorbic acid, and 20 mM H_2O_2. The decrease of absorbance values is recorded during 3 min at 290 nm in 2 mL cuvettes. Results are recorded as unit mg^{-1} protein.

2.5. Monodehydroascorbate reductase (MDHAR EC 1.6.5.4) is a flavinadenin dinucleotide (FAD) enzyme that is present as chloroplastic and cytosolic isozymes. MDHAR exhibits a high specificity for monodehydroasorbate (MDHA) as the electron acceptor, preferring NADH rather than NADPH as the electron donor (Asada 1999). MDHAR is also located in peroxisomes and mitochondria, where it scavenges H_2O_2 (del Rio et al. 2002).

1. The method for evaluating the MDRHAR activity is based on the oxidation of NADH measured at 340 nm (Hossain et al. 1984). The measurement of MDHAR activity is assayed at 25 °C in a reaction mixture containing 50 mM Tris (hydroxymethyl) aminomethane (Tris–HCl) buffer (pH 7.6), 0.125% Triton X-100, 0.2 mM NADH, 2.5 mM ascorbate, 5 μg ascorbate oxidase and enzyme extract. The decrease in absorbance at 340 nm due to NADH/H^+ oxidation

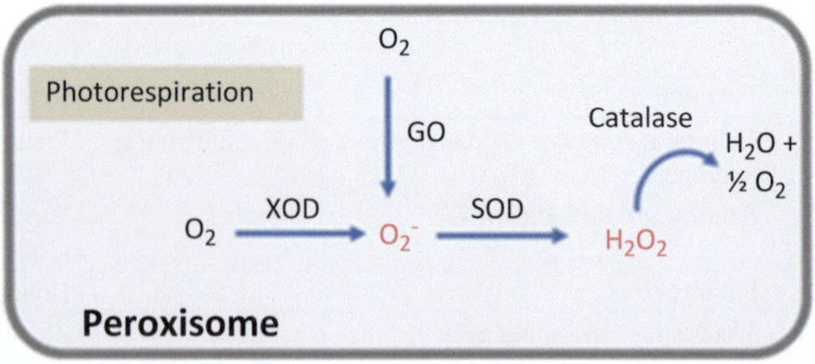

Fig. 10.6 Catalase activity and ROS detoxification in Peroxisome (Source: Houssain and Dietz 2016)

($\varepsilon = 6.2\ mM^{-1}\ cm^{-1}$) is monitored. The enzyme activity is measured in terms of µmol of NADH/H^+ oxidized $min^{-1}\ g^{-1}$ fresh weight at $25 \pm 2\ °C$.

2. Assay of MDHAR at 340 nm by a modification of the method of Miyake and Asada (1992). Monodehydroascorbate was generated via the action of ascorbate oxidase (0.4 unit; 1 unit = 1 µmol of ascorbate oxidized per min) in a reaction mixture (1 mL) containing 100 mm Hepes/KOH (pH 7.6), 25 µM NADPH, 2.5 mm ascorbate, and 100 µL of extract.

2.6. Dehydroascorbate reductase (DHAR, EC 1.8.5.1) regenerates ASC from the oxidized state and regulates the cellular ASC redox state, which is crucial for tolerance to various abiotic stresses leading to the production of ROS (Fig. 10.6). It has also been found that DHAR overexpression also enhances plant tolerance against various abiotic stresses (Kwon et al. 2003; Chen and Gallie 2005).

1. Evaluation of DHAR activity by monitoring the increase in absorbance at 265 nm due to ascorbate formation ($\varepsilon = 14\ mM^{-1}\ cm^{-1}$) (Nakano and Asada 1981). The reaction mixture contains 25 mM phosphate buffer (pH 7.0), 3.5 mM reduced glutathione (GSH), 0.4 mM dehydroascorbate and enzyme extract. The reaction rate is corrected for non-enzymatic reduction of dehydroascorbate by GSH. The enzyme activity is expressed as µmol of ascorbate $min^{-1}\ g^{-1}$ fresh weight at $25 \pm 2\ °C$.
2. Assay of DHAR activity by measuring the GSH-dependent reduction of dehydroascorbate as described by Murshed et al. (2008). The assay mixture contains 50 mM Tris-HCl buffer (pH 7.0), 0.1 mM EDTA, 2.5 mM GSH, 8 mM dehydroascorbate and 15 µL crude extract. An aliquot of 5 µL freshly prepared dehydroascorbate solution is added to initiate the reaction and the increase in absorbance is recorded at 265 nm for 5 min. Specific activity is calculated in terms of µmol dehydroascorbate per mg of protein per min, using extinction coefficient $\varepsilon = 14.15\ mM^{-1}\ cm^{-1}$.

2.7. Glutathione Reductase (GR EC 1.6.4.2) is a flavo-protein oxidoreductase localized predominantly in chloroplasts and in small amount in mitochondria and cytosol (Edwards et al. 1990). This enzyme catalizes GSSG reduction to GSH at the

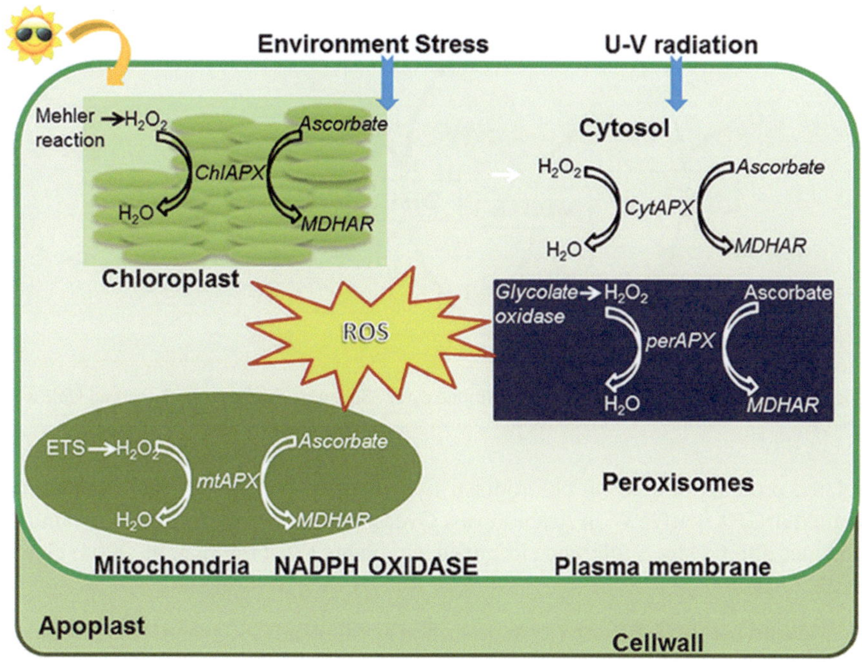

Fig. 10.7 Localization of APX enzymes and detoxification of ROS in subcellular compartments (Source: Pandey et al. 2017)

expense of NADPH. It maintains the balance between GSH and ascorbate pools (Fig. 10.7). GSH and GSSG represent a cellular redox buffer and their ratio decrease by oxidative stress. GR has a pivotal role to maintain a high GSH/GSSG ratio (Fig. 10.6). A number of abiotic stresses have been shown to affect the activity of GR in plants (Gill and Tuteja 2010).

1. Estimation of glutathione reductase (GR) according to Foyer and Halliwell (1976), by following the decrease in absorbance at 340 nm due to NADPH oxidation ($\varepsilon = 6.2\ mM^{-1}\ cm^{-1}$). The reaction mixture consists of 25 mM phosphate buffer (pH 7.8), 0.5 mM oxidized glutathione (GSSG), 0.12 mM NADPH and enzyme extracts. The enzyme activity is measured in terms of μmol of NADPH oxidized $min^{-1}\ g^{-1}$fresh weight at 25 ± 2 °C.
2. Determination of GR activity [U (mg protein)$^{-1}$] by the oxidation of NADPH at 340 nm with a molar absorption coefficient of $6.2\ mM^{-1}\ cm^{-1}$, as described by Nordhoff et al. (1993). The reaction mixture is composed of 100 mM potassium phosphate buffer (pH 7.8), 2 mM EDTA, 0.2 mM NADPH, 0.5 mM glutathione (oxidized form, GSSG) and 10 μL enzyme extract (total reaction mixture 1 mL). The reaction is initiated by the addition of NADPH at 25 °C.

2.8. Glutathione peroxidases (GPXs, EC 1.11.1.9) are a large family of diverse isozymes that use GSH to reduce H_2O_2 and organic and lipid hydroperoxides (Noctor et al. 2002) (Fig. 10.8). Millar et al. (2003) identified a family of seven

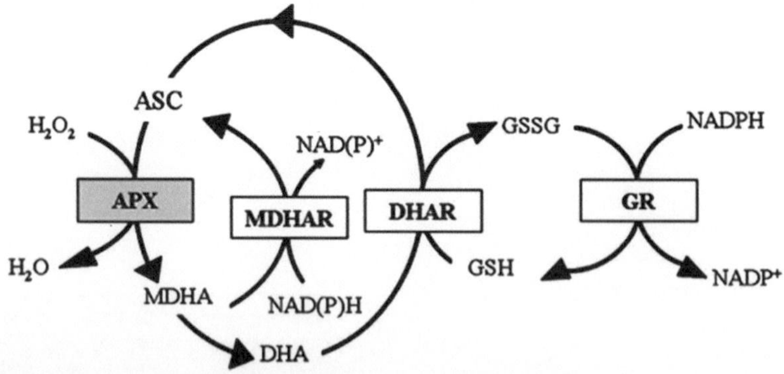

Fig. 10.8 The chloroplast ascorbate-glutathione cycle (Scheme adapted from Foyer and Halliwell 1976)

related proteins in cytosol, chloroplast, mitochondria and endoplasmic reticulum, named *AtGPX1-AtGPX7* in Arabidopsis. Overexpression of GPX has been found to enhance abiotic stress tolerance in transgenic plants (Yoshimura et al. 2004; Gaber et al. 2006).

1. Measurement of GPX activity as described by Elia et al. (2003) and Hasanuzzaman et al. (2011). The reaction mixture consists of 100 mM K-P buffer (pH 7.0), 1 mM EDTA, 1 mM sodium azide (NaN_3), 0.12 mM NADPH, 2 mM GSH, 1 unit GR, 0.6 mM H_2O_2 (as a substrate) and 20 μL of sample solution. The oxidation of NADPH is recorded at 340 nm for 1 min and the activity is calculated using the extinction coefficient $\varepsilon = 6.62\ mM^{-1}\ cm^{-1}$.
2. Determination of GPX activity using glutathione peroxidase cellular activity assay kit according to manufacturer's instruction (CGP1 Sensitivity: 0.005 units mL^{-1}, Sigma Aldrich) (Jincy et al. 2017). In brief, 10 μL of homogenate is added to a 990 μL reaction mixture containing 0.25 mM NADPH, 2.1 mM reduced glutathione, 0.5 units mL^{-1} glutathione reductase, and 300 μM tert-butyl hydroperoxide. The change in absorbance of the reaction product is kinetically quantified every 10 s at 340 nm for 1 min, and the activity is expressed as mol NADP$^+$ formed $min^{-1}\ kg^{-1}$ fresh weight.

2.9. Glutathione *S*-transferases (**GSTs**, EC 2.5.1.18) are a large and different group of enzymes that catalyze the conjugation of electrophilic xenobiotic substrates with glutathione (Fig. 10.8). Plant GSTs are known to function in herbicide detoxification, hormone homeostasis, vacuolar sequestration of anthocyanin, tyrosine metabolism, hydroxyperoxide detoxification, regulation of apoptosis and in plant responses to biotic and abiotic stresses (Dixon et al. 2010). GSTs can reduce peroxides with the help of GSH and produce scavengers of cytotoxic and genotoxic compounds (Noctor et al. 2002). These enzymes are generally cytoplasmic proteins, but microsomal, plastidic, nuclear and apoplastic isoforms has also been reported

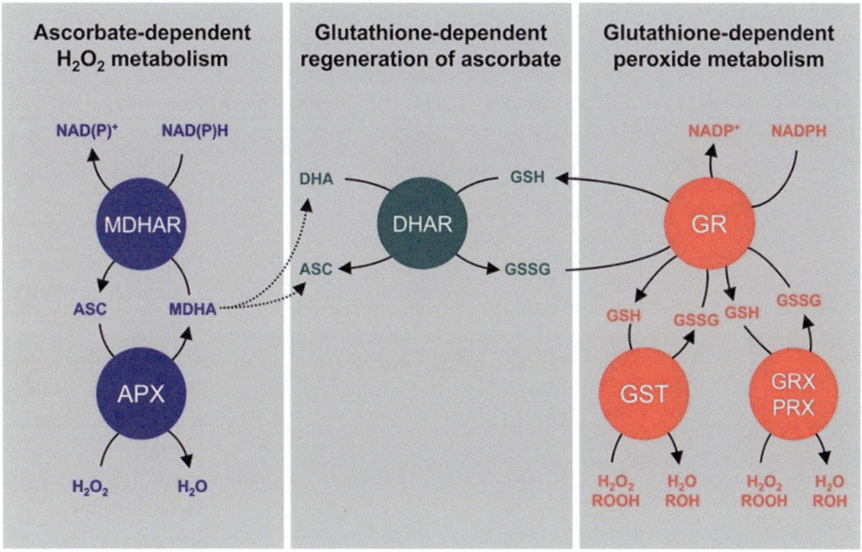

Fig. 10.9 Interdependence and independence of glutathione and ascorbate in peroxide metabolism (Source: Foyer and Noctor 2011)

(Frova 2003). GSTs are very abundant in some cases representing more than 1% of soluble proteins in plant cells (Edwards et al. 2000). It has been found that GST overexpression also enhances plant tolerance to various abiotic stresses (Zhao and Zhang 2006).

1. Measurement of GST activity (Hossain et al. 2006). Reaction mixture contains 100 mM Tris-HCl buffer (pH 6.5), 1.5 mM GSH, 1 mM 1-chloro-2,4- dinitrobenzene (CDNB); enzyme solution (700 μL final volume). Reaction is initiated by CDNB; increase in absorbance is measured at 340 nm, 1 min. Activity is calculated using extinction coefficient $\varepsilon = 9.6\ mM^{-1}\ cm^{-1}$.
2. GST assay. 1 mL reaction mixture contains 0.1 M sodium phosphate buffer (pH 6.5), 20 μL enzyme extract and 2% 1-chloro-2,4-dinitrobenzene (CDNB). The enzyme activity [U (mg protein)$^{-1}$] is calculated by monitoring the kinetic of reaction mixture for 180 s (60 s intervals) at 340 nm in a spectrophotometer, as described by Kumar et al. (2009) (Fig. 10.9).

During adaptation to a situation that leads to increased oxidative stress, antioxidant enzyme activities would be expected to increase. However, this response is variable and depends on plant species, stage of development, degree of the stress and cross-working of the different antioxidant enzymes that cooperate to alleviate the damaging effects of ROS and to develop tolerance against various environmental stress conditions (Table 10.2).

Table 10.2 Summary of the antioxidant enzyme changes in different wheat genotypes and different tissues type under different tested abiotic-stress conditions

Abiotic stress	SOD	CAT	APX	POX	GR	GPX/ GST	MDHAR	DHAR	References
Drought	↓	–	–	↓					Alexieva et al. (2001)
		↑							Luna et al. (2005)
	–	↑	↑	↑	↑				Devi et al. (2012)
	↑	↑	↑		↑				Wang et al. (2005)
						↓			Chen et al. (2004)
						↓			Lascano et al. (2001)
						↑			Guo et al. (2006)
						↑			Dhindsa (1991)
	↓	↓	↓		↓		↓	↑	Jiménez et al. (2002)
	↑	↓	↑	↑	↑		↑	↑	Sharma and Dubey (2005)
Salinity		↑	↑	↑					Barakat (2011)
		↑	↓	↓					Heidari (2009)
	↑	↑			↓				Esfandiari et al. (2007)
	↑	↑			↑				Sairam et al. (2002)
Cold	↑	↑	↑	↑					Janmohammadi et al. (2012)
	↑	–	↑	↑	↑				Turk et al. (2014)
	↓		↓				↓	↓	Fryer et al. (1998)
	↑	↑	↑	↑	↑	↑	↑	↑	Zhang et al. (2008)
Heat	↑	↑	↑		↑				Badawi et al. (2007)
	↑			↑					Ibrahim et al. (2013)
	↑	↑		↑					Gupta et al. (2013)
	↓↑	↓		↓	↓↑				Wang et al. (2014)
UV-B	↑			↑					Ibrahim et al. (2013)
	↑	↑		–					Alexieva et al. (2001)
	↓	↓		↓	↑				Barabás et al. (1998)

SOD Superoxide dismutase, *CAT* catalase, *APX* ascorbate peroxidase, *POX* guaiacol peroxidase, *GR* glutathione reductase, *DHAR* dehydro ascorbate reductase, *MDHAR* monodehydroascorbater-eductase, *GPX* glutathione peroxidase, *GST* glutathione S-transferase, ↑ increase, ↓ decline and – unchanged (Source: Caverzan et al. 2016)

3 Conclusion

ROS are produced by cellular metabolic activities and induced by environmental stress factors. They are molecules highly reactive because of unpaired electrons in their structure and react with several biological macromolecules in cell, such as carbohydrates, nucleic acids, lipids, and proteins, and alter their functions, damaging the integrity of the cell and ultimately leading to its death. ROS synthesis is widespread, with production sites being present in both intracellular and extracellular locations. The antioxidant enzymes work in conjunction to alleviate the damaging effects of ROS and their highly compartmentalized nature is well defined. The reaction mechanisms can be used to evaluate the antioxidant activity of various naturally occurring enzymes. Regulation of redox state is critical for cell viability, activation, proliferation, and organ function.

References

Aebi H (1984) Catalase in Vitro. In: Colowick SP, Kaplan NO (eds) Methods in enzymology, vol 105. Acad. Press, Florida, pp 114–121

Alexieva V, Sergiev I, Mapelli S, Karanov E (2001) The effect of drought and ultraviolet radiation on growth and stress markers in pea and wheat. Plant Cell Environ 24:1337–1344

Alscher RG, Erturk N, Heath LS (2002) Role of superoxide dismutases (SODs) in controlling oxidative stress in plants. J Exp Bot 53(372):1331–1341

Asada K (1999) The water–water cycle in chloroplasts: scavenging of active oxygens and dissipation of excess photons. Annu Rev Plant Physiol Plant Mol Biol 50:601–639

Badawi GH, Tahir ISA, Nakata N, Tanaka K (2007) Induction of some antioxidant enzymes in selected wheat genotypes. Afr Crop Sci Conf Proceed 8:841–848

Barabás KN, Szegletes Z, Pestenacz A, Fulop K, Erdei U (1998) Effects of excess UV-8 irradiation on the antioxidant defence mechanisms in wheat (*Triticum aestivum* L.) seedlings. Plant Physiol 153:146–153

Barakat NAM (2011) Oxidative stress markers and antioxidant potential of wheat treated with phytohormones under salinity stress. J Stress Physiol Biochem 7:250–267

Barka EA (2001) Protective enzymes against reactive oxygen species during ripening of tomato (Lycopersicon esculentum) fruits in response to low amounts of UV-C. Aust J Plant Physiol 28:785–791

Baxter A, Mittler R, Suzuki N (2014) ROS as key players in plant stress signalling. J Exp Bot 65:1229–1240

Caverzan A, Casassola A, Patussi Brammer S (2016) Antioxidant responses of wheat plants under stress. Genet Mol Biol 39(1):1–6

Chen Z, Gallie DR (2005) Increasing tolerance to ozone by elevating folia ascorbic acid confers greater protection against ozone than increasing avoidance. Plant Physiol 138:1673–1689

Chen KM, Gong HJ, Chen GC, Wang SM, Zhang CL (2004) Gradual drought under field conditions influences the glutathione metabolism, redox balance and energy supply in spring wheat. J Plant Grow Reg 23:20–28

Corban GJ, Hadjikakou SK, Tsipis AC, Kubicki M, Bakas T, Hadjiliadis N (2011) Inhibition of peroxidase-catalyzed iodination by thioamides: experimental and theoretical study of the antithyroid activity of thioamides. New J Chem 35:213–224

Das MK, Roychoudhury A (2014) ROS and responses of antioxidant as ROS-scavengers during environmental stress in plants. Front Env Sc 2(53):1–13

Devi R, Kaur N, Gupta AK (2012) Potential of antioxidant enzymes in depicting drought tolerance of wheat (Triticum aestivum L.). Indian J Biochem Biophys 49:257–265

Dhindsa RS (1991) Drought stress, enzymes of glutathione metabolism, oxidation injury, and protein synthesis in *Tortula ruralis*. Plant Physiol 95:648–651

Dhindsa RS, Plumb-Dhindsa P, Thorpe TA (1981) Leaf senescence: correlated with increased levels of membrane permeability and lipid peroxidation, and decreased levels of superoxide dismutase and catalase. J Exp Bot 126:93–101

Dietz KJ, Jacob S, Oelze ML, Laxa M, Tognetti V, De Miranda SM, Baier M, Finkemeier I (2006) The function of peroxiredoxins in plant organelle redox metabolism. J Exp Bot 57:1697–1709

Dixon DP, Skipsey M, Edwards R (2010) Roles for glutathione transferases in plant secondary metabolism. Phytochemistry 71:338–350

Edwards EA, Rawsthorne S, Mullineaux PM (1990) Subcellular distribution of multiple forms of glutathione reductase in leaves of pea (*Pisum sativum* L.). Planta 180:278–284

Edwards R, Dixon DP, Walbot V (2000) Plant glutathione S-transferases: enzymes with multiple functions in sickness and in health. Trends Plant Sci 5:193–198

Elia AC, Galarini R, Taticchi MI, Dorr AJM, Mantilacci L (2003) Antioxidant responses and bioaccumulation in *Ictalurus melas* under mercury exposure. Ecotoxicol Environ Saf 55:162–167

Esfandiari E, Shekari F, Shekari F, Esfandiari M (2007) The effect of salt stress on antioxidant enzymes activity and lipid peroxidation on the wheat seedling. Not Bot Hort Agrobot Cluj 35:48–56

Foyer CH, Halliwell B (1976) The presence of glutathione reductase in chloroplast: a proposed role in ascorbic acid metabolism. Planta 133:21–25

Foyer CH, Noctor G (2011) Ascorbate and glutathione: the heart of the redox hub. Plant Physiol 155:2–18

Foyer CH, Shigeoka S (2011) Understanding oxidative stress and antioxidant functions to enhance photosynthesis. Plant Physiol 155:93–100

Frova C (2003) The plant glutathione transferase gene family: genomic structure, functions, expression and evolution. Physiol Plant 119:469–479

Fryer MJ, Andrews JR, Oxborough K, Blowers DA, Baker NR (1998) Relationship between CO_2 assimilation, photosynthetic electron transport, and active O_2 metabolism in leaves of maize in the field during periods of low temperature. Plant Physiol 116(2):571–580

Gaber A, Yoshimura K, Yamamoto T, Yabuta Y, Takeda T, Miyasaka H, Nakano Y, Shigeoka S (2006) Glutathione peroxidase-like protein of Synechocystis PCC 6803 confers tolerance to oxidative and environmental stresses in transgenic Arabidopsis. Physiol Plant 128:251–262

Garciâ-Limones C, Hervâs A, Navas-Cortes JA, Jimeânez-Diâz RM, Tena M (2002) Induction of an antioxidant enzyme system and other oxidative stress markers associated with compatible and incompatible interactions between chickpea (*Cicerarietinum* L.) and *Fusarium oxysporum* f. sp. *Ciceris*. Physiol Mol Plant Pathol 61:325–337

Garg N, Manchanda G (2009) ROS generation in plants: boon or bane? Plant Biosys 143:8–96

Gill SS, Tuteja N (2010) Reactive oxygen species and antioxidant machinery in abiotic stress tolerance in crop plants. Plant Physiol Biochem 48:909–930

Guo Z, Ou W, Lu S, Zhong Q (2006) Differential responses of antioxidative system to chilling and drought in four rice cultivars differing in sensitivity. Plant Physiol Biochem 44:828–836

Gupta NK, Agarwal S, Agarwal VP, Nathawat NS, Gupta S, Singh G (2013) Effect of short-term heat stress on growth, physiology and antioxidativedefence system in wheat seedlings. Acta Physiol Plant 35:1837–1842

Hasanuzzaman M, Hossain MA, Fujita M (2011) Nitric oxide modulates antioxidant defense and the methylglyoxal detoxification system and reduces salinity-induced damage of wheat seedlings. Plant Biotechnol Rep 5:353–365

Heidari M (2009) Antioxidant activity and osmolyte concentration of sorghum (*Sorghum bicolor*) and wheat (*Triticum aestivum*) genotypes under salinity stress. Asian J Plant Sci 8:240–244

Hong CY, Hsu YT, Tsai YC, Kao CH (2007) Expression of ascorbate peroxidase 8 in roots of rice (*Oryza sativa* L.) seedlings in response to NaCl. J Exp Bot 58:3273–3283

Hossain MA, Nakano Y, Asada K (1984) Monodehydroascorbate reductase in spinach chloroplast and its participation in regeneration of ascorbate for scavenging hydrogen peroxide. Plant Cell Physiol 11:351–358

Hossain MZ, Hossain MD, Fujita M (2006) Induction of pumpkin glutathione S-transferase by different stresses and its possible mechanisms. Biol Plant 50:210–218

Houssain MS, Dietz KJ (2016) Tuning of redox regulatory mechanisms, reactive oxygen species and redox homeostasis under salinity stress. Front Plant Sci 7:1–15

Ibrahim MM, Alsahli AA, Al-Ghamdi AA (2013) Cumulative abiotic stresses and their effect on the antioxidant defense system in two species of wheat, *Triticum durum* desf and *Triticum aestivum* L. Arch Biol Sci 65:1423–1433

Jajic I, Sarna T, Strzalka K (2015) Senescence, stress, and reactive oxygen species. Plants 4:393–411

Janmohammadi M, Enayati V, Sabaghnia N (2012) Impact of cold acclimation, de-acclimation and re-acclimation on carbohydrate content and antioxidant enzyme activities in spring and winter wheat. Icel Agric Sci 25:3–11

Jiménez A, Creissen G, Kular B, Firmin J, Robinson S, Verhoeten M, Mullineaux P (2002) Changes in oxidative processes and components of the antioxidant system during tomato fruit ripening. Planta 214:751–758

Jincy M, Djanaguiraman M, JeyakumaraK P, Subramanian SK, Jayasankar S, Paliyath G (2017) Inhibition of phospholipase D enzyme activity through hexanal leads to delayed mango (*Mangifera indica* L.) fruit ripening through changes in oxidants and antioxidant enzymes activity. Scientia Hort 218:316–325

Kehrer JP (2000) The Haber–Weiss reaction and mechanisms of toxicity. Toxicology 149:43–50

Kumar M, Yadav V, Tuteja N, Kumar Johri A (2009) Antioxidant enzyme activities in maize plants colonized with *Piriformospora indica*. Microbiology 155:780–790

Kumar A, Dutt S, Bagler G, Ahuja PS, Kumar S (2012) Engineering a thermo-stable superoxide dismutase functional at subzero to >50°C, which also tolerates autoclaving. Sci Rep 2:387

Kwon SY, Choi SM, Ahn YO, Lee HS, Lee HB, Park YM, Kwak SS (2003) Enhanced stress-tolerance of transgenic tobacco plants expressing a human dehydroascorbate reductase gene. J Plant Physiol 160:347–353

del Río LA, Corpas FJ, Sandalio LM, Palma JM, Gómez M, Barroso J (2002) Reactive oxygen species, antioxidant systems and nitric oxide in peroxisomes. J Exp Bot 53:1255–1272

Lascano HR, Antonicelli GE, Luna CM, Melchiorre MN, Gomez LD, Racca RW, Trippi VS, Casano LM (2001) Antioxidant system response of different wheat cultivars under drought: field and in vitro studies. Aust J Plant Physiol 28:1095–1102

Lee YP, Kim SH, Bang JW, Lee HS, Kwak SS, Kwon SY (2007) Enhanced tolerance to oxidative stress in transgenic tobacco plants expressing three antioxidant enzymes in chloroplasts. Plant Cell Rep 26:591–598

Lin CC, Kao CH (1999) NaCl induced changes in ionically bounds peroxidase activity in roots of rice seedlings. Plant Soil 216:147–153

Luna C, Pastori GM, Driscoll S, Foyer CH (2005) Drought controls on H_2O_2 accumulation, catalase (CAT) activity and CAT gene expression in wheat. J Exp Bot 56:417–423

Marino D, Dunand C, Puppo A, Pauly N (2012) A burst of plant NADPH oxidases. Trends Plant Sci 17:9–15

Maxwell DP, Wang Y, Mcintosh L (1999) The alternative oxidase lowers mitochondrial reactive oxygen production in plant cells. Proc Natl Acad Sci U S A 96:8271–8276

Meyer Y, Belin C, Delorme-Hinoux V, Reichheld JP, Riondet C (2012) Thioredoxin and glutaredoxin systems in plants: molecular mechanisms, crosstalks, and functional significance. Antiox Redox Signal 17:1124–1160

Millar AH, Mittova V, Kiddle G, Heazlewood JL, Bartoli CG, Theodoulou FL, Foyer CH (2003) Control of ascorbate synthesis by respiration and its implication for stress responses. Plant Physiol 133:443–447

Mittler R (2002) Oxidative stress, antioxidants and stress tolerance. Trends Plant Sci 7:405–410

Mittler R, Vanderauwera S, Gollery M, van Breusegem F (2004) The reactive oxygen gene network of plants. Trends Plant Sci 9:490–498

Miyake C, Asada K (1992) Thylakoid-bound ascorbate peroxidase in spinach chloroplasts and photoreduction of its primary oxidation product monodehydroascorbate radicals in thylakoids. Plant Cell Physiol 33:541–553

Murshed R, Lopez-Lauri F, Sallanon H (2008) Microplate quantification of enzymes of the plant ascorbate–glutathione cycle. Anal Biochem 383:320–322

Nakano Y, Asada K (1981) Hydrogen peroxide is scavenged by ascorbate-specific peroxidase in spinach chloroplast. Plant Cell Physiol 22:867–880

Noctor G, Gomez L, Vanacker H, Foyer CH (2002) Interactions between biosynthesis, compartmentation, and transport in the control of glutathione homeostasis and signaling. J Exp Bot 53:1283–1304

Noctor G, Mhamdi A, Foyer CH (2014) The roles of reactive oxygen metabolism in drought: not so cut and dried. Plant Physiol 164:1636–1648

Noctor G, Reichheld JP, Foyer CH (2017) ROS-related redox regulation and signaling in plants. Semin Cell Dev Biol. https://doi.org/10.1016/j.semcdb.2017.07.013

Nordhoff A, Bucheler US, Werner D, Schirmer RH (1993) Folding of the four domains and dimerization are impaired by the *Gly*446→Glu exchange in human glutathione reductase. Implications for the design of antiparasitic drugs. Biochemistry 32:4060–4066

Oda T, Hashimoto H, Kuwabara N, Akashi S, Hayashi K, Kojima C, Wong HL, Kawasaki T, Shimamoto K, Sato M, Shimizu T (2010) Structure of the N-terminal regulatory domain of a plant NADPH oxidase and its functional implications. J Biol Chem 285:1435–1445

Onsa GH, Saari N, Selamat J, Bakar J (2004) Purification and characterization of membrane-bound peroxidases from *Metroxylon sagu*. Food Chem 85:365–376

Pandey S, Fartyal D, Agarwal A, Shukla T, James D, Kaul T, Negi YK, Arora S, Reddy MK (2017) Abiotic stress tolerance in plants: myriad roles of ascorbate peroxidase. Front Plant Sci 8:581. https://doi.org/10.3389/fpls.2017.00581

Polidoros NA, Scandalios JG (1999) Role of hydrogen peroxide and different classes of antioxidants in the regulation of catalase and glutathione S-transferase gene expression in maize (*Zea mays* L.). Physiol Plant 106:112–120

Sairam RK, Rao KV, Srivastava GC (2002) Differential response of wheat genotypes to long term salinity stress in relation to oxidative stress, antioxidant activity and osmolyte concentration. Plant Sci 163:1037–1046

Sharma P, Dubey RS (2005) Drought induces oxidative stress and enhances the activities of antioxidant enzymes in growing rice seedlings. Plant Growth Regul 46(3):209–221

Suzuki N, Miller G, Morales J, Shulaev V, Torres MA, Mittler R (2011) Respiratory burst oxidases: the engines of ROS signaling. Curr Opin Plant Biol 14:691–699

Teixeira FK, Menezes-Benavente L, Margis R, Margis-Pinheiro M (2004) Analysis of the molecular evolutionary history of the ascorbate peroxidase gene family: inferences from the rice genome. J Mol Evol 59:761–770

Teixeira FK, Menezes-Benavente L, Galvão VC, Margis R, Margis-Pinheiro M (2006) Rice ascorbate peroxidase gene family encodes functionally diverse isoforms localized in different subcellular compartments. Planta 224:300–314

Temizgul R, Kaplan M, Kara R, Yilmaz S (2016) Effect of salt concentrations on antioxidant enzyme activity of grain *Sorghum*. Curr Trends Natu Sci 5(9):171–178

Torres MA, Dangl JL (2005) Functions of the respiratory burst oxidase in biotic interactions, abiotic stress and development. Curr Opin Plant Biol 8:397–403

Turk H, Erdal S, Genisel M, Atici O, Demir Y, Yanmis D (2014) The regulatory effect of melatonin on physiological, biochemical and molecular parameters in cold-stressed wheat seedlings. Plant Growth Regul 74:139–152

Wang Y, Wisniewski M, Meilan R, Cui M, Webb R, Fuchigami L (2005) Overexpression of cytosolic ascorbate peroxidase in tomato confers tolerance to chilling and salt stress. J Am Soc Hortic Sci 130:167–173

Wang C, Wen D, Sun A, Han X, Zhang J, Wang Z, Yin Y (2014) Changes in antioxidant enzyme activity and gene expression in response to high temperature stress in wheat seedlings. J Cereal Sci 60:653–659

Yoshimura K, Miyao K, Gaber A, Takeda T, Kanaboshi H, Miyasaka H, Shigeoka S (2004) Enhancement of stress tolerance in transgenic tobacco plants overexpressing *Chlamydomonas* glutathione peroxidase in chloroplasts or cytosol. Plant J 37:21–33

Zhang Y, Luo Y, Hou YX, Jiang H, Chen Q, Tang RH (2008) Chilling acclimation induced changes in the distribution of H_2O_2 and antioxidant system of strawberry leaves. Agric J 3(4):286–291

Zhao F, Zhang H (2006) Salt and paraquat stress tolerance results from co-expression of the *Suaeda salsa* glutathione S-transferase and catalase in transgenic rice. Plant Cell Tissue Organ Cult 86:349–358

Zhu Y, Zuo M, Liang Y, Jiang M, Zhang J, Scheller HV, Tan M, Zhang A (2013) MAP 65-1a positively regulates H_2O_2 amplification and enhances brassinosteroid-induced antioxidant defense in maize. J Exp Bot 64:3787–3802

Chapter 11
Flow Cytometric Measurement of Different Physiological Parameters

Fabrizio Araniti, Teodoro Coba de la Peña, and Adela M. Sánchez-Moreiras

1 Introduction

A variety of physiological parameters involved in signal transduction pathways, enzymatic activities, ATP production, and many other physiological processes can be analyzed by flow cytometry. Parameters as intracellular pH, membrane potential, calcium concentration, reactive oxygen species (ROS) generation, or glutathione content can be estimated thanks to this technique. One of the principal advantages of flow cytometry is that it allows the measurement of these parameters in living and small cells, usually in real time during the physiological stimulation. Moreover, flow cytometry can detect not only the fluorescence of biochemically specific developed fluorescent dyes but also autofluorescence (such as the chlorophyll *a* fluorescence), which can be measured in parallel and can provide supplementary information about the physiological condition of the measured cells (Franklin et al. 2001).

For the measurement of physiological parameters with fluorescent probes, detailed observations for suitable loading and intracellular distribution of the fluorescent probe must be firstly done by fluorescence microscopy and flow cytometry. Secondly, the effects of different ionophores and inhibitors (specific in each case for the physiological parameter under study) on fluorescence are analyzed by flow cytometry. These observations allow us to verify that the fluorescence behavior of the probe is correct and really dependent on the parameter under study, evaluating

F. Araniti
Dipartimento di Agraria, Facoltà di Agraria, Università Mediterranea di Reggio Calabria, Reggio Calabria, Italy

T. Coba de la Peña
Centro de Estudios Avanzados en Zonas Áridas (CEAZA), La Serena, Chile

A. M. Sánchez-Moreiras (✉)
Department of Plant Biology and Soil Science, University of Vigo, Vigo, Spain
e-mail: adela@uvigo.es

© Springer International Publishing AG, part of Springer Nature 2018

A. M. Sánchez-Moreiras, M. J. Reigosa (eds.), *Advances in Plant Ecophysiology Techniques*, https://doi.org/10.1007/978-3-319-93233-0_11

and discarding possible artifacts. These observations also allow a calibration of the fluorescence and of the range variations of the physiological parameter. Thirdly, physiological effects (and thus, fluorescence variations) induced by different substances (hormones or other biological substances, chemicals) and environmental conditions (abiotic or biotic stress factors) under study are analyzed by flow cytometry. Finally, the recorded data and histograms are analyzed. Cell loading of the probes referred below is accomplished simply by adding the probe directly to the buffer or cell suspension, at different concentrations and incubation times (other loading techniques not discussed here, involve electroporation and microinjection).

Besides the fluorescent dyes already mentioned, an expanding family of fluorescent proteins (FPs), which have become essential tools for studies of cell biology and physiology and that can be detected by flow cytometry analyses, has been developed in the last years. These genetically encoded fluorescent markers can be measured without manipulating the sample and without adding external chemicals to the cell, which represents a great advantage in the analyses of living cells (Day and Davidson 2009). Certain of these fluorescent proteins are calcium, ROS or pH-sensitive fluorescent proteins that have become essential tools for studies of pH homeostasis and cell physiology (Benčina 2013).

2 Physiological Parameters

2.1 Intracellular pH

Almost every process of cell metabolism induces or is affected by apoplastic, cytosolic or intraorganellar (luminal) pH changes, including ATP generation, extensibility of the wall, membrane potential, movement of hormones, receptor-ligand interactions, cell growth and proliferation, movement of substances across membranes, etc. In the same way, changes in intracellular pH are a cell response to external applied agents, like hormones, growth factors, biotic or abiotic stress and others. Different cell organelles, as lysosomes, vacuoles or the Golgi apparatus, have different pH than cytosol, which is necessary to accomplish their functions; i.e. almost all vacuolar functions depend either on the acidic pH of the lumen or on the pH gradient across the membrane, and acidification is achieved through the action of the V-H^+-ATPase proton pump, located at the membrane (Rodrigues et al. 2013). As well, alkalinization of Golgi pH can cause delayed transport, immature glycosylation, and altered Golgi morphology (Maeda et al. 2008), among others. Therefore, cells have developed different mechanisms for regulating intracellular pH (Smith 1979; Felle 1988; Kurkdjian and Guern 1989; Sakano 2001).

Internal pH can be measured by flow cytometry in single cells or isolated organelles in a generally simple procedure that can be done using any cytometer equipped with a 488-nm argon laser. Fluorescence properties (intensity, emission and excitation spectrum) of several available probes vary depending on the H^+ concentration of their environment, because hydrogen ion binding changes the electronic structure of the probe (Haugland et al. 1996). Maximum response of the probe will occur for pH values near its *pKa.* The sensitivity of these probes is around 0.1–0.2 pH units.

On the other hand, the development of genetically encoded fluorescent pH-sensitive sensors (GFP mutants) with a range of *pKa* values, such as Cameleon (Miyawaki et al. 1999), pHluorin (Miesenbock et al. 1998) or pHusion (Gjetting et al. 2012), which offer information about pH dynamics without dye loading or sample manipulation, appears in the last years as a good option for non-invasive intracellular pH detection in those cells that can be transformed with pH-sensing GFPs (Benčina 2013). pH affects the equilibrium between protonated and deprotonated forms of these fluorescent proteins appearing as good pH sensors to detect changes in pH_i. An expanding family of pH sensors for apoplastic (Gao et al. 2004), cytosolic (Moseyko and Feldman 2001), mitochondrial (Li and Tsien 2012), Golgi network (Lam et al. 2012) and endoplasmic reticulum (Kneen et al. 1998) pH_i measurement in living cells has been developed. Flow cytometry can easily record the fluorescence emitted by these non-invasive pH indicators.

2.1.1 Measurement of Intracellular pH by Fluorescent Probes

Two of the most used pH sensitive probes to measure cytosolic pH are the weak acids BCECF (2′,7′-bis-(2 carboxyethyl)-5-(and 6) carboxyfluorescein acetoxymethyl ester) and SNARF1 (Semi naphtho rhoda fluorine acetoxymethyl ester), which have protonated and free base forms with different emission spectra that allow to get a signal proportional to pH_i (Gonugunta et al. 2008). Both have *pK* values close to 7.0, although SNARF-1 is a more sensitive indicator than BCECF in the physiological range (Robinson et al. 1997). Covalent binding with an acetoxymethyl (AM) residue allows probes to be permeable across biological membranes. Acetoxymethyl-ester forms of these probes are commercially available. Once into the cell, cellular esterases remove acetoxymethyl residue, and the probes become negatively charged, and thereby membrane impermeable, trapped into the cell or into a cell compartment.

BCECF is optimally excited at 488 nm, and its maximal fluorescence emission is 520 nm. Its *pKa* is 6.98, and thereby it is very suitable for the study of cytosolic pH (6.5–7.5). In case of acidification, BCECF becomes more protonated, and its fluorescence intensity decreases. Cellular alkalinization induces an increase in intensity. BCECF is a fast-response probe, allowing kinetic studies of pH changes in real time. One example of the use of BCECF and flow cytometry in plant material is shown in Giglioli-Givarc’h et al. (1996), where activation of a phosphoenolpyruvate kinase after cytosolic alkalinization in *Digitaria sanguinalis* protoplasts is characterized.

SNARF1 is excited at 488 nm and is a 'ratiometric dye', that is, its emission maximum shifts upon pH changes in the microenvironment. The protonated form of the fluorescent probe has a maximum emission at 540 nm, and the maximum of the deprotonated form is at 630 nm. It is possible to record continuously the fluorescence intensity at both wavelengths by flow cytometry, using two detectors. In fact, the ratio of both fluorescence intensities is a very reliable and specific measure, because it discards fluorescence intensity variations induced by several unspecific factors, like differential individual loading among cells (Haugland et al. 1996).

Some ionophores and substances used in the validation and calibration of pH fluorescent probes are:

- Nigericine induces a permeabilization of cell membrane to proton ions, so the extracellular and intracellular proton concentrations make equal, if extracellular and intracellular K^+ concentrations are the same.
- Propionic is a weak acid that induces intracellular acidification. NH_4Cl is a weak base that induces intracellular alkalinization. Both are used for monitoring changes on fluorescence intensity.

The 'null point method' is used for calibrating and converting fluorescence intensity values in pH units. Dye-loaded cells are incubated in a series of buffers at different pHs in presence of Nigericine. Intracellular pH equals extracellular pH, and a direct correspondence between known intracellular pH and fluorescence intensity is established (Haugland et al. 1996).

Regarding luminal pH, vacuolar lumen acidity has been assessed by Rodrigues et al. (2013) with two pH-sensitive probes, Acridine Orange (AO) and LysoSensor Green DND-189, by staining the vacuoles with 30 μM AO or 5 μM LysoSensor Green and incubating them for 10 min in the dark at room temperature.

2.1.2 Measurement of Intracellular pH by Fluorescent Proteins (FPs)

Different fluorescent proteins, such as pHlourins or Pt-GFP, have been increasingly used in the last years facilitating the detection of cytoplasmic pH in plants (Swanson et al. 2011). The ability of flow cytometry to do rapid analysis and acquisition of multiparameter data at the single-cell level for each cell in a population makes this technique very appropriate to detect the pH sensors based on fluorescent proteins (FPs). This is the case of root and leaf intracellular pH measurements in *Arabidopsis* under abiotic stress (Gao et al. 2004; Schulte et al. 2006) or under growth monitoring (Monshausen et al. 2007).

As previously showed by Valkonen et al. in 2013 for pHluorin detection, a ratiometric flow cytometer equipped with 405- and 488-nm light paths for the dual excitation of all pHluorin-based ratiometric pH probes can be used for pHi analyses. Green fluorescence is detected after excitation with 488-nm and 405-nm laser light, and the ratios of these two fluorescence parameters (F405-nm and F488-nm) are calculated for every cell. Finally, to correlate ratios with pH, a calibration curve that links fluorescence intensity ratios to pH is generated (Benčina 2013).

2.2 Cytosolic Ca^{2+} Concentration

Calcium concentration is a critical factor in the control of many cellular responses, being usually among the most rapid responses, in the range of nanoseconds, in plant cells. Calcium is a second messenger for a broad variety of stimuli, regulating metabolism and gene expression. Tuteja and Mahajan (2007), Riveras et al. (2015), and Edel et al. (2017) have recently reviewed the role of calcium as second messenger in different signaling pathways of plants.

Knowing the changes of intracellular calcium concentrations is important when the response of plant cells to stress factors is being studied, and fast and sensitive techniques are necessary for accurate measurements of this parameter. Continuous monitoring of thousands of cells is now possible thanks to the new generation of flow cytometers, which provide a new method for dynamic Ca^{2+} measurements of the entire population (Vines et al. 2010).

Grynkiewicz et al. (1985) and Haugland (2003) have described several fluorescent probes for measuring cytosolic Ca^{2+} changes. A nice review about the characteristics and use of high-affinity (Calcium-Green-1, Fluo-3, Fluo-4, Fura-2, Indo-1, Oregon Green 488 BAPTA, Ca^{2+} Yellow, Ca^{2+} Orange, Ca^{2+} Crimson, and X-Rhod/Rhod-2), and low-affinity (Mag-Fura-2, Mag-Fluo-4, Mag-Indo-1, Mag-Fura-5, Mag-Fura-Red, Fura-2-ff, Fluo-5N, Oregon Green BAPTA-5N, Rhod-5N, Rhod-FF, X-rhod-5F, X-rhod-FF) calcium indicators has been also published by Paredes et al. in 2008.

Examples of specific dyes excited in the visible range of the spectrum are Calcium-Green 2 and Fluo-1. An 80-fold increase in the fluorescence intensity of Fluo-1 can be recorded upon binding to Ca^{2+}. Zottini and Zannoni (1993) also reported the first measurement of $[Ca^{2+}]$ in plant mitochondria using the fluorescent Ca^{2+} indicator fura-2/AM (Tsien 1981), which can be successfully trapped into the matrix of mitochondria. As well, Huang et al. (1997) uses also Fura-2 to detect increases in cytosolic Ca^{2+} in parsley mesophyll senescent cells. Subbaiah et al. (1998) investigated the relationship between mitochondrial and cytosolic Ca^{2+} changes in anoxic maize cells using the positive charged dye Rhod-2 AM, which has a dissociation constant (K_d) of 570 nm for Ca^{2+} and can be accumulated within the matrix of the mitochondria. The non-ratiometric dyes Fluo-3 with a K_d of ~390 nm (Minta et al. 1989) and Fluo-4, which binds to calcium with similar affinity but with a substantially higher fluorescence output (Gee et al. 2000), allow the flow cytometric measurement of calcium on instruments that are not equipped with a UV light source (June and Moore 2004). However, due to the difficulties on calibration, June and Moore (2004) recommended the combined use of Fluo-3 and Fura Red. One of the most suitable fluorescent probes for the study of calcium by flow cytometry is Indo-1. This is a ratiometric dye, excited in the ultraviolet (338 nm), and its emission spectra shifts following calcium binding (maximum emission for Indo-1 in the absence of Ca^{2+} is 490 nm, and 405 nm if bound to Ca^{2+}). Ratio measurements (405/490) allow accurate quantifying of Ca^{2+} concentrations by flow cytometry. Indo-1 is a fast response-dye, and its K_d is 230 nM. Loading can be performed simply by addition of the acetoxymethyl ester form of Indo-1 in the extracellular medium. Darjania et al. (1993) have measured calcium concentration in *Vicia faba* protoplasts using indo-1 and fluorometry. Bush and Jones (1987, 1990) have developed a methodology for measuring calcium changes in aleurone protoplasts by fluorometry, using this dye, as well as Allen et al. (1999) in *Arabidopsis* guard cells.

However, as previously commented for intracellular pH measurement, the non-invasive detection of cytosolic and intraorganellar calcium has been increasingly used in the last years, displacing the use of fluorescent dyes in flow cytometry. The use of bioluminescent protein aequorin, a genetically encoded Ca^{2+} sensor with three calcium-binding sites, as well as the fusion of aequorin with GFP (green fluorescent protein) or other photoproteins, highly improved the measurement of mitochondrial, chloroplastic and endoplasmic reticulum [Ca^{2+}], among others (recently reviewed in Bakayan et al. 2017). As well, the recent development of genetically encoded fluorescent indicators, in which Ca^{2+} modifies the fluorescence of a circularly permutated GFP (camgaroos and pericams) or stimulates the reversible association of two GFP mutants of different colors (cameleons), has allowed a faster, more accurate and non-invasive measurement of intracellular calcium (Demaurex and Frieden 2003). The genetically encoded fluorescent Ca^{2+} indicators Yellow Cameleons (YCs), which have cyan and yellow fluorescent proteins (CFP and YFP), have been successfully used in the last years to measure [Ca^{2+}] in guard cells in response to abscisic acid and methyl jasmonate, in roots to analyze responses to salt, hormones, membrane hyperpolarization, or mechanical stimulation, and in leaves to detect calcium in response to extra-cellular ATP, touch, cold, and hydrogen peroxide (Kudla et al. 2010; Swanson et al. 2011; Bonza et al. 2013; Martí et al. 2013; Behera et al. 2015; Loro et al. 2016). Recently, Doucette and collaborators (2016) recommended the use of flow cytometry in the detection of these genetically encoded sensors, based on Förster resonance energy transfer (FRET) between fluorescent proteins (FPs), to avoid heterogeneity in the FRET ratio and the variability of microscopic methods. The use of a cytometer with laser capable of exciting CFP (cyan fluorescent protein) can allow the measurement of CFP-YFP FRET, as that from yellow cameleons (YC). Intermolecular and intramolecular FRET can be measured through flow cytometry, as previously demonstrated by Dye (2005), Adachi and Tsubata (2008), and Doucette et al. (2016). Although different limitations are still in discussion to be overcome with this method (i.e. accurate FRET ratio measurements can only be made for cells expressing relatively high levels of the reporter, and can only provide information about population average behavior, not the relationship between different parameters in a single cell), Doucette et al. (2016) suggest that multiplexed cytometric analysis of intramolecular FP FRET signals could be successfully used to investigate signal transduction cascades, ion fluxes, and metabolism, or to screen compound collections.

Some ionophores and other substances used for validation and calibration are Ionomycin, which increases the permeability of biological membranes to calcium allowing the concentration-dependent flux of this ion across membranes, and has a higher affinity by calcium at neutral and alkaline pH; Ionophore 4-bromo-A23187, which also binds to Ca^{2+} but its affinity is higher at acidic pH; $CaCl_2$, which induces a massive entry of Ca^{2+} into the cell; $MnCl_2$, which induces fluorescence quenching of all calcium-specific probes; and some chelator agents like BAPTA [1,2-bis(2-aminophenoxy), ethane-N,N,N′,N′-tetraacetic acid] and EGTA [ethylene glycol bis (β-aminoethyl ether)], which bind free calcium ion and are used, in presence of ionophore, for reducing or regulating extracellular (or even intracellular) free-Ca^{2+} concentrations.

Easily reproducible protocols for flow cytometric measurements of intracellular Ca^{2+} concentrations can be found in June and Moore (2004), Vines et al. (2010), Posey et al. (2015), and Doucette et al. (2016).

2.3 Reactive Oxygen Species Generation

The incomplete reduction of the molecular oxygen in plants originates molecules largely known as Reactive Oxygen Species (ROS) and their derivatives. Among the derivatives we can distinguish three types (Das and Roychoudhury 2014):

- Non-radicals such as singlet oxygen ($^1O_2$), hydrogen peroxide (H_2O_2), and ozone (O_3);
- Free radicals characterized by one or more unpaired electrons e.g. alkoxyl ($RO^·$), superoxide ($O_2^{·-}$), peroxyl ($ROO^·$), hydroxyl ($OH^·$), hydroperoxyl ($HO_2^·$);
- Peroxynitrite ($ONOO^-$), which derivates from the reaction of superoxide and nitric oxide ($^·NO$), acting both as reactive nitrogen species (RNS) and ROS.

In plants, ROS play a pivotal role in several biological processes such as plant development, plant signaling, stress response, etc. (Sharma et al. 2012).

ROS can be generated in several cell compartments and organelles, especially in those characterized by high electron transport rates (e.g. mitochondria, chloroplasts and peroxisomes) (Apel and Hirt 2004). Their accumulation could induce several damages to those organelles but plants possess ROS-scavenging enzymatic and non-enzymatic mechanisms that in normal growing conditions are able to prevent ROS-mediated toxicity (Ortega-Villasante et al. 2005; Demidchik 2015; Petrov et al. 2015). During stress conditions, these defense mechanisms could be overridden or inhibited resulting in ROS accumulation that, as a consequence, could induce cell death through protein degradation/denaturation, lipid peroxidation and nucleotides degradation (Ortega-Villasante et al. 2005; Demidchik 2015; Petrov et al. 2015).

Several ways have been developed for characterizing and quantifying ROS and their radical scavenging enzymes (Cakmak and Marschner 1992; Kruk et al. 2005; Jambunathan 2010). Among them, spectrophotometric methods and in situ localization using dyes are among the most historically used (Zhou et al. 2006; Ortega-Villasante et al. 2016). In the last decades, the use of flow cytometry is quickly growing and several specific probes have been developed for selective ROS detection (O'Brien et al. 1997; Walrand et al. 2003; Eruslanov and Kusmartsev 2010).

For singlet oxygen detection both Singlet Oxygen Sensor Green reagent (SOSG) and trans-1-(2-Methoxyvinyl)pyrene are the most widely used probes (Driever et al. 2009; Flors et al. 2006; Tang et al. 2009; Thompson et al. 1986a). Both are extremely selective for singlet oxygen but whereas the SOSG reagent is the most used for singlet oxygen detection, the trans-1-(2-Methoxyvinyl) pyrene is extremely sensitive allowing the detection of picomole quantities of this ROS (Posner et al. 1984; Thompson et al. 1986b). The SOSG, before reacting with singlet oxygen, is characterized by a slightly blue fluorescence with excitation peaks at 372 and 393 nm, and emission peaks at 395 and 416 nm. Once exposed to singlet oxygen, it emits a green fluorescence with a maximum of excitation/emission around 504/525 nm (Wiederschain 2011). Care should be taken during the experiments, since the fluorescence could quickly degrade in some solutions and alkaline pH could stimulate green fluorescence in absence of the ROS (Burns et al. 2012). Therefore, it is really important to have a control in order to correlate the intensity of the green fluorescence with singlet oxygen concentration. Moreover, could be useful to have a positive control, using chemicals such as Hypericin, to induce in health cells and/or tissues singlet oxygen production (Thomas et al. 1992; Triantaphylidès and Havaux 2009).

Concerning the applications of these probes in plant science, Pattanayak et al. (2012) used the SOSG probe to demonstrate that the accelerated cell death 2 protein (ACD2) localizes dynamically during infection to protect cells from pro-death mobile substrate molecules, some of which may originate in chloroplasts, but have major effects on mitochondria. Moreover, Wang et al. (2015) used this probe to demonstrate that the enhanced transcription of CYP38, mediated by ROS, increases plant tolerance to high light stress.

Superoxide radical has been implicated in the plant response to a wide number of stress conditions (Alscher et al. 2002; Mittler 2002; Wang et al. 2003). Since its lifetime is extremely short (in the order of nanoseconds) it generally induces peroxidation only when it is produced in close proximity to their targets (Georgiou et al. 2008). Superoxide radical detection could be achieved using the Dihydroethidium and its cationic derivative known as red mitochondrial superoxide indicator (Mito SOX™). Both chemicals are extensively used in superoxide detection even if Mito SOX was designed for a highly selective detection of superoxide in the mitochondria of living cells (Wiederschain 2011).

Dihydroethidium, also called hydroethidine, is characterized by a blue-fluorescence in the cytosol, once it intercalates within the cell's DNA it oxidizes staining the organelles (nuclei, mitochondria, etc.) with a bright red fluorescence. Moreover, when intracellular peroxidases, in combination with reactive oxygen species such as superoxide, catalyze the oxidative reaction, a highly red fluorescent product should be observed due to ethidium production.

This compound stains the cytoplasm of living cells in blue with an excitation/emission at 370/420 nm and chromatin of living cells in red with an excitation/emission at 535/610 nm (Wiederschain 2011).

As previously said, Mito SOX™ is extremely selective for mitochondria. Therefore, it should be the first choice if a selective localization of superoxide radical is needed. The high selectivity of this chemical is due to the cationic triphenylphosphonium substituent, which is responsible for the uptake of the probe in actively respiring mitochondria. Superoxide-driven Mito SOX oxidation leads to the production of 2-hydroxyethidium that exhibits a fluorescent excitation peak at 400 nm, which is absent in the excitation spectrum of the ethidium oxidation product generated by others ROS, thus conferring to this chemical its high selectivity (Wiederschain 2011).

Cid et al. (1996) used dihydroethidium in flow cytometry experiments to evaluate copper toxicity on the membrane system of a marine diatom, observing a time dependent peroxidase activity in response to copper treatment. Moreover, Bradner and Nevalainen (2003) used the dihydroethidium coupled to flow cytometry to evaluate the oxidative status on spores and mycelial growth of antarctic microfungi exposed to prohibitive temperatures.

Concerning H_2O_2 quantification, the most used probes in flow cytometry applied to plants are both 2′,7′-dichlorodihydrofluorescin-diacetate (H_2DCFDA excitation/emission ~492–495/517–527 nm) and Dihydrorhodamine 123 (DRH123 excitation/emission ~488/560 nm).

DRH123 is an uncharged, non-fluorescent ROS indicator that, once passively diffused across membranes, is oxidized by peroxidase to cationic rhodamine 123, which exhibits green fluorescence, mainly localized in the mitochondrial inner membrane and has a sensitivity to H_2O_2 that is extremely higher than H_2DCFDA.

On the contrary, H_2DCFDA, after diffusion into the cell, is firstly deacetylated to a non-fluorescent compound by cellular esterases and peroxidases and then oxidized by ROS to 2′,7′–dichlorofluorescein (DCF), which is a highly fluorescent chemical and its localization is not limited to mitochondria. The higher is the concentration of H_2O_2 the greater is the fluorescent signal emitted by the probe. As reported by Tsuchiya et al. (1994) in animal tissues, the use of H_2DCFDA could be coupled with propidium iodide to simultaneously monitor oxidant production and cell injury. This technique could be also applied to plant cells.

As reported by Haugland et al. (1996), the inconvenience of both probes is that they can be oxidized by cytochrome c, and by the oxidative phosphorylation occurring in the mitochondria, making difficult to perform a quantitative estimation of ROS activity and allowing only a comparison of relative fluorescence intensities.

Concerning the potential applications of these probes in plant studies, Cronjé et al. (2004) used H_2DCFDA probe to evaluate, in tobacco protoplasts cells, if potentiation of heat-induced Hsp70 mediated by salicylic acid somministration could contribute to a reduction in apoptosis.

Saison et al. (2010), in green algal cultures treated with core–shell copper oxide nanoparticles, observed, using H_2DCFDA, an induction of cellular aggregation processes and a ROS-mediated deteriorative effect on chlorophyll by inducing the photoinhibition of photosystem II. Moreover, Joo et al. (2001) used flow cytometry and H_2DCFDA to evaluate the role of ROS in mediating the gravitropic response of maize roots trough the induction of auxin production. In particular, they observed that the scavenging of ROS by antioxidants inhibited root gravitropism, concluding that ROS generation is pivotal in root gravitropism.

2.4 Glutathione Levels

Glutathione (GSH), and its functionally homologous thiol, is a metabolite essential for several animal and plant functions. Its pivotal role in mammals metabolism has been largely discussed. In fact, it has been demonstrated that the inhibition, in newborn rats, of GSH production caused multi-organ failures followed by a rapid death (Meister 1994).

In plant cells, numerous physiological functions have been attributed to GSH (Rennenberg 1995; Noctor et al. 2012) and one of the most known is its role in protecting cells against damages incurred by free radicals and oxidants. In the past, H_2O_2 scavenging was mainly attributed to ascorbic acid, whereas the important role of GSH was les apparent. Nowadays, it is well known that in presence of both, GSH and H_2O_2, the enzyme glutathione peroxidase reduces free hydrogen peroxide to water according to the following formula:

$$2\text{GSH} + \text{H}_2\text{O}_2 \rightarrow \text{GS} - \text{SG}(\text{glutathione disulfide}) + 2\text{H}_2\text{O}$$

Recent studies carried on Arabidopsis knockout mutants, characterized by GSH depletion due to a lack of the first enzyme involved in GSH synthesis, pointed out embryonic death during seed maturation (Cairns et al. 2006) a seedling-lethal phenotype (Pasternak et al. 2008). Moreover, transgenic Arabidopsis plants, where the tomato glutathione S-transferase was highly expressed, were characterized by a high resistance to both, drought and salt stress (Xu et al. 2015). In addition, it has been demonstrated that GSH is involved in various physiological processes, such as the regulation of sulfur assimilation (Maruyama-Nakashita and Ohkama-Ohtsu 2017), formaldehyde metabolism and detoxification (Haslam et al. 2002; Achkor et al. 2003), in defense against biotic stress (Ball et al. 2004; Parisy et al. 2007; Maughan et al. 2010) as well as in plant signaling and development (Noctor et al. 2012).

Due to the high importance of this chemical in plant metabolism and defense several techniques have been employed for its quantification. Among them, especially in human and animal field, flow cytometry is one of the most important. Several fluorescent probes have been developed to determine cellular levels of GSH and glutathione S-transferase (GST). But kinetic measurements under saturating substrate conditions is quite difficult because of the high and variable levels of intracellular glutathione, the multitude of glutathione S-transferase isozymes (Hedley and Chow 1994; Vanderven et al. 1994), and because of the reaction of the reagents used for GSH quantification with intracellular thiols other than glutathione (Bakker et al. 1991). Therefore, during GSH and GST quantification, is important to do preliminary experiments testing several dyes under controlled conditions, in which glutathione is depleted, to verify which dye fits better for the experiments purposes (Tauskela et al. 2000).

Among the dyes used for GSH detection, Bimanes was one of the most used although its use was recently substituted by ThiolTracker™ Violet reagent (excitation/emission maxima ~404/526 nm), which reacts with reduced thiols, and is more than tenfold brighter. This compound can cross living cell membranes, becoming cell-impermeable after reacting with cellular thiols (Held 2010). Therefore, living cells staining should be carried out in a thiol free buffer and then they could be immediately observed with a conventional xenon or mercury arc lamps or cells can be fixed with aldehyde before imaging. Although this compound is largely used for GSH detection and quantifications in human cells and other organisms (Rubio et al. 2011; Benson et al. 2015; Deorukhkar et al. 2015), no scientific articles are available regarding its use on plants.

Another probe used for GSH and thiols quantification is represented by the ortho-phthaldialdehyde (OPA) reagent. This is a non-fluorescent compound that, once loaded into the cells, forms two different fluorescent conjugates with both thiols and GSH, which could be observed using an excitation of 350 nm and an emission of 450, 525 and 575 nm for OPT-GSH conjugates and 405 nm for OPT-thiols conjugates. As previously reported by several authors, to get a reliable estimation of GSH content it is extremely important to have a ratiometric measurement of both types of conjugates (Treumer and Valet 1986; Haugland et al. 1996; Coba de la Peña 2001).

The thiol-reactive probes CellTracker™ Green CMFDA (based on 5-chloromethyl fluorescein diacetate) is another interesting probe, largely used in both flow cytometry and laser scanning microscopy, for intracellular GSH and intracellular thiols quantification using visible light with an argon-ion laser (Lilius et al. 1996). The conjugated products obtained from the reaction of this probe with GSH have higher fluorescence than monochlorobimane. After cells loading, the dye is well retained and it can be observed using an excitation/emission of 492/517 nm. This probe is largely used in plant field, both in flow cytometry and confocal microscopy studies (Vivancos et al. 2010; De Simone et al. 2015; Munoz et al. 2016). In particular, Munoz (2014) and Munoz et al. (2016) used this probe coupled to flow cytometry on the microalga *Chlorella vulgaris* to understand the mechanisms of detoxification of various forms of arsenic and dimethylarsenic acid.

Finally, a probe commonly used for GSH quantitation and GST activity is the non-fluorescent cell-permanent monochlorobimane, which once conjugated with thiols emits a strong blue fluorescence (Coleman et al. 1997). This probe could be used in both living cells as well as in tissue homogenates, and the glutathione conjugate of monochlorobimane has absorption/emission maxima ~394/490 nm (Kamencic et al. 2000). The possibility to work with tissue homogenates allows GSH quantification also through fluorescence spectrophotometry. In fact, this technique was used by Meyer et al. (2001) to quantify GSH level in Arabidopsis cells. Moreover, it has been used to analyze tracheary element differentiation in *Zinnia elegans* cells (Weir et al. 2005) and in studies focused on the reversibility of early stages of apoptosis in plant cells (O'Brien et al. 1998).

2.5 Membrane Potential

Electric potential differences across membranes of prokaryotic and eukaryotic cells reflect the differential distribution and activity of ions such as Na^+, Cl^-, H^+ and especially K^+ across these biological membranes. Diverse membrane electrogenic pumps generate these ionic gradients, with a contribution from the intrinsic membrane permeability for each ion. Membrane potential plays a major role in the processes involving external stimulation of the cell, photosynthesis, nutrient and ion transport across the membrane, and signal transduction. In eukaryotic cells, major examples are cytoplasmic, mitochondrial (inner membrane) and lysosome membrane potential, negative inside the cell (or inside the organelles) relative to the external medium. In chloroplasts, the thylakoid potential is relatively more positive inside, but here the major electrochemical gradient is due to protons, the lumen being acid and the stroma alkaline. The mean potential values in eukaryotic cells are between −10 and −100 mV. In the mitochondria, the potential values are around −100 mV, and −50 mV in lysosomal membranes (Shechter 1984).

Membrane potential changes involve either depolarization (that is, a decrease in transmembrane potential) or hyperpolarization (an increase in the potential difference across the membrane).

Many excellent reviews are available concerning the fluorometric methods and probes developed for estimation of membrane potential, especially in organelles or in cells that are too small to allow the use of microelectrodes (Montana et al. 1989; Gross and Loew 1989; Loew 1993; Haugland et al. 1996; Roy and Hajnóczky 2009; Sabnis 2015).Dyes usually used in flow cytometry are molecules with a single negative or positive net charge, are highly hydrophobic and their partition across the membrane is a function of the Nernst equation:

$$[C]_{in} / [C]_{out} = e^{-nFE/RT}$$

where n = net electric charge of the indicator; [C] = intra and extracellular concentration of the indicator; F = Faraday constant; R = gas constant; T = temperature (°K); E = membrane potential.

These dyes are excited at the visible range of the spectrum, and with slow response to environmental changes in membrane potential ($\Delta\Psi$; Plásek and Sigler 1996). In general, fluorescence dyes' detection by flow cytometry is useful for finding differences in $\Delta\Psi$ within or between populations rather than for assigning specific values of $\Delta\Psi$ (Shapiro 2004).

Oxonol dyes have one net negative electric charge, and they will accumulate principally in the external volume of a negatively charged membrane, a lesser portion of the dye being retained in the internal compartment. A hyperpolarization of the membrane produces dye redistribution to the external medium. These dyes are excluded from the negatively charged inner mitochondrial compartment and their fluorescence reflects mainly the plasmic membrane potential. One example is the Oxonol dye $DiBaC_4(3)$, commonly used in cytometry, because it can be excited at

488 nm. These dyes are very sensitive to the variation of external ionic concentrations.

Cyanine dyes have one net positive electric charge at physiological pH, so their cellular partitioning is the contrary of oxonol dyes. These dyes are also partially accumulated in some organelles with negative inner membrane potential, like mitochondria and endoplasmic reticulum, and they are relatively toxic to cells. The cellular fluorescence intensity reflects membrane potential from the plasma membrane and also mitochondrial and endoplasmic reticulum membranes. This class of dye is the most used in flow cytometry. Two examples are the Carbocyanines $DiOC_6(3)$ and $DiOC_5(3)$. These dyes can undergo quenching when they are at high local concentration and polymerize.

The Carbocyanine dye JC-1 can be used for the study of mitochondrial potential. At low local dye concentration (low potential), the molecule is in the monomeric state with green fluorescence emission (527 nm) when excited at 490 nm. When the mitochondria are hyperpolarized, the local dye concentration increases and it forms polymer conjugates (J-conjugates) with a shifted red fluorescence (590 nm). This property makes possible ratiometric red/green fluorescence measurements in flow cytometry.

Merocyanine dyes undergo molecular reorientations with membrane depolarization, forming fluorescent dimers with altered absorption spectra. Merocyanine 540, principally associated with unsaturated lipids, is a common example.

Finally, we can cite Rhodamine 123. Its incorporation depends on the voltage gradient of the mitochondrial inner membrane, and it is less toxic that Carbocyanine dyes. This dye is used in tests for early modifications of energy metabolism.

Validation of the specificity of the dye fluorescence is done using some ionophores that modify membrane potential:

- Valinomycin facilitates the passages of K+ ions down their concentration gradient across the membrane.
- Gramicidine D makes pores in the membrane, facilitating the free passage of mono and divalent ions, and it is generally used for membrane depolarization.
- Vanadate inhibits the ATPase proteins susceptible to phosphorylation, that are mainly the cytoplasmic membrane ATPases.
- Regarding mitochondrial potential, FCCP (carbonyl cyanide *p*-trifluoro methoxyphenyl hydrazone) induces an increase in mitochondrial membrane permeability for H^+ passage, producing depolarisation.
- Nigericine is used to cancel pH gradient, and thereby to hyperpolarize the mitochondrial membrane because there is an interchange that is electrically neutral between K^+ and H^+ ions.

Examples of calibration curves and flow cytometric determinations of absolute membrane potential are shown in Krasznai et al. (1995). A good protocol to give an estimation of membrane potential ($\Delta\Psi$) by flow cytometry, as a function of the distribution of fluorescent lipophilic dye between cells and surrounding medium is provided in Shapiro (2004).

Acknowledgement This chapter is an update of 'Coba de la Peña (2001) Use of flow cytometry to measure physiological parameters. In: Reigosa MJ (Ed), Handbook of Plant Ecophysiology Techniques. Kluwer Academic Publishers, The Netherlands, pp. 53–64.'

References

Achkor H, Díaz M, Fernández MR, Biosca JA, Parés X, Martínez MC (2003) Enhanced formaldehyde detoxification by overexpression of glutathione-dependent formaldehyde dehydrogenase from Arabidopsis. Plant Physiol 132:2248–2255

Adachi T, Tsubata T (2008) Fret-based Ca^{2+} measurement in B lymphocyte by flow cytometry and confocal microscopy. Biochem Biophys Res Commun 367:377–382

Allan AC, Fluhr R (1997) Two distinct sources of elicited reactive oxygen species in tobacco epidermal cells. Plant Cell 9:1559–1572

Allen GJ, Kuchitsu K, Chu SP, Murata Y, Schroeder JI (1999) Arabidopsis abi1-1 and abi2-1 phosphatase mutations reduce abscisic acid-induced cytoplasmic calcium rises in guard cells. Plant Cell 11(9):1785

Alscher RG, Erturk N, Heath LS (2002) Role of superoxide dismutases (SODs) in controlling oxidative stress in plants. J Exp Bot 53:1331–1341

Apel K, Hirt H (2004) Reactive oxygen species: metabolism, oxidative stress, and signal transduction. Annu Rev Plant Biol 55:373–399

Bakayan A, Domingo B, Vaquero CF, Peyrieras N, Llopis J (2017) Fluorescent protein–photoprotein fusions and their applications in calcium imaging. Photochem Photobiol 93:448–465

Bakker P, Stap J, Tukker C, Van Oven C, Veenhof C, Aten J (1991) An indirect immunofluorescence double staining procedure for the simultaneous flow cytometric measurement of iodo- and chlorodeoxyuridine incorporated into DNA. Cytometry A 12:366–372

Ball L, Accotto GP, Bechtold U, Creissen G, Funck D, Jimenez A, Kular B, Leyland N, Mejia-Carranza J, Reynolds H, Karpinski S, Mullineaux PM (2004) Evidence for a direct link between glutathione biosynthesis and stress defense gene expression in Arabidopsis. Plant Cell 16:2448–2462

Behera S, Wang N, Zhang C, Schmitz-Thom I, Strohkamp S, Schültke S, Hashimoto K, Xiong L, Kudla J (2015) Analyses of Ca^{2+} dynamics using a ubiquitin-10 promoter-driven Yellow Cameleon 3.6 indicator reveal reliable transgene expression and differences in cytoplasmic Ca^{2+} responses in Arabidopsis and rice (*Oryza sativa*) roots. New Phytol l206:751–760

Benčina M (2013) Illumination of the spatial order of intracellular pH by genetically encoded pH-sensitive sensors. Sensors 13:16736–16758

Benson KF, Newman RA, Jensen GS (2015) Antioxidant, anti-inflammatory, anti-apoptotic, and skin regenerative properties of an *Aloe vera*-based extract of *Nerium oleander* leaves (NAE-8®). Clin Cosmet Investig Dermatol 8:239

Bonza MC, Loro G, Behera S, Wong A, Kudla J, Costa A (2013) Analyses of Ca^{2+} accumulation and dynamics in the endoplasmic reticulum of Arabidopsis root cells using a genetically encoded Cameleon sensor. Plant Physiol 163:1230–1241

Bradner J, Nevalainen K (2003) Metabolic activity in filamentous fungi can be analysed by flow cytometry. J Microbiol Met 54:193–201

Burns JM, Cooper WJ, Ferry JL, King W, DiMento BP, McNeill K, Miller CJ, Miller WL, Peake BM, Rusak SA, Rose AL, Waite TD (2012) Methods for reactive oxygen species (ROS) detection in aqueous environments. Aquatic Sci 74:683–734

Bush DS, Jones RL (1987) Measurement of cytoplasmic calcium in aleurone protoplasts using indo-1 and fura-2. Cell Calcium 8:455–472

Bush DS, Jones RL (1990) Measuring intracellular levels in plant cells using the fluorescent probes, indo-1 and fura-2. Plant Physiol 93:841–845

Cairns NG, Pasternak M, Wachter A, Cobbett CS, Meyer AJ (2006) Maturation of Arabidopsis seeds is dependent on glutathione biosynthesis within the embryo. Plant Physiol 141:446–455

Cakmak I, Marschner H (1992) Magnesium deficiency and high light intensity enhance activities of superoxide dismutase, ascorbate peroxidase, and glutathione reductase in bean leaves. Plant Physiol 98:1222–1227

Cid A, Fidalgo Paredes P, Herrero C, Abalde J (1996) Toxic action of copper on the membrane system of a marine diatom measured by flow cytometry. Cytometry 25:32–36

Coba de la Peña TC (2001) Use of flow cytometry to measure physiological parameters. In: Handbook of plant ecophysiology techniques. Springer, p 53–64

Coleman J, Randall R, Blake-Kalff M (1997) Detoxification of xenobiotics in plant cells by glutathione conjugation and vacuolar compartmentalization: a fluorescent assay using monochlorobimane. Plant Cell Environ 20:449–460

Cronjé MJ, Weir IE, Bornman L (2004) Salicylic acid-mediated potentiation of Hsp70 induction correlates with reduced apoptosis in tobacco protoplasts. Cytometry A 61:76–87

Darjania J, Curvetto N, Delmastro S (1993) Loading of *Vicia faba* guard cell protoplasts with indo-1 to measure cytosolic calcium concentration. Plant Physiol Biochem 31:793–798

Das K, Roychoudhury A (2014) Reactive oxygen species (ROS) and response of antioxidants as ROS-scavengers during environmental stress in plants. Front Environ Sci 2:53

Day RN, Davidson MW (2009) The fluorescent protein palette: tools for cellular imaging. Chem Soc Rev 38:2887–2921

De Simone A, Dong Y, Vivancos PD, Foyer CH (2015) GSH partitioning between the nucleus and cytosol in *Arabidopsis thaliana*. In: De Kok LJ, Hawkesford MJ, Rennenberg H, Saito K, Schnug E (eds) Molecular physiology and ecophysiology of sulfur. Springer, Cham, pp 37–48

Demaurex N, Frieden M (2003) Measurements of the free luminal ER Ca^{2+} concentration with targeted "cameleon" fluorescent proteins. Cell Calcium 34:109–119

Demidchik V (2015) Mechanisms of oxidative stress in plants: from classical chemistry to cell biology. Environ Exp Bot 109:212–228

Deorukhkar A, Ahuja N, Mercado AL, Diagaradjane P, Raju U, Patel N, Mohindra P, Diep N, Guha S, Krishnan S (2015) Zerumbone increases oxidative stress in a thiol-dependent ROS-independent manner to increase DNA damage and sensitize colorectal cancer cells to radiation. Cancer Med 4:278–292

Doucette J, Zhao Z, Geyer RJ, Barra M, Balunas MJ, Zweifach A (2016) Flow cytometry enables multiplexed measurements of genetically encoded intramolecular FRET sensors suitable for screening. J Biomol Screen 21:535–547

Driever SM, Fryer MJ, Mullineaux PM, Baker NR (2009) Imaging of reactive oxygen species in vivo. In: Pfannschmidt T (ed) Plant signal transduction, Methods in Molecular Biology, vol 479. Humana Press, Totowa, pp 109–116

Dye BT (2005) Flow cytometric analysis of CFP–YFP FRET as a marker for in vivo protein–protein interaction. Clin Appl Immunol Rev 5:307–324

Edel KH, Marchadier E, Brownlee C, Kudla J, Hetherington AM (2017) The evolution of calcium-based signalling in plants. Curr Biol 27:667–679

Eruslanov E, Kusmartsev S (2010) Identification of ROS using oxidized DCFDA and flow-cytometry. In: Advanced protocols in oxidative stress II. Springer, p 57–72

Felle H (1988) Short-term pH regulation in plants. Physiol Plant 74:583–591

Flors C, Fryer MJ, Waring J, Reeder B, Bechtold U, Mullineaux PM, Nonell S, Wilson MT, Baker NR (2006) Imaging the production of singlet oxygen in vivo using a new fluorescent sensor, Singlet Oxygen Sensor Green®. J Exp Bot 57:1725–1734

Franklin NM, Stauber JL, Lim RP (2001) Development of flow cytometry-based algal bioassays for assessing toxicity of copper in natural waters. Environ Toxicol Chem 20:160–170

Gao D, Knight MR, Trewavas AJ, Sattelmacher B, Plieth C (2004) Self-reporting Arabidopsis expressing pH and [Ca^{2+}] indicators unveil ion dynamics in the cytoplasm and in the apoplast under abiotic stress. Plant Physiol 134:898–908

Gee KR, Brown KA, Chen WN, Bishop-Stewart J, Gray D, Johnson I (2000) Chemical and physiological characterization of fluo-4 Ca(2+)-indicator dyes. Cell Calcium 27:97–106

Georgiou CD, Papapostolou I, Grintzalis K (2008) Superoxide radical detection in cells, tissues, organisms (animals, plants, insects, microorganisms) and soils. Nat Protoc 3:1679

Giglioli-Guivarch N, Pierre JN, Vidal J, Brown S (1996) Flow cytometric analysis of cytosolic pH of mesophyll cell protoplasts from the crabgrass *Digitaria sanguinalis*. Cytometry 23:241–249
Gjetting SK, Ytting CK, Schulz A, Fuglsang AT (2012) Live imaging of intra- and extracellular pH in plants using pHusion, a novel genetically encoded biosensor. J Exp Bot 63:3207–3218
Gonugunta VK, Srivastava N, Puli MR, Raghavendra AS (2008) Nitric oxide production occurs after cytosolic alkalinization during stomatal closure induced by abscisic acid. Plant Cell Environ 31:1717–1724
Gross D, Loew LM (1989) Fluorescent indicators of membrane potential: micro-spectrofluorometry and imaging. Meth Cell Biol 30:193–218
Grynkiewicz G, Poenie M, Tsien RY (1985) A new generation of indicators with greatly improved fluorescence properties. J Biol Chem 260:3440–3450
Haslam R, Rust S, Pallett K, Cole D, Coleman J (2002) Cloning and characterisation of S-formylglutathione hydrolase from *Arabidopsis thaliana*: a pathway for formaldehyde detoxification. Plant Physiol Biochem 40:281–288
Haugland RP (2003) Indicators for Ca^{2+}, Mg^{2+}, Zn^{2+} and other metal ions. In: Larison KD (ed) Handbook of fluorescent probes and research products, 9th edn. Molecular Probes, Eugene Chapter 20
Haugland RP, Spence MT, Johnson ID (1996) Handbook of fluorescent probes and research chemicals. Molecular Probes, Eugene 390 pp
Hedley DW, Chow S (1994) Evaluation of methods for measuring cellular glutathione content using flow cytometry. Cytometry A 15:349–358
Held P (2010) An introduction to reactive oxygen species: measurement of ros in cells. BioTek Instruments, Winooski
Huang FY, Philosoph-Hadas S, Meir S, Callaham DA, Hepler PK (1997) Increases in cytosolic Ca^{2+} in parsley mesophyll cells correlate with leaf senescence. Plant Physiol 115:51–60
Jambunathan N (2010) Determination and detection of reactive oxygen species (ROS), lipid peroxidation, and electrolyte leakage in plants. Methods Mol Biol 639:292–298
Joo JH, Bae YS, Lee JS (2001) Role of auxin-induced reactive oxygen species in root gravitropism. Plant Physiol 126:1055–1060
June CH, Moore JS (2004) Measurement of intracellular ions by flow cytometry. Curr Protocol Immunol 5:5.5.1–5.5.20
Kamencic H, Lyon A, Paterson PG, Juurlink BH (2000) Monochlorobimane fluorometric method to measure tissue glutathione. Anal Biochem 286:35–37
Kneen M, Farinas J, Li Y, Verkman AS (1998) Green fluorescent protein as a noninvasive intracellular pH indicator. Biophys J 74:1591–1599
Krasznai Z, Márián T, Balkay L, Emri M, Trón L (1995) Flow cytometric determination of absolute membrane potential of cells. J Photoch Photobio 28:93–99
Kruk I, Aboul-Enein HY, Michalska T, Lichszteld K, Kładna A (2005) Scavenging of reactive oxygen species by the plant phenols genistein and oleuropein. Luminescence 20:81–89
Kudla J, Batistic O, Hashimoto K (2010) Calcium signals: the lead currency of plant information processing. Plant Cell 22:541–563
Kurkdjian A, Guern J (1989) Intracellular pH: measurement and importance in cell activity. Annu Rev Plant Physiol Plant Mol Biol 40:271–303
Lam AJ, St-Pierre F, Gong Y, Marshall JD, Cranfill PJ, Baird MA, McKeown MR, Wiedenmann J, Davidson MW, Schnitzer MJ (2012) Improving FRET dynamic range with bright green and red fluorescent proteins. Nat Methods 9:1005–1012
Li Y, Tsien RW (2012) pHTomato, a red, genetically encoded indicator that enables multiplex interrogation of synaptic activity. Nat Neurosci 15:1047–1053
Lilius H, Haestbacka T, Isomaa B (1996) A combination of fluorescent probes for evaluation of cytotoxicity and toxic mechanisms in isolated rainbow trout hepatocytes. Toxicol In Vitro 10:341–348
Loew LM (1993) Potentiometric membrane dyes. In: Masson WT (ed) Fluorescent and luminiscent probes for biological activity, A Practical Guide to Technology for Quantitative Real-Time Analysis. Academic, London

Loro G, Wagner S, Doccula FG, Behera S, Weinl S, Kudla J, Schwarzländer M, Costa A, Zottini M (2016) Chloroplast-specific in vivo Ca^{2+} imaging using Yellow Cameleon fluorescent protein sensors reveals organelle-autonomous Ca^{2+} signatures in the stroma. Plant Physiol 171:2317–2330

Maeda Y, Ide T, Koike M, Uchiyama Y, Kinoshita T (2008) GPHR is a novel anion channel critical for acidification and functions of the Golgi apparatus. Nat Cell Biol 10:1135–1145

Martí MC, Stancombe MA, Webb AAR (2013) Cell- and stimulus type-specific intracellular free Ca2+ signals in Arabidopsis. Plant Physiol 163:625–634

Maruyama-Nakashita A, Ohkama-Ohtsu N (2017) Sulfur assimilation and glutathione metabolism in plants. In: Hossain M, Mostofa M, Diaz-Vivancos P, Burritt D, Fujita M, Tran LS (eds) Glutathione in plant growth, development, and stress tolerance. Springer, Cham, pp 287–308

Maughan SC, Pasternak M, Cairns N, Kiddle G, Brach T, Jarvis R, Haas F, Nieuwland J, Lim B, Müller C, Salcedo-Sora E, Kruse C, Orsel M, Hell R, Miller AJ, Bray P, Foyer CH, Murray JA, Meyer AJ, Cobbett CS (2010) Plant homologs of the *Plasmodium falciparum* chloroquine-resistance transporter, PfCRT, are required for glutathione homeostasis and stress responses. P Natl A Sci 107:2331–2336

Meister A (1994) Glutathione-ascorbic acid antioxidant system in animals. J Biol Chem Paper Edition 269:9397–9400

Meyer AJ, May MJ, Fricker M (2001) Quantitative in vivo measurement of glutathione in Arabidopsis cells. Plant J 27:67–78

Miesenbock G, De Angelis DA, Rothman JE (1998) Visualizing secretion and synaptic transmission with pH-sensitive green fluorescent proteins. Nature 394:192–195

Minta A, Kao JP, Tsien RY (1989) Fluorescent indicators for cytosolic calcium based on rhodamine and fluorescein chromophores. J Biol Chem 264:8171–8178

Mittler R (2002) Oxidative stress, antioxidants and stress tolerance. Trends Plant Sci 7:405–410

Miyawaki A, Griesbeck O, Heim R, Tsien RY (1999) Dynamic and quantitative Ca^{2+} measurements using improved cameleons. Proc Nat Acad Sci USA 96:2135–2140

Monshausen GB, Bibikova TN, Messerli MA, Shi C, Gilroy S (2007) Oscillations in extracellular pH and reactive oxygen species modulate tip growth of Arabidopsis root hairs. Proc Natl Acad Sci U S A 104:20996–21001

Montana V, Farkas DL, Loew LM (1989) Dual wavelength radiometric fluorescence measurements of membrane potential. Biochemistry 28:4536

Moseyko N, Feldman LJ (2001) Expression of pH-sensitive green fluorescent protein in *Arabidopsis thaliana*. Plant Cell Environ 24:557–563

Munoz LP (2014) The mechanisms of arsenic detoxification by the green microalgae *Chlorella vulgaris*. Middlesex University Dissertation

Munoz LP, Purchase D, Jones H, Raab A, Urgast D, Feldmann J, Garelick H (2016) The mechanisms of detoxification of As (III), dimethylarsinic acid (DMA) and As (V) in the microalga *Chlorella vulgaris*. Aquat Toxicol 175:56–72

Noctor G, Mhamdi A, Chaouch S, Han Y, Neukermans J, Marquez-Garcia B, Queval G, Foyer CH (2012) Glutathione in plants: an integrated overview. Plant Cell Environ 35:454–484

Nomura H, Komori T, Kobori M, Nakahira Y, Shiina T (2008) Evidence for chloroplast control of external Ca^{2+}-induced cytosolic Ca^{2+} transients and stomatal closure. Plant J 53:988–998

O'Brien IE, Baguley BC, Murray BG, Morris BA, Ferguson IB (1998) Early stages of the apoptotic pathway in plant cells are reversible. Plant J 13:803–814

O'Brien IE, Reutelingsperger CP, Holdaway KM (1997) Annexin-V and TUNEL use in monitoring the progression of apoptosis in plants. Cytometry A 29:28–33

Ortega-Villasante C, Rellán-Alvarez R, Del Campo FF, Carpena-Ruiz RO, Hernández LE (2005) Cellular damage induced by cadmium and mercury in *Medicago sativa*. J Exp Bot 56:2239–2251

Ortega-Villasante C, Burén S, Barón-Sola Á, Martínez F, Hernández LE (2016) In vivo ROS and redox potential fluorescent detection in plants: present approaches and future perspectives. Meth 109:92–104

Paredes RM, Etzler JC, Watts LT, Lechleiter JD (2008) Chemical calcium indicators. Methods 46:143–151

Parisy V, Poinssot B, Owsianowski L, Buchala A, Glazebrook J, Mauch F (2007) Identification of PAD2 as a γ-glutamylcysteine synthetase highlights the importance of glutathione in disease resistance of Arabidopsis. Plant J 49:159–172

Pasternak M, Lim B, Wirtz M, Hell R, Cobbett CS, Meyer AJ (2008) Restricting glutathione biosynthesis to the cytosol is sufficient for normal plant development. Plant J 53:999–1012

Pattanayak GK, Venkataramani S, Hortensteiner S, Kunz L, Christ B, Moulin M, Smith AG, Okamoto Y, Tamiaki H, Sugishima M, Greenberg JT (2012) Accelerated cell death 2 suppresses mitochondrial oxidative bursts and modulates cell death in Arabidopsis. Plant J 69:589–600

Petrov V, Hille J, Mueller-Roeber B, Gechev TS (2015) ROS-mediated abiotic stress-induced programmed cell death in plants. Front Plant Sci 6:69

Plasek J, Sigler K (1996) Slow fluorescent indicators of membrane potential: a survey of different approaches to probe response analysis. J Photoch Photobio 33:101–124

Posey AD Jr, Kawalekar OU, June CH (2015) Measurement of intracellular ions by flow cytometry. Curr Protoc Cytom 72:9.8.1–9.8.21

Posner GH, Lever JR, Miura K, Lisek C, Seliger HH, Thompson A (1984) A chemiluminescent probe specific for singlet oxygen. Biochem Bioph Res Co 123:869–873

Rennenberg H (1995) Process involved in glutathione metabolism. In: Wallsgrove RM (ed) Amino acids and their derivatives in higher plants. Institute of arable Crops Research (IACR), Rothamsted, pp 155–171

Riveras E, Alvarez JM, Vidal EA, Oses C, Vega A, Gutiérrez RA (2015) The calcium ion is a second messenger in the nitrate signaling pathway of Arabidopsis. Plant Physiol 169:1397–1404

Robinson JP, Darzynkiewicz Z, Dean PN, Hibbs AR, Orfao A, Rabinovitch PS, Wheeless LL, Chow S, Hedley D (1997) Flow cytometric measurement of intracellular pH. Curr Protoc Cytom 00(1):9.3.1–9.3.10

Rodrigues R, Silva RD, Noronha H, Pedras A, Gerós H, Côrte-Real M (2013) Flow cytometry as a novel tool for structural and functional characterization of isolated yeast vacuoles. Microbiology 159:848–856

Roy SS, Hajnóczky G (2009) Fluorometric methods for detection of mitochondrial membrane permeabilization in apoptosis. Methods Mol Biol 559:173–190

Rubio V, Zhang J, Valverde M, Rojas E, Shi Z-Z (2011) Essential role of Nrf2 in protection against hydroquinone-and benzoquinone-induced cytotoxicity. Toxicol In Vitro 25:521–529

Sabnis RW (2015) Handbook of fluorescent dyes and probes. Wiley, New Jersey

Saison C, Perreault F, Daigle J-C, Fortin C, Claverie J, Morin M, Popovic R (2010) Effect of core–shell copper oxide nanoparticles on cell culture morphology and photosynthesis (photosystem II energy distribution) in the green alga, *Chlamydomonas reinhardtii*. Aquat Toxicol 96:109–114

Sakano K (2001) Metabolic regulation of pH in plant cells: role of cytoplasmic pH in defense reaction and secondary metabolism. Int Rev Cytol 206:1–44

Schulte A, Lorenzen I, Böttcher M, Plieth C (2006) A novel fluorescent pH probe for expression in plants. Plant Methods 2:7–21

Shapiro HM (2004) Estimation of membrane potential by flow cytometry. Curr Protoc Cytom 9(6):9.6.1–9.6.12

Sharma P, Jha AB, Dubey RS, Pessarakli M (2012) Reactive oxygen species, oxidative damage, and antioxidative defense mechanism in plants under stressful conditions. J Bot 2012:1–26

Shechter E (1984) In: Masson (ed) Membranes biologiques. Masson S.A., Paris

Smith FA (1979) Intracellular pH and its regulation. Annu Rev Plant Physiol 30:289–311

Subbaiah CC, Bush DS, Sachs MM (1998) Mitochondrial contribution to the anoxic Ca^{2+} signal in maize suspension-cultured cells. Plant Physiol 118:759–771

Swanson SJ, Choi WG, Chanoca A, Gilroy S (2011) In vivo imaging of Ca^{2+}, pH, and reactive oxygen species using fluorescent probes in plants. Annu Rev Plant Biol 62:273–297

Tang PM-K, Liu X-Z, Zhang D-M, Fong W-P, Fung K-P (2009) Pheophorbide a based photodynamic therapy induces apoptosis via mitochondrial-mediated pathway in human uterine carcinosarcoma. Cancer Biol Ther 8:533–539

Tauskela JS, Hewitt K, Kang LP, Comas T, Gendron T, Hakim A, Hogan M, Durkin J, Morley P (2000) Evaluation of glutathione-sensitive fluorescent dyes in cortical culture. Glia 30:329–341

Thomas C, MacGill RS, Miller GC, Pardini RS (1992) Photoactivation of hypericin generates singlet oxygen in mitochondria and inhibits succinoxidase. Photochem Photobiol 55:47–53

Thompson A, Biggley W, Posner G, Lever J, Seliger H (1986a) Microsomal chemiluminescence of benzo [a] pyrene-7, 8-dihydrodiol and its synthetic analogues trans-and cis-1-methoxyvinylpyrene. Biochim Biophys Acta 882:210–219

Thompson A, Seliger HH, Posner GH (1986b) Chemiluminescent probes for singlet oxygen in biological reactions. In: Methods in enzymology, vol 133. Elsevier, p 569–584

Treumer J, Valet G (1986) Flow-cytometric determination of glutathione alterations in vital cells by o-phthaldialdehyde (OPT) staining. Exp Cell Res 163:518–524

Triantaphylidès C, Havaux M (2009) Singlet oxygen in plants: production, detoxification and signaling. Trends Plant Sci 14:219–228

Tsien RY (1981) A non-disruptive technique for loading calcium buffers and indicators into cells. Nature 290:527–528

Tsuchiya M, Suematsu M, Suzuki H (1994) [12] In vivo visualization of oxygen radical-dependent photoemission. In: Methods in enzymology, vol 233. Elsevier, p 128–140

Tuteja N, Mahajan S (2007) Calcium signaling network in plants. Plant Signal Behav 2:79–85

Valkonen M, Mojzita D, Penttilä M, Benčina M (2013) Noninvasive high-throughput single-cell analysis of the intracellular pH of *Saccharomyces cerevisiae* by ratiometric flow cytometry. Appl Environ Microbiol 79:7179–7187

Vanderven AJ, Mier P, Peters WH, Dolstra H, Vanerp P, Koopmans PP, Vandermeer J (1994) Monochlorobimane does not selectively label glutathione in peripheral blood mononuclear cells. Anal Biochem 217:41–47

Vines A, McBean GJ, Blanco-Fernández A (2010) A flow-cytometric method for continuous measurement of intracellular Ca(2+) concentration. Cytometry A 77:1091–1907

Vivancos PD, Dong Y, Ziegler K, Markovic J, Pallardó FV, Pellny TK, Verrier PJ, Foyer CH (2010) Recruitment of glutathione into the nucleus during cell proliferation adjusts whole-cell redox homeostasis in *Arabidopsis thaliana* and lowers the oxidative defence shield. Plant J 64:825–838

Walrand S, Valeix S, Rodriguez C, Ligot P, Chassagne J, Vasson M-P (2003) Flow cytometry study of polymorphonuclear neutrophil oxidative burst: a comparison of three fluorescent probes. Clin Chim Acta 331:103–110

Wang W, Vinocur B, Altman A (2003) Plant responses to drought, salinity and extreme temperatures: towards genetic engineering for stress tolerance. Planta 218:1–14

Wang Y, Zeng L, Xing D (2015) ROS-mediated enhanced transcription of CYP38 promotes the plant toleran. Tauskela JS et al. (2000) Evaluation of glutathione-sensitive fluorescent dyes in cortical culture. Glia 30:329–341

Weir IE, Maddumage R, Allan AC, Ferguson IB (2005) Flow cytometric analysis of tracheary element differentiation in *Zinnia elegans* cells. Cytometry A 68:81–91

Wiederschain GY (2011) The molecular probes handbook. A guide to fluorescent probes and labeling technologies. Biochem Mosc 76:1276–1276

Xu J, Xing X-J, Tian Y-S, Peng R-H, Xue Y, Zhao W, Yao Q-H (2015) Transgenic Arabidopsis plants expressing tomato glutathione S-transferase showed enhanced resistance to salt and drought stress. PLoS One 10:e0136960

Zhou B, Wang J, Guo Z, Tan H, Zhu X (2006) A simple colorimetric method for determination of hydrogen peroxide in plant tissues. Plant Growth Regul 49:113–118

Zottini M, Zannoni D (1993) The use of Fura-2 fluorescence to monitor the movement of free calcium ions into the matrix of plant mitochondria. (*Pisum sativum* and *Helianthus tuberosus*). Plant Physiol 102:573–578

Chapter 12
Flow Cytometry: Cell Cycle

Teodoro Coba de la Peña and Adela M. Sánchez-Moreiras

1 The Plant Cell Cycle

Quantitative analyses of cell cycle can give essential information about the response of plants to short- or long-term abiotic or biotic stress, as most species alter leaf expansion or root growth as one of the first responses to cope with adverse environmental conditions (Boyer 1982). Tardieu and Granier (2000) observed a reduction of leaf area under water and light deficits due to partial blockage of nuclei in G1, which increased cell cycle duration and decreased final cell number. This effect can be detected shortly after the application of the stress and, sometimes, does not alter the photosynthetic rate, as is independent of carbon metabolism. Something similar happens with root development, where increasingly more works are focusing the interest on the study of auxin-regulated gene expression, the role of protein kinases as key regulators in plant growth and development, and the cell cycle rate and dynamic measurements in stressed tissues (Sánchez-Moreiras et al. 2006).

Flow cytometry makes possible a fine approach to the study of these events, including basic mechanisms of the cell cycle (rates of proliferating and quiescent cells, characterization of cell subsets and states upon cell cycle length and progression), and also study of effects of different putative modulators and inhibitors (hormones, growth factors, toxins, maybe allelochemicals, etc.) and environmental conditions (including stress) on the cell cycle.

This chapter is an update of 'Coba de la Peña T, Sánchez-Moreiras AM (2001) Flow cytometry: cell cycle. In: Reigosa MJ (Ed), Handbook of Plant Ecophysiology Techniques. Kluwer Academic Publishers, The Netherlands, pp. 65–80'

T. Coba de la Peña
Centro de Estudios Avanzados en Zonas Áridas (CEAZA), La Serena, Chile

A. M. Sánchez-Moreiras (✉)
Department of Plant Biology and Soil Science, University of Vigo, Vigo, Spain
e-mail: adela@uvigo.es

© Springer International Publishing AG, part of Springer Nature 2018

A. M. Sánchez-Moreiras, M. J. Reigosa (eds.), *Advances in Plant Ecophysiology Techniques*, https://doi.org/10.1007/978-3-319-93233-0_12

Cell cycle rate and dynamics can be affected due to very diverse environmental conditions. As extensively reviewed by Granier and collaborators (2007), several factors can spatial and temporally affect cell division rate in the different organs of the plant in very different ways. In this respect, water deficit or reduced incident light cause a fall in mitotic activity decreasing the cell division rate in roots and shoots (Chiatante et al. 1997; Schuppler et al. 1998; Cookson and Granier 2006), while high temperature can reduce or increase the cell division rate depending on the range of temperatures and the equilibrium with the duration of the cell cycle (Granier et al. 2007).

Besides abiotic conditions, also contaminants (Wonisch et al. 1999); different pesticides, such as acetochlor or diquat (Chauhan et al. 1999; Freeman and Rayburn 2006); and other chemical compounds, such as secondary metabolites (Sánchez-Moreiras et al. 2008), were found to have a great impact on cell cycle, reducing the number of cells in division. Wisniewska and Chelkowski (1994) and Packa (1997, 1998) studied the potential genotoxicity of *Fusarium* mycotoxins on wheat cells, finding decreased mitotic index after the treatment, with excessive condensation of prophasic and metaphasic chromosomes, accumulation of metaphases and a significantly increase of the percentage of cells with chromosomal aberrations. As well, the secondary metabolite artemisinin, a highly phytotoxic compound produced by *Artemisia annua*, was also found to show abnormal metaphase and anaphase configurations (Dayan et al. 1999), and aryltetralin plant lignans induced inhibition at all phases of mitosis with abnormal star anaphase chromosomal configurations. The exact mechanisms of action of these compounds are still unknown, but a primary effect seems to be the alteration of the formation of the spindle microtubular organization centers, resulting in the formation of multiple spindle poles and an asymmetrical convergence of the chromosomes (Oliva et al. 2002). More recently, flow cytometric analyses and mitotic index showed a retard of cell cycle in lettuce meristems treated with Benzoxazolin-2(3H)-one (BOA), with selective activity at G2/M checkpoint (Sánchez-Moreiras et al. 2008). Blocking and delay of mitosis was also found on Arabidopsis root meristems after some minutes of citral treatment (Graña et al. 2013).

Although the plant cell cycle can be regulated at multiple points, biotic and abiotic stress seems to predominantly operate at the G1 to S and G2 to M transitions (Granier et al. 2007). Cells (and, in particular, the nuclei) can be at different possible states or phases (Marie and Brown 1993; Francis 2009; Scofield et al. 2014): G0, G1, S, G2 and M. The whole of G1, S and G2 phases is termed 'interphase'.

- Cells in G0 phase (or Gap 0): cells in quiescent state after mitosis, i.e. differentiated or undifferentiated cells that do not divide and are not involved in active (proliferating) cell cycle events. This quiescent state can be reversible, and then cells enter in G1 phase.
- Cells in G1 phase: involved in cell growth and active cell cycle. They are characterized by a 2C nuclear DNA content (that is, with double DNA amount than that of gametes). This interval precedes nuclear DNA synthesis.

- Cells in S (Synthesis) phase: DNA synthesis takes place, and cell can duplicate progressively their nuclear DNA content.
- Cells in G2 phase, which is an interval between the end of DNA synthesis and the beginning of mitosis. They are characterized by a 4C nuclear DNA content.
- Cells in Mitosis (M): chromatin condenses, becoming chromosomes. Nuclear envelope disappears. Later on, chromosome segregation occurs, appearing new nuclear membranes, originating two daughter nuclei. This event is usually followed by cell division (cytokinesis). Thereby, this presently 4C cell divides in two 2C daughter cells.

Daughter cells can enter in G0 phase for a time, or enter directly in the G1 phase of a new cell cycle. In every described phase, cell cycle progression can stop and cell entries in a new quiescent phase. By this way, cells in quiescent G1, S and G2 phases (called $G1_Q$, S_Q and $G2_Q$, respectively) appear. Transition phases between quiescent and proliferating cells have also been described, and they are called $G1_T$, S_T and $G2_T$. G0 and $G2_Q$ can be followed by irreversible differentiation of the tissue cells, that do not divide anymore, although regression to undifferentiated and newly proliferating cells has sometimes been observed in mesophyll cells (Marie and Brown 1993).

If anomalous mitosis occurs (endomitosis, characterized by no formation of mitotic spindle and no attainment of chromosome segregation), a single nucleus with double number of chromosomes (corresponding to a 4C DNA content) becomes permanent. This event can take place several times, originating cells with a DNA content of 8C, 16C, 32C, 64C, etc., that is, with different ploidy levels. Endopolyploidy is originated by this way. In fact, this phenomenon is common in plants (Barow and Jovtchev 2007), and different tissues of a given plant organism can show different ploidy levels (polysomaty).

Length of each phase varies upon species, tissues and cell physiology. In a sample of proliferating cells (as is the case of plant meristems or some plant cell suspensions), most of them are in G1 phase, because this is the longest phase. The higher part of plant tissues is composed by fully differentiated (quiescent) cells, which do not divide anymore.

2 Flow Cytometry for Cell Cycle Analyses in Plants

In this section, we will describe the simplest modality of cell cycle analysis performed by flow cytometry.

Moreover, using simultaneously other fluorescent dyes and fluorescent-labeled monoclonal antibodies, RNA and protein content and synthesis, identification of antigens and molecular markers that are specific and/or critical of a cell cycle subphase can also be analyzed at the same time that the cell cycle.

Samples to be analyzed by flow cytometry are prepared from plant suspensions or meristems. Briefly, using a Petri dish, intact plant tissue is chopped with a razor

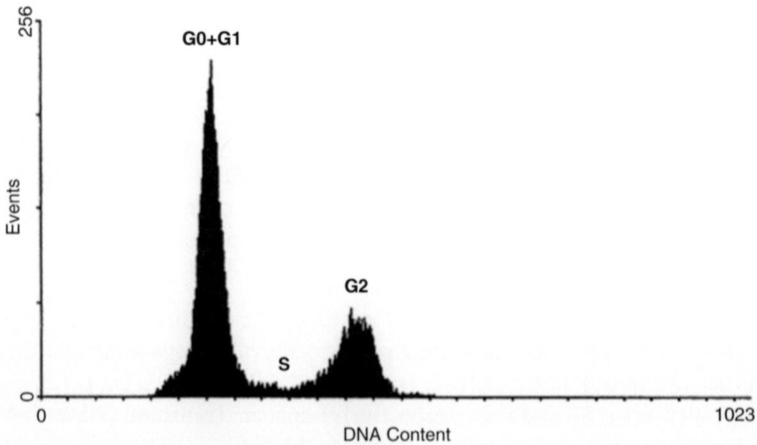

Fig. 12.1 Monoparametric linear histogram for cell cycle analysis of *Lactuca sativa* root meristems. (From Coba de la Peña and Sánchez-Moreiras 2001)

blade, into a nuclear buffer. This suspension is filtered through nylon filters 30 μm pore size). If an intercalating dye is used, RNase treatment is necessary, previous to dye addition. Then, a nucleic acid-specific dye, like Ethidium Bromide (EtBr) for instance, is added for nuclei labeling. After an incubation of 30 min, labeled plant nuclei suspension is analyzed in a flow cytometer.

Of course, a mixture of nuclei in different cell cycle (active or quiescent) phases is present in this asynchronous suspension. There is a lineal correlation between fluorescence intensity of EtBr-labeled nuclei and DNA content.

One example of monoparametric histogram obtained from a labeled nuclei suspension from root meristems of *Lactuca sativa* by flow cytometry is shown in Fig. 12.1. Both axes are linear scales. Relative fluorescence intensity (proportional to the DNA content, in the X-axis of up to 1024 channels) is represented versus the number of analyzed nuclei ('Events' in the Y-axis).

In this simple case, three nuclear populations are shown:

- The first one (G0+G1) corresponds to 2C nuclei, and it includes quiescent G0, $G1_Q$ and proliferating G1 undifferentiated cells, and also 2C differentiated cells. We cannot distinguish among these different types of nuclei on the only basis of this monoparametric DNA content-depending fluorescence analysis, and all of them are placed in the same peak.
- The second peak (G2) corresponds to 4C nuclei, and it includes G2 cycling nuclei that have finished DNA replication, but also quiescent $G2_Q$ nuclei nuclei and differentiated 4C cells, not involved in the proliferating cell cycle. The mean fluorescence intensity of this 4C peak is approximately double than that of 2C peak. Usually, fluorescence intensity ratio 4C/2C is not 2, but 1.8 or 1.9, owing to labeling irregularities due to differences in chromatin condensation state (Galbraith 1989).

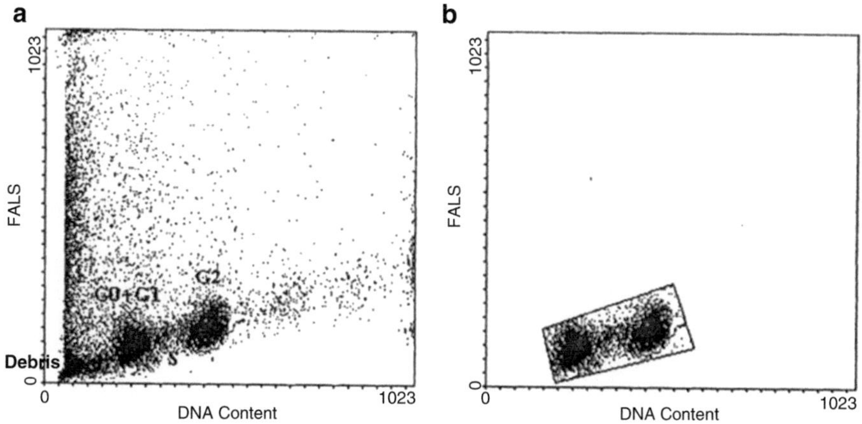

Fig. 12.2 Bi-parametric cytogram where Forward Angle Light Scatter (FALS corresponding to particle size) is represented *versus* EtBr-fluorescence intensity (corresponding to DNA content). (**a**) Total events, including the three interesting nuclei populations (G0+G1, S and G2), and population debris. (**b**) The three populations referred above are selected using gates, and only these selected nuclei are taken into account for monoparametric histogram display and analysis. (From Coba de la Peña and Sánchez-Moreiras 2001)

- Finally, the third population (S, between both peaks) is recorded as a strip connecting the first and the second peak population. This population is constituted by S-phase nuclei, in different stages of DNA replication. This is the reason of the strip shape of this population in the flow cytometry histogram. These nuclei include, of course, cycling S and non-cycling S_Q cells.

In fact, the very first analysis of nuclei suspension that must be performed by flow cytometry is a biparametric one: nuclei and debris are identified recording simultaneously EtBr-specific fluorescence intensity and particle size (FALS), as it is shown in Fig. 12.2.

In this cytogram, several principal nuclei populations are clearly identified: G0+G1 corresponds to 2C nuclei. G2 corresponds to 4C nuclei, and their size and fluorescence intensity are approximately the double than in the case of G0+G1 population. A little S population is placed between. Another population has small size and weak fluorescence intensity, and it corresponds to cellular and nuclear debris (broken nuclei, cell and membrane fragments, etc., weakly labeled with EtBr). This debris can be gated and eliminated, using discriminating windows of the cytometer software (Fig. 12.2b). The histogram showed in Fig. 12.1 results from gating and projecting the EtBr-fluorescence intensity parameter from Fig. 12.2, where debris has yet been virtually removed.

When mitosis takes place, nuclear envelope disappears, and the dispersed chromosomes (of different sizes and weak fluorescence intensities) cannot be detected or distinguished from debris in this experimental approach. Thereby, mitotic cells are lost and not detected by flow cytometry in these conditions, and this population (M) is not recorded in the histograms. In fact, for a correct evaluation of G2, M and

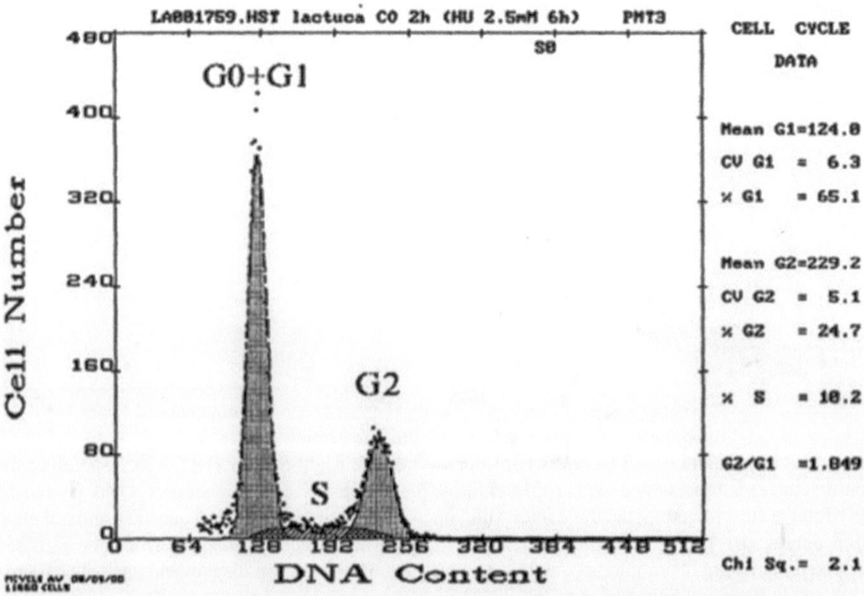

Fig. 12.3 Histogram resulting from application of computer program Multicycle (Flow System, San Diego) on raw data histogram shown in Fig. 12.1. (From Coba de la Peña and Sánchez-Moreiras 2001)

G1 lengths, mitotic indices (percent of mitosis) must be evaluated complementarily to flow cytometry using other techniques, as it is exposed in chapter about mitotic index.

This type of monoparametric analysis by flow cytometry allows simple and useful cell cycle analysis, as it will be shown below.

Usually, it is necessary to apply specific software to the flow cytometry-obtained histograms to perform a suitable cell cycle analysis from the raw data of the initial histograms. Several programs are commercially available, and each one uses different algorithms (Gray et al. 1990). These programs allow a suitable estimation of peak shape, CV of each nuclei population, %G1, %S, %G2, background subtraction, and chi-square (χ^2) estimation of fitting between raw data and estimated data. In our laboratory, the computer program Multicycle (Phoenix Flow Systems, San Diego) is used. Figure 12.3 shows Multicycle estimation of the monoparametric histogram shown above (Fig. 12.1).

Asynchronous cell populations from different tissues, meristems, and cell suspensions can be analyzed, so the percentage of cells in each cell cycle phase can be estimated. But if we are interested in obtaining metaphase chromosomes (for ulterior sorting and characterization), or in testing the effect of putative cell cycle modulators, a previous synchronization is required.

There are some commercially available inhibitors that block or stop the cell cycle in a specific phase (Planchais et al. 2000 and references therein). To avoid malformations in the following phases, the inhibitor has to be rapidly efficient at low concentrations (Planchais et al. 2000). Cyclin-dependent kinase (CDK) inhibitors, commonly used in plant cells, are olomoucine, which inhibits at G1 to S and G2 to M transitions (Glab et al. 1994), and roscovitine and bohemine, which has been found to block the cell cycle in G1, G1/S, and G2/M in tobacco cell suspensions (Planchais et al. 1997). DNA synthesis inhibitors commonly used in plant preparations are hydroxyurea (HU) and aphidicolin. Aphidicolin causes a specific and reversible inhibition of the DNA polymerase α, leading to a removable cell cycle block at the G1/S transition (Cuq et al. 1995 and references therein), while Hydroxyurea (HU) reversibly inhibits the enzyme ribonucleotide reductase, and therefore the production of deoxyribonucleotides. Treatment with this inhibitor induces the accumulation of cells in G1 and early S phase (Doležel et al. 1999 and references therein). Anti-tubulin drugs (colchicine, oryzalin, propyzamide, etc.) and proteasome inhibitors (MG132, lactacystin) are also used to block cells at early and late mitosis (Planchais et al. 2000). Finally, starvation and physical methods have also been used for inducing partial cell cycle synchronization, principally in cell suspensions, but chemicals are more specific tool.

Once the commercial inhibitor is added, cycling cells continue the cell cycle progression up to the cycle phase point where that inhibitor has a specific effect, and all the cells will arrest the cell cycle at that phase after an incubation time. After some time, inhibitor is removed from the medium by washing and whole cycling cell population re-starts and goes on the cycle simultaneously, and this synchronous cell population progression can be acutely analyzed. In the same way, the specific effect of a putative cell cycle modulator under study can be finely analyzed. By adding the tested substance at different times after inhibitor removing, cell cycle phase and subphase-specific effects can be detected. For example, monocerin (benzopyran toxin produced by the fungus *Exserophilum turcicum*) induces a delay in the cell cycle progression of synchronized root meristems of maize, specifically in S and G2 phases, as it was revealed in a study where synchronization was performed with aphidicolin (Cuq et al. 1995). Lee et al. (1996) used hydroxyurea for root tip synchronization and subsequent metaphase chromosome isolation from maize, and Sánchez-Moreiras et al. (2008) used also this inhibitor for the first time to block cell cycle in G1 phase and analyze the effects of the secondary metabolite 2-benzoxazolinone on *Lactuca sativa* root meristems. After inhibitor removal, a synchronous cell population of about 25% of total recorded nuclei was detected in progression through S and G2 phases (%S was about 10–12% in asynchronous lettuce meristems; Sánchez-Moreiras et al. 2008).

In synchronized plant cell cultures, S nuclei can represent more than 50% of total population. At present, we are testing the putative effects of some allelochemicals on this synchronized cell cycle. Figure 12.4 shows some steps of synchronous cell cycle progression at different times after HU removal, showing both raw flow cytometry histograms (left) and the corresponding Multicycle-treated data (right). Immediately after HU inhibitor removal, 79% of detected nuclei were at G0+G1 phase, 8.4% in G2 phase, and 12.4% in S phase (Fig. 12.4a). HU has induced blockage and accumulation of nuclei in G1 phase. Start and advance of synchronized

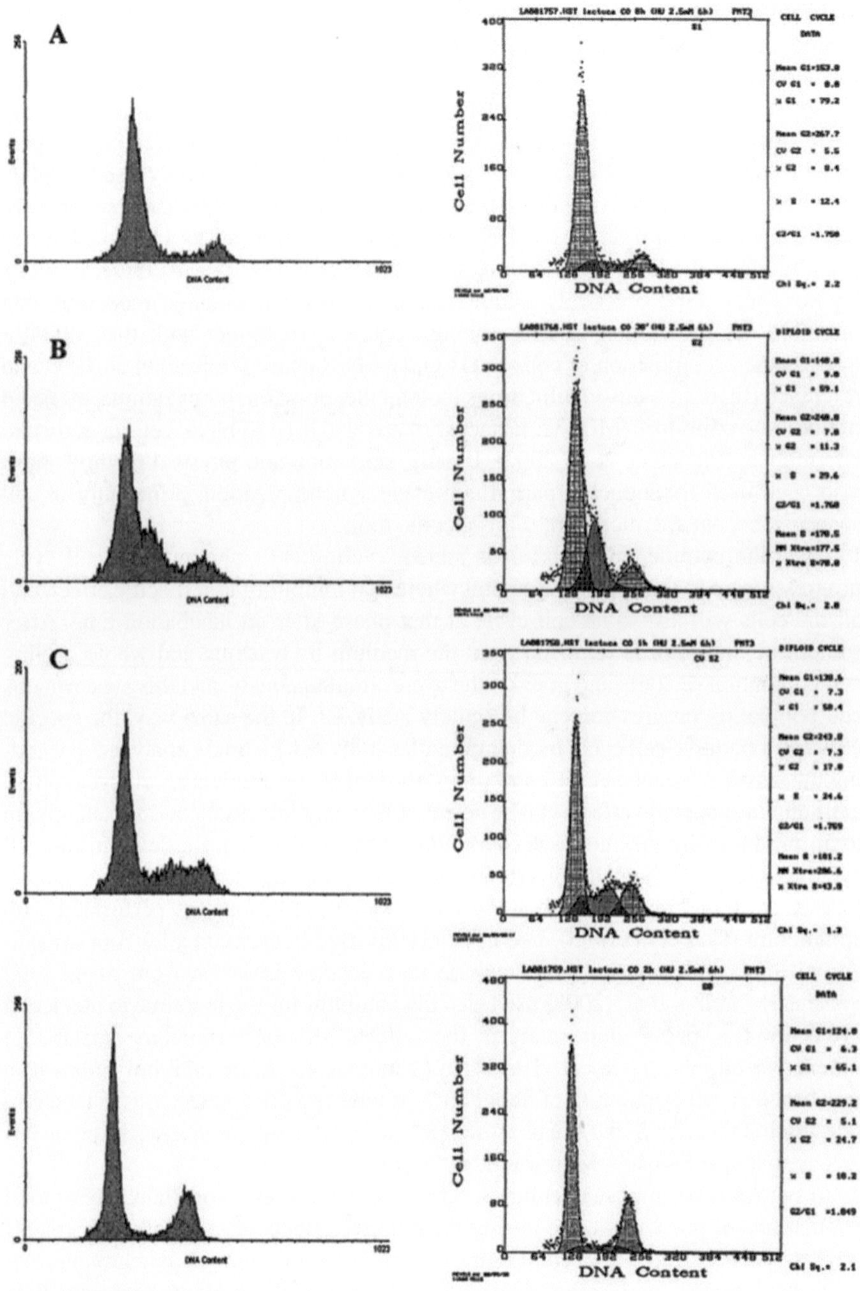

Fig. 12.4 Comparative cell cycle analysis after release of hydroxyurea-synchronized root cell meristems. (From Coba de la Peña and Sánchez-Moreiras 2001)

nuclei in S phase (29.6% of detected nuclei) is observed 30 min after (Fig. 12.4b). One hour after HU release, synchronized nuclei population begins to incorporate into G2 phase (Fig. 12.4c). Finally, all synchronized nuclei are incorporated in G2 phase of the cell cycle 2 h after HU release (Fig. 12.4d). G2 population, normally about 10% in non-synchronized meristems, reaches 24.7% in this situation. After this step, synchronized nuclei entries into mitosis and the samples have abundant metaphasic chromosomes. In this particular experimental system, a new synchronized G1 phase is not observed.

3 BI-OR Multiparametric Analysis of the Cell Cycle

Cell cycle analysis by flow cytometry can be performed measuring simultaneously other parameters, like RNA, protein contents and a wide range of antigens, using several fluorescent probes and fluorescent-labeled monoclonal antibodies. These measures allow a fine characterization and discrimination between cycling and non-cycling cells. Cell cycle can be analyzed after 5-bromodeoxyuridine (BrdUrd) incorporation. This is a thymidine analogue that is incorporated in the DNA of S-phase cells. Incorporation of Hoechst 33258 (HO), an AT-binding dye, will be reduced upon the degree of BrdUrd incorporation, owing to Hoechst do not bind DNA if BrdUrd is present instead of thymidine. Propidium Iodide (PI), an intercalating dye that is not affected by BrdUrd incorporation, is also added in this system. Simultaneous analysis of both fluorescent intensities will provide information on relative DNA content (PI) and relative fluorescence quenching (loss of HO intensity) due to DNA synthesis in presence of BrdUrd. Biparametric histograms are analyzed. By this way, it is possible to distinguish quiescent from proliferating cells, and to estimate the number of cycles they have progressed. Alternatively, anti-BrdUrd monoclonal antibodies can be used (Coba de la Peña and Brown 2001; Kim and Sederstrom 2015).

RNA levels can be detected and analyzed simultaneously to DNA (Bergounioux et al. 1988) using Acridine Orange. This is a metachromatic dye that stains differentially doubled stranded from single stranded nucleic acids. Acridine Orange fluoresces green in the first case, and red in the second, when excited in blue light (Grunwald 1993). The resulting DNA-RNA biparametric histograms allow identifying $G1_Q$, S_Q, $G2_Q$, $G1_T$, S_T, and $G2_T$ populations. Total cellular proteins can be estimated by flow cytometry using Sulphorhodamine 101 (SR 101) or Fluorescent Isothiocyanate (FITC). Moreover, a wide variety of fluorescent-labeled monoclonal antibodies against cellular antigens are available, and they can be used simultaneously with DNA-specific fluorescent dyes (Petit et al. 1993). However, as DNA denaturation is required for BrdU detection, and most protein epitopes can be destroyed avoiding classical antibody staining techniques for multiplex analysis, a novel method that overcomes DNA denaturation but still allows detection of BrdU has been developed by Cappella et al. (2008). This new procedure is based on Click chemistry detection of the thymidine analog 5-Ethynyl-2′-deoxyuridine (EdU) and increases the options for analyzing cell cycle by flow cytometry (O'Donnell et al. 2013). This can be performed in fixed (or even living) cells or protoplasts. DNA-binding vital fluorescent dyes have been recently developed (Haugland 1996).

Thereby, DNA, RNA, total protein, and even other parameters, can be estimated simultaneously in a flow cytometer, using simultaneously several fluorescent probes, several detectors, up to three lasers for excitation, and multiparametric histograms. For instance, Onelli et al. (1997) have performed immune-characterization of PCNA (Proliferating Cell Nuclear Antigen) in synchronized root meristems of *Pisum sativum* by flow cytometry. An example of combination of this technique with molecular biology is shown in Segers et al. (1996), where it was observed that a cycling-dependent kinase gene is preferentially expressed during S and G phases in meristematic cells of *Arabidopsis thaliana*.

Different examples, more information and details about these experimental approaches are exposed in Bergounioux and Brown (1990), Bergounioux et al. (1992), Robinson et al. (1997) (for fresh plant tissues analyses), Suda and Trávníček (2006) (for dehydrated plant tissues), and Marie et al. (2000) (for phytoplankton analyses).

4 Protocol for Analyzing the Cell Cycle of Synchronized Lettuce Root Meristem by Flow Cytometry

Briefly, young lettuce plants are incubated with the cell inhibitor hydroxyurea (HU) for 6 h. After washing with distilled water, plants are immediately incubated with the treatment or the control. Samples (nuclear suspensions) are prepared from root meristems and analyzed by flow cytometry every 2 h during 12–14 h after HU removal, comparing the synchronized-cell cycle progression of treated plants with that of corresponding controls. By this way, partial or total inhibition of cell cycle can be detected (Sánchez-Moreiras et al. 2008).

A schematic representation of a general protocol for sample preparation and cell cycle analysis by flow cytometry can be shown in Fig. 12.5.

4.1 Equipment and reagents

- Flow cytometer with VIS (visible) excitation source
- Seeds of *Lactuca sativa* cv. Great Lakes, California (Fitó, S.A.) Hydroxyurea (Sigma H 8627, 2.5 mM in water, pH 6.0)
- Galbraith nuclear buffer: 45 mM $MgCl_2$, 30 mM sodium citrate, 20 mM MOPS pH 7.0, 0.1% (w/v) Triton X-100, supplemented with 100% beta-mercaptoethanol and Tween 20.
- Ehidium Bromide (Sigma product E 8751, stock 10 mg mL^{-1} in water)

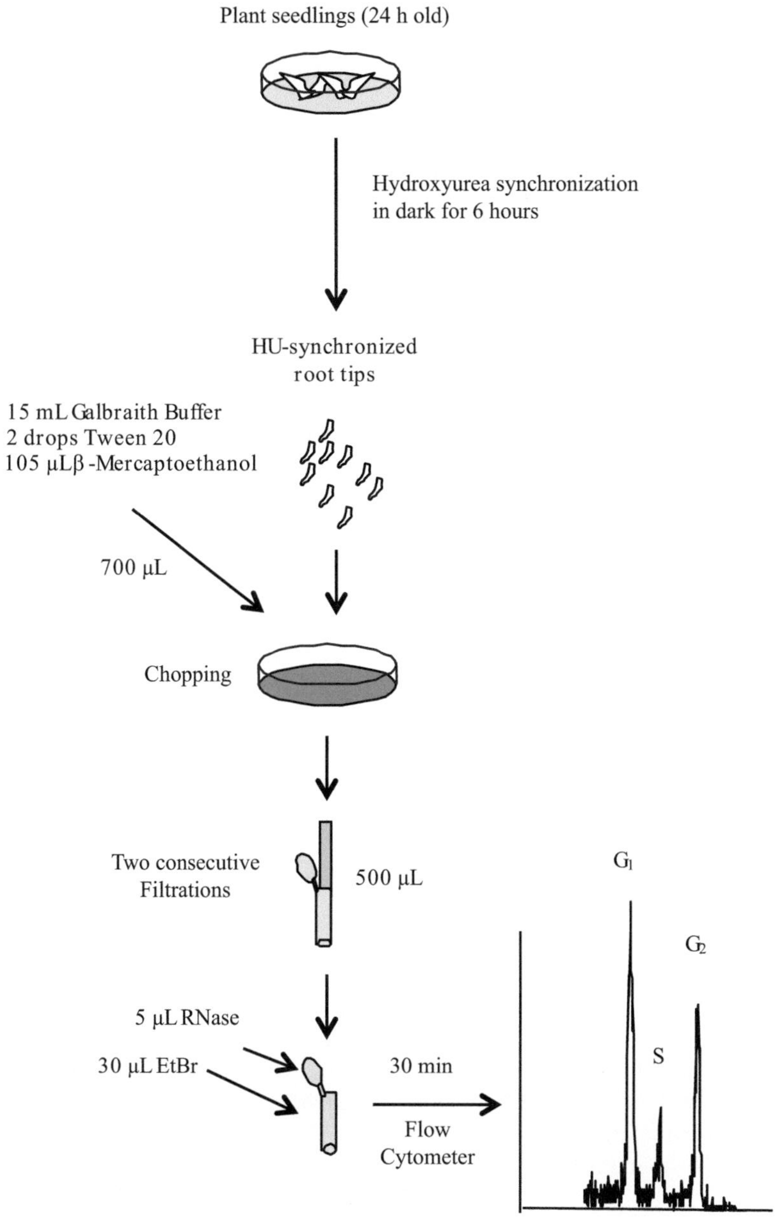

Fig. 12.5 Schematic representation of sample preparation and cell cycle analysis procedure by flow cytometry. (From Coba de la Peña and Sánchez-Moreiras 2001)

- RNase A (Boheringer Mannheim 85340024-78, stock 1% solution in Tris-HCl, NaCl and glycerol, pH 7.6)
- Heat chamber with a fixed temperature of 26 °C
- Petri dishes
- Razor blades
- 30 µm diameter nylon filters
- Micropipettes
- Plastic trays

4.2 Method

1. *Lactuca sativa* seeds are placed on moistened filter paper, into plastic trays covered with cooking foil. Seeds are germinated at 27 °C and dark for 20 h.
2. 1–3 mm-root length plants are transferred to Petri dishes containing filter paper that has been moistened with 5 mL of 2.5 mM hydroxyurea, pH 6.0. Twenty plants are placed in each Petri dish and incubated for 6 h at 27 °C in the dark.
3. HU is removed by washing twice with distilled water pH 6.0. Immediately after, plants are transferred to other Petri dishes with filter papers that have been moistened with 4 mL of either treatment (treated plants) or distilled water pH 6.0 (control plants). These plants are incubated at 27 °C and dark.
4. From this moment, and every 2 h, samples of treated plants and corresponding controls are processed simultaneously for flow cytometry analysis. The 1 mm-apical tips of root meristems from forty treated plants (that is, the content of two Petri dishes) are chopped with a razor blade on another Petri dish containing 700 µL of Galbraith buffer, supplemented with 100% Tween 20 (2 drops in 15 mL buffer) and 100% beta-mercaptoethanol (7 µL in 1 mL buffer). The obtained suspension is filtered twice through 30 µm-nylon filters, and 500 µL of filtered nuclei suspension are obtained into Eppendorf tubes. Control plants must be submitted simultaneously to the same process. The product of forty meristems constitutes one sample for flow cytometry.
5. 5 µL of 1% RNase solution are added to the nuclei suspension and, immediately after, 30 µL of 10 mg/ml Ethidium Bromide (EtBr) are added. Incubation with EtBr is for 30 min at room temperature and dark.
6. Set flow cytometer with the laser turned on 488 nm excitation wavelength. Five types of histograms (or cytograms) must be displayed in the cytometer screen:

   (a) FALS versus DNA-specific fluorescence (biparametric, see Fig. 12.3): it allows to gate debris and to eliminate it from analysis.
   (b) Peak signal versus integral signal of DNA-specific fluorescence (biparametric): it allows discarding between single nuclei and doublets.
   (c) Red signal (chlorophyll) versus yellow signal (EtBr-labeled DNA): it allows discarding stained nuclei from pigments and debris with red fluorescence.

(d) DNA fluorescence in log scale (monoparametric): it allows visualizing all peak populations.
(e) DNA fluorescence in linear scale (monoparametric, see Fig. 12.1): these are the data for cell cycle analysis.

7. Cell cycle histograms are recorded for treated and control plants every 2 h, up to arrive to 12 or 14 h of monitoring. At least 10,000 nuclei from each sample must be analyzed in the flow cytometer.
8. Data processing begins: clean histograms on a linear scale are obtained by previous gating on the other histograms.
9. Histogram profiles are analyzed using the computer program Multicycle (Flow Systems, San Diego), and G0+G1, S and G2 populations are estimated comparatively in control and BOA-treated plants.

References

Barow M, Jovtchev G (2007) Endopolyploidy in plants and its analysis by flow cytometry. In: Doležel J, Greilhuber J, Suda J (eds) Flow cytometry with plant cells. Wiley-VCH Verlag GmbH, Weinheim, pp 349–372

Bergounioux C, Brown SC (1990) Plant cell cycle analysis with isolated nuclei. Methods Cell Biol 33:563–573

Bergounioux C, Perennes C, Brown SC, Sarda C, Gadal P (1988) Nuclear RNA quantification in protoplast cell cycle phases. Cytometry 9:84–87

Bergounioux C, Brown SC, Petit P (1992) Flow cytometry and plant protoplast cell biology. Physiol Plant 85:374–386

Boyer JS (1982) Plant productivity and environment. Science 218:443–448

Cappella P, Gasparri F, Pulici M, Moll J (2008) A novel method based on click chemistry, which overcomes limitations of cell cycle analysis by classical determination of BrdU incorporation, allowing multiplex antibody staining. Cytometry A 73:626–636

Chauhan LKS, Saxena PN, Gupta SK (1999) Cytogenetic effects of cypermethrin and fenvalerate on the root meristem cells of *Allium cepa*. Environ Exp Bot 42:181–189

Chiatante D, Rocco M, Maiuro L, Scippa GS, Di Martino C, Bryant JA (1997) Cell division and DNA topoisomerase I activity in root meristems of pea seedlings during water stress. Plant Biosyst 131:163–173

Coba de la Peña T, Brown S (2001) Flow cytometry. In: Hawes C, Satiat-Jeunemaitre B (eds) Plant cell biology: a practical approach. Oxford University Press, Oxford

Coba de la Peña T, Sánchez-Moreiras AM (2001) Flow cytometry: cell cycle. In: Reigosa MJ (ed) Handbook of plant ecophysiology techniques. Kluwer Academic Publishers, The Netherlands, pp 65–80

Cookson SJ, Granier C (2006) A dynamic analysis of the shade-induced plasticity in *Arabidopsis thaliana* rosette leaf development reveals new components of the shade-adaptative response. Ann Bot 97:443–452

Cuq F, Brown SC, Petitprez M, Alibert G (1995) Effects of monocerin on cell cycle progression in maize root meristems synchronised with aphidicolin. Plant Cell Rep 15:138–142

Dayan FE, Hernández A, Allen SN, Moraes RM, Vroman JA, Avery MA, Duke SO (1999) Comparative phytotoxicity of artemisinin and several sesquiterpene analogues. Phytochemistry 50:607–614

Doležel J, Číhalíková J, Weiserová J, Lucretti S (1999) Cell cycle synchronisation in plant root meristems. Methods Cell Sci 21:95–107

Francis D (2009) What's new in the plant cell cycle? In: Lüttge U, Beyschlag W, Büdel B, Francis D (eds) Progress in botany. Springer-Verlag, Berlin, pp 33–49

Freeman JL, Rayburn AL (2006) Aquatic herbicides and herbicide contaminants: in vitro cytotoxicity and cell-cycle analysis. Environ Toxicol 21:256–263

Galbraith DW (1989) Analysis of higher plants by flow cytometry and cell sorting. Int Rev Cytol 116:165–228

Glab N, Labidi B, Qin L-X, Trehin C, Bergounioux C, Meijer L (1994) Olomoucine, an inhibitor of the cdc2/cdk2 kinases activity, blocks plant cells at the G1 to S and G2 to M cell cycle transitions. FEBS Lett 353:207–211

Graña E, Sotelo T, Díaz-Tielas C, Araniti F, Krasuska U, Bogatek R, Reigosa MJ, Sánchez-Moreiras AM (2013) Citral induces auxin-mediated malformations and arrests cell division in *Arabidopsis thaliana* roots. J Chem Ecol 39:271–282

Granier C, Cookson SJ, Tardieu F (2007) Cell cycle and environmental stresses. In: Inzé D (ed) Cell cycle control and development. Blackwell Publishing Ltd, Oxford, pp 335–355

Gray JW, Dolbeare F, Pallavicini MG (1990) Quantitative cell-cycle analysis. In: Melamed MR, Lindmo T, Mendelshon ML (eds) Flow cytometry and sorting. Wiley, New York

Grunwald D (1993) Flow cytometry and RNA studies. Biol Cell 78:27–30

Haugland RP (1996) Handbook of fluorescent probes and research chemicals. Molecular Probes Inc., Eugene

Kim KH, Sederstrom JM (2015) Assaying cell cycle status using flow cytometry. Curr Protoc Mol Biol 111:28.6.1–28.6.11

Lee JH, Arumuganathan K, Kaepler SM, Kaepler HF, Papa CM (1996) Cell synchronisation and isolation of metaphase chromosomes from maize (*Zea mays* L.) root tips for flow cytometry analysis and sorting. Genome 39:697–703

Marie D, Brown SC (1993) A cytometric exercise in plant DNA histograms, with 2C values for 70 species. Biol Cell 78:41–51

Marie D, Simon N, Guillou L, Partensky F, Vaulot D (2000) DNA/RNA analysis of phytoplankton by flow cytometry. In: Robinson JP, Darzynkiewicz Z, Dobrucki J, Hyun W, Nolan J et al (eds) Current protocols in cytometry, pp 11.12.1–11.12.14

O'Donnell EA, Ernst DN, Hingorani R (2013) Multiparameter flow cytometry: advances in high resolution analysis. Immune Netw 13:43–54

Oliva A, Moraes RM, Watson SB, Duke SO, Dayan FE (2002) Aryltetralin lignans inhibit plant growth by affecting the formation of mitotic microtubular organizing centers. Pest Biochem Physiol 72:45–54

Onelli E, Citterio S, O'Connor JE, Levi M, Sgorbati S (1997) Flow cytometry and immunocharacterization with proliferating cell nuclear antigen of cycling and non-cycling cells in synchronised pea root tips. Planta 202:188–195

Packa D (1997) Cytogenetic effects of *Fusarium* mycotoxin on root tip cells of rye (*Secale cereale* L.), wheat (*Triticum aestivum* L.) and fields bean (*Vicia faba* L. var. Minor). J Appl Genet 38:259–272

Packa D (1998) Potential genotoxicity of *Fusarium* mycotoxins in *Vicia* and *Pisum* cytogenetic tests. J Appl Genet 39:171–192

Petit JM, Denis-Gay M, Ratinaud MH (1993) Assessment of fluorochromes for cellular structure and function studies by flow cytometry. Biol Cell 78:1–13

Planchais S, Glab N, Tréhin C, Perennes C, Bureau J-M, Meijer L, Bergounioux C (1997) Roscovitine, a novel cyclin-dependent kinase inhibitor, characterizes restriction point and G2/M transition in tobacco BY-2 cell suspension. Plant J 12:191–202

Planchais S, Glab N, Inzé D, Bergounioux C (2000) Chemical inhibitors: a tool for plant cell cycle studies. FEBS Lett 476:78–83

Robinson JP, Darzynkiewicz Z, Dean PN, Hibbs AR, Orfao A, Rabinovitch PS, Wheeless LL, Galbraith DW, Lambert GM, Macas J, Dolezel J (1997) Analysis of nuclear DNA content and ploidy in higher plants. Curr Protoc Cytom 2(1):7.6.1–7.6.22

Sánchez-Moreiras AM, Coba de la Peña T, Reigosa MJ (2006) Cell cycle analyses for understanding growth inhibition. In: Reigosa MJ, Pedrol N, González L (eds) Allelopathy. A physiological process with ecological implications. Springer Academic Publishers, Dordrecht, pp 451–463

Sánchez-Moreiras AM, Coba de la Peña T, Reigosa MJ (2008) The natural compound benzoxazolin-2(3H)-one selectively retards cell cycle in lettuce root meristems. Phytochemistry 69:2172–2179

Schuppler U, He PH, John PCL, Munns R (1998) Effect of water stress on cell division and cell-division-cycle 2-like cell cycle kinase activity in wheat leaves. Plant Physiol 117:667–678

Scofield S, Jones A, Murray JAH (2014) The plant cell cycle in context. J Exp Bot 65:2557–2562

Segers G, Gadisseur I, Bergounioux C, De AEJ, Jacqmard A, Montagu MW, Inzé D (1996) The Arabidopsis cyclin-dependent kinase gene cdc2bAt is preferentially expressed during S and G sub (2) phases of the cell cycle. Plant J 10:601–612

Suda J, Trávníček P (2006) Reliable DNA ploidy determination in dehydrated tissues of vascular plants by DAPI flow cytometry—new prospects for plant research. Cytometry A 69A:273–280

Tardieu F, Granier C (2000) Quantitative analysis of cell division in leaves: methods, developmental patterns and effects of environmental conditions. Plant Mol Biol 43:555–567

Wisniewska H, Chelkowski J (1994) Influence of deoxynivalenol on mitosis of root tip cells of wheat seedlings. Acta Physiol 16:159–162

Wonisch A, Tausz M, Müller M, Weidner W, De Kok LJ, Grill D (1999) Treatment of young spruce shoots with SO_2 and H_2S: effects on fine root chromosomes in relation to changes in the thiol content and redox state. Water Air Soil Poll 116:423–428

Chapter 13
Mitotic Index

Elisa Graña

1 Introduction

The cell division cycle is a highly controlled process, essential for plant growth, whose purpose is to generate two identical daughter cells. Vegetative cell division, or mitosis, encompasses four sequential steps: two gap (G) phases separate the DNA replication (S phase) and chromosome segregation (M or mitosis) (Fig. 13.1). The first gap (G1) is the first step of cell division. It is located between cell division and DNA synthesis, and at this stage, each chromosome appears as a single chromatid with a single DNA molecule. G1 ends when cell moves into S phase (or synthesis phase), which corresponds to DNA duplication, obtaining the double of genetic material ready to be distributed between the two new daughter cells. Once achieved, cell enters in G2, the second gap previous to mitosis that differs from G1 in showing the double of DNA (two identical chromatids). Finally, cell division occurs in M phase or mitosis, which usually ends in cytokinesis. Resulting cells can continue to divide, remaining at meristematic zones, or leaving the cell cycle to undergo differentiation. In plants, mitosis specifically occurs in meristems, localized in leaves, stems and roots (Sánchez-Moreiras et al. 2008; Dewitte and Murray 2003; de Souza Junior et al. 2016).

Successful progression of cell division requires of different checkpoints, especially at the G1/S and G2/M transitions, to ensure that the previous phase has been correctly completed (Van't Hof 1985). These controlling checkpoints, which regulate the order and timing of cell division, are crucial for maintaining genomic integrity and a balanced growth and division.

Determination of mitotic index (or cell division rate) in meristematic zones results very useful to know the health status and meristematic activity of the cells

E. Graña (✉)
Plant Ecophysiology Laboratory, Department of Plant Biology and Soil Science, University of Vigo, Vigo, Spain
e-mail: eli-grana@uvigo.es

© Springer International Publishing AG, part of Springer Nature 2018

A. M. Sánchez-Moreiras, M. J. Reigosa (eds.), *Advances in Plant Ecophysiology Techniques*, https://doi.org/10.1007/978-3-319-93233-0_13

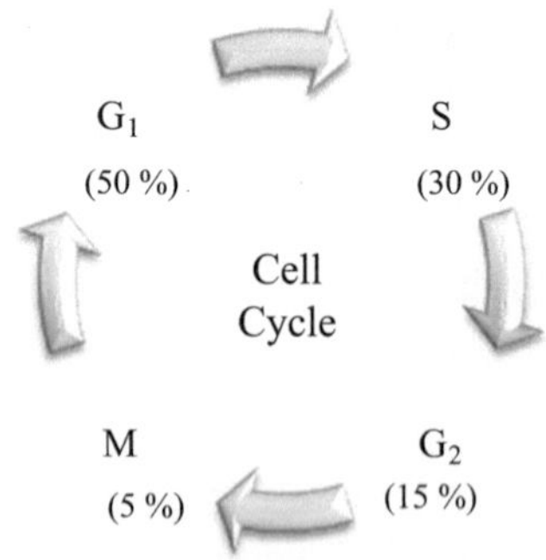

Fig. 13.1 Schematic representation of mitotic cell cycle. Data between brackets indicate the percentage of time spent by one cell in each phase. (Redrawn from Sánchez-Moreiras et al. 2001)

(Fiskesjö 1985). That is the main reason why this simple method has been widely used, especially when root growth inhibition is observed (Dayan et al. 2000), although as has been said, it can be also used to measure the mitotic activity of other organs.

Mitotic index is used to measure cytotoxicity in living organisms (Smaka-Kincl et al. 1996), based on the increase/decrease of the rate of cell division (Debnath et al. 2016; Jain et al. 2016). It can be simply calculated as follows and is given in percentage:

$$\mathrm{MI} = \frac{\text{Prophase} + \text{Metaphase} + \text{Anaphase} + \text{Telophase}}{\text{Total No. of cells}} \times 100$$

As a measure to trace cytotoxic substances, significant decrease in mitotic activity is an evidence of genotoxic potential. When mitotic index reaches values below **50%** of a negative control are considered sub-lethal effects (Sharma and Vig 2012; Jain et al. 2016), and this point is known as **cytotoxic limit value** (Panda and Sahu 1985; Sharma et al. 2012); while lethal effects are considered when inhibition decreases below **22%** (Jain et al. 2016; Sharma and Vig 2012). Reductions in mitotic index are usually attributed to an inhibition in DNA synthesis or a stop in the G2 phase (Sudhakar et al. 2001).

On the other hand, an increase in mitotic index can be the consequence of a reduction of the time necessary for DNA repair (Evseeva et al. 2003). It can reveal an uncontrolled cell proliferation being able to end in tumor formation (Hoshina 2002), or be indicative of shortening of the duration of the mitotic cycle (Al-Ahmadi 2013). Any of these options are characterized as being detrimental to cells.

In addition to Mitotic Index, it can be also calculated the Active Mitotic Index, which gives additional information about the percentage of actively dividing cells (cells at metaphase and anaphase) (Borah and Talukdar 2002; Madaan and Mudgal 2011):

$$\mathrm{AMI} = \frac{\text{Metaphase} + \text{Anaphase}}{\text{Total No. of cells observed}} \times 100$$

2 Mitotic Index and Its Applications

Mitotic index is widely used to evaluate the genotoxic potential of many substances in studies of environmental biomonitoring (Smaka-Kincl et al. 1996), manly using *Allium cepa* or *Lactuca sativa* as testing organisms due to the big size of their cells, which makes easier the visualization of different mitotic phases. Besides, higher plant bioassays are quick, cheap and easy to handle. This test is especially successful for screening, monitoring and detection of clastogenicity of environmental mutagens, including atmospheric, water and soil pollutants (Fiskesjö 1993; Ma et al. 1995).

Thus, MI has been used to test the toxic potential of many pollutants: overexposure to chromium (Rai and Dayal 2016), zinc oxide nanoparticles (Kumari et al. 2011), copper chloride (Can et al. 2016), aluminum (Salabert de Campos and Viccini 2003), insecticides (Panda and Sahu 1985), radioisotope-contaminated air in Chernobyl area (Cebulska-Wasilewska 1992; Ichikawa et al. 1996), chlorpyrifos, benzene, nitrogen oxide, nitric oxide, ozone or sulfur dioxide (Schairer et al. 1978).

Another area where the calculation of mitotic index is also very useful is allelopathy. Many of the allelopathic studies are focused on evaluating the ability of plant extracts or plant naturally-occurring isolated compounds to act as plant growth regulators. In this way, mitotic index is used to test the anti-proliferative profile of different plant extracts such as *Terminalia arjuna* and *Moringa oleifera* (Debnath et al. 2016); *Brassica juncea* (Sharma et al. 2012), *Zanthoxylum limonella* (Charoenying et al. 2010) or *Schinus* spp. (Pawlowski et al. 2012). Besides, mitotic index is also inhibited by plant isolated compounds like BOA (Sánchez-Moreiras et al. 2008), citral (Graña et al. 2013), 1,8-cineole (Romagni et al. 2000), or cinmethylin (El-Deek and Hess 1968). All these works are just an example, as there are many works of this type in the bibliography.

As well, this technique has been also used to test the suitability of cell culture media, and to be sure that it does not compromise cell viability (Maisch et al. 2016), or to verify that applied electric fields can enhance apical root regeneration (Kral et al. 2016).

3 Chromosomal Aberrations

At the same time that cells are visualized using a brightfield microscope, chromosomal aberrations can be easily observed, especially when *Allium* or *Lactuca* specimens are used. Alterations in mitotic index are usually accompanied by cytogenetic instabilities, and most of them are lethal and may cause genetic disorders (Debnath et al. 2016). Atypical number of chromosomes or structural abnormalities in chromosomes are very common. The most common chromosomal aberrations are summarized below and are also schematically represented in Fig. 13.2:

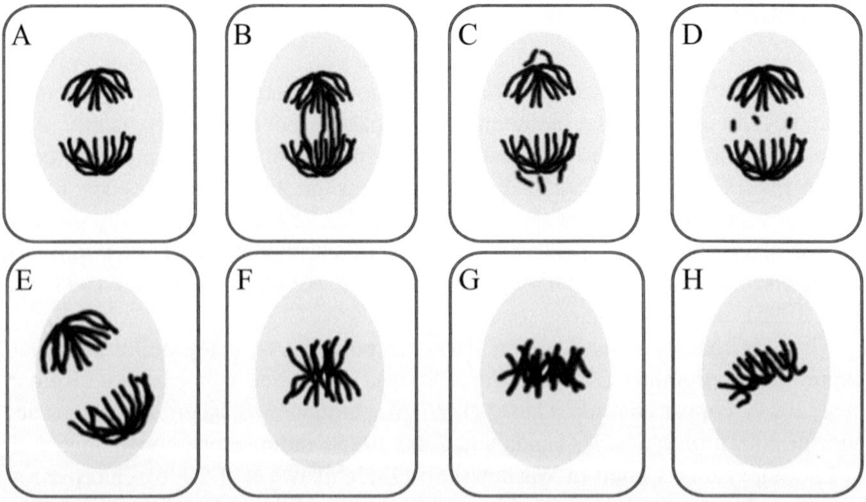

Fig. 13.2 Schematic representation of most common chromosomal aberrations: (**a**) Normal anaphase; (**b**) Chromosomal bridges; (**c**) Laggard chromosomes; (**d**) Fragmented chromosomes; (**e**) Diagonal anaphase; (**f**) Normal metaphase; (**g**) Sticky chromosomes at metaphase; (**h**) c-Mitosis

- **Chromosomal bridges**. The twin chromatids fail to separate (or there is a delay in separation); and as a result, chromosomes are subjected to an abnormal separation. Bridges are the result of stickiness of the chromosomes or due to the presence of dicentric chromosomes (Vorobjev et al. 1993; Debnath et al. 2016). Chromosomal bridges are usually observed in anaphase.
- **Laggard chromosomes** are the result of the inhibition of chromosome movement in anaphase, mainly due to a disturbance on the microtubule arrangement. Part of genetic material is 'left behind', causing an unequal distribution of chromosomes in the daughter cells. Besides, in some cases laggard chromosomes lead to micronuclei formation, since them reorganize later than other chromosomes (Kozgar 2014; de Storme and Mason 2014; Debnath et al. 2016). Micronuclei are the simplest indicator of DNA damage (Migid et al. 2007).
- **Fragmented chromosomes**, also known as banded chromosomes, are small chromosome pieces, mainly observed as fragments in anaphase (Fiskesjö 1988). Chromosome fragmentation causes unbalanced chromosome patterns and uneven distribution of chromatids (Nordenskiöld 1963). It is believed that fragments originate from fragmentation of terminal regions of regular chromosomes (Sheikh et al. 1995), or that are the result of the rupture of chromosome bridges (Liu and Makaroff 2006).

- **Diagonal anaphase** refers to the abnormal location of the spindle fibers, located in opposite corners instead of parallel to the cell division plate. It is also known as polar shifting (El-Bayoumi et al. 1979; Pandey and Sakya 2009).

– **Sticky chromosomes** are characterized by clustering at any phase of the cell cycle. Chromosomes lose their sharpness and tend to clump (Salabert de Campos and Viccini 2003; Rai and Dayal 2016). Apparently, they lose their individuality and can be observed as a mass made of chromatic substance (Kabarity and Malallah 1980).
– **c-Mitosis**. It happens when all chromosomes remain at cell division plate level, but no separation of chromatids occurs. It can be observed as collapsed chromosomes in the central part of the cell (Vorobjev et al. 1993). It is consequence of the inactivation of the spindle apparatus together with a delay in centromere division (Mann and Storey 1966; Jackson 1969; Shehab 1980).

4 The Method

The method here presented consists on an adaptation of the classical squash techniques from Armbruster and collaborators (1991), and can be used for plant seedlings, but also for cell suspension cultures. It is based on division synchronization and nuclear staining to study the effect of a given substance on the division behavior of actively dividing plant cells (Fig. 13.3).

Cycling cells are characterized by dividing asynchronously in meristems. To obtain a large number of cells in the same phase of the cell cycle, it is necessary to induce synchronization through the use of chemical agents. This kind of compounds generally act preventing the formation of the mitotic spindle or inhibiting the synthesis of DNA to block cell cycle progression (Sánchez-Moreiras et al. 2001). Since this method is widely used in genotoxicity studies or to assess the potential as cell cycle inhibitor of a given substance, the most common procedure is to compare the data from a control treatment with those obtained for the tested compound. Staining cells at different mitotic phases can be seen in Fig. 13.4.

The method for mitotic index calculation uses hydroxiurea (HU) as cell cycle arresting agent, it is focused on the analysis of seedlings root apexes and it consists in the following steps (Protocol is summarized in Fig. 13.3):

1. Treat plant seedlings with 2 mM HU during 14 hours to obtain cell cycle synchronization. HU is applied especially to radicles, ensuring that they are completely covered with treatment.
2. Remove HU through three consecutive washes with distilled water. Synchronization progressively decreases once HU treatment is released.
3. Immediately, apply the treatment of the studied compound. At this time, the control treatment (usually distilled water) is also treated. For a better asses of cell division under a potentially genotoxic substance it is advisable to do treatments at different times to observe progression. A powerful cell division blocker can act in minutes, but it is also interesting to know what happens after several hours.

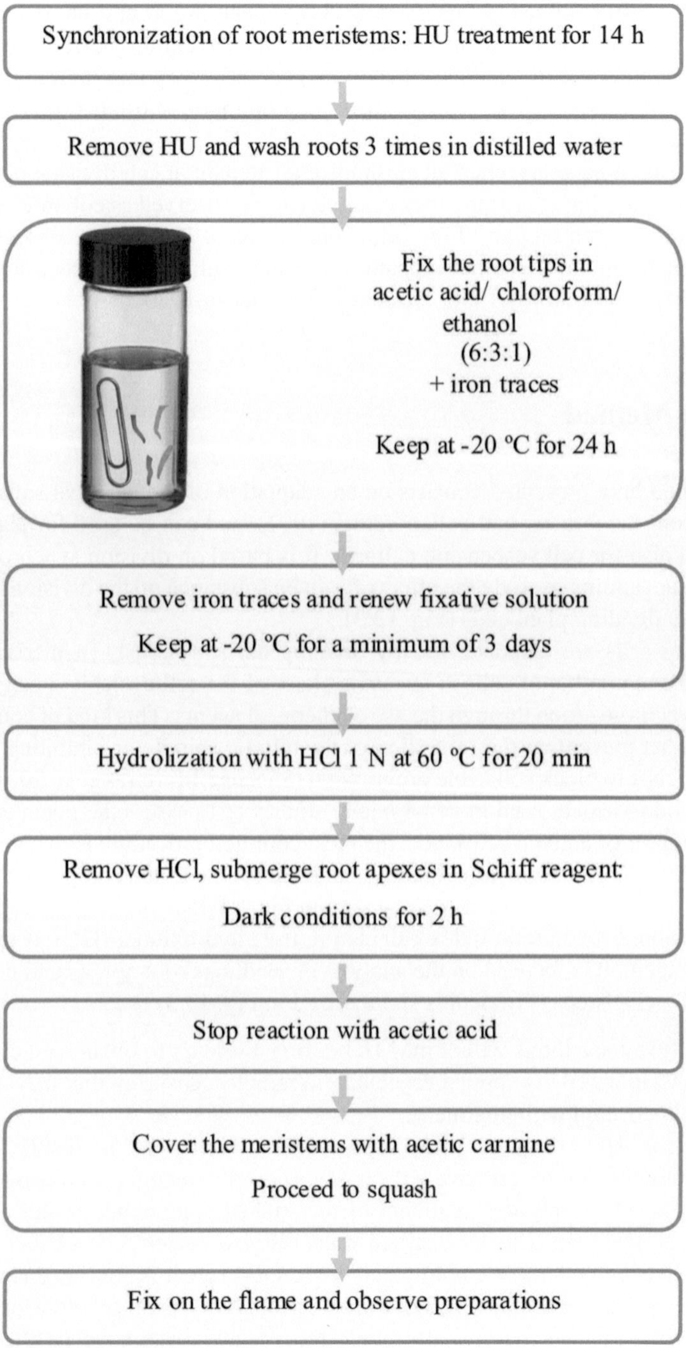

Fig. 13.3 Representative scheme of the mitotic index protocol

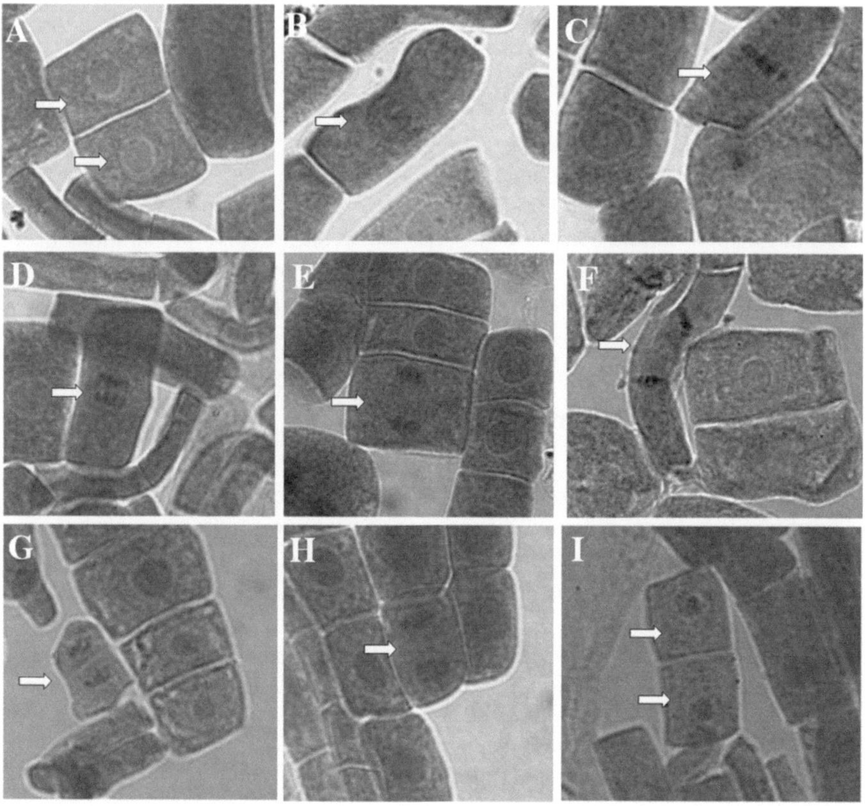

Fig. 13.4 Microphotographs of different phases of mitosis on *Arabidopsis* seedlings meristems. (**a**) Interphase cells; (**b**) Nucleus at incipient prophase; (**c**) metaphase; (**d–f**) different stages of anaphase; (**g** and **h**) telophases; (**i**) recently separated daughter cells. (Images belong to control-treated meristems. Experiment done by Graña et al. 2013)

4. Collect plant material. Root apexes are excised about 1 cm above the end of the roots. Then, they are submerged in a fixative medium consisting of acetic acid/chloroform/ethanol (6:3:1). This step is done in small crystal vials (5–10 mL) with hermetic lids to avoid evaporation, introducing some iron traces in them. This metal acts as a mordant and changes the isoelectric point and accelerates the nuclear staining (Sánchez-Moreiras et al. 2001).
5. Store vials at −20 °C for 24 h.
6. Remove iron traces, renew fixative solution and store again at −20 °C for a minimum of 3 days.
7. After that time, hydrolyze plant material with 1 N HCl at 60 °C for 20 min. Ensure that lids are well closed to avoid the evaporation of HCl.
8. Remove HCl and submerge root apexes in Schiff reagent in dark conditions for 2 h. Chromosomes will be stained with pink to violet color.

9. Stop reaction: put plant material on a slide and cover it with a drop of acetic acid. At this point, root apexes can be cut with a razor blade.
10. Put a drop of acetic carmine over the meristems and squash them. This step will separate cells and form a monolayer, essential to observe the nuclei of all the cells.
11. Set the preparation by passing the slide over a flame for 1–3 s. It has to be done very carefully avoiding t burn it.
12. Fix the slide with the cover using nail polish and observe preparations using bright-field microscope.
13. Score the number of mitotic cells and also the total number of cells.

5 Tricks and Recommendations

- Step 4: A very easy way to get iron traces is just using a clip, a staple or an iron pushpin. Besides, at this step, it is not necessary to prepare big amounts of fixative solution: just prepare the enough solution to cover all radicles.
- Step 10: A good idea to uniformly squash the meristems is pressing the cover with an eraser.
- Step 11: The time of exposure to flame is relative; it depends on the thickness of the tissue. It is very common to overexpose, burn the meristems and leave them useless. I recommend do it in small fractions of time and view the meristems to check fixation.
- Step 13: It is generally recommended to score about 1000 cells per replicate, which can represent 3–5 meristems depending on their size.

References

Al-Ahmadi MS (2013) Effects of organic insecticides, Kingbo and Azdar 10 EC, on mitotic chromosomes in root tip cells of *Allium cepa*. Int J Gen Mol Biol 5(5):64–70

Armbruster BL, Molin WT, Bugg MW (1991) Effects of the herbicide dithiopyr on cell division in wheat root tips. Pest Biochem Physiol 39:110–120

Borah SP, Talukdar J (2002) Studies on the phytotoxic effects of extract of castor seed (*Ricinus communis* L.). Cytologia 67:235–243

Can AA, Isik G, Yucel E (2016) The effects of copper ($CuCl_2$) on mitotic cell division of Lebanon cedar (*Cedrus libani*). Fresenius Environ Bull 25(1):4324–4326

Cebulska-Wasilewska A (1992) Tradescantia stamen-hair mutation bioassay on the mutagenicity of radioisotope-contaminated air following the Chernobyl nuclear accident and one year later. Mutat Res 270:23–29

Charoenying P, Teerarak M, Laosinwattana C (2010) An allelopathic substance isolated from *Zanthoxylum limonella* Alston fruit. Sci Hort 125:411–416

Dayan FE, Romagni JG, Duke SO (2000) Investigating the mode of action of natural phytotoxins. J Chem Ecol 26(9):2079–2094 de Souza Junior JDA, Grossi de Sa MF, Engler G, de Almeida Engler J (2016) Imaging nuclear morphology and organization in cleared plant tissues treated with cell cycle inhibitors. In: Caillaud M-C (ed) Plant cell division – methods and protocols, Springer protocols. Humana Press, New York, pp 59–60

de Storme N, Mason A (2014) Plant speciation through chromosome instability and ploidy change: cellular mechanisms, molecular factors and evolutionary relevance. Curr Plant Biol 1:10–33

Debnath B, Paul C, Debnath A, Saha D (2016) Evaluation of cytotoxicity of *Terminalia arjuna* (Roxb.) Wight & Arn. and *Moringa oleifera* Lam. in root meristems cells of *Allium cepa* L. J Med Plant Stud 4(3):107–110

Dewitte W, Murray JAH (2003) The plant cell cycle. Annu Rev Plant Biol 54:235–264

El-Bayoumi AS, Kabarity A, Habib A (1979) Cytological effects of papaverine hydrochloride on root tips of *Allium cepa* L. Cytologia 44:745–755

El-Deek MH, Hess FD (1968) Inhibited mitotic entry is the cause of growth inhibition by cinmethylin. Weed Sci 34:684–688

Evseeva TI, Stanislav A, Geras'kin I, Shuktomova I (2003) Genotoxicity and toxicity assay of water sampled from a radium production industry storage cell territory by means of Allium-test. J Environ Radioact 68:235–248

Fiskesjö G (1985) The Allium test as a standard in environmental monitoring. Hereditas 102:99–112

Fiskesjö G (1988) The Allium-test an alternative in environmental studies: the relative toxicity of metal ions. Mutat Res 197:243–260

Fiskesjö G (1993) Allium test I: a 2–3 day plant test for toxicity assessment by measuring the mean root growth of onions (*Allium cepa* L.). Environ Toxicol Water Qual 8:461–470

Graña E, Sotelo T, Díaz-Tielas C, Araniti F, Krasuska U, Bogatek R, Reigosa MJ, Sánchez-Moreiras AM (2013) Citral induces auxin and ethylene-mediated malformations and arrests cell division in *Arabidopsis thaliana* roots. J Chem Ecol 39:271–282

Hoshina MM (2002) Evaluation of a possible contamination of the waters of the Claro River-Municipality of Rio Claro, part of the Corumbataí River Basin, with the mutagenicity tests using *Allium cepa*. State University of São Paulo, Rio Claro (in Portuguese)

Ichikawa S, Nakano A, Kenmochi M, Yamamoto I, Murai M, Takahashi E, Yamaguchi A, Watanabe K, Tomiyama M, Sugiyama K, Yogo A, Yazaki T, Okumura M, Shima N, Satoh M, Yoshimoto M, Xiao LZ (1996) Yearly variation of spontaneous somatic mutation frequency in the stamen hairs of *Tradescantia* clone KU 9 grown outdoors, which showed a significant increase after the Chernobyl accident. Mutat Res 349(2):249–259

Jackson WT (1969) Regulation of mitosis. II. Interaction of isopropyl N-phenyl-carbamate and melatonin. J Cell Sci 5:745–755

Jain P, Singh P, Sharma HP (2016) Anti-proliferative activity of some medicinal plants. Int J Pharmacol Pharm Sci 3(2):46–52

Kabarity A, Malallah G (1980) Mitodepressive effects of Khat extract in the meristematic region of *Allium cepa* root tips. Cytologia 45:733–738

Kozgar I (2014) Mutation breeding in chickpea: perspectives and prospects for food security. De Gruyter Open, Berlin, p 28

Kral N, Ougolnikova AH, Sena G (2016) Externally imposed electric field enhances plant root tip regeneration. Regeneration 3(3):156–167

Kumari M, Khan SS, Pakrashi S, Mukherjee A, Chandrasekaran N (2011) Cytogenetic and genotoxic effects of zinc oxide nanoparticles on root cells of *Allium cepa*. J Hazard Mat 190:613–621

Liu Z, Makaroff CA (2006) Arabidopsis separase AESP is essential for embryo development and the release of cohesion during meiosis. Plant Cell 18:1213–1225

Ma TH, Xu Z, Xu C, McConnell H, Rabago EV, Arreola GA, Zhang H (1995) The improved *Allium/Vicia* root tip micronucleus assay for clastogenicity of environmental pollutants. Mutat Res 334:185–195

Madaan N, Mudgal V (2011) Phytotoxic effects of selenium on the accessions of wheat and safflower. Res J Environ Sci 5(1):82–87

Maisch J, Kreppenhofer K, Büchler S, Merle C, Sobich S, Görling B, Luy B, Ahrens R, Guber AE, Nick P (2016) Time-resolved NMR metabolomics of plant cells based on a microfluidic chip. J Plant Physiol 200:28–34

Mann JD, Storey WB (1966) Rapid action of carbamate herbicides upon plant cell nuclei. Cytologia 31:203–207

Migid HMA, Azab YA, Ibrahim WM (2007) Use of plant genotoxicity bioassay for the evaluation of efficiency of algal biofilters in bioremediation of toxic industrial effluent. Ecotoxicol Environ Safe 66:57–64

Nordenskiöld H (1963) A study of meiosis in the progeny of x-irradiated *Luzula purpurea*. Hereditas 49:33–47

Panda BB, Sahu UK (1985) Induction of abnormal spindle function and cytokinesis inhibition in mitotic cells of *Allium cepa* by the organophosphorus insecticide fensulfothion. Cytobios 42:147–155

Pandey A, Sakya SR (2009) Effect of triazophos on mitotic activity and chromosomal behavior in root meristems of *Allium cepa* L. Bot Orient 6:4–7

Pawlowski Â, Kaltchuk-Santos E, Zini CA, Caramão EB, Soares GLG (2012) Essential oils of *Schinus terebinthifolius* and *S. molle* (Anacardiaceae): Mitodepressive and aneugenic inducers in onion and lettuce root meristems. S Afr J Bot 80:96–103

Rai P, Dayal S (2016) Evaluating genotoxic potential of chromium on *Pisum sativum*. Chromos Bot 11(2):44–47

Romagni JG, Allen SN, Dayan FE (2000) Allelopathic effects of volatile cineoles on two weedy plant species. J Chem Ecol 26:303–313

Salabert de Campos JM, Viccini LF (2003) Cytotoxicity of aluminum on meristematic cells of *Zea mays* and *Allium cepa*. Caryologia 56(1):65–73

Sánchez-Moreiras AM, Coba de la Peña T, Martínez Otero A, Blanco Fernández A (2001) Mitotic index. In: Reigosa Roger MJ (ed) Handbook of plant ecophysiology techniques. Kluwer Academic Publishers, Dordrecht, p 83

Sánchez-Moreiras AM, Coba de la Peña T, Reigosa MJ (2008) The natural compound benzoxazolin-2(3H)-one selectively retards cell cycle in lettuce root meristems. Phytochemistry 69(11):2172–2179

Schairer LA, Van't Hof J, Hayes CG, Burton RM, de Serres FJ (1978) Exploratory monitoring of air pollutants for mutagenicity activity with the *Tradescantia* stamen hair system. Environ Health Perspect 27:51–60

Sharma S, Vig AP (2012) Antigenotoxic effects of Indian mustard *Brassica juncea* (L.) Czern aqueous seeds extract against mercury (Hg) induced genotoxicity. Sci Res Essays 7(13):1385–1392

Sharma S, Nagpal A, Vig AP (2012) Genoprotective potential of *Brassica juncea* (L.) Czern against mercury-induced genotoxicity in *Allium cepa* L. Turk J Biol 36:622–629

Shehab AS (1980) Cytological effects of medicinal plants in Qatar. II. Mitotic effect of water extract of *Teucrium pilosum* on *Allium cepa*. Cytologia 45:57–64

Sheikh SA, Kondo K, Hoshi Y (1995) Study on diffused centromeric nature of *Drosera* chromosomes. Cytologia 60:43–47

Smaka-Kincl V, Stegner P, Lovka M, Toman MJ (1996) The evaluation of waste, surface and ground water quality using the Allium test procedure. Mutat Res 368:171–179

Sudhakar R, Gowda KN, Venu G (2001) Mitotic abnormalities induced by silk dyeing industry effluents in the cells of *Allium cepa*. Cytologia 66(3):235–239

Van't Hof J (1985) Control points within the cell cycle. In: Bryant JA, Francis D (eds) The cell division cycle in plants. Cambridge University Press, Cambridge, pp 1–13

Vorobjev IA, Liang H, Berns MW (1993) Optical trapping for chromosome manipulation: a wavelength dependence of induced chromosome bridges. Biophys J 64(2):533–538

Chapter 14
Fluorescent Probes and Live Imaging of Plant Cells

Elfrieda Fodor and Ferhan Ayaydin

1 Introduction

Fluorescent probes are used in almost all areas of plant research ranging from molecular biology to ecophysiology studies. Development of novel fluorochromes and fluorescent proteins in combination with advanced microscopy techniques allow us to analyze cells, tissues, organs and whole plants in great detail. Live fluorescence microscopy imaging of plants is of particular importance for ecophysiology studies where complex interactions of plants and their environment need to be understood at molecular, cellular and organismal level. Here we present an overview of fluorescent probes and live cell microscopy setup for plants and provide a detailed protocol for fluorescent live-dead viability assay using fluorescein diacetate and propidium iodide fluorescent dyes.

1.1 Natural Fluorophores and Autofluorescence in Plants

Naturally occurring fluorescence or intrinsic fluorescence is common to many molecules of living organisms, which in turn confer auto-fluorescence properties to the tissue or cell compartment they reside in. The excitation and emission wavelengths of the autofluorescence encountered in cells can be diverse, spanning the whole visible spectrum and beyond. Common examples of molecules with intrinsic fluorescence are the aromatic residues, phenylalanine, tyrosine and tryptophan of proteins, with

E. Fodor
Institute of Biochemistry, Biological Research Centre, Szeged, Hungary

F. Ayaydin (✉)
Cellular Imaging Laboratory, Institute of Plant Biology, Biological Research Centre, Szeged, Hungary
e-mail: ayaydin.ferhan@brc.mta.hu

© Springer International Publishing AG, part of Springer Nature 2018
A. M. Sánchez-Moreiras, M. J. Reigosa (eds.), *Advances in Plant Ecophysiology Techniques*, https://doi.org/10.1007/978-3-319-93233-0_14

fluorescence maxima at 282, 303 and 348 nm, respectively, where tryptophan accounts for about 90% of the fluorescence of the proteins. Certain cofactors such as NAD, FAD, FMN and porphyrins exhibit fluorescence, as well (Lakowicz 2007). There are several plant-specific molecules bearing intrinsic fluorescence. For example, plant cell walls rich in lignin fluoresce blue with emission peak around 360 nm (Chapman et al. 2005) and the chlorophyll of green plants emits red fluorescence between 650 and 750 nm range (Agati 1998). There are also several other plant-specific molecules, which fluoresce in blue and blue-green (certain alkaloids such as colchicine, terpene and flavonoids) or in yellow and orange (acridone, the polyacetylenes and isoquinoline) or in red (anthocyanins and azulenes) spectral ranges (Roshchina 2012).

During fluorescence imaging of plants, autofluorescence can be either advantageous or disadvantageous depending on the application. Red chlorophyll autofluorescence of plants, for example, can be used as an intrinsic probe to locate chloroplasts in green tissues during fluorescence microscopy imaging. Intrinsic chlorophyll fluorescence also allows assessing changes of photosynthetic apparatus, state of plant health, stress tolerance, disease onset and nutrient deficiency (Buschmann 2007). On the other hand, intense red chlorophyll fluorescence often interferes with red and far-red colored fluorescent probes in multicolor labeling experiments. In such cases, either a specific bandpass emission filter or spectral unmixing approach can be used to isolate the signal of interest (Berg 2004; Mylle et al. 2013). Similarly, the cyan and blue colored cell wall autofluorescence is suitable to mark cell and tissue borders. While being advantageous, cell wall autofluorescence may also cause interference in experiments involving blue and cyan colored exogenous fluorescent probes. Autofluorescence can also be induced by mechanical stress and wounding of plants, such as during sampling of plant parts for microscopy analysis. Fluorescent phenolic compounds are being formed when contents of burst vacuoles at the cut site mix with cytoplasmic enzymes. This may cause cut or injured sites of living plant samples to fluoresce. Although this property can be exploited in wounding response studies, often it creates unwanted background fluorescence during imaging. Therefore, mechanical stress and physical injury should be minimized during live analysis of plant samples.

Figure 14.1a, b show a setup for direct live analysis of a leaf of a potted maize plant analyzed with confocal fluorescence microscope without detaching the leaf. Using this setup, chlorophyll fluorescence is captured to locate chloroplasts of parenchyma tissue (Fig. 14.1c). Similarly, violet light-induced intrinsic fluorescence is used to capture guard cells of epidermis (Fig. 14.1d) and trichomes of the leaf midrib region (Fig. 14.1e).

Fig. 14.1 (continued) detection range. (**e**) Trichome autofluorescence is captured at the leaf midrib region using 405 nm laser excitation (emission: 425–525 nm). Multiple confocal optical sections are merged to obtain extended depth of focus image. (**f**) Timelapse imaging of tobacco BY2 suspension cells using differential interference contrast (DIC) technique. Arrow indicates translocation of a cytoplasmic cargo. Nuclei (N), nucleoli (n) and one of the vacoules (V) are marked on the first image. (**g**) Fluorescein diacetate (FDA, green, live cells) and propidium iodide (PI, red, dead cells) live/dead staining of Arabidopsis (Col) suspension culture. (**h**) FDA/PI viability analysis of phosphinotricin (PPT) resistant transgenic maize suspension culture after treatment with PPT (15 mg L^{-1}) for 5 days

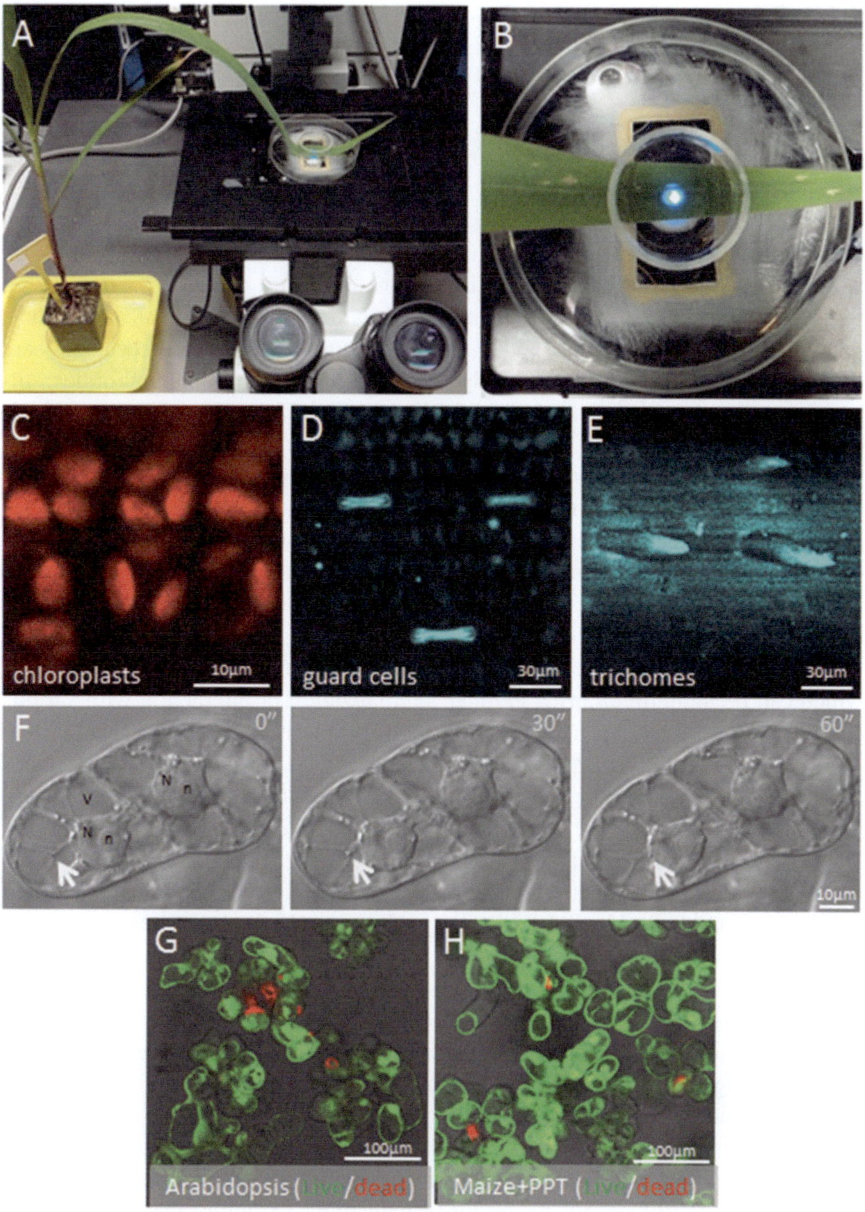

Fig. 14.1 Laser scanning confocal microscopy imaging of live plant samples. (**a**) A potted maize plant is placed next to the microscope stage and one of the leaves is layered onto a coverslip-bottom Petri dish (9 cm diameter) to image intrinsic leaf fluorescence using 40× oil immersion objective. (**b**) Closeup view of the same sample on the microscope stage. The leaf is immobilized on top of the larger Petri dish by using a smaller glass Petri dish (3 cm diameter). (**c**) Using the mounted leaf sample, maize leaf chloroplasts of parenchyma tissue are captured using chlorophyll fluorescence (emission: 650–750 nm) excited by 543 nm laser beam. (**d**) Guard cell autofluorescence of three stomata at the leaf surface is captured using 405 nm laser excitation and 425–525 nm emission

1.2 Organic Fluorescent Dyes and Reporter Molecules

During the last decades, a wide variety of synthetic fluorescent dyes and fluorescent reporter molecules with diverse spectral characteristics had been developed for various specific applications (Johnson and Spence 2010). According to the mode of attachment to the target to be studied, they can be divided into two broad categories, (i) probes that target the molecule of interest by *affinity binding*, (e.g. the nucleic acid dyes DAPI and propidium iodide) and (ii) fluorescent dyes that possess an active group through which the labeling is achieved in a *covalent* manner, either directly to the target molecule, or to an intermediate molecule – such as an antibody, that in turn will target the molecule of interest. Examples of this second category are the reactive fluoresceins, rhodamine dyes, acridines, fluorenes, naphthalimides, stilbenes, reactive AlexaFluor dyes and their derivatives with specific activities towards specific chemical groups.

Covalently attached fluorophores are used for generation of *fluorescent reporter molecules*. Such reporter molecules available are numerous, since just about any molecule can be engineered as covalently-modified synthetic fluorophore. When such molecules are reintroduced into the cell, they can be monitored by their fluorescence and their localization and behavior could report on the endogenous molecules that they mimic.

Examples of fluorescent reporter molecule types are: polysaccharide reporters, which can be monitored during cellular internalization or interaction with various cellular molecules during trafficking; labeled metabolites; substrates for different enzymes; labeled fatty acids and lipids, which can be partitioned into cell membranes and can therefore be used to report on their environment. Also in this category and of special interest are the labelled immunoreagents such as fluorescently labelled primary and secondary antibodies. Such labelled antibodies are usually available with various functional groups and are available in various colors. They can be used to probe location or activity of diverse biological molecules or the molecular interactions they participate in. Labeling technologies were further advanced by the commercial efforts of several companies that developed novel fluorochromes (e.g: AlexaFluor, Bodipy, CyDyes, ATTO, Chromis, CAL fluor, Quasar and IR Dyes), which all possess several unique properties and advantages (e.g: photostability, high quantum yield and brightness) as compared to traditionally used fluorescein and rhodamine-based dyes.

Among the most common applications of fluorescent dyes is the visualization of cell structures and components. For almost all intracellular organelles diverse stable and bright dyes had been developed for visualization. Several of them can be used for live analysis in plants such as probes for mitochondria (mitotracker green, nonyl acridine orange), nucleus (DAPI, Hoechst 33242), intracellular membranes ($DiOC_6$), cell walls (calcofluor white), vacuoles (Carboxy DCFDA), lipid droplets (Nile Red, Ac-201) and plasma membrane (FM 4-64) (Johnson and Spence 2010; Kuntam et al. 2015; Lovy-Wheeler et al. 2007; Schoor et al. 2015)

Another common application of the probes is to monitor biological processes in cells. Several fluorescent labels were developed for signal transduction studies; for tracking lipid metabolism, protein kinases, phosphatases and tracking nucleotide binding of proteins. Special probes are developed for studying ion channels and receptor binding, as well as endocytosis and exocytosis. A number of unique cellular functions can also be studied by specialized fluoroprobes, such as cell proliferation, cell cycle and apoptosis (Johnson and Spence 2010).

Probes called fluorescent tracers and indicators (or sensors) are also used in imaging studies. Various types of tracers had been developed (Kumar and Gilula 1996; Nashmi et al. 2002; Sukhorukov et al. 1995; Vercelli et al. 2000) to trace cell morphology and cell lineage for example. Such probes are usually dextran conjugates or peptide and protein conjugates, or fluorescent microspheres and they are tracers for either lipophilic membranes or polar, cell-injectable cytoplasmic tracers. Requirement for tracers is to be biologically inert and non-toxic for the host, while allowing prolonged tracking.

The general procedure for using organic dyes in plant live cell imaging involves preparing the fluorescent probe in a live cell compatible physiological solution and its delivery to plant cells. In plant cells, one particular limitation for dye delivery is the plant cell wall. Cell wall impermeable but plasma membrane permeable probes can be delivered to plant cells by preparing protoplasts. In case of plasma membrane impermeable probes, electroporation, microinjection, ester loading or low pH loading techniques can be used as an alternative (Fricker et al. 2001).

1.3 Fluorescent Proteins in Plant Cell Imaging

The cloning of the green fluorescent protein (GFP) of *Aequorea victoria* in 1992 (Prasher et al. 1992) and its first use for genetically tagging cellular proteins (Chalfie et al. 1994) represented a major turning point for fluorescence imaging of biological samples and initiated an exponential increase of its applications for biological studies. The possibility of genetically tagging proteins of living cells with fluorescent proteins has opened the gates for studies of cellular proteins in their natural environment by fluorescence microscopy. Ever since, great efforts are being made to search for other varieties of either naturally occurring fluorescent proteins, or improving and diversifying the existing ones by evolving them through mutations. Hence, their application possibilities are being considerably widened (Chudakov et al. 2010).

Structurally, GFP-like proteins possess β-barrel structures with a short helix tethered inside in the middle of the barrel, harboring the posttranslationally acquired chromophore. The diversity of the structures of the chromophores and of the nearby amino acid residues in these proteins leads to a range of emission colors and varied spectral properties for these proteins. Exploiting these, numerous efforts had been devoted to generate mutants with new, refined or enhanced spectral properties (Rizzo et al. 2009; Shu et al. 2006), and at present, a wide range of fluorescent proteins are available. Their color palette covers the entire spectral range from deep-blue to far-red region, allowing not only for single but also for multicolor labeling of different targets simultaneously in a living cell, hence, opening doors to versatile applications.

By using fluorescent proteins to tag different endogenous proteins, not only structural organizations can be monitored but also molecular interactions and dynamic processes within the cell, by using live cell imaging (Shaw and Ehrhardt 2013).

The basic procedure for using fluorescent proteins in plant cell imaging involves construction of the gene of interest in fusion with a fluorescent protein gene in a suitable vector, its delivery into plant cells (e.g: using Agrobacterium mediated delivery or microprojectile bombardment) to express the protein either transiently or stably and visualization of the expressed protein using live imaging microscopy setup (Cui et al. 2016; Groover and Jackson 2007). In green plant tissues, the possible intereference of red chlorophyll fluorescence should be taken into consideration when choosing fluorescent proteins for tagging and imaging. For green/red dual labelling and colocalization experiments, enhanced green fluorescent protein (EGFP) and mCherry combination can be used with appropriate filter sets to prevent interference of chlorophyll fluorescence. For triple labelling, enhanced cyan fluorescent protein (ECFP), enhanced yellow fluorescent protein (EYFP) (or its brighter derivative Venus) and mCherry can be used with appropriate filters. Beside protein localization studies, fluorescent protein fusions are also successfully used in plants to assess protein-protein interactions by using fluorescence resonance energy transfer (FRET) technique. ECFP and EYFP fused proteins are often used as FRET pairs, however, several novel alternative FRET pairs are being developed with better spectral properties (Müller et al. 2013). Another in vivo technique for protein interaction studies is the so called bimolecular fluorescence complementation (BiFC) where a fluorescent protein is split into carboxy-terminal (C-terminal) and amino-terminal (N-terminal) fragments, each of which is fused with a partner protein whose interactions are being assessed (Walter et al. 2004). In BiFC method, if the two proteins of interest interact with each other, the C-terminal and N-terminal parts of the split fluorescent protein will be brought close together such that they form a functional fluorescent protein. If the proteins do not interact, no fluorescence is observed. However, it is essential to include appropriate internal controls especially in BiFC type protein-protein interaction experiments, as the fluorescent protein halves are prone to self-assembly independently of protein-protein interactions (Horstman et al. 2014).

1.4 Live Cell Imaging of Plants

Basic live cell imaging of plant cells can be performed even with a simple light microscope equipped with a detector. Keeping the cells alive and healthy is the first and foremost concern during live cell imaging, since living cells are responsive to minute changes in environmental conditions. Therefore, it is vital to keep cells in an environment close to their natural or *in vitro* growth conditions during the entire duration of imaging. Parameters such as temperature, light, pH, humidity, osmolarity, nutrient and oxygen supply are among the most important factors that need to be regulated.

For long term observations, special environmental chamber inserts or whole-microscope enclosures are available from several manufacturers that allow for keeping cells, seedlings or plants under constant temperature and humidity with options to precisely regulate oxygen and carbon dioxide content of the chamber, as well. For short term observations, simpler observation chambers and imaging setups can be used. For *in vitro* grown suspension cultures, standard plastic Petri dishes provide the simplest solution for low magnification imaging. On the other hand, majority of the high magnification, high numerical aperture objectives are of oil-immersion type and require very short working distances. These objectives are often designed to be used with 0.17 mm thick, No 1.5 size coverslips. Therefore, for high resolution imaging, coverslip-bottom containers and invert microscopes should be preferred. For simultaneous handling of multiple samples, coverslip-bottom 4-compartment Petri dishes or 8-compartment chambered coverglass systems can be used. For high-throughput imaging, there are also 96 well, 384 well or even 1536 well coverslip-bottom plates available. However, all of these coverslip-bottom containers designed specifically for high resolution imaging applications, are significantly more expensive than standard plastic-bottom ones. As a low cost alternative, it is also possible to prepare imaging chambers by cutting out part of the bottom of regular plastic containers and attaching a coverslip at the bottom using a non-toxic adhesive such as pure silicon rubber, which is also used for building aquariums. (For sterilization: After complete curing of adhesive, dip the container and lid in 70% ethanol for 10 min, rinse with absolute ethanol and let it dry face down on a sterile tissue paper in laminar hood).

Figure 14.1b shows an example of a custom-made coverslip-bottom live cell imaging chamber for observation of plant cells using high resolution oil immersion objectives. For imaging of deeper layers, water immersion objectives should be preferred to better match the refractive index of the living plant tissue. In addition to bottom, inserting a coverslip to the lid of the container allows the use of a contrast enhancement technique called differential interference contrast (DIC), which benefits from all-glass light pathway. As opposed to standard bright field imaging, DIC imaging offers higher contrast to better resolve cellular and intracellular details in case of transparent, unstained samples such as the one shown in Fig. 14.1f. Note that nuclei (N), nucleoli (n), vacuoles (V) and cytoplasmic strands are clearly visible in the cultured tobacco BY-2 cells shown in Fig. 14.1f, where time-lapse live DIC imaging shows translocation of a cytoplasmic cargo to the vicinity of nucleus (arrows). In addition to DIC, there are several other contrast enhancement techniques such as phase contrast, Hoffmann modulation contrast, darkfield illumination or polarized illumination (Dawe et al. 2006). Depending on the nature of the sample, these non-intrusive techniques can provide significantly enhanced image for unstained samples during live observation of plant cells. All of these methods, however, require special microscope accessories attached to the basic light microscope. This is also the case for fluorescence imaging, which requires additional hardware to excite and detect fluorescent signals. Unlike transmitted light imaging modes, fluorescence imaging techniques do not require transparent samples. Majority of modern fluorescence microscopes work in reflection mode, hence non-transparent samples can also be conveniently observed.

Fluorescence microscopy, however, requires fluorescent molecules such as endogenously produced lignin or chlorophyll, or externally delivered fluorescent dyes or transgene-expressed fluorescent proteins. Detection of the reflected fluorescence emission allows to image thick leaves, stems or roots alive under fluorescence microscope, however, the quality of the image provided by conventional wide-field fluorescence microscopy deteriorates significantly when focused deep into the sample, due to a blur originating from the out-of-focus fluorescence emission.

The technique called confocal imaging greatly enhances the image quality by eliminating fluorescence emission originating from out-of-focus regions. This filtering out in confocal imaging is achieved by using a pinhole or slit-based system in front of the detector, so that it allows only the emission arising from focal point to enter the detector. These systems require sensitive detectors and also high illumination intensities such as those provided by lasers, high-power light-emitting diodes or high performance arc-discharge lamps. Using these powerful light sources during live analysis requires extra care as high intensity illumination can cause phototoxicity and photodamage especially for chlorophyll containing green tissues. A laser scanning confocal fluorescence microscope imaging setup for live observation of a leaf of potted maize plant is shown on Fig. 14.1a, b. In this configuration, a maize leaf is layered on a custom-made coverslip-bottom Petri dish and it is observed from below using an inverted microscope and an oil-immersion objective. A similar setup can be used for live observation of roots of a potted plant by using a pot with a coverslip/coverglass window at the bottom. Live, high resolution cellular imaging of intact plants is particularly important for ecophysiology studies where minimal disturbance of plant growth and metabolism is essential. To assess the health and viability of plant cells and tissues, below we present an example protocol based on using dual color fluorescence and fluorescence microscopy imaging that enables determination of live-dead cell ratio for the specimens.

2 Protocol: Live-Dead Viability Assay in Plant Cells Using Propidium Iodide and Fluorescein Diacetate

Determination of plant cell viability is essential to optimize growth conditions and to assess cell health following various experimental procedures and treatments (e.g: protoplastation, stress treatment, genetic modification). Fluorescence-based viability assays employ fluorescent stains and/or dye-conjugated substrates. One of the widely used approaches is dual staining with propidium iodide and fluorescein diacetate. Propidium iodide (PI) is a nucleic acid intercalating dye, which has very low plasma membrane permeability hence unable to penetrate living cells. It penetrates to plasma membrane compromised, dead cells and stains nuclear DNA as well as cytoplasmic and nuclear RNA. When excited by green light (e.g: 543 nm HeNe laser or mercury lamp with Texas red/Rhodamine filter set), propidium iodide fluoresces red and marks dead cells. Contrary to PI, the live cell staining agent fluorescein diacetate (FDA) is cell permeable, but nonfluorescent. However, when the acetyl groups of FDA molecules are cleaved by intracellular esterases, fluorescein molecules are being released inside the living cells. When excited by blue light (e.g: using 488 nm laser or mercury lamp with fluorescein filter set), the bright green fluorescence of the fluorescein molecule serves as a marker of the cellular viability. Hence co-incubation of cells with FDA/PI can be used as a convenient dual-color viability assay. As an example protocol, below we present detailed steps for viability assay of a non-transgenic Arabidopsis culture (Fig. 14.1g) and of a herbicide phosphinotricin (PPT)-resistant transgenic maize culture following treatment with PPT (Fig. 14.1h).

2.1 Equipment and Reagents

1. *Arabidopsis thaliana* cell line (ecotype Columbia) in Murashige and Skoog (MS) medium with 0.24 mg L^{-1} 2,4-dichlorophenoxyacetic acid (2,4-D) and 0.014 mg L^{-1} Kinetin (Murashige and Skoog 1962).
2. Transgenic maize cell line (*Zea mays* 'cv. H1233') with resistance to herbicide phosphinotricin (PPT) (Tiricz et al. 2018) in N6M medium with 0.5 mg L^{-1} 2,4-D and 15 mg/l PPT (Morocz et al. 1990).
3. Phosphinotricin (Duchefa Biochemie BV, The Netherlands) for herbicide treatment. 10 mg mL^{-1} in water. Sterile filtered, kept frozen in aliquots.
4. Propidium iodide (PI) 1000× stock solution: Prepare 2 mg mL^{-1} PI solution in water. Caution: PI is a nucleic acid intercalating dye; hence it should be handled with extreme care due to health hazards.
5. Fluorescein diacetate (FDA) 1000× stock solution: Prepare 1 mg mL^{-1} FDA stock solution in dimethyl sulfoxide (DMSO)
6. Consumables: sterile 1.5 mL microfuge tubes, pipettes, sterile pipette tips, microscope slides and coverslips.
7. Equipment: laminar flow hood, desktop centrifuge with swing-out rotor, conventional or confocal fluorescence microscope with fluorescein and Texas Red (or tetramethylrhodamine, TRITC) filter sets.

2.2 Method

1. Under a laminar flow hood, sample 0.5 mL from plant cultures (Arabidopsis and maize) into a 1.5 mL microfuge tube and let the cells to settle. (For PPT treatment, maize cells were incubated for 5 days prior imaging in 15 mg L^{-1} PPT containing growth medium). Alternatively, cells can be centrifuged briefly (3 min, 100 g, swing-out rotor) to aid settling. Care must be taken not to mechanically stress the cultures. Using wide bore pipettes or cutting the tip of plastic pipettes with a sterile scalpel can help reduce mechanical stress during handling of cells.
2. Replace culture supernatant with 1.5 mL fresh culture medium and resettle the cells. This washing step helps minimize residual esterase activity present in old culture medium. On the other hand, performing FDA/PI labeling in unwashed culture medium may help identify possible bacterial, yeast or fungi contamination present in the culture.

3. Freshly prepare 0.5 mL solution of 1 mg mL^{-1} FDA and 2 mg mL^{-1} PI in plant growth medium or in phosphate buffered saline (PBS). If fresh culture medium contains components with esterase activity, preparation of labeling reagents in PBS (or in MS macro salts) is preferred to prevent premature cleavage of FDA, which results in increased background fluorescence due to free fluorescein. When working with protoplasts, proper osmotic adjustment (e.g: using 0.3–0.5 M Sorbitol, depending on cell type) is essential to prevent bursting of cells during assay.
4. Incubate 0.5 mL of labeling reagent with settled cells for 2–5 min by gently inverting the tube.
5. Take 50–100 μL from stained samples onto a microscope slide, coverslip or coverslip-bottom Petri dish and observe cells using a fluorescence microscope configured for imaging fluorescein (green fluorescent living cells) and propidium iodide (red fluorescent dead cells).
6. Count at least 500 cells in 3 replicates to assess the ratio of green and red fluorescent cells.

Acknowledgements We thank Dr. Sándor Mórocz (Cereal Research Non-Profit Ltd., Szeged, Hungary) for providing non-transgenic Zea mays (H1233) cultures, Prof. Dénes Dudits (BRC, Szeged, Hungary) for providing *Arabidopsis thaliana* (Col) cultures; Béatrice Satiat-Jeunemaitre (CNRS, Gif sur Yvette, France) for providing *Nicotiana tabacum* (BY-2) cultures; Katalin Török and Ildikó Válkony for maintaining maize plants and suspension cultures. This work was supported by the National Research, Development and Innovation Office, NKFIH (Grant Number K116318) and by GINOP-2.3.2-15-2016-00001 grant.

References

Agati G (1998) Response of the in vivo chlorophyll fluorescence spectrum to environmental factors and laser excitation wavelength. Pure Appl Opt: J Eur Opt Soc Part A 7:797

Berg R (2004) Evaluation of spectral imaging for plant cell analysis. J Microsc 214:174–181

Buschmann C (2007) Variability and application of the chlorophyll fluorescence emission ratio red/far-red of leaves. Photosynth Res 92:261–271

Chalfie M, Tu Y, Euskirchen G, Ward WW, Prasher DC (1994) Green fluorescent protein as a marker for gene expression. Science 263:802–805

Chapman S, Oparka KJ, Roberts AG (2005) New tools for in vivo fluorescence tagging. Curr Opin Plant Biol 8:565–573

Chudakov DM, Matz MV, Lukyanov S, Lukyanov KA (2010) Fluorescent proteins and their applications in imaging living cells and tissues. Physiol Rev 90:1103–1163

Cui Y, Gao C, Zhao Q, Jiang L (2016) Using fluorescent protein fusions to study protein subcellular localization and dynamics in plant cells. In: Schwartzbach SD, Skalli O, Schikorski T (eds) High-resolution imaging of cellular proteins, Methods in molecular biology book series. Humana Press, New York, pp 113–123

Dawe GS, Schantz J-T, Abramowitz M, Davidson MW, Hutmacher DW (2006) Light microscopy. In: Dokland T, Hutmacher DW, Lee Ng MM, Shantz J-T (eds) Techniques in microscopy for biomedical applications. World Scientific, Singapore, pp 9–54

Fricker M, Parsons A, Tlalka M, Blancaflor E, Gilroy S, Meyer A, Plieth C (2001) Fluorescent probes for living plant cells. In: Hawes C, Satiat-Jeunemaitre B (eds) Plant cell biology: a practical approach, 2nd edn. Oxford University Press, Oxford, pp 35–84

Groover A, Jackson D (2007) Live-cell imaging of GFP in plants. Cold Spring Harb Protocol 2007:pdb.ip31

Horstman A, Tonaco IAN, Boutilier K, Immink RG (2014) A cautionary note on the use of split-YFP/BiFC in plant protein-protein interaction studies. Int J Mol Sci 15:9628–9643

Johnson I, Spence M (2010) The molecular probes handbook. A guide to fluorescent probes and labeling technologies, 11th edn. Life Technologies, Carlsbad 1060 pp

Kumar NM, Gilula NB (1996) The gap junction communication channel. Cell 84:381–388

Kuntam S, Puskás LG, Ayaydin F (2015) Characterization of a new class of blue-fluorescent lipid droplet markers for live-cell imaging in plants. Plant Cell Rep 34:655–665

Lakowicz JR (2007) Principles of fluorescence spectroscopy. Springer, New York

Lovy-Wheeler A, Cárdenas L, Kunkel JG, Hepler PK (2007) Differential organelle movement on the actin cytoskeleton in lily pollen tubes. Cytoskeleton 64:217–232

Morocz S, Donn G, Nérneth J, Dudits D (1990) An improved system to obtain fertile regenerants via maize protoplasts isolated from a highly embryogenic suspension culture. TAG 80:721–726

Müller SM, Galliardt H, Schneider J, Barisas BG, Seidel T (2013) Quantification of Förster resonance energy transfer by monitoring sensitized emission in living plant cells. Front Plant Sci 29:413

Murashige T, Skoog F (1962) A revised medium for rapid growth and bio assays with tobacco tissue cultures. Physiol Plant 15:473–497

Mylle E, Codreanu M-C, Boruc J, Russinova E (2013) Emission spectra profiling of fluorescent proteins in living plant cells. Plant Methods 9:10

Nashmi R, Velumian AA, Chung I, Zhang L, Agrawal SK, Fehlings MG (2002) Patch-clamp recordings from white matter glia in thin longitudinal slices of adult rat spinal cord. J Neurosci Methods 117:159–166

Prasher DC, Eckenrode VK, Ward WW, Prendergast FG, Cormier MJ (1992) Primary structure of the *Aequorea victoria* green-fluorescent protein. Gene 111:229–233

Rizzo MA, Davidson MW, Piston DW (2009) Fluorescent protein tracking and detection: fluorescent protein structure and color variants. Cold Spring Harb Protocol 2009:pdb.top63

Roshchina VV (2012) Vital autofluorescence: application to the study of plant living cells. Int J Spectr 2012:124672

Schoor S, Lung S-C, Sigurdson D, Chuong SD (2015) Fluorescent staining of living plant cells. In: ECT Y, Stasolla C, Sumner MJ, Huang BQ (eds) Plant microtechniques and protocols. Springer, Cham, pp 153–165

Shaw SL, Ehrhardt DW (2013) Smaller, faster, brighter: advances in optical imaging of living plant cells. Annu Rev Plant Biol 64:351–375

Shu X, Shaner NC, Yarbrough CA, Tsien RY, Remington SJ (2006) Novel chromophores and buried charges control color in fruits. Biochemistry 45:9639–9647

Sukhorukov VL, Djuzenova CS, Frank H, Arnold WM, Zimmermann U (1995) Electropermeabilization and fluorescent tracer exchange: the role of whole cell capacitance. Cytometry A 21:230–240

Tiricz H, Nagy B, Ferenc G, Török K, Nagy I, Dudits D, Ayaydin F (2018) Relaxed chromatin induced by histone deacetylase inhibitors improves the oligonucleotide-directed gene editing in plant cells. J Plant Res 131(1):179–189

Vercelli A, Repici M, Garbossa D, Grimaldi A (2000) Recent techniques for tracing pathways in the central nervous system of developing and adult mammals. Brain Res Bull 51:11–28

Walter M et al (2004) Visualization of protein interactions in living plant cells using bimolecular fluorescence complementation. Plant J 40:428–438

Chapter 15
Confocal and Transmission Electron Microscopy for Plant Studies

Adela M. Sánchez-Moreiras, Marianna Pacenza, Fabrizio Araniti, and Leonardo Bruno

1 Transmission Electron Microscopy

1.1 The Significance of Structural and Ultra-Structural Studies

The development of high-sensitive microscopic techniques, together with the development of methods for the conservation of cells and the improvement of methods for obtaining ultra-fine sections, gave rise to the knowledge of the cellular structure in the second half of the twentieth century. The combination of these techniques with immune-cytological techniques and the use of computational systems of image analysis have increased the interest in the use of electron and confocal microscopy in biological research. In fact, the use of transmission electron microscopy is very useful as a first approach for the characterization of the plant response to biotic or abiotic stress conditions. Knowing which areas of the cell morphology are being altered after treatment is essential to establish new physiological or biochemical measures for the in detail study of the effects of stressing factors on plants. A general view of the root or shoot tissues examined under microscopy will allow the identification of cellular organization, where alterations can be detected as erroneous division patterns, loss of cell identity or cellular disorganization (Rost et al. 1996; Zhu and Rost 2000). In particular, root tips act as finely tuned sensor for

A. M. Sánchez-Moreiras
Department of Plant Biology and Soil Science, University of Vigo, Vigo, Spain

M. Pacenza · L. Bruno (✉)
Department of Biology, Ecology and Earth Science, University of Calabria, Arcavacata di Rende, CS, Italy
e-mail: leonardo.bruno@unical.it

F. Araniti
Department Agraria, Mediterranea University, Reggio Calabria, Italy

© Springer International Publishing AG, part of Springer Nature 2018

A. M. Sánchez-Moreiras, M. J. Reigosa (eds.), *Advances in Plant Ecophysiology Techniques*, https://doi.org/10.1007/978-3-319-93233-0_15

different kinds of stress, which makes their study especially interesting for interpreting the plant stress response (Colmer et al. 1994; Koyro 2002).

Root cells are generated, from a small number of stem cells at the apex of the root, by division and continuous cell expansion and differentiation, so that it is possible to find all stages of cell development at the same time (Scheres et al. 2002). Roots have four types of stem cells or initial cells (Dolan et al. 1993): (i) the initial cells of the lateral layer or of the epidermis of the root, which give rise to the epidermis and to the lateral layer; (ii) the central zone of the calyptra, the columella, which has its own initial cells; (iii) the vascular tissue and the pericycle, which also have their own initial cells; and, finally, (iv) the cortex and the endodermis, which are generated by divisions of the initial cells of the cortex and the endodermis (Scheres et al. 2002). Internally, and in contact with all these undifferentiated cells, is the quiescent center, a group of mitotically inactive cells that are responsible for maintaining the activity of the stem cells (Scheres et al. 2002). The initial cells of the columella divide only anticline (perpendicular to the axis of growth) giving rise to a single row of cells, and their progeny undergo rapid cell expansion and differentiation producing amyloplasts, which play a fundamental role in gravitropism. The other three initial cell types are divided by anticlinal and periclinal divisions (parallel to the growth axis) giving rise to several layers of cells that acquire different identities (Dolan et al. 1993). Any change on the division, expansion or differentiation of initial cells can lead to totally malformed tissues.

Transmission electron microscopy study allows obtaining information about the morphology, disposition and size of these cells and their cellular organelles, which are usually affected when plants are under stress. Moreover, ultra-structural changes induced by a specific type of stress are usually quite consistent among species, which gives an advantage for the interpretation of the relevance of biotic or abiotic stress impact. The most commonly affected cell parts are the nucleus, the mitochondria, the vacuoles and the chloroplasts in photosynthetic cells, although cell wall alterations and Golgi activity can also give essential information to get a whole picture of the cellular response to stress (Fig. 15.1).

1.1.1 Nucleus

Abiotic stress, as produced by toxins, metals, herbicides, pollutants, reactive oxygen species, deficit or excess water, high or low temperatures, salinity or high light, frequently results in the induction of cell death processes (Joseph and Jini 2010). Actually, programmed cell death can be part, under these conditions, of an adaptive mechanism to survive stress (Palavan-Unsal et al. 2005). Nuclei alterations, typical of apoptotic-like PCD (AL-PCD), such as chromatin condensation and DNA fragmentation, formation of herniae in the nuclear membrane or migration of the nucleus to the cellular periphery, have been found in presence of abiotic stress (Tao et al. 2000; Palavan-Unsal et al. 2005; Díaz-Tielas et al. 2012).

Besides PCD-related changes, nuclei can also show other alterations that will give interesting information about other kind of plant responses to stress; i.e. amorphous and irregular nuclei and bi- or multi-nucleate cells can be associated

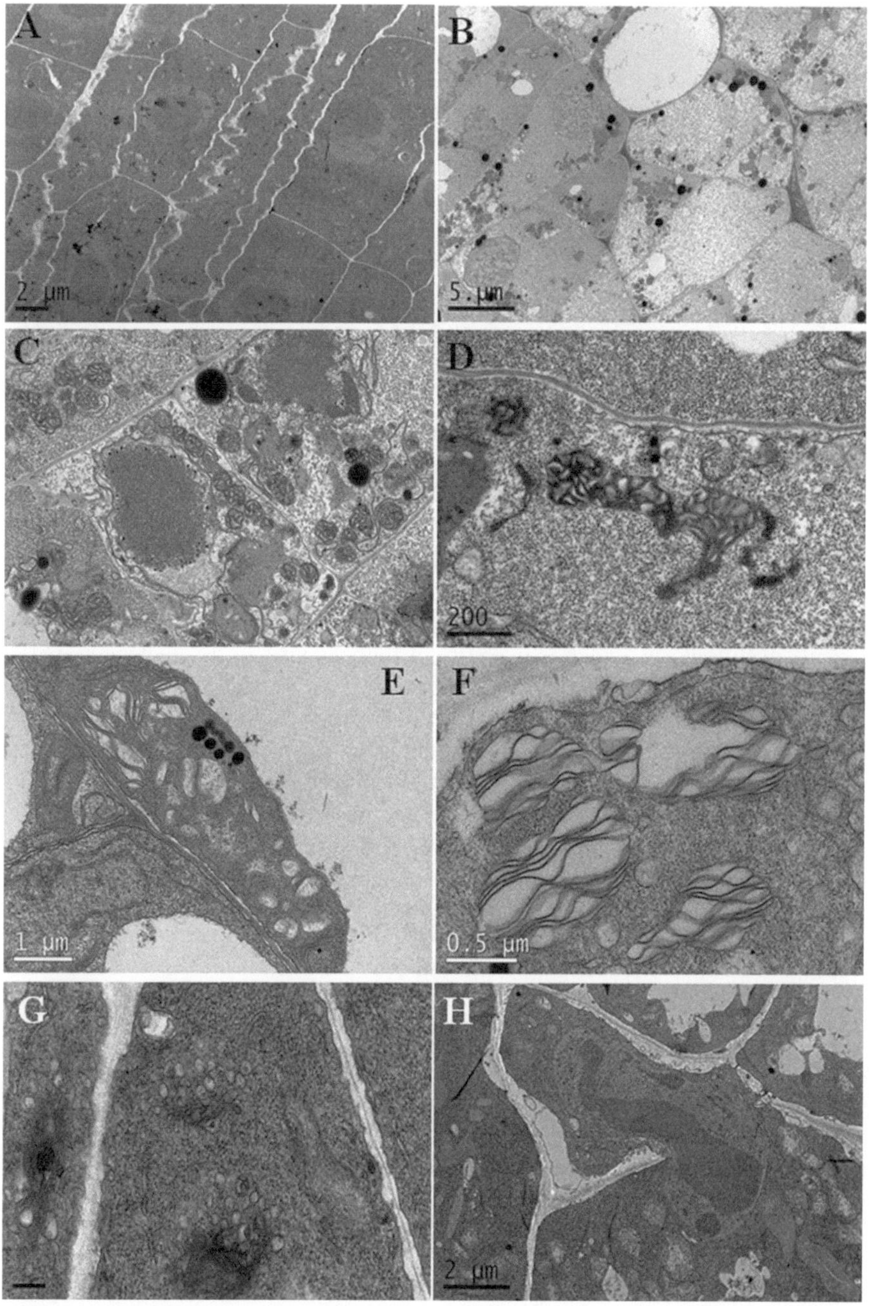

Fig. 15.1 TEM images of *Arabidopsis* treated cells. (**a**) Zigzag cell walls in citral-treated roots; (**b**) Disorganized and degraded cells in chalcone-treated roots; (**c**) Condensed mitochondria and distorted nuclei with nuclear membrane hernia and fragmented chromatin in chalcone-treated cells; (**d**) Condensed, irregular and broken mitochondria in chalcone-treated cell; (**e–f**) Swollen thylakoid in chalcone-treated shoots; (**g**) Increased Golgi-activity in eugenol-treated cells; (**h**) Incomplete cell wall formation in farnesene-treated cell

with cell wall malformations and, in particular, with erratically arranged microtubules (MTs), and shorter and fewer MTs in the arrays that can lead to aberrant chromosomal arrangements, misaligned or incomplete cell plates and distorted cell division (Kawamura et al. 2006; Araniti et al. 2016; Graña et al. 2017).

1.1.2 Mitochondria

Mitochondria play also a key role in AL-PCD response to stress, as apoptotic-like signal is usually initiated in these organelles. The intracellular machinery responsible for AL-PCD depends on the mitochondrial release of cytochrome c and other apoptogenic factors (Reape et al. 2008). Permeability transition pores are formed in the outer mitochondrial membrane, leading to a decrease in mitochondrial membrane potential or to the opening of a voltage-dependent anionic channel that results in the release of these apoptogenic factors from the space between the two mitochondrial membranes (Yao et al. 2004; Reape et al. 2008), which will lead to the cleavage of specific intracellular substrates that lead to cell death. Highly condensed and swollen mitochondria showing electron-dense matrix and a large number of cristae are typical of depolarized mitochondria involved in AL-PCD (Kiechle and Zhang 2002; Díaz-Tielas et al. 2012). Mitotic role in AL-PCD can include also the process of mitoptosis, which is the entry of mitochondria in the nucleus in which apoptogenic factors are released (Skulachev et al. 2004) and that has been previously shown for Arabidopsis stressed cells (Díaz-Tielas et al. 2012).

Increased number and size of mitochondria is also an interesting signal of plant response to stress. Peroxidation of mitochondrial membranes by ROS can result in membrane depolarization and irreversible loss of mitochondrial functions such as mitochondrial respiration, oxidative phosphorylation and ion transport (Vladimirov et al. 1980; Arpagaus et al. 2002), with the concomitant ATP deficit in the cell, which will increase mitochondrial number as a strategy of root cell to compensate mitochondrial dysfunction satisfying cellular energy requirements (Hardie 2011; Suksungworn et al. 2017; Araniti et al. 2016).

1.1.3 Chloroplasts

As key organelles in plant physiology due to the significance of photosynthesis for plant development, chloroplasts are one of the most common early target organelles of biotic and/or abiotic stress (Kratsch and Wise 2000; Giełwanowska et al. 2015; Díaz-Tielas et al. 2017). In particular, chloroplasts are especially susceptible to oxidative stress due to the high concentration of unsaturated fatty acids present in the thylakoids. Chloroplast swelling, thylakoid dilation and disintegration, plastoglobuli formation, reduced size and number of starch granules, unstacking of grana, intergranal vacuolation, disintegration of chloroplast envelope, darkening of the stroma and accumulation of lipid droplets, are among the most common changes observed in chloroplasts under cold, salinity, high light, herbicide, heavy metal, high temperature, or drought stress (Kratsch and Wise 2000; Stefanowska et al. 2002; Zhang et al. 2010; Meng et al. 2016; Díaz-Tielas et al. 2017). Disruption of chloroplast ultra-structure is associated with an early detriment of photosynthesis affecting the correct development of plants.

1.1.4 Golgi, Vacuoles and Other Cellular Structures

The increase in the activity and number of Golgi complexes and endoplasmic reticulum has been related to active detoxification processes, since they participate in the release of secretory vesicles and in the synthesis of lipid globules (Kaur et al. 2005). Detoxified bioproducts are compartmentalized in vacuoles or deposited in the cell wall, which is usually related to an increase in the number of vacuoles and the presence of thickened cell walls. It has also been shown that treatment with compounds that inhibit mitochondrial respiration results in associations between the Golgi apparatus and the endoplasmic reticulum as a result of a decrease in vesicle production due to energy depletion (Robinson and Kristen 1982), in addition to changes in the morphology of the Golgi complex, which acquires a cup-shaped arrangement (Kandasamy and Kristen 1987).

Moreover, vacuolation has been previously related to PCD process in plant cells. In particular, the rupture and disappearance of large central vacuoles commonly found in healthy plant cells has been found to occur prior to the degradation of the nucleus and the programmed death of cells (Zheng et al. 2017). The protoplast of several vacuoles can also break and release its hydrolytic contents to the cytosol (Díaz-Tielas et al. 2012).

Finally, enlargement of the vacuoles has also been found as a response to salt, drought or chemical stress due to accumulation of proline and other osmolytes in the vacuole for osmotic adjustment (Patakas et al. 2002) or to the sequestration of cations, heavy metals or other undesired compounds (Sauge-Merle et al. 2003).

Other cellular structures, such as peroxisomes (Fahy et al. 2017); autophagosomes (Minibayeva et al. 2012; Araniti et al. 2016); amyloplasts (Graña et al. 2013); or ribosomes can be also altered or just appear/disappear under mild or severe plant stress.

1.1.5 Plant Cell Wall

Stressing factors can lead to the appearance of incomplete cell walls due to problems in the formation of the phragmoplast and alterations in its thickness (Hepler and Bonsignore 1990; Vaughn et al. 1996). In many cases, the underlying cause of these wall dysfunctions are problems in cytokinesis and formation of the phragmoplast for the physical separation of the cytoplasm during cell division, since a successful cytokinesis is essential not only for correct morphogenesis, but also for an appropriate architecture of the body of the plant (Verma 2001). Moreover, the abnormal cell plates formed after microtubule disrupter can be enriched in callose thickening cell walls (Vaughn 2006).

Biosynthesis of wall polysaccharides shows enormous plasticity to several environmental factors (His et al. 2001), with different types of stress causing different balance of deposition of the different components, principally affecting the cellulose-xyloglucan network and compensating the reduction in cellulose content by increasing the pectin content (Burton et al. 2000; Aouar et al. 2010; Graña et al. 2013). Thickened and multilamellar cell walls with strong deposit content and irregularities in phragmoplast formation have been previously reported in plants under stress (García-Angulo et al. 2009; Graña et al. 2013; Araniti et al. 2016). Moreover, the thickening of the cell wall can greatly difficult the formation of plasmodesmata, impeding correct cell-cell communication and consequently causing the loss of tissue identity (Graña et al. 2013).

1.2 The Technique of Transmission Electron Microscopy (TEM)

Electron microscopy is mainly based on the observation under vacuum of fixed and dehydrated cellular and subcellular structures in stained ultrathin sections. The use of electron microscopy for the study of plant preparations started with the observation of isolated organelles (such as chloroplasts) and the visualization of cell walls (Roland and Vian 1991), mainly due to the difficulties that the presence of large vacuoles (soft compartments) and rigid cell walls (hard structures) represented in the preparation of electron microscopy sections from plant samples. Considering that the tonoplast (the membrane surrounding the vacuole) is highly sensitive to chemicals, strong care should be taken during the preparation of EM samples to avoid vacuole rupture and release of potentially damaging enzymes or compounds to the cytosol. Development of specific cytochemistry techniques, more specific reagents and a plurality of embedding materials allowed overpassing these difficulties and the heterogeneity of plant specimens and permitted to obtain high quality preparations. Although a slow and more complex microscopic technique, when compared to light or confocal microscopy, transmission electron microscopy (TEM) offers strong resolution and accuracy – to a resolution of 0.2 nm – when fine details are under study (Wilson and Bacic 2012).

Moreover, nowadays, we have new branches of transmission electron microscopy that enhance preparation and imaging of the samples; e.g. TEM tomography can generate semi-3D reconstructions (±60/70°) of plant cells and tissues at nanometer resolutions, while cryo-TEM allows visualization of frozen unstained samples (no reagents added during the process) avoiding alteration of ultrastructural details. TEM tomography and cryo-TEM can be coupled, although the necessity of sophisticated machinery strongly increases the cost of the preparations (Wilson and Bacic 2012).

There is no model recipe that can be used to prepare all kinds of plant tissues, therefore we show below a procedure to fix and stain root meristems or young shoots of *Arabidopsis* seedlings (Fig. 15.1). Meristematic zones are usually preferred, especially for beginners, when visualizing a sample due to the presence of smaller vacuoles and densely packed cytoplasm. When analyzing the effects of a treatment on cell ultra-structure, the numerical expression of morphological information (i.e. number of mitochondria, size of nuclei, number of autophagosomes, amount of swelling thylakoids, etc.) can be useful in the assessment of the effects of that treatment on cells, tissues and organs (Steer 1991).

1.3 Experimental Procedure for TEM Analyses

1. For roots: cut the apical meristem or the area of interest (1–2 mm) and introduce it immediately in 0.1 M cacodylate buffer (pH 7.2) until the fixation step
For shoots: cut the areas of interest of the living shoots into pieces of 1–2 mm, eliminate any superficial humidity and introduce them in an eppendorf tube (4–5 pieces per tube). Immediately after, fill the tubes with 0.1 M cacodylate buffer (pH 7.2), and introduce these tubes in a vacuum chamber for 15 min
2. Fix the root or shoot sample in 5% glutaraldehyde in 0.1 M cacodylate buffer (pH 7.2) at 4 °C for 4 h. To avoid artifacts or undesired modifications proceed as soon as possible with this step after sample excision
3. Wash in 0.1 M cacodylate buffer at 4 °C for 12 h (3 × 4 h)
4. After washing, incubate the samples in 2% osmium tetroxide in 0.1 M cacodylate buffer at 4 °C for 3 h
5. Incubate then in 2% uranyl acetate in 10% acetone at 4 °C for 1 h
6. Dehydrate with the following increasing acetone solutions (do all steps at 4 °C):
50% Acetone, 2 × 30 min
75% Acetone, 2 × 1 h[a]
80% Acetone, 2 × 1 h
95% Acetone, 2 × 1 h
100% Acetone, 2 × 2 h[b]
7. Infiltrate the sample in Spurr's resin at 4 °C as follows:
Spurr: acetone (1:3 v/v) (3 × 2 h)
Spurr: acetone (1:1 v/v) (3 × 2 h)[b]
Spurr: acetone (3:1 v/v) (2 × 2 h plus 1 × 3 h)
8. Embed the sample in 100% Spurr's resin for 2 × 3 h and left it overnight at room temperature
9. Embed the sample again in 100% resin (2 × 3 h)
10. Place the samples in pure resin in the molds and let them polymerize at 60 °C for 48 h
11. As the same fixation and embedding medium can be used for light and electron microscopy, check samples in semi-thin sections with light microscopy prior to continue with electron microscopy. For that, prepare semithin sections (0.7 μm) for light microscopy and then ultrathin sections (50–70 nm) for electron microscopy (ultrathin sections are mandatory to allow electrons correctly passing through the sample)

12. Contrast the samples as follows: incubate with uranyl acetate for 30 min; wash in boiled distilled water for 2 min; incubate with lead citrate for 10 min, and wash again in boiled distilled water for 2 min
13. Assemble ultrathin sections on copper grids of 100/200 mesh and examine by TEM using a JEOL JEM-1010 transmission electron microscope (at 100 kV) (Peabody, MA, USA) equipped with a CCD Orius-Digital Montage Plug-in camera (Gatan Inc., Gatan, CA, USA) and Gatan Digital Micrograph software (Gatan Inc.)

[a]The process can be stopped for a certain period
[b]It can stay overnight

2 Confocal Microscopy

The optical imaging technique based on confocal microscopy, also known as confocal laser scanning microscopy, aims to increase both contrast and optical resolution of a given sample micrograph using a spatial pinhole that blocks out-of-focus light during the formation of images (Matsumoto 2003). Unlike conventional microscopy, where the light passes through the sample, as far as it can penetrate, in confocal microscopy a beam of light is focused at one narrow depth level at a time. This technique, achieving a controlled and highly limited depth of focus, allows the reconstruction of 3D structures, of both biological and non-biological samples, through the acquisition at different depths of multiple two-dimensional images that are successively elaborated by a software (Matsumoto 2003). The ability to recreate three-dimensional images of microscopic samples allowed its diffusion in several research fields such as material science, semiconductor inspection and life science (Bruchez et al. 1998; Cardinale 2014; He et al. 2015; Fuchs et al. 2015; Bertani et al. 2017; Ben-Tov et al. 2018).

Concerning life science, confocal microscopy has revolutionized our knowledge about cells. It has became a pivotal tool for imaging both cell function and structure as well as the complexities of the morphology and dynamics of single cells and entire tissues (Amin et al. 2017). Moreover, since confocal microscope optically sections the specimen in a relatively non-invasive manner, it can be used to observe fixed as well as living samples increasing the potential of this technique in the understanding of several molecular and physiological dynamics (Araniti et al. 2017; Luo and Russinova 2017; Dinh et al. 2018).

In the past, the research through confocal microscopy was mainly focused on the study of cell structure and on the observation of spatial distribution of the organelles, whereas nowadays scientists continuously try to push to its limits confocal technology. For example, Shargil et al. (2015) have been able to detect and localize the cucumber green mottle mosaic virus (CGMMV) in both vegetative and reproductive tissues of cucumber and melon through in-situ hybridization technique. In particular, they observed that all tissues were infected by the virus, whereas the pollen was virus-free, highlighting that virulence transmission was mainly due to vectors such as insect instead of gamic transmission.

Zhao et al. (2012), studying soils contaminated with zinc oxide and nanoparticles demonstrated that both zinc oxide and nanoparticles penetrate root epidermis and cortex through the apoplastic way, whereas penetrate endodermis through the symplast. In addition, confocal microscopy technique has been largely used to study both plant nutrition and signaling in plants, and ion fluxes propagation could be followed through both entire living cells and tissues (Larrieu et al. 2015; Tian et al. 2016). For example, recent studies demonstrated that plants are able to quickly activate a signaling system based on Ca^{2+} waves that propagate in plant through the cortex and endodermal cell layers. In particular, it has been observed that Ca^{2+} movement depends on the vacuolar ion channel TPC1 and this Ca^{2+}/TCP1 system, eliciting specific responses in target organs, could contribute to increased stress-tolerance by whole plant (Choi et al. 2014).

Nowadays, there are available on the market probes for any kind of research activity, and this technological development has opened a new world to scientists interested in the dynamic complexities of cells and tissues (Wiederschain 2011). For example, the fluorescent labeled actin is a probe largely used to investigate the structural dynamics of the cytoskeleton (Araniti et al. 2016; Vaškebová et al. 2017), whereas Mito-Traker, Er-Traker, Bodi-Py ceramide, Lyso-Traker and DAPI probes are used to stain cellular organelle such as mitochondria, endoplasmic reticulum, Golgi complex, lysosomes and nuclei, respectively (Wiederschain 2011; Zhou et al. 2014; Chen and Zhang 2015; Chumak et al. 2015; Lee and Back 2017; Ibl et al. 2018). Fluorophores for probing cell function and structure (from ion flux to cell viability and from organelles and membrane to whole cells), for detecting biomolecules, for detecting and localizing oxidative stress, etc., are largely available and they have a contained cost considering the information they give (Wiederschain 2011).

Nowadays, using a chimeric fluorescent protein as an endogenous probe, such as GFP (green fluorescent protein), YFP (yellow fluorescent protein), etc., labeled proteins can be expressed within cells (specific tissues organelle and entire tissues) (Chalfie et al. 1994). Indicator molecules, which change their fluorescent characteristics as they sense changes in hormones concentration, pH, ions concentration, membrane potential etc., have been introduced into model species (e.g. *Arabidopsis thaliana*) and their mutants allowing to optically monitor the physiological state and its changes of entire tissues, as well as of individual cells (D'Angelo et al. 2006; Monshausen et al. 2007; Liu et al. 2015; Liu and Müller 2017). For example, Araniti et al. (2017) used several PIN*s*::GFP mutants to identify the mode of action of the natural compound farnesene, whereas Bruno et al. (2017) demonstrated that cadmium impacted growth of Arabidopsis primary root, altering auxin-cytokinin cross-talk and scarecrow expression. Those results highlight the potential of confocal microscopy, which is a robust and versatile technique with a variety of fields of application.

Two confocal microscopy techniques used to evaluate Arabidopsis root anatomy and protein immune-localization will be described below.

2.1 mPS-PI Staining

The use of sections obtained by cutting embedded tissues into paraffin blocks or resin and mounted onto microscope slides is always been the method of choice for the observation and the study, in plants, of the organization of tissues and their cells (Fig. 15.2).

Thanks to the advent of confocal microscopy (CLSM, Confocal Laser Scanner Microscopy), it became possible to obtain optical sections of the tissues without having to resort to actual sectioning of the samples, which implies a substantial reduction of the workload and, consequently, of the experimental timing. Moreover, the CLSM allows collection of sequential images on z-axis thanks to the use of specific software, and the subsequent reconstruction 3D of the samples. Unfortunately, the use of confocal microscopy on fresh samples is limited to thin and semi-transparent organs, like roots of seedlings of *Arabidopsis thaliana* (Helariutta et al. 2000; Kurup et al. 2005; Laplaze et al. 2005; Stadler et al. 2005).

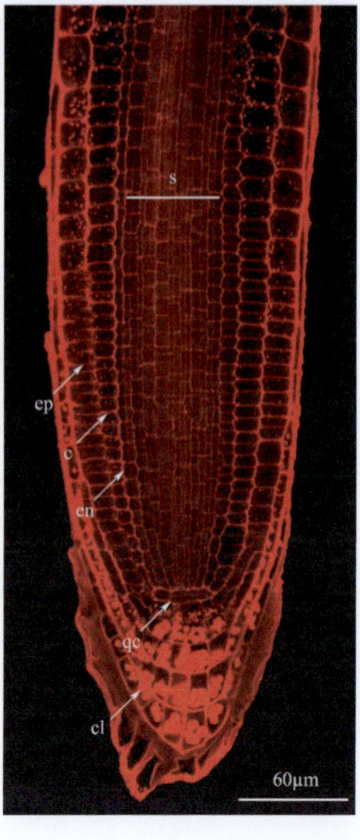

Fig. 15.2 Confocal laser image of primary root tip in *Arabidopsis thaliana* seedling at 7 days after germination (DAG), stained by using mPS-staining protocol (Truernit et al. 2008). *cl* columella, *c* cortex, *en* endodermis, *ep* epidermis, *M* meristematic zone, *pl* cap peripheral layers, *qc* quiescent center, *s* stele, *TZ* transition zone. Scale bar 46 μm

In fact, the presence in the aerial parts of plants of substances suitable to prevent the excessive penetration of light radiation, in addition to the presence of cell walls, cause spherical aberrations and dispersion of the luminous laser beam of the microscope, preventing penetration through the sample. These factors compromise the quality of the final images, especially during the observation of deeper layers of tissues (from ≥50 to 100 μm) (Haseloff 2003; Moreno et al. 2006).

To obtain optimal results in the observation of deeper layers of tissues (0.3 × 0.3 × 0.5 μm) by using confocal microscopy, it is therefore advisable to resort to the use of clarifying agents with high refractive indexes on correctly fixed samples (Truernit et al. 2008). The mPS-PI (Pseudo-Schiff Propidium Iodide) staining technique, published by Truernit et al. (2008), allows obtaining, in relatively short times, high-resolution images at cellular level of *Arabidopsis thaliana* tissues, in all organs and in all developmental stages. The staining is based on aldehyde groups' creation in cell walls carbohydrates of the samples following periodic acid treatment. These aldehyde groups, later, react covalently with pseudo-Schiff reagents such as propidium iodide. The formation of these covalent bonds on the walls makes them highly fluorescent, allowing the analysis of tissues at the cellular level through confocal microscopy in a detailed manner (Truernit et al. 2008). This method can also be combined with gene expression studies using β-glucuronidase (GUS).

2.1.1 Materials

- Distilled water;
- Methanol (CH_4O; CAS 67-56-1; FW 32.04);
- Acetic acid (CH_3COOH; CAS 64-19-7; FW 60.05);
- Ethanol (C_2H_6O; CAS 64-17-5; FW 46.07);
- Periodic Acid (HIO_4; CAS 10450-60-9; FW 227.94);
- Sodium metabisulphite ($Na_2S_2O_5$; CAS 7681-57-4; FW 190.10);
- Chloridric acid (HCl; CAS 7647-01-0; FW 36.46);
- Propidium iodide ($C_{27}H_{34}I_2N_4$; CAS 25535-16-4; FW 668.39);
- Chloral hydrate ($C_2H_3Cl_3O_2$; CAS 302-17-0; FW 165.40);
- Sodium dodecyl sulfate (SDS) ($NaC1_2H_{25}SO4$; CAS 151-21-3; FW 288.37);
- Sodium hydroxide (NaOH; CAS 1310-73-2; FW 39.99);
- Arabic gum (CAS 9000-01-5; FW).

2.1.2 Experimental Protocol

1. Fix the samples in the fixative (50% methanol, 10% acetic acid in distilled water), making sure that they are completely covered, at 4 °C for at least 12 h.
   The samples can thus be stored in the fixative up to a month.
2. Transfer the tissue in 80% (v/v) ethanol and incubate at 80 °C from 1 to 5 min, depending on the type of tissue (example: 1 min for leaves, 5 min for flower stems).

Ethanol treatment is important to increase the efficiency of staining in some organs. As far as root and root primordial staining is concerned, the step with ethanol can be omitted, while in ovules and seeds staining this step is replaced by overnight treatment of the samples with 1% SDS and 0.2 N NaOH at room temperature (RT), followed by washing in water and incubation in 25% bleach solution (2.5% of Cl^- active) from 1 to 5 min. Then, perform another washing in water and proceed with treatment with 1% periodic acid (w/v) (point 4).

3. Transfer samples in fixative and incubate at RT for 1 h.
4. Wash samples in distilled water and incubate in 1% periodic acid (w/v) at RT for 40 min.
5. Wash samples in distilled water.
6. Prepare Schiff's reagent (100 mM sodium metabisulphite, 0.15 N HCl in distilled water), and freshly add the propidium iodide to the final concentration of 100 μg mL^{-1}. Incubate the samples with Schiff's reagent at RT in the dark until the samples are visibly stained (around 2 h).
7. Transfer samples on slides and cover them with chloral hydrate solution (4 g chloral hydrate, 1 mL glycerol and 2 mL water).
8. Keep the slides overnight at RT in a closed container, to prevent them from drying out.
9. Remove the excess of chloral hydrate solution and coat the samples with Hoyer solution (30 g Arabic gum, 200 g chloral hydrate, 20 g glycerol and 50 mL water).
10. Cover the samples with the coverslip and allow them to stabilize for at least 3 days before the observation with a confocal microscope (excitation wavelength: 488 nm; reflection signal between 520 and 720 nm).

2.2 *Whole Mount Immunolocalization in Plant*

Despite the enormous progresses achieved in recent years in gene expression analyses, which allow to define the expression patterns of entire pathways in short times within plant (the so-called high-throughput techniques), in some cases, the in-situ localization of proteins at the cellular and sub-cellular level is still today the most efficient method for the study of functions, mechanisms of action and possible interactions between proteins and/or other cellular components.

Direct visualization by using recombinant DNA techniques with Green Fluorescent Protein (GFP) is one of the most used techniques for *in situ* localization of proteins, for its high reproducibility and the very short experimental times required for observation once obtained the specific construct for the proteins of interest; however, this approach in some cases does not accurately reflect the localization of the wild type (non-recombinant) protein, since the molecular properties of the recombinant protein are altered by GFP presence. In these cases, if the specific antibody is available, it is possible to resort to the immunolocalization of endogenous proteins (Sauer et al. 2006).

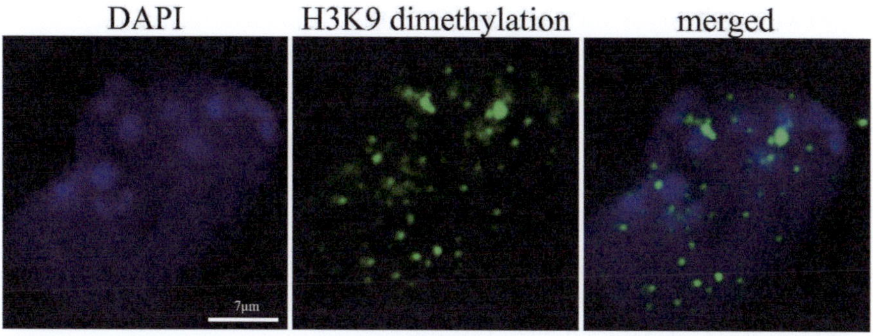

Fig. 15.3 Immunolocalization of histone H3 dimethylation at lysine position 9 on nuclei of the primary root of *Arabidopsis thaliana* seedlings growth at 8 days after germination (DAG)

In literature, there are many immunolocalization methods suitable for plants, however these methods require a prior work of inclusion and dissection of the tissues of interest, which makes them very laborious and unsuitable for a large number of samples, due to the large amount of time required to produce the necessary sections. One of the quickest and best performing methods is the one published by Sauer et al. (2006), which uses whole mount samples. A slightly modified version of this protocol is proposed here. This technique was optimized for *Arabidopsis thaliana* (Fig. 15.3), but good results were also obtained using the roots of tobacco plants (*Nicotiana tabacum*) and tomatoes (*Lycopersicon esculentum*).

The protocol consists of five main steps: (1) Tissue fixation, (2) Tissue permeation, (3) Incubation with the primary antibody, (4) Incubation with the secondary antibody, (5) Staining with DAPI (optional).

2.2.1 Materials

- Distilled water
- Potassium chloride (KCl; CAS 7447-40-7; FW 74.55)
- Sodium chloride (NaCl; CAS 7647-14-5; FW 58.44)
- Sodium phosphate dibasic dihydrate ($Na_2HPO_4 \cdot 2H_2O$; CAS 10028-24-7; FW 177.99)
- Potassium dihydrogen phosphate (KH_2PO_4; CAS 7778-77-0; FW 136.09)
- 1.4-Piperazinediethanesulfonic acid (PIPES; $C_8H_{18}N_2O_6S_2$; CAS 5625-37-6; FW 302.37)
- Ethylene glycol diacetate (EGTA; $HOCH_2CH_2OH$; CAS 107-21-1; FW 62.07)
- Magnesium sulfate heptahydrate ($MgSO_4 \cdot 7\ H_2O$; CAS 7487-88-9; FW 120.37)
- Potassium hydroxide (KOH; CAS 1310-58-3; FW 56.11)
- Paraformaldehyde ($HO(CH_2O)_nH$; CAS 30525-89-4; FW 30.03 (monomer))
- Triton X-100 (-Oct-C_6H_4-$(OCH_2CH_2)_xOH$ x=9-10; CAS 9002931)
- Driselase (CAS 85186-71-6; EC number: 286-055-3)
- IGEPAL CA-630 ($(C_2H_4O)_nC_{14}H_{22}O$; CAS 9002-93-1)

- Dimethyl sulfoxide (DMSO; $(CH_3)_2SO$; CAS 67-68-5; FW 78.13)
- Bovine serum albumin (BSA; CAS 9048-46-8; EC number 232-936-2)
- Primary antibody
- Secondary antibody
- Polyvinyl alcohol mounting medium with DABCO, antifading ($[-CH_2CHOH-]_n$).

To detect proteins in hypocotyls. Cotyledons or young leaves:

- Methanol (CH_4O; CAS 67-56-1; FW 32.04);
- Ethanol (C_2H_6O; CAS 64-17-5; FW 46.07);
- Xylene (C_8H_{10}; CAS 1330-20-7; FW 106.16).

Optional:

- 4′,6-Diamidino-2-phenylindole dihydrochloride (DAPI; $C_{16}H_{15}N_5 \cdot 2HCl$; CAS 28718-90-3; FW 350.25).

2.2.2 Experimental Protocol

Prepare the phosphate-buffered saline (PBS) or the microtubule-stabilizing buffer (MTSB).

10X PBS: Weigh and add 2 g KCl, 80 g NaCl, 17.8 g $Na_2HPO_4 \cdot 2\ H_2O$ and 2.4 g KH_2PO_4 in one liter solution. After diluting 10X PBS with water to obtain 1X PBS, adjust pH to 7.4 with the addition of KOH.

1X MTSB: 15 g PIPES, 1.9 g EGTA, 1.32 g $MgSO_4 \cdot 7\ H_2O$, 5 g KOH in one liter solution. If necessary, adjust pH to 6.8–7.0 with KOH.

1. Fix the samples in the fixative (4% paraformaldehyde, 0.1% Triton X-100 in 1X PBS). It is advisable to collect the ovules with the embryos in ice, since this operation takes a long time.
2. Incubate the samples in a vacuum pump with desiccator for 1 h at RT. For coriaceous organs, such as hypocotyls, it is recommended to prolong the treatment in order to guarantee the penetration of the fixative into the tissues.
3. Remove the fixative and perform two washes with 1X PBS for 5–10 min at RT.
4. To detect proteins in root apexes, in lateral roots and in embryos, go directly to point 5.

   To detect proteins in hypocotyls, cotyledons or young leaves:

   (a) Remove the 1X PBS and wash in methanol for 10 min at 37 °C. Repeat this wash if necessary until the chlorophyll is completely removed.
   (b) Remove the methanol and incubate the samples with a solution of 50% ethanol and 50% xylene for 10 min at 37 °C. Repeat this wash twice.

(c) Remove the ethanol and xylene solution and incubate the samples in xylene for 10 min at 37 °C. Repeat this wash twice.
(d) Remove xylene and incubate samples with a 50% ethanol and 50% xylene solution for 10 min at RT. Repeat this wash once.
(e) Remove the ethanol and xylene solution and incubate the samples in absolute ethanol for 10 min at RT. Repeat this wash once.
(f) Remove ethanol and incubate with 90% ethanol in 1X PBS for 5 min at RT.
(g) Remove ethanol and incubate with 75% ethanol in 1X PBS for 5 min at RT.
(h) Remove ethanol and incubate with 50% ethanol in 1X PBS for 5 min at RT.
(i) Remove ethanol and incubate with 25% ethanol in 1X PBS for 5 min at RT.

5. Remove the liquid and wash twice with distilled water for 5 min at RT.
6. Incubate samples with 2% Driselase in 1X PBS at 37 °C for 30–60 min.
7. Remove the Driselase and perform six washes with 1X PBS for 5–10 min at RT.
8. Incubate the samples with 3% IGEPAL CA-630 and 10% DMSO in 1X PBS.
9. Remove the IGEPAL CA-630 and DMSO solution and perform five to ten washes with 1X PBS for 5–10 min at RT.
10. Incubate the samples with 3% BSA in 1X PBS for 60 min at RT.
11. Remove the BSA solution and incubate the samples with the solution containing the primary antibody (2% BSA in 1X PBS, the concentration of primary antibody must be determined experimentally) for at least 4 h at 37 °C. Incubation with the primary antibody can be prolonged overnight at 4 °C.
12. Remove the primary antibody and perform six washes with 1X PBS at RT.
13. Remove the 1X PBS and incubate the samples with the secondary antibody (2% BSA in 1X PBS, the recommended secondary antibody concentration ranges from 1: 600 to 1: 2000, but concentration may need to be optimized) for at least 3 h at 37 °C.
14. Remove the secondary antibody and perform six washes with 1X PBS at RT.
15. Prepare the DAPI stock solution (1 mg mL^{-1} DAPI in 1X PBS) (optional).
16. Prepare the DAPI solution (by diluting the DAPI stock solution in water at a concentration of 1: 1000) and incubate the samples for 30 min at RT (optional).
17. Remove the DAPI solution and perform four washes with distilled water for 10 min at RT (optional).
18. Remove the water, mount the samples on the slides with a drop of antifade mounting medium and cover the samples with a coverslip.
19. Observe samples under a confocal microscope or store slides at 4 °C in the dark to preserve fluorescence. Depending on the type of fluorophore used, the samples can be preserved from few days to weeks after staining.

Acknowledgments The implementation of these techniques was possible thanks to the invaluable assistance of Inés Pazos and Jesús Méndez from the Central Research Services (CACTI) of the University of Vigo.

References

Amin MA, Nandi S, Mondal P, Mahata T, Ghosh S, Bhattacharyya K (2017) Physical chemistry in a single live cell: confocal microscopy. Phys Chem Chem Phys 19:12620–12627

Aouar L, Chebli Y, Geitmann A (2010) Morphogenesis of complex plant cell shapes: the mechanical role of crystalline cellulose in growing pollen tubes. Sex Plant Reprod 23:15–27

Araniti F, Graña E, Krasuska U, Bogatek R, Reigosa MJ, Abenavoli MR, Sánchez-Moreiras AM (2016) Loss of gravitropism in farnesene-treated arabidopsis is due to microtubule malformations related to hormonal and ROS unbalance. PloS one 11(8):e0160202

Araniti F, Bruno L, Sunseri F, Pacenza M, Forgione I, Bitonti MB, Abenavoli MR (2017) The allelochemical farnesene affects *Arabidopsis thaliana* root meristem altering auxin distribution. Plant Physiol Biochem 121:14–20

Arpagaus S, Rawyler A, Braendle R (2002) Occurrence and characteristics of the mitochondrial permeability transition in plants. J Biol Chem 277:1780–1787

Ben-Tov D, Idan-Molakandov A, Hugger A, Ben-Shlush I, Günl M, Yang B, Usadel B, Harpaz-Saad S (2018) The role of COBRA-LIKE 2 function, as part of the complex network of interacting pathways regulating Arabidopsis seed mucilage polysaccharide matrix organization. Plant J 94(3):497–512. https://doi.org/10.1111/tpj.13871

Bertani FR, Mozetic P, Fioramonti M, Iuliani M, Ribelli G, Pantano F, Santini D, Tonini G, Trombetta M, Businaro L, Selci S, Rainer A (2017) Classification of M1/M2-polarized human macrophages by label-free hyperspectral reflectance confocal microscopy and multivariate analysis. Sci Rep 7:8965

Bruchez M, Moronne M, Gin P, Weiss S, Alivisatos AP (1998) Semiconductor nanocrystals as fluorescent biological labels. Science 281:2013–2016

Bruno L, Pacenza M, Forgione I, Lamerton LR, Greco M, Chiappetta A, Bitonti MB (2017) In *Arabidopsis thaliana* cadmium impact on the growth of primary root by altering SCR expression and auxin-cytokinin cross-talk. Front Plant Sci 8:1323

Burton RA, Gibeaut DM, Bacic A, Findlay K, Roberts K, Hamilton A, Baulcombe DC, Fincher GB (2000) Virusinduced silencing of a plant cellulose synthase gene. Plant Cell 12:691–706

Cardinale M (2014) Scanning a microhabitat: plant-microbe interactions revealed by confocal laser microscopy. Front Microbiol 5:94

Chalfie M, Tu Y, Euskirchen G, Ward WW, Prasher DC (1994) Green fluorescent protein as a marker for gene expression. Science 263:802–805

Chen H, Zhang X (2015) Subcellular localization of CAX proteins in plants. Mol Soil Biol 6:1–5

Choi W-G, Toyota M, Kim S-H, Hilleary R, Gilroy S (2014) Salt stress-induced Ca^{2+} waves are associated with rapid, long-distance root-to-shoot signaling in plants. PNAS 111:6497–6502

Chumak N, Mosiolek M, Schoft VK (2015) Sample preparation and fractionation of *Arabidopsis thaliana* sperm and vegetative cell nuclei by FACS. Bio Protocol 5(22):e1664

Colmer TD, Fan TWM, Higashi RM, Läuchli A (1994) Interactions of Ca^{2+} and NaCl stress on the relations and intracellular pH of *Sorghum bicolor* root tips. An in vivo ^{32}P-NMR study. J Exp Bot 45:1037–1044

D'Angelo C, Weinl S, Batistic O, Pandey GK, Cheong YH, Schültke S, Albrecht V, Ehlert B, Schulz B, Harter K, Luan S, Bock R, Kudla J (2006) Alternative complex formation of the Ca^{2+}-regulated protein kinase CIPK1 controls abscisic acid-dependent and independent stress responses in Arabidopsis. Plant J 48:857–872

Díaz-Tielas C, Graña E, Sotelo T, Reigosa MJ, Sánchez-Moreiras AM (2012) The natural compound trans-chalcone induces programmed cell death in *Arabidopsis thaliana* roots. Plant Cell Environ 35:1500–1517

Díaz-Tielas C, Graña E, Maffei M, Reigosa MJ, Sánchez-Moreiras AM (2017) Plasma membrane depolarization precedes photosynthesis damage and long-term leaf bleaching in (*E*)-chalcone-treated *Arabidopsis* shoots. J Plant Physiol 218:56–65

Dinh N, van der Ent A, Mulligan DR, Nguyen AV (2018) Zinc and lead accumulation characteristics and in vivo distribution of Zn^{2+} in the hyperaccumulator *Noccaea caerulescens* elucidated with fluorescent probes and laser confocal microscopy. Environ Exp Bot 147:1–12

Dolan L, Janmaat K, Willemsen V, Linstead P, Poethig S, Roberts K, Scheres B (1993) Cellular organization of the *Arabidopsis thaliana* root. Development 119:71–84

Fahy D, Sanad MNME, Duscha K, Lyons M, Liu F, Bozhkov P, Kunz H-H, Hu J, Neuhaus HE, Steel PG, Smertenko A (2017) Impact of salt stress, cell death, and autophagy on peroxisomes: quantitative and morphological analyses using small fluorescent probe N-BODIPY. Sci Rep 7:39069

Fuchs N, Krajewski P, Bernhard C (2015) In situ observation of austenite grain growth in plain carbon steels by means of high-temperature laser scanning confocal microscopy. BHM Berg- und Hüttenmännische Monatshefte 160:214–220

Giełwanowska I, Pastorczyk M, Kellmann-Sopyła W, Górniak D, Górecki RJ (2015) Morphological and ultrastructural changes of organelles in leaf mesophyll cells of the arctic and antarctic plants of Poaceae family under cold influence. Arct Antarct Alp Res 47:17–25

Graña E, Sotelo T, Díaz-Tielas C, Araniti F, Krasuska U, Bogatek R, Reigosa MJ, Sánchez-Moreiras AM (2013) Citral induces auxin-mediated malformations and arrests cell division in *Arabidopsis thaliana* roots. J Chem Ecol 39:271–282

Graña E, Costas-Gil A, Longueira S, Celeiro M, Teijeira M, Reigosa MJ, Sánchez-Moreiras AM (2017) Auxin-like effects of the natural coumarin scopoletin on Arabidopsis cell structure and morphology. J Plant Physiol 218:45–55

Hardie G (2011) AMP-activated protein kinase—an energy sensor that regulates all aspects of cell function. Genes Dev 25:1895–1908

Haseloff J (2003) Old botanical techniques for new microscopes. Biotechniques 34:1174–1183

He YM, Clark G, Schaibley JR, He Y, Chen MG, Wie YJ, Ding X, Zhang Q, Yao W, Xu X, Lu CY, Pan JW (2015) Single quantum emitters in monolayer semiconductors. Nat Nanotechnol 10:497

Helariutta Y, Fukaki H, Wysocka-Diller J, Nakajima K, Jung J, Sena G, Hauser MT, Benfey PN (2000) The SHORT-ROOT gene controls radial patterning of the Arabidopsis root through radial signaling. Cell 101:555–567

Hepler PK, Bonsignore CL (1990) Caffeine inhibition of cytokinesis: ultrastructure of cell plate formation/degradation. Protoplasma 157:182–192

His I, Driouich A, Nicol F, Jauneau A, Höfte H (2001) Altered pectin composition in primary walls of Korrigan, a dwarf mutant of Arabidopsis deficient in membrane-bound endo-1,4-β-glucanase. Planta 212:348–358

Ibl V, Peters J, Stöger E, Arcalís E (2018) Imaging the ER and endomembrane system in cereal endosperm. In: Hawes C, Kriechbaumer V (eds) The plant endoplasmic reticulum, Methods in Molecular Biology, vol 1691. Humana Press, New York, pp 251–262

Joseph B, Jini D (2010) Salinity induced programmed cell death in plants: challenges and opportunities for salt-tolerant plants. J Plant Sci 5:376–390

Kandasamy MK, Kristen U (1987) Pentachlorophenol affects mitochondria and induces formation of Golgi apparatus-endoplasmic reticulum hybrids in tobacco pollen tubes. Protoplasma 141:112–120

Kaur H, Inderjit, Kaushik S (2005) Cellular evidence of allelopathic interference of benzoic acid to mustard (*Brassica juncea* L.) seedling growth. Plant Physiol Biochem 43:77–81

Kawamura E, Himmelspach R, Rashbrooke MC, Whittington AT, Gale KR, Collings DA, Wasteneys GO (2006) MICROTUBULE ORGANIZATION 1 regulates structure and function of microtubule arrays during mitosis and cytokinesis in the Arabidopsis root. Plant Physiol 140:102–114

Kiechle FL, Zhang X (2002) Apoptosis: biochemical aspects and clinical implications. Clin Chim Acta 326:27–45

Koyro HW (2002) Ultrastructural effects of salinity in higher plants. In: Läuchli A, Lüttge U (eds) Salinity: environment – plants – molecules. Springer, Dordrecht, pp 139–157

Kratsch HA, Wise RR (2000) The ultra structure of chilling stress. Plant Cell Environ 23:337–350

Kurup S, Runions J, Köhler U, Laplaze L, Hodge S, Haseloff J (2005) Marking cell lineages in living tissues. Plant J 42:444–453

Laplaze L, Parizot B, Baker A, Ricaud L, Martiniere A, Auguy F, Franch C, Nussaume L, Bogusz D, Haseloff J (2005) GAL4-GFP enhancer trap lines for genetic manipulation of lateral root development in *Arabidopsis thaliana*. J Exp Bot 56:2433–2442

Larrieu A, Champion A, Legrand J, Lavenus J, Mast D, Brunoud G, Oh J, Guyomarc'h S, Pizot M, Farmer EE, Turnbull C, Vernoux T, Bennett MJ, Laplaze L (2015) A fluorescent hormone biosensor reveals the dynamics of jasmonate signalling in plants. Nat Commun 6:6043

Lee H-Y, Back K (2017) Cadmium disrupts subcellular organelles, including chloroplasts, resulting in melatonin induction in plants. Molecules 22:1791

Liu J, Müller B (2017) Imaging TCSn::GFP, a synthetic cytokinin reporter. In: Kleine-Vehn J, Sauer M (eds) *Arabidopsis thaliana*, Plant Hormones. Methods in Molecular Biology, vol 497. Humana Press, New York, pp 81–90

Liu J, Yang L, Luan M, Wang Y, Zhang C, Zhang B, Shi J, Zhao F, Lan W, Luan S (2015) A vacuolar phosphate transporter essential for phosphate homeostasis in Arabidopsis. PNAS 112:E6571–E6578

Luo Y, Russinova E (2017) Quantitative microscopic analysis of plasma membrane receptor dynamics in living plant cells. In: Russinova E, Caño-Delgado A (eds) Brassinosteroids, Methods in Molecular Biology, vol 1564. Humana Press, New York, pp 121–132

Matsumoto B (2003) Cell biological applications of confocal microscopy, vol 70. Academic, Cambridge, MA

Meng F, Luo Q, Wang Q, Zhang X, Qi Z, Xu F, Lei X, Cao Y, Chow WS, Sun G (2016) Physiological and proteomic responses to salt stress in chloroplasts of diploid and tetraploid black locust (*Robinia pseudoacacia* L.). Sci Rep 6:23098

Minibayeva F, Dmitrieva S, Ponomareva A, Ryabovol V (2012) Oxidative stress-induced autophagy in plants: the role of mitochondria. Plant Physiol Biochem 59:11–19

Monshausen G, Bibikova T, Messerli M, Shi C, Gilroy S (2007) Oscillations in extracellular pH and reactive oxygen species modulate tip growth of Arabidopsis root hairs. PNAS 104:20996–21001

Moreno N, Bougourd S, Haseloff J, Feijó JA (2006) Imaging plant cells. In: Pawley J (ed) Handbook of biological confocal microscopy. Springer, Boston, pp 769–787

Palavan-Unsal N, Buyuktuncer ED, Tufekci MA (2005) Programmed cell death in plants. J Cell Mol Biol 4:9–23

Patakas A, Nikolaou N, Zioziou E, Radoglou K, Noitsakis B (2002) The role of organic solute and ion accumulation in osmotic adjustment in drought-stressed grapevines. Plant Sci 163:361–367

Reape TJ, Molony EM, McCabe PF (2008) Programmed cell death in plants: distinguishing between different modes. J Exp Bot 59:435–444

Robinson E, Kristen U (1982) Membrane flow via the Golgi apparatus of higher plant cells. Int Rev Cytol 77:89–127

Roland JC, Vian B (1991) General preparation and staining of thin sections. In: Hall JL, Hawes C (eds) Electron microscopy of plant cells. Academic, London, pp 1–66

Rost TL, Baum SF, Nichol S (1996) Root apical organization in *Arabidopsis thaliana* ecotype "WS" and a comment on root cap structure. Plant Soil 187:91–95

Sauer M, Paciorek T, Benková E, Friml J (2006) Immunocytochemical techniques for whole-mount in situ protein localization in plants. Nat Protoc 1:98

Sauge-Merle S, Cuine S, Carrier P, Lecomte-Pradines C, Luu DT, Peltier G (2003) Enhanced toxic metal accumulation in engineered bacterial cells expressing *Arabidopsis thaliana*. Appl Environ Microbiol 69:490–494

Scheres B, Benfey P, Dolan L (2002) Root development. In: The *Arabidopsis* book, vol 1. American Society of Plant Biologists, Rockville

Shargil D, Zemach H, Belausov E, Lachman O, Kamenetsky R, Dombrovsky A (2015) Development of a fluorescent in situ hybridization (FISH) technique for visualizing CGMMV in plant tissues. J Virol Methods 223:55–60

Skulachev VP, Bakeeva LE, Chernyak BV et al (2004) Thread-grain transition of mitochondrial reticulum as a step of mitoptosis and apoptosis. Mol Cell Biochem 256/257:341–358

Stadler R, Wright KM, Lauterbach C, Amon G, Gahrtz M, Feuerstein A, Oparka KJ, Sauer N (2005) Expression of GFP-fusions in Arabidopsis companion cells reveals non-specific protein trafficking into sieve elements and identifies a novel post-phloem domain in roots. Plant J 41:319–331

Steer MW (1991) Quantitative morphological analyses. In: Hall JL, Hawes C (eds) Electron microscopy of plant cells. Academic, London, pp 85–104

Stefanowska M, Kuras M, Kacperska A (2002) Low temperature induced modifications in cell ultrastructure and localization of phenolics in winter oilseed rape (*Brassica napus* L. var. oleifera L.) leaves. Ann Bot 90:637–645

Suksungworn R, Srisombat N, Bapia S, Soun-Udom M, Sanevas N, Wongkantrakorn N, Kermanee P, Vajrodaya S, Duangsrisai S (2017) Coumarins from *Haldina cordifolia* lead to programmed cell death in giant mimosa: potential bio-herbicides. Pak J Bot 49:1173–1183

Tao L, van Staden J, Cress WA (2000) Salinity induced nuclear and DNA degradation in meristematic of soybean (*Glycine max* (L.)) roots. Plant Growth Regul 30:49–54

Tian Q, Zhang X, Yang A, Wang T, Zhang W-H (2016) CIPK23 is involved in iron acquisition of Arabidopsis by affecting ferric chelate reductase activity. Plant Sci 246:70–79

Truernit E, Bauby H, Dubreucq B, Grandjean O, Runions J, Barthélémy J, Palauqui J-C (2008) High-resolution whole-mount imaging of three-dimensional tissue organization and gene expression enables the study of phloem development and structure in Arabidopsis. Plant Cell 20:1494–1503

Vaškebová L, Šamaj J, Ovečka M (2017) Single-point ACT2 gene mutation in the Arabidopsis root hair mutant der1–3 affects overall actin organization, root growth and plant development. Ann Bot 00:1–13

Vaughn KC (2006) The abnormal cell plates formed after microtubule disrupter herbicide treatment are enriched in callose. Pest Biochem Physiol 84:63–71

Vaughn KC, Hoffman JC, Hahn MG, Staehelin LA (1996) The herbicide dichlobenil disrupts cell plate formation: inmunogold characterization. Protoplasma 194:117–132

Verma DP (2001) Cytokinesis and building of the cell plate in plants. Annu Rev Plant Physiol Plant Mol Biol 52:751–784

Vladimirov YA, Olenev VI, Suslova TB, Cheremisina ZP (1980) Lipid peroxidation in mitochondrial membranes. Adv Lipid Res 17:173–249

Wiederschain GY (2011) A guide to fluorescent probes and labeling technologies. In: Johnson I, Spence M (eds) The molecular probes handbook, vol 76. Biochemistry, Moscow, pp 1276–1276

Wilson SM, Bacic A (2012) Preparation of plant cells for transmission electron microscopy to optimize immunogold labeling of carbohydrate and protein epitopes. Nat Protoc 7:1716–1727

Yao N, Eisfelder BJ, Marvin J, Greenberg JT (2004) The mitochondrion – an organelle commonly involved in programmed cell death in *Arabidopsis thaliana*. Plant J 40:596–610

Zhang R, Wise RR, Struck KR, Sharkey TD (2010) Moderate heat stress of *Arabidopsis thaliana* leaves causes chloroplast swelling and plastoglobule formation. Photosynth Res 105:123–134

Zhao L, Peralta-Videa JR, Ren M, Varela-Ramirez A, Li C, Hernandez-Viezcas JA, Aguilera RJ, Gardea-Torresdey JL (2012) Transport of Zn in a sandy loam soil treated with ZnO NPs and uptake by corn plants: Electron microprobe and confocal microscopy studies. Chem Eng J 184:1–8

Zheng Y, Zhang H, Deng X, Liu J, Chen H (2017) The relationship between vacuolation and initiation of PCD in rice (*Oryza sativa*) aleurone cells. Sci Rep 7:41245

Zhou J, Wang J, Yu J-Q, Chen Z (2014) Role and regulation of autophagy in heat stress responses of tomato plants. Front Plant Sci 5:174

Zhu T, Rost TL (2000) Directional cell-to-cell communication in the *Arabidopsis* root apical meristem. III. Plasmodesmata turnover and apoptosis in meristem and root cap cells during four weeks after germination. Protoplasma 213:99–107

Chapter 16
Plant Programmed Cell Death (PCD): Using Cell Morphology as a Tool to Investigate Plant PCD

T. John Conway and Paul F. McCabe

1 Introduction

Programmed cell death (PCD) is a form of cellular suicide that serves to eliminate damaged or unwanted cells. In plants PCD operates from embryogenesis to the death of the whole plant and is an integral component of both plant defence and development.

Activation of a PCD process can lead to the death of a few cells or instead whole plants can senesce, die and transfer nutrients to the next generation. PCD in the flower offers an insight into the various roles and outcomes of specific PCD events. Following pollination, at the whole organ level, petals and sepals will undergo PCD with resources from these organs being diverted to the rest of the plant. In the case of pollen development, a small layer of tissue surrounding the developing pollen – known as the tapetum – undergoes PCD and deposits a coat of lipids (pollen kitt) on the maturing pollen surface. PCD in flowers also includes megaspore and antipodal cell degeneration, nucellular degeneration and pollen tube death during self-incompatibility (SI) (Rogers 2006). Further examples of PCD occur in seed maturation and seed ripening. The timing and intensity of death is crucial as, for example, suboptimal tapetal PCD results in sterile pollen and a condition known as cytoplasmic male sterility (CMS) (Bohra et al. 2016).

Plants are sessile. Because of this they must be able to alter their metabolic processes to an ever-changing environment. When plants experience stressful environmental conditions such as drought, flooding, high UV light, salt, heavy metal or heat stress they can no longer continue normal metabolic processes and PCD is induced. As developmental cues can induce PCD in flowers; so environmental stimuli can

T. John Conway · P. F. McCabe (✉)
School of Biology and Environmental Science, University College Dublin, Dublin, Ireland
e-mail: paul.mccabe@ucd.ie

© Springer International Publishing AG, part of Springer Nature 2018

A. M. Sánchez-Moreiras, M. J. Reigosa (eds.), *Advances in Plant Ecophysiology Techniques*, https://doi.org/10.1007/978-3-319-93233-0_16

also result in activation of PCD. Cell death may occur in only a few cells, for example following leaf damage, the cells surrounding the damaged area will die, thereby sealing the wound and preventing water loss and pathogen entry. Alternatively, PCD may lead to significant organ reconstruction such as aerenchyma formation. Aerenchyma is tissue composed of interconnected gas conducting spaces formed by collapse and PCD of cells within the cortex region of oxygen-starved roots or leaves. Aerenchyma formation is an important response of many crop species to waterlogging which is a major problem in many parts of the world.

Plants do not possess an animal-like immune system. Instead, in response to pathogen attack, plants will initiate a hypersensitive response (HR), where cells surrounding a site of infection undergo PCD, a process which functions to isolate the pathogen, starve it of resources and prevent the spread of the infection (Morel and Dangl 1997). Misregulation of the HR can lead to unnecessary excessive loss of tissue, or alternatively, fail to prevent the spread of disease.

These examples of PCD in plants highlight the central role PCD plays in plant development, defence and response to environmental conditions. While individual PCD events are often described as being either developmentally or environmentally activated it is most likely that all PCD events are driven by a combination of environmental and developmental factors with each PCD event being influenced to a greater or lesser extent by either primary stimulation. So, while execution of PCD is genetically driven, the activation of the process is influenced by significant contributions from environmental stimuli. An appreciation of the environmental factors that impact PCD regulation is therefore crucial to fully understanding the control of PCD in crops and plants. The aforementioned examples of PCD also illustrate how suboptimal PCD can detrimentally affect plant biology and significantly impact plant fitness – resulting in crop sterility, excessive biomass loss in environmentally challenging conditions, or a compromised HR defence. Given the essential role PCD plays in plant biology and the impact suboptimal PCD can have on yield and fitness it is understandable that research into the genetic and environmental aspects of the core regulation of PCD has increased over the years.

Technically, a crucial component of experimental research into genetic or environmental components of PCD, is to be able to distinguish between living cells, cells which have undergone a PCD process, and cells which have died necrotically (unorganised rapid cell death). Many experimental protocols use hallmark characteristics to identify PCD events (Kacprzyk et al. 2016). These hallmark characteristics have often been informed by research into animal apoptosis and there are several hallmark characteristics common to both such as, nDNA fragmentation, chromatin condensation, mitochondrial membrane rupture, release of cytochrome c, contraction of the cytoplasm and elevation of signalling molecules such as reactive oxygen species and Ca^{2+} (Kacprzyk et al. 2011).

Studying the regulatory mechanisms of PCD in vivo can be difficult as often cells undergoing PCD are buried beneath living tissue or it is only possible to observe the cells post mortem. For many of the hallmark characteristics there is only a finite window of opportunity to detect their presence and even then, the presence of these markers is only enough to give a qualitative and not a quantitative measure of PCD. To examine the subtle role environmental factors may have on altering PCD activation thresholds it is necessary to quantitatively assess rates and timing of PCD events. One way to do this is by scoring a hallmark morphology that is associated with PCD.

2 Using Morphology to Investigate Plant PCD

In our laboratory we use cellular morphology to quantify rates of plant PCD. When plant cells die they often undergo a marked condensation of the cytoplasm, which results in a visible gap between the plasma membrane and cell wall (Reape and McCabe 2008). Morphological retraction also has been shown to occur in animal cells and been shown to be an integral part of the PCD process and is in fact a driver of PCD (apoptosis) where volume decrease is regulated by ion flux. Our group has shown that in plants this morphology is the result of an active retraction of an intact membrane. Further evidence that this morphology is an active PCD process as distinct from a necrotic collapse of a ruptured membrane, is the requirement to have active ATP metabolism to be initiated properly in order to drive the retraction process (Kacprzyk et al. 2017). Examples of plant PCD events occurring with protoplast retraction include death induced by environmental or abiotic stress, pathogen attack or developmental PCD, such as suspensor elimination, leaf morphogenesis or PCD during the final stages of senescence.

This characteristic morphology marker was used to evaluate rates of activation of a programmed cell death pathway in carrot cells in response to stress and the absence of survival signals (McCabe et al. 1997). McCabe and colleagues used a key concept of cell survival and initiation of PCD in response to stress – cells are constantly monitoring pro- and anti- PCD signals and when cells are stressed pro-PCD signals increase. It was found that at low stress levels cells acclimate to stress and remain alive; at moderate stress levels cells trigger a cell death programme (PCD) and, at higher stress levels still, cells are overwhelmed and death occurs without any activation of a death programme (necrosis). Figure 16.1 shows the PCD response curve of 5 days old *Arabidopsis thaliana* seedling root hair cells when subjected to an increasing heat shock.

Figure 16.1 shows that as the temperature increases the percentage of each cell type changes. Above 44 °C cells begin to die. At 48 °C most cells are able to initiate a PCD programme. Beyond 52 °C cells are no longer capable of initiating a PCD programme. At exposure to higher temperatures cells begin to lose membrane integrity, are no longer capable of ATP synthesis and are generally no longer capable of completing the necessary metabolic processes required for PCD initiation.

We have used this morphological trait to develop PCD assay protocols using cell cultures, seedling root hairs and fern gametophytes. In each system material can be prepared quickly and cheaply. Also in each system sufficient numbers of individual cells can be observed using light microscopy and easily categorised as being viable, having undergone PCD or necrotic by using simple staining techniques. These sys-

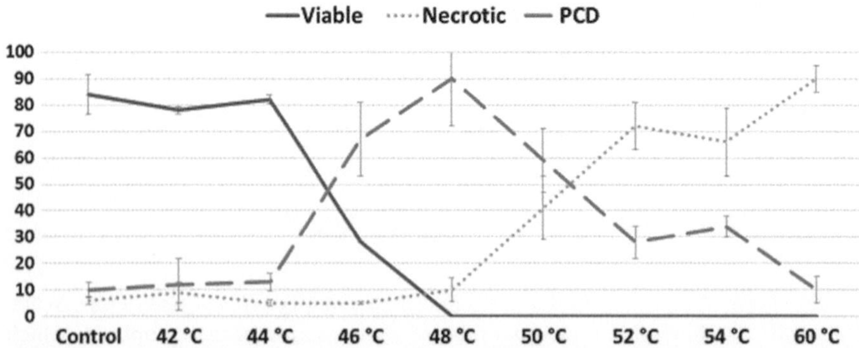

Fig. 16.1 The effect on cell type numbers by increasing the exposure to heat shock. Five days old seedlings of Arabidopsis thaliana are exposed to a range of heat shocks from 42 to 60 °C for 10 min. Root hairs are scored 24 h later. Root hairs that fluoresce under UV light and after staining with FDA are deemed to be viable. Non-fluorescing cells are examined for morphology and those with contracted cytoplasm are scored as having undergone PCD. Those with no fluorescence and no retraction are scored as necrotic

tems can then be used to study environmental factors that influence, or induce, PCD by subjecting cells to specific environmental conditions and using morphology to quantify rates of cellular PCD.

2.1 *Cell Cultures*

Cell cultures are maintained on a 7-day cycle by adding 10 mL of culture to 100 mL of medium containing 30 g L^{-1} sucrose, 4.3 g L^{-1} Murashige and Skoog (M&S) basal salt mixture, 500 μL L^{-1} naphthaleneacetic acid, and 50 μL L^{-1} kinetin in distilled water, pH 5.8. Cultures are maintained at a constant temperature of 23 °C at 100 rpm. In our laboratory we typically use *Arabidopsis thaliana* cultures, but the system works equally well with any species which can be maintained as a cell culture.

2.2 *Root Hair Assay*

Root hairs are single cells, which extend out at right angles from the main root. Each seedling root contains numerous root hairs, which can be readily observed by light microscopy. In our laboratory we typically use *Arabidopsis thaliana* seedlings. Preparation of material begins with sterilisation of seeds with 20% (V/V) commercial bleach (final concentration of NaOCL is 1%) followed by washing four times in sterile distilled water (SDW). After sterilisation seeds are plated on ½ strength M&S (basal salts, 2.15 g L^{-1}) medium, 1% sucrose and 1.5% agar in 9 cm petri dishes and stratified at 4 °C in the dark for 24 h before being placed vertically at 22 °C, 16 h light, 8 h dark. Five-day old seedlings are suitable for experiments.

Again, as with the cell culture system, the root hair protocol can be utilised with a wide range of species, which satisfy a few necessary criteria. Synchronised germination is required, there needs to be a suitable number of root hairs, too few and the number of seedlings needed becomes very large and too many and it becomes impossible to discriminate individual root hair cells (Hogg et al. 2011).

2.3 *Fern Gametophyte*

In order to expand our research to include non-spermatophyte species we have assessed a range of tissue types from lower plant species for their suitability to facilitate large scale scoring of PCD in whole plants. An important criterion for their suitability being that a large number of visible cells exhibits PCD morphology when subjected to a known inducer of PCD. The thallus of fern gametophytes was found to be ideal. Fern gametophytes are planar single cell thick prothalli, which can be propagated easily in liquid cultures. Hundreds of suitable gametophytes can be acquired within 14 days after simple propagation in liquid media. As with the root hair system the first step is spore sterilisation in 20% commercial bleach, washed four times in SDW then placed in ½ strength M&S media (without sucrose), pH 6.0. Cultures are maintained at a constant temperature of 23 °C at 100 rpm.

In each of the systems described (cell culture, root hairs, fern gametophyte) an identical methodology is used to analyse the percentage of cell that are viable, have undergone PCD, or have died necrotically, after exposure to a treatment. Viable cells will fluoresce when stained with FDA and viewed under UV light. Cells which do not fluoresce are examined for PCD morphology. Cells exhibiting the morphology are deemed to have undergone PCD and cells which lack both the morphology and fluorescence are categorised as necrotic.

Figure 16.2 shows an example of a viable, PCD and necrotic cell from each of the three systems described.

After observation and categorisation of the cells a simple statistical analysis can be carried out where each cell type is expressed as a percentage of the total. In each system a sufficient number of replications is required depending on the variability of results obtained in pilot studies. Typically, we score three seedlings for each root hair experiment with each seedling having more than one hundred root hair cells scored. Three replicate experiments are run to ensure consistency of results. For cell cultures we count a minimum one hundred cells from each of three Erlenmeyer flasks, and for fern gametophytes we examine three gametophytes in each experiment and run three replicates to ensure consistency.

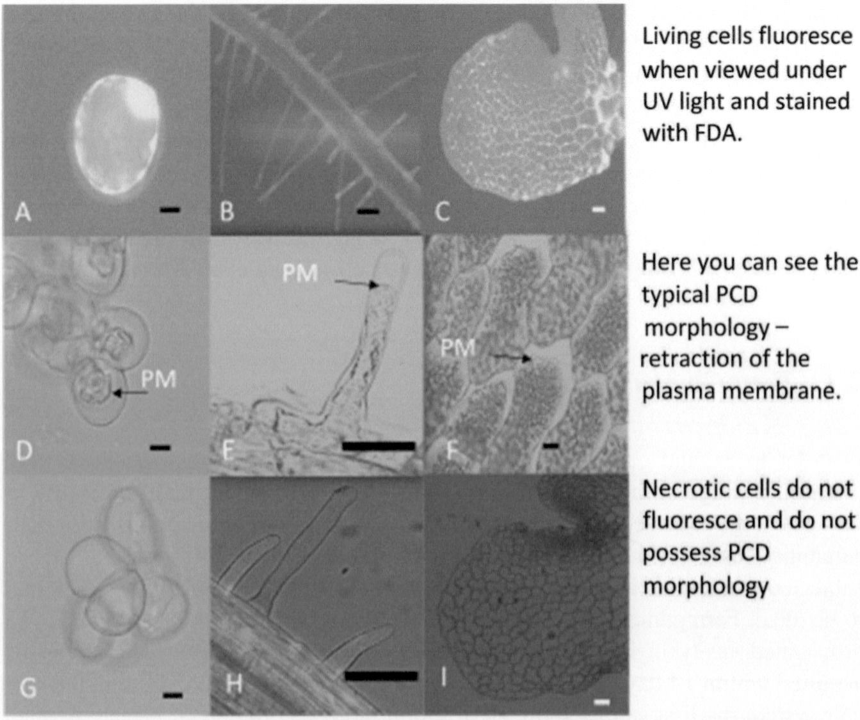

Fig. 16.2 Induction of PCD morphology and necrosis in *Arabidopsis thaliana* cells (D&G); *A. thaliana* root hairs (E&H); and fern gametophytes (F&I). *PM* plasma membrane. Scale bar 10 μm (**a**, **b**, **d**–**h**) and 50 μm (**c**, i)

3 Work Flow

Using morphology as a tool to investigate PCD constitutes a rapid and adaptable way to quantify and therefore investigate PCD in plants. Figure 16.3 describes a workflow for a typical experiment. The first step is plant material preparation. In each of the systems described enough suitable material can be obtained cheaply and quickly. The next step, frequently involves incubating the plant material in known or putative inhibitors or regulators of PCD. For example, we have previously examined the role played in PCD regulation by the signalling molecules ROS and Ca^{2+} and used antioxidants and calcium channel blockers to investigate the operation of these signalling pathways. The penultimate step is to expose the cells to a stress and finally record the percentage of cells of each category, viable, PCD and necrotic. Analysis of the data then allows us to infer the contribution either growth conditions, or incubation in chemicals, has made to the PCD sensitivity thresholds.

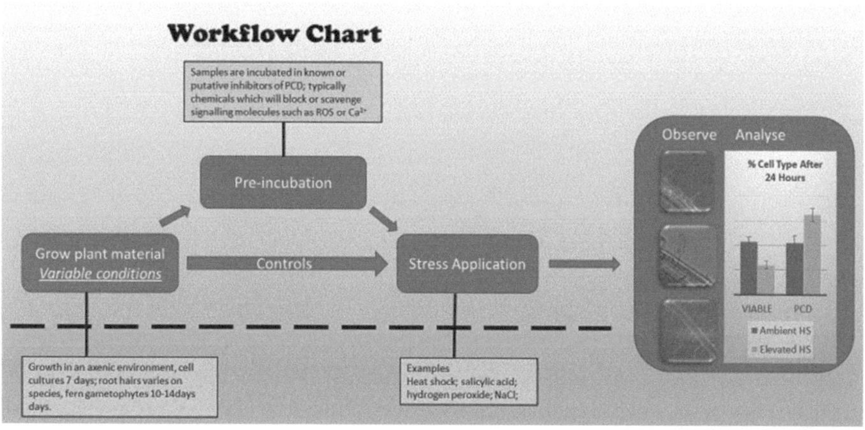

Fig. 16.3 A simplification of the workflow for using morphology to investigate the core regulatory mechanisms involved in plant PCD. This workflow is simple, quick and flexible. Grow the required plant material, incubate in regulators of PCD, apply a stress and the use morphology to quantify the rate and timing of PCD events in individual cells

4 Examining the Impact Environmental Growth Conditions Have on PCD Sensitivity Thresholds

A significant advantage to using morphology to quantify rates of PCD is that this system lends itself to directly examining how environment can alter PCD sensitivity thresholds. Our group used PCD morphology in a cell culture system to explore the contribution chloroplasts and light play in regulating PCD. Doyle and colleagues (2010), used cell cultures to demonstrate that the rate of PCD induced by heat stress in dark grown cultures (without functioning chloroplasts) and light grown cultures (whose chloroplasts have been chemically retarded) was similar, in contrast to light grown cultures (with functional chloroplasts) which had reduced levels of PCD. Antioxidant treatment of light grown cultures produced a PCD rate profile comparable to cultures with non-functioning chloroplasts suggesting that chloroplast generated ROS and exposure to light were active regulators in PCD regulation.

To further explore the impact environment plays on PCD sensitivity thresholds we have used the root hair system to compare induced PCD rates between seedlings grown in ambient (400 ppm) and super elevated (1900 ppm) CO_2. We found a significant difference between the rates of PCD observed in ambient and elevated grown seedlings when examined 24 h after a 10 min heat shock (Fig. 16.4). Seedlings which were not exposed to a heat shock showed no difference in rates of viable, PCD or necrotic cells.

The *Arabidopsis thaliana* seedlings were germinated in CONVIRON BDW40 walk-in plant growth chambers. The seedlings were placed in magenta culture boxes

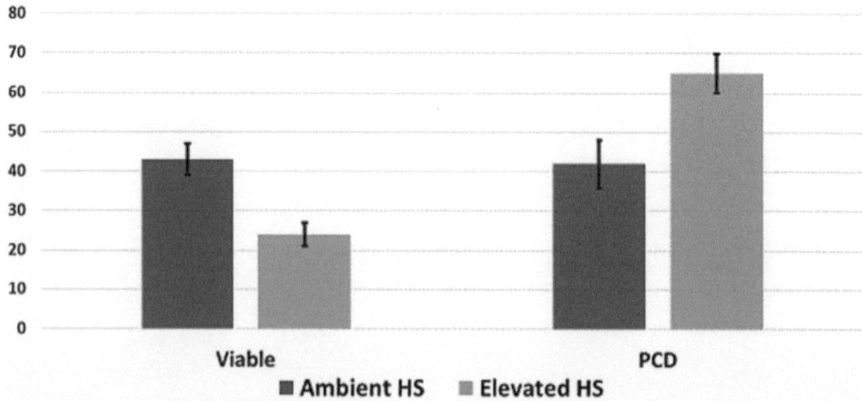

Fig. 16.4 Cell viability and PCD rates of ambient (400 ppm) and elevated grown (1900 ppm) *Arabidopsis thaliana* seedlings 24 h after a 10 min 46 °C heat shock (HS). Results shown are the average of three seedlings, where each seedling has in-excess of 100 root hairs scored. The error bars show standard error

fitted with air filters. Light conditions were set at 16 h light and 8 h dark. Light intensity rises from 300 to 600 μmol m^{-2} s^{-1} over 3 h when changing from dark to light and falls from 600 to 300 μmol m^{-2} s^{-1} over 3 h from light to dark. Temperature was set at 15 °C in the dark, rising to 21 °C in light. Relative humidity was 70%. CO_2 concentration was 1900 ppm in the elevated chamber and 400 ppm in the ambient chamber.

A recurring observation in plant PCD studies is the contribution ROS play in the regulation of PCD and contemporary investigations into plant stresses at elevated CO_2 have shown a consistent alteration in reactive oxygen species (ROS) homeostasis (Xu et al. 2015). It is possible that changes in ROS homeostasis are responsible for the changes in PCD rates we observed in the above experiment. Given that the root hair assay is suitable for use with a variety of inducers of PCD and can be used with a range of other species, we feel it is ideally suited for use in further exploring the role changing atmospheric CO_2 concentrations have on PCD.

5 Summary

1. Plant PCD is a normal and necessary response to excessive abiotic or biotic stress. PCD also plays a ubiquitous and important role in plant developmental processes, from embryogenesis to the death of the whole plant.
2. Environmental conditions affect regulation of PCD and suboptimal PCD negatively impacts plant fitness and reduces crop yields.
3. Cells undergoing PCD exhibit a distinctive morphology – condensation of the cytoplasm and retraction of the plasma membrane.

4. This easily identifiable marker of PCD has been utilised to develop assay protocols using cell cultures, seedling root hairs and fern gametophytes. These protocols constitute a rapid and adaptable way to accurately assess rates of PCD in plants.
5. These protocols offer a unique way to investigate how environmental factors affect PCD sensitivity thresholds. For example *A. thaliana* cell cultures have been used to investigate the role of light and chloroplasts in regulating PCD and *A. thaliana* seedlings have been used to show that atmospheric CO_2 concentrations alter PCD sensitivity thresholds.
6. Plant research is increasingly concerned with how plants will respond to future climatic changes, not least the impact rising CO_2 will have on plant biology. The protocols described here offer a unique way to examine how environment influences a central aspect of plant cell biology – namely the decision 'to die or not to die'.

References

Bohra A, Jha UC, Adhimoolam P, Bisht D, Singh NP (2016) Cytoplasmic male sterility (CMS) in hybrid breeding in field crops. Plant Cell Rep 35:967–993

Doyle SM, Diamond M, McCabe PF (2010) Chloroplast and reactive oxygen species involvement in apoptotic-like programmed cell death in Arabidopsis suspension cultures. J Exp Bot 61:473–482

Hogg BV, Kacprzyk J, Molony EM, O'Reilly C, Gallagher TF, Gallois P, McCabe PF (2011) An in vivo root hair assay for determining rates of apoptotic-like programmed cell death in plants. Plant Methods 7:45

Kacprzyk J, Daly CT, McCabe PF (2011) The botanical dance of death. Programmed cell death in plants. Adv Bot Res 60:169–261

Kacprzyk J, Dauphinee AN, Gallois P, Gunawardena A, McCabe PF (2016) Methods to study plant programmed cell death. In: Puthalakath H, Hawkins C (eds) Programmed cell death: methods and protocols. Methods in molecular biology, vol 1419. Springer, New York, pp 145–160

Kacprzyk J, Brogan NP, Daly CT, Doyle SM, Diamond M, Molony EM, McCabe PF (2017) The retraction of the protoplast during PCD is an active, and interruptible, calcium-flux driven process. Plant Sci 260:50–59

McCabe PF, Levine A, Meijer P-J, Tapon NA, Pennell RI (1997) A programmed cell death pathway activated in carrot cells cultured at low cell density. Plant J 12:267–280

Morel JB, Dangl JL (1997) The hypersensitive response and the induction of cell death in plants. Cell Death Differ 4:671–683

Reape TJ, McCabe PF (2008) Apoptotic-like programmed cell death in plants. New Phytol 180:13–26

Rogers HJ (2006) Programmed cell death in floral organs: how and why do flowers die? Ann Bot 97:309–315

Xu Z, Jiang Y, Zhou G (2015) Response and adaptation of photosynthesis, respiration, and antioxidant systems to elevated CO_2 with environmental stress in plants. Front Plant Sci 6:1–17

Chapter 17
Visualization of Plant Microtubules

Elisa Graña

1 Introduction

Microtubules (MTs) are highly dynamic components of the cell cytoskeleton that are involved in many important processes such as cell division (chromosome movement, formation of preprophase band, phragmoplast, cortical band before preprophase, etc.), cellular transport (endocytosis, exocytosis, organelle movement: nuclei, chloroplasts, amyloplasts, etc.), and growth and differentiation (transport of cellulose precursors to the cell wall to form cellulose microfibrils, transition from division to expansion, stomata movement, etc.) (Marc 1997; Wasteneys 2004; Alberts 2008; Nick 2008a; Celler et al. 2016).

Microtubules were first clearly described by Ledbetter and Porter in 1963, who also named them as microtubules. At that time, they were still trying to find out the best fixing method to preserve their experimental samples in TEM (transmission electron microscopy), by testing different combinations of fixatives. Ledbetter and Potter described MTs after examining cortex cells from *Phleum pretense*, *Spirodela oligorrhiza* and *Juniperus chinensis* in interphase. Microtubules were parallel to each other, clustered in small groups of 5–6 units and circumferentially arranged to the long axis of the cells, like 'hoops around a barrel'. These authors observed also MTs as hollow cylinders in cross-sections. As well, MTs were also seen in dividing cells (specifically in telophase), although authors recognized that was much more difficult to find them in dividing cells. Ledbetter and Porter stated that 'thin lines oriented normal to the cell plate representing the tubules, are apparent in the interzone of the spindle'. Curiously, in that work they also inferred a possible relation between microtubules and cellulose microfibrils: 'the tubules or sub-units of them act as primers for cellulose deposition'.

E. Graña (✉)
Plant Ecophysiology Laboratory, Department of Plant Biology and Soil Science, University of Vigo, Vigo, Spain
e-mail: eli-grana@uvigo.es

© Springer International Publishing AG, part of Springer Nature 2018

A. M. Sánchez-Moreiras, M. J. Reigosa (eds.), *Advances in Plant Ecophysiology Techniques*, https://doi.org/10.1007/978-3-319-93233-0_17

Nowadays, we know that MTs are formed thanks to the binding of certain globular proteins, α- and β-tubulin, which associate to form heterodimers that form linear rows of tubulin dimmers named protofilaments. One single microtubule consists of the union of 13 of these protofilaments (Hyams and Lloyd 1994; Kwiatkowska 2006; Donhauser et al. 2010). Its polymerization needs the presence of a chaperonin, which folds α- and β-tubulin (Gao et al. 1993). In addition to the structural proteins tubulins, MTs are accompanied by another kind of proteins, known as microtubule-associated proteins or MAPs, whose function is related to the organization of MTs (Mao et al. 2005).

Microtubules are defined as semi-rigid and polarized structures, with a plus (+) or growing end, where polarization predominates over depolarization; and a minus (−) or depolymerizing end (Alberts 2008; Guo et al. 2009). Microtubules are continuously being assembled and disassembled, in a property that is called *dynamic instability* (Mitchison and Kirschner 1984). Generally, the rate of assembly and disassembly is balanced, but in case that polymerization stops, microtubule is disassembled and it may disappear (Alberts 2008). It is specifically the dynamic nature of MTs the one responsible for their flexibility to reorganize themselves into diverse arrays, allowing changes in growth depending on different environmental or chemical signals (Chen et al. 2014). One of the most relevant MAPs is the motor protein kinesin, which joins MTs with the help of adenosine triphosphate (ATP). This joint allows kinesins to transport traffic vesicles, organelles and another proteins from minus to plus microtubule ends (Reddy 2001).

As MTs are direct or indirect targets of numerous signaling pathways, and also participate in signal transduction itself (Nick 2008b), their configuration is variable depending on the needs of the cell (Goddard et al. 1994). In fact, Yuan et al. (1994) described the microtubule dynamics like 'an adaptation whereby sessile plants can continually and rapidly alter their direction of cell expansion in response to external and internal stimuli'.

Wasteneys (2004) postulates that one of MTs main roles is the modulation of signaling pathways that allow them to face environmental changes, being able to act as 'sensors or transducers' for inputs that regulate plant growth (Landrein and Hamant 2013; Nick 2013; Bhaskara et al. 2016). Different signaling molecules could directly or indirectly bind to microtubules thanks to protein complexes, and could be freed to the cytoplasm and be activated once microtubules run into depolymerization (Wasteneys 2004). Different signal triggers described to induce cytoskeleton rearrangement are osmotic stress, cold, exposure to heavy metals, pathogens, hormones, gravity, light, or high molecular weight molecules such as PEG (Westeneys 2004; Wang et al. 2011; Mei et al. 2012; Chen et al. 2014). Interestingly, many of these signals converge in auxin.

Wasteneys also described MTs to be 'heavily congested places': nucleotides, ions and specially proteins (MAPs and their regulatory elements, motor proteins), could use them as the central place from where assembly, coupling and stability is regulated and organized.

Assembly starts on the microtubules organizing centers, also known as MTOCs. The minus end is associated with the MTOC, and microtubule grows to the plus direction. In animal cells, MTOCs are well identified as centrosomes, but in plants the assembly and organization sites have been under controversy, and it is thought that are located at the nuclear surface. As γ-tubulin predominates in the MTOCs from animal cells, many studies have focused on finding γ-tubulin clusters in plant cells, but no conclusive results were obtained (Goddard et al. 1994; Liu et al. 1994; Vaughn and Harper 1998).

The regulation of MTOCs involves assembling, dynamics, interactions with other cell elements, and association with motor and structural proteins (Marc 1997). In this way, the MTs associated proteins, or MAPs, cooperate with the tubulin dimmers assembly, join adjacent MTs or link MTs to other cellular structures (Marc 1997).

2 Typical Microtubule Arrangement

Cortical microtubules can be typically found arranged in three different conformations: transverse, oblique and longitudinal respect to the direction of cell growth. In interphase root tip cells, cortical MTs have been reported to be transverse (Collings and Wasteneys 2005), since transversal orientation is established early in the meristematic cells (Panteris et al. 2013). This orientation is constant in the cortex, in the endodermis and also in the stele (Panteris et al. 2013).

Microtubule reorientation starts when cells stop growing. MTs are found arranged in an oblique way, until they reach the longitudinal orientation, typical of elongation zone. Suppression of cell growth is traditionally related to longitudinal alignment (Panteris et al. 2013; Chen et al. 2014) (see Fig. 17.1).

3 Why Study Microtubules?

Studying these cytoskeleton components is not only useful for understanding the mechanisms of cellular organization, but for understanding the response of cells to different stimuli that are known to change the microtubule array configuration. The knowledge of the microtubule dynamics opens the door to novel technical applications. For example, some studies have been focused to improve the quality of wood and its products, based on the role of MTs in orientation and organization of cellulose microfibrils during the formation of the secondary cell wall, determining the mechanical properties of the wood (Funada 2008). On the other hand, several compounds have been described as typical antimicrotubule drugs: oryzalin, trifularin, colchicines, paclitaxel (better known as taxol), etc. Their antimicrotubule effects can be included in two groups: **mitotic disrupters** and **microtubule assembly inhibitors**. Both have been further studied due to their huge practical use. Besides being used as anticancer o anthelmintics agents (Jordan et al. 1998), these

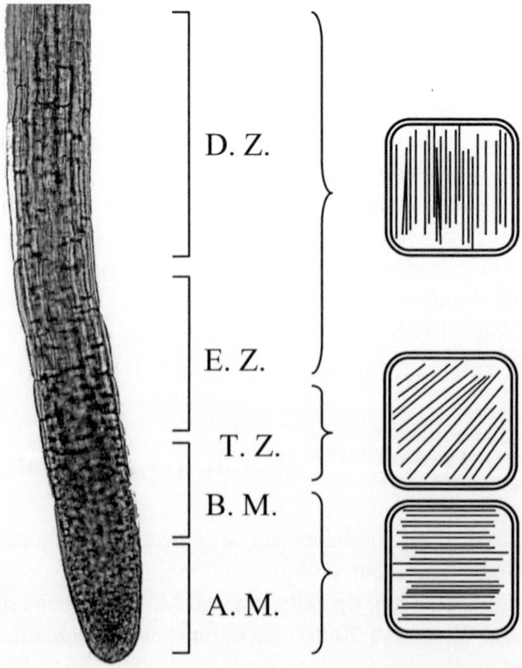

Fig. 17.1 Typical microtubule arrangement, depending on different root zones (*A.M.* apical meristem, *B.M.* basal meristems, *T.Z.* transition zone, *E.Z.* elongation zone, *D.Z.* differentiation zone)

compounds are source, for example, of potential herbicides, also known as anticytoskeletal herbicides.

One of the most studied microtubule assembly inhibitors is colchicine. This drug inhibits the assembly of new tubulin dimmers by blocking the union to the microtubule cap, preventing their polymerization by joining and blocking the binding sites of tubulin (Dayan et al. 2010; Oliva et al. 2002). Consequently, colchicine induces loss of MTs during cell division, leading to mitotic aberrations like blocked metaphases or polymorphic nuclei (Bartels and Hilton 1973; Vaughn et al. 1987).

Compounds with similar activities are podophyllotoxin (which inhibits the microtubule assembly, resulting in mitotic arrest at prometaphase, thereby decreasing the number of cells entering mitosis) (Vaughn and Vaughan 1988); dithiopyr (its target are the MAP proteins, avoiding the stabilization of the MTs involved in cell division) (Senseman 2007); or trifluralin (both dithiopyr and trifluralin cause irregularly formed nuclei, branched and undulating phragmoplast and incomplete and reticulate cell walls) (Lehnen and Vaughn 1991). Mitotic disrupters cause easily identifiable effects on plants. The most typical macroscopic effect is the 'club root' morphology, which is due to isodiametric cell growth in the elongation zone (Vaughn 2006). As result, roots appear massively thickened and distorted, accompanied by a decrease in hair root density. Besides, plant growth is generally retarded.

Among the microscopic effects, the most representative are branched and abnormally oriented cell plates, which do not divide the daughter cytoplasms in a proper way. Abnormalities in the movement of chromosomes have also been observed.

As result, it is possible to find multiple nuclei or small nuclear fragments, since the new nuclear envelopes are reformed around disoriented chromosomes (Vaughn 2006). This effect is due to an irregular organization of the spindle microtubules, appearing in clusters radiating from the same center. As consequence, abnormal anaphases are formed, known as 'star anaphases' (Lehnen et al. 1990).

Well-known mitotic disrupters are artemisin (Dayan et al. 1999), terbutol (Lehnen et al. 1990), or oryzalin, which induces the loss of cortical and spindle microtubules, causing mitotic aberrations as lobed nuclei and multinuclei cells (Bartels and Hilton 1973). It has been observed that oryzalin binds the tubulin dimer, co-polymerizing with free tubulin and decreasing the microtubule assembly. As well, left-handed helical growth in *Arabidopsis thaliana* oryzalin-treated seedlings has been also described (Nakamura et al. 2004).

In this chapter two different protocols to visualize plant MTs, one by immunostaining (using fluorescence or confocal microscopy) and other by transmission electron microscopy (TEM), are in detail described.

4 Immunostaining of Plant Microtubules

This method is carried out in two consecutive days and is based on plant tissue fixation, cell wall digestion and immunolocalization with specific antibodies. It has been specifically developed to observe cortical microtubules from *Arabidopsis thaliana* radicles, and it is based on the protocol of Holzinger and collaborators (2009), with some modifications. The steps to follow are detailed below, and are also summarized in Fig. 17.2. Examples of microtubule immunolabeling can be seen on Fig. 17.3.

Day 1

1. Roots from *Arabidopsis thaliana* seedlings are fixed for 45 min at room temperature in freshly prepared buffer containing 0.5% glutaraldehyde and 1.5% formaldehyde prepared in microtubule-stabilizing buffer or MSB (50 mM PIPES, 2 mM EGTA, 2 nM $MnSO_4$, pH 7.2) with 0.1% Triton X-100.
2. Wash samples in MSB buffer with 0.1% Triton X-100, for 20 min.
3. Wash samples again in MSB buffer until no foaming.
4. Chop the plant material with a razor blade to approximately 3 mm length.
5. Digest the cell walls to allow the entry of the antibodies in the cytoplasm, with MSB containing 1% cellulase and 1% pectolyase Y-23, pH 5.5. Digest at room temperature for 30 min.
6. Rinse in MSB, pH 7.2.
7. Permeabilize the root samples in methanol at −20 °C for 10 min.
8. Rehydrate samples by washes with PBS buffer, pH 7.4.
9. Incubate with 1 mg mL^{-1} $Na_2B_4O_7$ in PBS buffer for 20 min. The goal of this step is to reduce aldehyde-induced autofluorescence: $Na_2B_4O_7$ blocks free aldehyde groups.

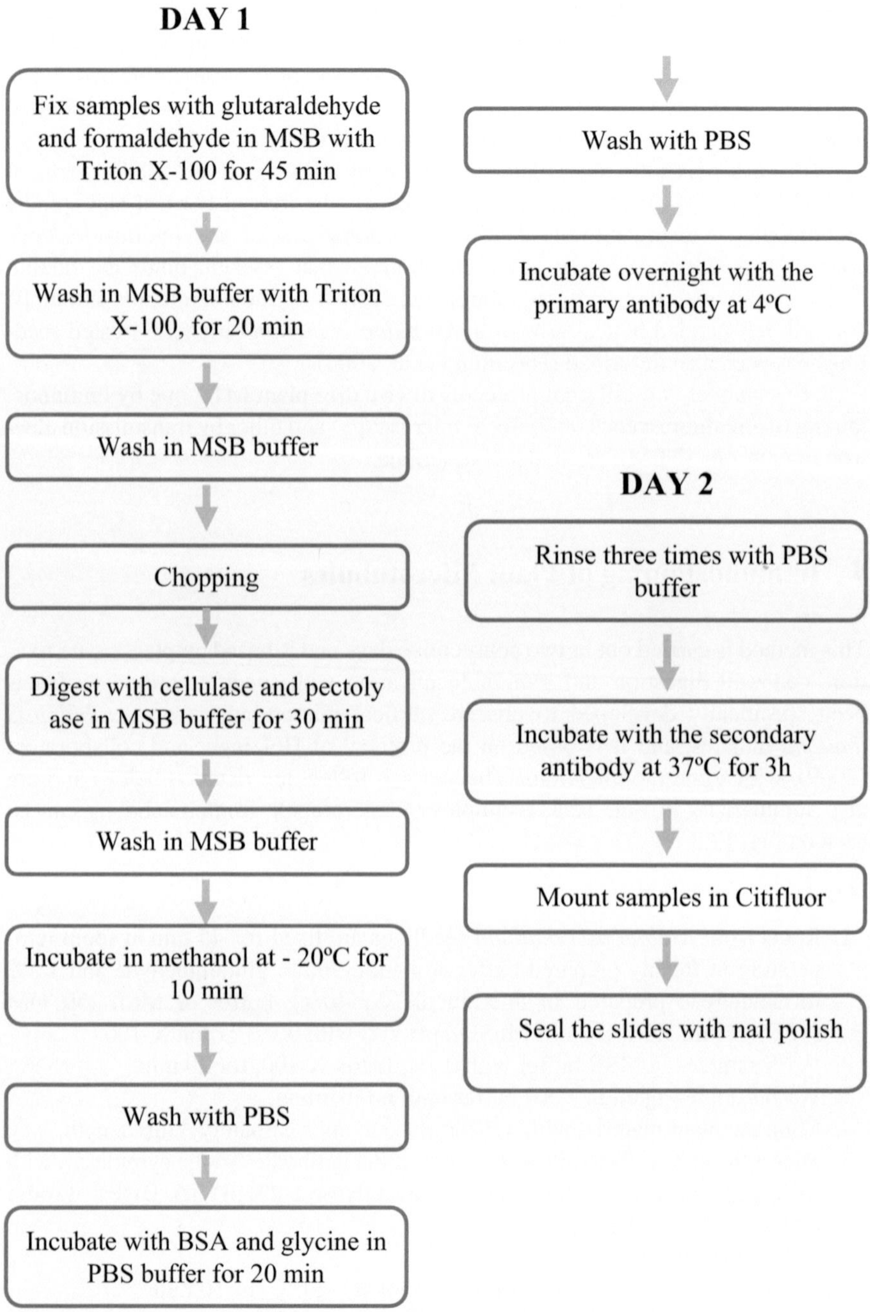

Fig. 17.2 Summarized scheme of microtubule immunolabeling procedure

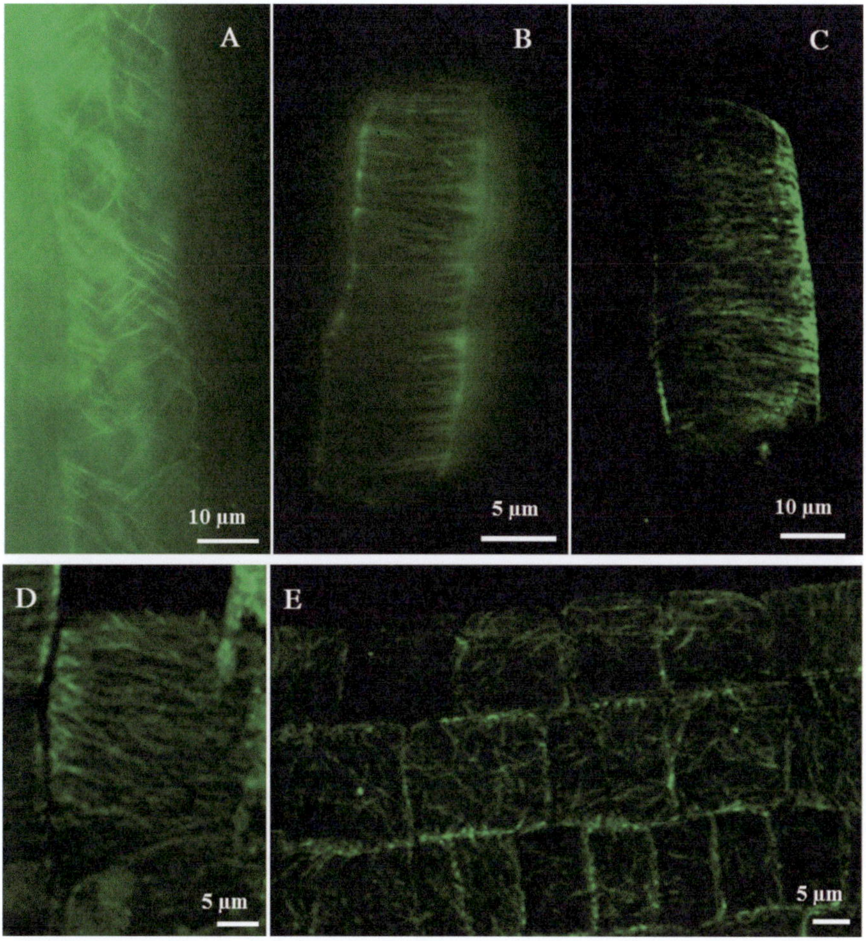

Fig. 17.3 Immunofluorescence of plant microtubules under fluorescence (**a**, **b**) or confocal microscopy (**c**–**e**; Z-stack images). Images (**a**, **d**) show obliquely oriented microtubules at transition zone. Images (**b**, **c**) show single cells with microtubules oriented 90° respect to the cell elongation plane. Image (**e**) shows erratically organized microtubules after farnesene treatment, a sesquiterpene known to cause microtubular alterations. (Araniti et al. 2016)

10. Wash plant material in PBS buffer.
11. Incubate with 1% BSA (bovine serum albumin) and 50 mM glycine in PBS buffer for 20 min.
12. Wash in PBS buffer.
13. Incubate overnight with the primary antibody (Sigma B512 anti-α tubulin; 1:1000 prepared in PBS buffer).

Day 2

14. Rinse three times with PBS buffer to remove properly the primary antibody. This prevents false positives.
15. Incubate samples with the secondary antibody (Alexa 488-conjugated goat anti-mouse IgG; 1:200 in PBS buffer) at 37 °C for 3 h.
16. Mount samples in Citifluor AF1 antifade agent to protect samples from the incident light, and seal covers to slides with nail polish.
17. Generate excitation at 488 nm and collect emission at 515 nm.

4.1 Tricks and Recommendations

- MSB buffer, without Triton, can be prepared in advanced and stored at 4 °C (Step 1).
- It is especially important that Step 1 is carried out at room temperature (including working buffers), since microtubules depolymerize in cold and nothing would be seen in the preparation.
- In order to facilitate samples management, steps 1–3 must be done in 2 mL eppendorfs; step 4 in 9 cm diameter Petri dishes; and steps 5–15 in six-well plates, using one well per step and treatment.
- Successive washing steps cause material loss in a very easy way. I strongly recommend collecting the chopped plant material using filters typical for cell and/or nuclei isolation protocols (Steps 5–15, see Fig. 17.2).
- If the immunostaining fails, I recommend increasing the amount of cellulose, since antibodies cannot access into the cell when cellulose microfibrils are very tight.

5 Visualization of Microtubules by TEM

The visualization of MTs by Transmission Electron Microscopy (TEM) gives additional information about microtubule arrangement; such as distance between microtubules, distance between microtubules and the cell wall, or number of microtubule clusters. Any variation of these parameters could be indicating abnormalities on microtubule organization due to plant stress (Araniti et al. 2016).

TEM visualization is especially useful for studying individual MTs, but also when other cellular structures need to also be visualized (Celler et al. 2016), which is specially interesting when the plant response to biotic and/or abiotic stress is being studied. So we will be able to see whether effects on MTs are also related to effects on cell wall disposition, presence of multi-nucleated cells, nuclei morphology or tissue organization (Araniti et al. 2016). Moreover, the high resolution of this technique allows to have a static picture of microtubule arrangement but gives also essential information about the presence of cross-linking proteins or about the interaction of MTs with other cellular structures and endomembrane components in roots under stress conditions (Celler et al. 2016).

5.1 TEM Procedure

Microtubules visualization is conducted according to Holzinger et al. (2007) with approximately 25–50 root tips per sample. This protocol has been improved together with the members of the Central Research Services of the University of Vigo (CACTI).

1. Cut the tissue in 1–2 mm
When working with thin root tips, cuts are done with a sharp blade on agar or a wax plate directly in the fixative to avoid further damages to the roots
2. Fix the tissue in 50 mM sodium cacodylate buffer (pH 7.0) containing 2.5% glutaraldehyde fixative at room temperature for 2 h
3. Wash 2 times (1 h each) with 50 mM cacodylate buffer (pH 7.0) at 4 °C
4. Immerse samples in 50 mM sodium cacodylate buffer (pH 7.0) with 1% osmium tetroxide at 4 °C for 12 h
5. Wash 2 times (1 h each) with 50 mM cacodylate buffer (pH 7.0) at 4 °C
6. Perform sample dehydration in increasing ethanol dilutions (at 4 °C): 10%, 20%, 40%, 60%[a], 80%, 90%, 95%, 100%, each for 20 min, except the last one for 40 min; and lately in propylene oxide (2×15 min)
7. Infiltrate the sample in Spurr's resin at 4 °C as follows:
Spurr: propylene oxide (1:3 v/v) (3×2 h)
Spurr: propylene oxide (1:1 v/v) (3×2 h)[a]
Spurr: propylene oxide (3:1 v/v) (2×2 h plus 1×3 h)
8. Embed the sample in 100% Spurr's resin for 2×3 h and left it overnight at room temperature
9. Embed the sample again in 100% resin (2×3 h)
10. Place the sample in molds with pure resin to allow polymerization at 60 °C for 2–3 days
11. Prepare semithin sections (0.7 μm) for light microscopy and ultrathin sections (50–70 nm) for electron microscopy
12. Contrast the sections as follows:
Uranyl acetate for 30 min
Wash with Milli-Q water for 2 min
Lead citrate for 12 min
Wash with Milli-Q water for 2 min
13. Assemble ultrathin sections on copper grids of 100/200 mesh and examine by TEM using a JEOL JEM-1010 transmission electron microscope (at 100 kV) (Peabody, MA, USA) equipped with a CCD Orius-Digital Montage Plug-in camera (Gatan Inc., Gatan, CA, USA) and Gatan Digital Micrograph software (Gatan Inc.)

[a]It can stay overnight

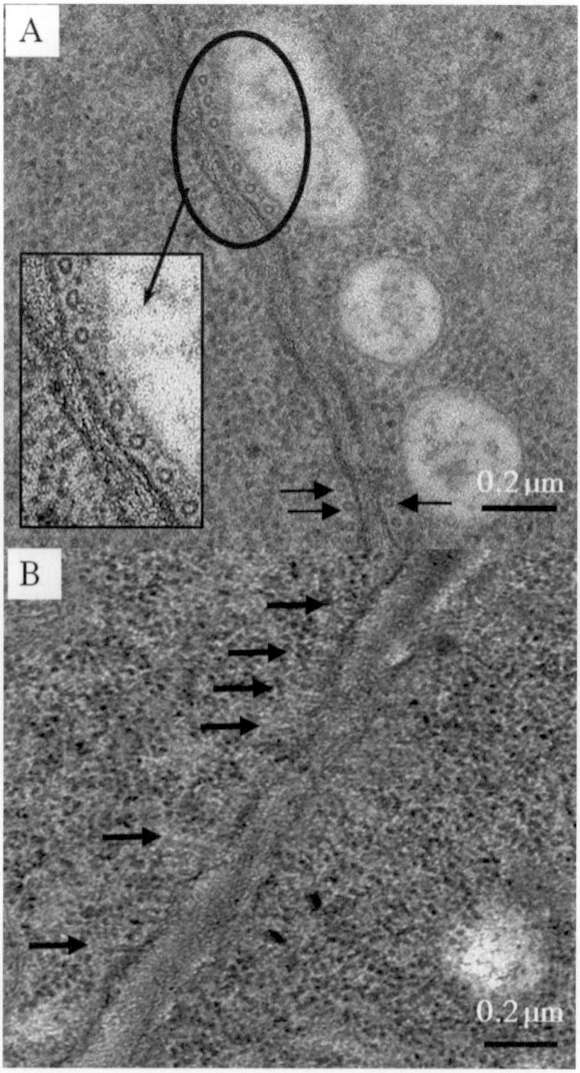

Fig. 17.4 Images show microtubules in (**a**) transversal section; and (**b**) longitudinal section. Arrows indicate microtubule localization

In so prepared ultra-thin sections, microtubules can be seen by TEM as small circles, aligned close to the cell wall in transversal sections (Fig. 17.4a) or as straight sticks in longitudinal sections (Fig. 17.4b).

Acknowledgments The implementation of these techniques was possible thanks to the invaluable assistance of Inés Pazos and Jesús Méndez from the Central Research Services (CACTI) of the University of Vigo.

References

Alberts B (2008) Molecular biology of the cell, 5th edn. Garland Science, New York

Araniti F, Graña E, Krasuska U, Bogatek R, Reigosa MJ, Abenavoli MR, Sánchez-Moreiras AM (2016) Loss of gravitropism in farnesene-treated *Arabidopsis* is due to microtubule malformations related to hormonal and ROS unbalance. PLoS One 11(8):e0160202

Bartels PG, Hilton JL (1973) Comparison of trifularin, oryzalin, pronamide, prophan and colchicines treatments on microtubules. Pestic Biochem Physiol 3:462–472

Bhaskara GB, Wen T-N, Nguyen TT, Versules PE (2016) Protein phosphatase 2Cs and microtubule-associated stress protein 1 control microtubule stability, plant growth, and drought response. Plant Cell 29:169–191

Celler K, Fujita M, Kawamura E, Ambrose C, Herburger K, Holzinger A, Wasteneys GO (2016) Microtubules in plant cells: strategies and methods for immunofluorescence, transmission electron microscopy and live cell imaging. In: Gavin RH (ed) Cytoskeleton: methods and protocols, methods in molecular biology, vol 1365. pp 155–184

Chen X, Grandont L, Li H, Hauschild R, Paque S, Abuzeineh A, Rakusová H, Benkova E, Perrot-Rechenmann C, Friml J (2014) Inhibition of cell expansion by rapid ABP-1 mediated auxin effect of microtubules. Nature 516:90–93

Collings DA, Wasteneys GO (2005) Actin microfilament and microtubule distribution in the expanding root of *Arabidopsis thaliana*. Can J Bot 83:579–590

Dayan FE, Hernandez A, Allen SN, Moraes RM, Vroman JA, Avery MA, Duke SO (1999) Comparative phytotoxicity of artemisin and several sesquiterpene analogues. Phytochemistry 50(607):614

Dayan FE, Duke SO, Grossmann K (2010) Herbicides as probes in plant biology. Weed Sci 58(3):340–350

Donhauser ZJ, Jobs WB, Binka EC (2010) Mechanics of microtubules: effects of protofilament orientation. Biophys J 99(5):1668–1675

Funada R (2008) Microtubules and the control of wood formation. In: Nick P (ed) Plant microtubules. Plant cell monographs, vol 11. Springer, Berlin, pp 83–119

Gao Y, Valnberg IE, Chow RL, Cowan NJ (1993) Two cofactors and cytoplasmic chaperonin are required for the folding of α- and β-tubulin. Mol Cell Biol 13:2478–2485

Goddard RH, Wick SM, Silflow CD, Snustad DP (1994) Microtubule components of the plant cell cytoskeleton. Plant Physiol 104:1–6

Guo L, Ho CK, Kong Z, Lee YR, Qian Q, Liu B (2009) Evaluating the microtubule cytoskeleton and its interacting proteins in monocots by mining the rice genome. Ann Bot 103:387–402

Holzinger A, Wasteneys G, Lütz C (2007) Investigating cytoskeletal function in chloroplast protrusion formation in the arctic-alpine plant *Oxyria digyna*. Plant Biol 9(3):400–410

Holzinger A, Kawamura E, Wasteneys GO (2009) Strategies for imaging microtubules in plant cells. Methods Mol Biol 586:243–262

Hyams JS, Lloyd CW (1994) Microtubules. Wiley, New York

Jordan A, Hadfield JA, Lawrence NJ, McGown AT (1998) Tubulin as a target for anticancer drugs: agents which interact with the mitotic spindle. Med Res Rev 18:259–296

Kwiatkowska D (2006) Flower primordium formation at the *Arabidopsis* shoot apex: quantitative analysis of surface geometry and growth. J Exp Bot 57:571–580

Landrein B, Hamant O (2013) How mechanical stress controls microtubule behavior and morphogenesis in plants: history, experiments and revisited theories. Plant J 75:324–338

Ledbetter MC, Porter KR (1963) A 'microtubule' in plant cell fine structure. J Cell Biol 19:239–250

Lehnen LP Jr, Vaughn KC (1991) Immunofluorescence and electron microscopic investigations of the effects of ditiopyr on onion root tips. Pestic Biochem Physiol 40(1):58–67

Lehnen LP Jr, Vaughan MA, Vaughn KC (1990) Terbutol affects spindle microtubule organizing centers. J Exp Bot 41(5):537–546

Liu B, Joshi HC, Wilson TJ, Silflow CD, Palevitz BA, Snustad DP (1994) γ-tubulin in *Arabidopsis*: gene sequence, immunoblot, and immunofluorescence studies. Plant Cell 6:303–314

Mao T, Jin L, Li H, Liu B, Yuan M (2005) Two microtubule-associated proteins of the *Arabidopsis* MAP65 family function differently on microtubules. Plant Physiol 138:654–662

Marc J (1997) Microtubule-organizing center in plants. Trends Plant Sci 2(6):223–230

Mei Y, Gao HB, Yuan M, Xue HW (2012) The *Arabidopsis* ARCP protein, CSI1, which is required for microtubule stability, is necessary for root and anther development. Plant Cell 24:1066–1080

Mitchison TJ, Kirschner MW (1984) Dynamic instability of microtubule growth. Nature 312:237–242

Nakamura M, Naoi K, Shoji T, Hashimoto T (2004) Low concentrations of propyzamide and oryzalin alter microtubule dynamics in *Arabidopsis* epidermal cells. Plant Cell Physiol 45(9):1330–1334

Nick P (2008a) Control of cell axis. In: Nick P (ed) Plant microtubules. Plant cell monographs, vol 11. Springer, Berlin, pp 3–46

Nick P (2008b) Microtubules as sensors for abiotic stimuli. In: Nick P (ed) Plant microtubules. Plant cell monographs, vol 11. Springer, Berlin, pp 175–203

Nick P (2013) Microtubules, signalling and abiotic stress. Plant J 75:309–323

Oliva A, Moraes RM, Watson SB, Duke SO, Dayan FE (2002) Aryltetralin lignans inhibit plant growth by affecting the formation of mitotic microtubular organizing centers. Pestic Biochem Physiol 72:45–54

Panteris E, Adamakis I-DS, Daras G, Hatzopoulos P, Rigas S (2013) Differential responsiveness of cortical microtubule orientation to suppression of cell expansion among the developmental zones of *Arabidopsis thaliana* root apex. PLoS One 8(12):e 82442

Reddy AS (2001) Molecular motors and their functions in plants. Int Rev Cytol 204:97–178

Senseman SA (2007) In: Senseman SA (ed) Herbicide handbook, 9th edn. Weed Science Society of America, Lawrence

Vaughn KC (2006) The abnormal cell plates formed after microtubule disrupter herbicide treatment are enriched in callose. Pestic Biochem Physiol 84:63–71

Vaughn KC, Harper JDI (1998) Microtubule organizing centers and nucleating sites in land plants. Int Rev Cytol 181:75–149

Vaughn KC, Vaughan MA (1988) Mitotic disrupters from higher plants. In: Cutler HG (ed) Biologically active natural products: potential use in agriculture. ACS symposium series 380, Washington, DC. pp 273–293

Vaughn KC, Marks MD, Weeks DP (1987) A dinitroaniline-resistant mutant of *Eleusine indica* exhibits cross-resistance and supersensitivity to antimicrotubule herbicides and drugs. Plant Physiol 83:956–964

Wang SH, Kurepa J, Hashimoto T, Smalle JA (2011) Salt stress-induced disassembly of *Arabidopsis* cortical microtubule arrays involves 26S proteasome dependent degradation of SPIRAL1. Plant Cell 23:3412–3427

Wasteneys GO (2004) Progress in understanding the role of microtubules in plant cells. Curr Opin Plant Biol 7(6):651–660

Yuan M, Shaw PJ, Warn RM, Lloyd CW (1994) Dynamic reorientation of cortical microtubules, from transverse to longitudinal, in living plant cells. Proc Natl Acad Sci U S A 91:6050–6053

Chapter 18
Multiprobe In-Situ Hybridization to Whole Mount *Arabidopsis* Seedlings

Leonardo Bruno, Fabrizio Araniti, and Olimpia Gagliardi

1 Introduction

In the multicellular organisms to know the temporal and spatial expression of genes gives specific information about the putative sites of activity of the corresponding proteins and also their function related to processes of growth and development as well as in response to biotic and abiotic stress conditions (Bruno et al. 2011). In-situ RNA-RNA hybridization (ISH) is a powerful technique that enables the localization of gene transcripts at the cellular level by using labeled probes complementary to nucleic acid of interest, followed by probe specific detection. Original methods for ISH have been developed in the 1980s (Harrison et al. 1974). Since that time, the methodologies associated with it have undergone a continuous evolution (Harrison et al. 1974).

The ISH technique is complementary to Northern blotting and RT-PCR (reverse transcriptase-polymerase chain reaction) where the RNA extraction procedure invariably results in the loss of spatial information (Bruno et al. 2011). It also complements DNA microarrays, a genome-wide expression profiling technique at seedling or organ level (Chuaqui et al. 2002; Wellmer et al. 2004). A resolution at the cellular level is possible when combined with cell sorting of fluorescent cells derived from transgenic reporter lines. However, it requires advanced equipment and is expensive (Birnbaum et al. 2005).

In addition to ISH, the analysis of promoter-reporter gene fusions is one of the most widely used techniques to investigate the spatial and temporal regulation of

L. Bruno (✉) · O. Gagliardi
Department of Biology, Ecology and Earth Science, Laboratory of Plant Biology, University of Calabria, Arcavacata di Rende, CS, Italy
e-mail: leonardo.bruno@unical.it

F. Araniti
Department Agraria, Mediterranea University, Reggio Calabria, Italy

© Springer International Publishing AG, part of Springer Nature 2018
A. M. Sánchez-Moreiras, M. J. Reigosa (eds.), *Advances in Plant Ecophysiology Techniques*, https://doi.org/10.1007/978-3-319-93233-0_18

gene expression as well as the protein subcellular localization (Taylor 1997). The promoter of the gene of interest is fused to the coding sequence of a reporter gene such as β-Glucuronidase (GUS), Green Fluorescent Protein (GFP) or Luciferase (LUC). The expression of the reporter gene in transgenic lines measures the promoter activity of the gene of interest. However, promoter activity does not necessarily correspond to the in vivo expression of the investigated gene (Taylor 1997). Regulatory elements, that are usually located in the promoter and are responsible for cell-specificity or inducibility, sometimes are located in introns or coding parts of the gene. A well-known example is the *AGAMOUS (AG)* gene in *Arabidopsis thaliana* with regulatory elements in its large second intron (Mizukami et al. 1996), 3′ regions, which are rarely included in promoter reporter constructs can also contribute to the control of gene expression. For example, *GL1* contains a transcriptional enhancer in its 3′ noncoding region that is required, together with promoter sequences from the 5′ region, for the normal *GL1* expression (Larkin et al. 1993).

Limitations to the use of reporter genes are the long half-live of their products. As a consequence, these cannot monitor rapid changes in the transcription, and diffusion of reporter gene products to neighboring cells resulting in artifacts (Drews et al. 1992). Finally, reporter gene analysis requires the construction of transgenic plants, which is time–consuming, expensive and demanding specific expertise.

An alternative way to determine the expression of genes is to look at the localization of the corresponding protein using specific antibodies (Christine and Bernard 2008). The main disadvantage of this method is that, for each protein, a specific antibody has to be generated and the conditions for immunohistochemistry have to be defined each time (fixation, dilution, detergents, washes). This is a slow process and therefore cannot be done for many probes at the same time.

Multi-probe whole mount mRNA in situ hybridization allows one to quickly and simultaneously define the expression of several genes at high resolution. For example, in *Arabidopsis thaliana* we employed this methodology to locate simultaneously three different genes involved in different physiological functions and/or processes (Bruno et al. 2011).

MISH as well as ISH are largely used not only in developmental studies but also in eco-physiological approaches. For example, Kim et al. (2008), using in-situ hybridization method observed that in *Arabidopsis thaliana*, the GRP7protein regulate stomata opening and closing in response to abiotic stresses. Lefebvre et al. (2006) used this methodology to study the involvement of NCED6 and NCED9 genes in ABA biosynthesis during seed development, whereas Giannino et al. (2004) demonstrated by using in-situ hybridization that the genes encoding for geranylgeranyl reductase locally increase both in cells mechanically wounded and naturally injured by the pathogenic fungus *Taphrina deformans*. Finally, Bruno et al. (2009) observed a local increase of transcript of geranylgeranyl reductase in specific cell domains in fruits of olive severely damaged by the insect *Bactroceraoleae*.

In conclusion, this new versatility methodology confirms its usefulness in eco-physiological studies.

2 Materials

2.1 Reagents for MISH

- 10× Biotin (BIO) RNA labeling mix: Roche cat. N. 1 685 597
- 10× Digoxigenin (DIG) RNA labeling mix: Roche cat. N. 1277073
- 10× Fluorescein (FITC) RNA labeling mix: Roche cat. N. 1685619
- 10× PBS (see reagent set up) dilution to 1× PBS is needed
- 1× PBS plus 0.1% Tween-20
- 2-(N-morpholino) ethane sulfonic acid (MES)
- 20× saline-sodium citrate buffer (SSC) (see reagent set up)
- 4 M NH_4Ac
- Plant Agar
- Agarose
- Alexa Fluor dyes 488 Donkey Anti-Mouse INVITROGEN
- Alexa Fluor dyes 555 Donkey Anti-Sheep INVITROGEN
- Alexa Fluor dyes 647 Donkey Anti-Rabbit INVITROGEN
- Boehring block reagent: Roche
- Bovine Serum Albumine (BSA) Roche
- Chloroform ($CHCl_3$)
- Diethylprocarbonate (DEPC) water! CAUTION Carcinogenic; work in fume hood
- DNase I recombinant, RNase-free (Roche)
- DTT (Roche)
- Ethanol, 90%, 85%, 75% (v/v) in DEPC water, 50%, 25% (v/v) in 1× Phosphate Buffer solution (PBS) and 100% ethanol
- Fixative solution (see reagent set up)
- Glycerol ≥99% (Sigma-Aldrich)
- HCl (FW 36.461)
- Hybridization solution and supplemented hybridization solution (see reagent set up)
- KCl (FW 74.55)
- KH_2PO_4 (FW 136.09)
- Methanol
- Mouse anti-Biotin: (Roche)
- Murashige Skoog basal salt
- Myo-Inositol
- Na_2HPO_4 (FW 177.96)
- NaCl (FW 58.44)
- NaOH
- NH_4Ac (FW 77.08)
- Paraformaldehyde
- Phenol
- Protector RNase Inhibitor (Roche)

- Proteinase K, stock 25 mg mL^{-1} (Roche)
- Rabbit anti-fluorescein: (Molecular Probes)
- RNA marker
- RNase-free water (see reagent set up)
- Salmon sperm (Sigma)
- Sheep anti-digoxigenin (Roche)
- Sodium acetate solution (NaOAc) 3 M, pH (5.2)
- SP6 RNA Polymerase (+transcription buffer) (Roche)
- Sucrose
- Suitable restriction endonuclease enzyme and buffer
- T7 RNA Polymerase (+transcription buffer) (Roche)
- Tris (FW 121.14)
- tRNA, from backer's yeast (20 mg mL^{-1}; Roche)
- Tween-20
- Xylene

2.2 Reagents for TSA

The reagents used to perform TSA method are the same of MISH method, which are reported in the paragraph 2.1. But, in addition, specific extra reagents are required for making TSA, such as:

- 1% (v/v) H_2O_2
- HRP-goat anti-mouse IgG
- HRP-goat anti-rabbit IgG (anti-dig)
- Alexa 488-conjugated tyramide (INVITROGEN)
- Alexa 555-conjugated tyramide (INVITROGEN)
- Alexa Fluor dyes 647 Donkey ant-rabbit (INVITROGEN)
- Rabbit anti-denatured GFP antibody (gentle gift from professor Takeshi Kaneko)
- HRP-goat anti mouse

2.3 Equipment

- Standard plastic, laboratory glassware and instrumentation (eppendorf 1.5–2 mL, falcon 15–50 mL, beakers, cylinders, shaker, fine tweezers, petri dishes, handleds, slides, etc.;
- Multi-well plates;
- Vacuum pump;
- Stereo microscope (e.g. Leica);
- DNA electrophoresis equipment (e.g. Biorad);
- Thermoblock (e.g. Eppendorf);

- Autoclave;
- Thermocycler (e.g. Eppendorf), required temperatures 50–55 °C;
- NanoDropSpectrophotomer (e.g. Celbio ND-1000);
- Micropipette (e.g. Gilson or Eppendorf);
- Confocal Microscope laser scanning (e.g. Leica TCSP 8);

Δ **CRITICAL STEP** It is pivotal to choose the fluorophore of the secondary antibody only after considering the available excitation and emission spectra of your specific microscope.

! CAUTION Before starting the experiments, benches and all instruments should be cleaned with absolute ethanol (96%). In order to degrade RNases, laboratory glassware and metal tools should be autoclaved at 121 °C for 21 min or baked at 180 °C for 2 h.

2.4 *Reagent Setup*

RNase-free water preparation: Add 1 mL L^{-1} DEPC in distilled water, shake the solution and incubate overnight under a fume hood. Autoclave properly to inactivate remaining DEPC.

! CAUTION DEPC should be handled under a fume hood since it is a carcinogenic compound.

10× PBS preparation: 1.3 M NaCl, 70 mM Na_2HPO_4* $2H_2O$ and 30 mM KH_2PO_4, pH to 7.4 with 1 M KCl.

PBT: 1XPBS plus 0.1% (v/v) Tween-20

Fixative solution: The fixative solution is prepared by solving paraformaldehyde (4% w/v) in PBS (1×).

Weight 4 g of paraformaldehyde and dissolve it in 100 mL of 1× PBS. The solution should be heated to 60–70 °C (use a glass beaker wrap up with aluminum foil) and stir continuously. Once reached the temperature, few drops of NaOH (1 N) should be added until the solution becomes clear. The solution should be cooled down before use.

Δ **CRITICAL STEP** Fixative solution must be freshly prepared or, alternatively, small aliquots should be prepared and stocked at −20 °C up to 3 months.

! CAUTION: The fixative solution should be prepared under a fume hood, since paraformaldehyde is extremely toxic.

20× SSC: 3 M NaCl, 300 mM sodium citrate, pH 7.0 with 1 M HCl. Other SSC dilutions needed 5×, 2× and 0.2×.

Hybridization solution: 50% (v/v) formamide in 5× SSC, 0.1% (v/v) Tween-20 and 0.1 mg mL^{-1} Heparin.

Supplemented hybridization solution: Add 10 μg mL^{-1} Salmon sperm to the hybridization solution.

50% Glycerol (v/v): Glycerol solution is prepared combining 5 mL of 1× PBS with 5 mL of glycerol and well mixed.

3 MISH Method

3.1 Seed Sterilization and Medium Preparation

Seed sterilization is a pivotal process that aims to both avoid contamination and synchronize germination. Seed sterilization could be carried out as previously described by Araniti et al. (2016, 2017).

Sterilize *Arabidopsis thaliana* L. (Heyn.) seeds, ecotype Columbia (Col-0), by incubation in EtOH (96%) for 2 min and successively in NaOCl: Triton X-100 (1.75%: 0.01%) solution for 12 min (Triton X-100 should be used to improve membrane permeability).

After sterilization, rinse *Arabidopsis thaliana* seeds three times × 5 min in sterilized distilled water and then vernalize them in 0.1% agar solution at 4 °C for 48 h in order to promote the synchronization of seed germination. Sown sterilized and vernalized seeds on petri plates (10 × 10) filled with 40 mL of agar medium (8 g L^{-1}) containing Murashige–Skoog medium (Sigma–Aldrich) (4.4 g L^{-1}), sucrose (10 g L^{-1}), Myo-Inositol (0.1 g L^{-1}) and MES (0.5 g L^{-1}) Buffer, adjust the pH at 6.0 using KOH. Transfer the sown plates to a growth chamber settled with a temperature of 21 °C, under 16 h (150 μmol m^{-2} s^{-1}) light and 8 h dark and 60% relative humidity.

3.2 DNA Template Preparation

Δ CRITICAL STEP: From this step the use of RNase-free experimental tips, tubes and reagents is recommended.

1. Linearize 10 μg of plasmid carrying a fragment of a gene specific cDNA with appropriate polymerase (a final volume 200 μL should be reached using DEPC water). Use an excess of enzyme/time (e.g. 5×, 4 h) and then check linearization on agarose gel.

Δ CRITICAL STEP: Plasmid should be completely cut since supercoiled DNA is a very efficient substrate for RNA polymerase and will generate high molecular weight RNA (mainly vector sequence). Do not use enzymes that leave a 3′ overhang since the RNA polymerase may drop onto and transcribe the bottom strand.

2. Run a small portion of plasmid DNA (≈200 ng of DNA) on agarose gel to check the complete restriction digestion.
3. Perform a phenol/chloroform extraction on the linearized plasmid. Resuspend the linearized plasmid in DEPC water (0.5 μg μL^{-1}).

Δ CRITICAL STEP: During the experiment all the solutions must be prepared using RNase free/DEPC.

4. Precipitate DNA using 2× volume of absolute ethanol, 0.1× volume 3 M sodium acetate, pH 5.2.
5. Keep DNA at −20 °C for at least 2 h (or −80 °C for 30 min), then centrifuge for 25 min at 13,000 rpm in a microfuge.
6. Wash the pellet with cold ethanol (70%), centrifuge for 15 min at 13,000 rpm in a microfuge, drain and air dry.
7. Resuspend in 40 μL DEPC water.
8. Check 5 μL of linearized plasmid DNA (pDNA) on 1% agarose gel. A single band should be visible. For accurate quantification and quality, 1.5 μL of linearized pDNA should be controlled with a NanoDropSpectrophotomer ND-1000.

PAUSE POINT Linearized plasmid DNA can be stored at −80 °C for up to 1 year.

3.3 *Synthesizing Labeled RNA Probe/(It Could Be Done Also by PCR)*

9. The transcription reaction should be settled up on ice in an RNase-free tube (20 μL total volume) using the following ratio:

DNA (~2 μg template DNA)	X μL
10× transcription buffer	2.0 μL
10× RNA labeling mix (DIG, BIO, FITC, or DNP)	2.0 μL
RNase inhibitor	1.0 μL
RNA polymerase (T7/SP6)	2.0 μL
DTT 10 mM	2.5 μL
DEPC water	Y μL
Total volume	**20 μL**

10. Incubate for 2 h at 37 °C.
11. Run 2 μL on a 1% agarose gel with a RNA marker. (A single or double band should be visible).

Δ **CRITICAL STEP** Look at the gel after 15 min, since RNA degrades quickly (Before the use, in order to eliminate RNases, the RNA gel box should be washed with 0.2 M NaOH for a period >30 min).

12. Add 75 μL of H_2O, 1 μL tRNA (100 mg mL^{-1}), 5u. of DNase I recombinant RNase-free, then incubate 10 min at 37 °C.
13. Add equal volume of NH_4Ac (4 M), two volumes of 96% ethanol and put at −20 °C for 2 h (or −20 °C overnight)
14. Centrifuge for 30 min at 13,000 rpm at 4 °C.

15. Wash the pellet carefully with cold ethanol (70%) and then centrifuge for 10 min at 13,000 rpm at 4 °C.
16. Drain completely, air dry the pellet and then dissolve it in a 100 μL DEPC water.

PAUSE POINT Labeled probe should be stored at −20 °C; stock probe in aliquots and avoid re-freezing and thawing.

3.4 *Plant Material Fixation and Dehydration*

17. Cut the tissue (root and apical organs) into small pieces with a razor blade and put it in a 15 ml falcon containing the fixative solution. To improve the infiltration of fixative, fix under vacuum until the tissues precipitate at the bottom of the tube (~20 min).

Δ **CRITICAL STEP** The dissected material must be fixed immediately since RNA degrades quickly. Prepare and use paraformaldehyde in a fume hood.

18. Transfer the tissue sample to plastic tubes or glass vials containing fresh fixative. Cap the vial, tape it on its side to an orbital platform shaker, and shake it gently, at 60–80 rpm, for 1 h at 4 °C.
19. Remove the fixative, and add the methanol 2×5 min shaking at 4 °C and 3×5 min in 96% ethanol.
20. Store in 96% ethanol overnight at −20 °C (if it is necessary seedlings can be stored for several days at −20 °C).

3.5 *Sample Treatment and Hybridization*

Δ **CRITICAL STEP** All steps, except the hybridization, are performed with a gentle shaking.

21. Incubate the samples in a 1:1 mixture of ethanol and xylene for 30 min;
22. Wash twice in ethanol for 5 min;
23. Hydrate the samples using a 75% ethanol solution (v/v in water × 10 min). Successively, remove the solution and replace it with a 50% (v/v 1× PBS × 10 min) ethanol solution. Finally, remove the 50% ethanol solution and replace it with a 25% ethanol solution (v/v in 1× PBS × 10 min).
24. Refix samples in fixative solution for 20 min at room temperature (RT).
25. Wash twice in PBT for 10 min.
26. Digest with proteinase K buffer (use a final concentration 125 μg mL^{-1} in water) for 15 min.
27. Stop the digestion by incubating the samples for 5 min in 1× PBS plus 0.2% glycine.

28. Wash twice in PBT for 10 min.
29. Refix material in fixative solution for 20 min at RT.
30. Wash twice in PBT for 10 min.
31. Wash in hybridization solution for 10 min.
32. Pre-hybridize the samples in hybridization solution for 1 h at 50 °C.
33. Hybridize overnight at 50–55 °C in supplemented hybridization solution (supplemented with Herring sperm DNA). The hybridization solution contains 20–100 ng of denatured probe/ml (denaturation is carried out in a PCR machine at 80 °C for 2 min and quickly cooled for 2 min before use).

Δ **CRITICAL STEP** Concentration of probe requires certain optimization.

3.6 Post-hybridization and Fluorescent Detection

34. Wash samples three times (10 min, 60 min and 20 min) in 50% (v/v) formamide, 2× SSC and 0.1% (v/v) Tween-20 at 50–55 °C (depending on the probe).
35. Wash samples for 20 min in 2× SSC, 0.1% (v/v) Tween-20 at 50–55 °C (depending on the probe).
36. Wash samples twice for 20 min in 0.2× SSC, 0.1% (v/v) Tween-20 at 50–55 °C (depending on the probe).
37. Wash samples three times for 10 min in PBT at RT.
38. Wash the samples one time for 30 min in PBT plus 1% BSA.

Δ **CRITICAL STEP** At this step, choose which primary and secondary detection reagents are required to detect the probes. Incubate the samples with a mixture of primary antibodies selected (anti-digoxigenin sheep, Roche, anti-biotin mouse, Roche); diluted (1:100) in (PBT + BSA), for 2 h at RT under gentle shaking.

39. Wash three times for 10 min in PBT.
40. Wash for 30 min in PBT plus BSA.
41. Incubate with a mixture of secondary antibodies (Alexa Fluor 555 dye Donkey Anti-Sheep, Invitrogen, Alexa Fluor 488 dye Donkey Anti-Mouse, Invitrogen) diluted (1: 100) in PBT plus BSA kept in the dark overnight at RT.

Δ **CRITICAL STEP** Protect the multi-well plate from light during this incubation.

3.7 Washing and Fluorescent Detection

42. Wash twice for 15 min in PBT under gentle shaking.

Δ **CRITICAL STEP** protects the multi-well plate from light.

43. Mount tissues in 50% (v/v) glycerol on a microscope slide for viewing.

Δ **CRITICAL STEP** this medium will not protect against photobleaching, but performs well for most experiments.

4 TSA-MISH Method

The TSA method is an improvement of the MISH method, which allows not only to detect low expressed genes but also to detect in the same sample genes with different expression level (Fig. 18.1). In order to detect both the transcript (mRNA) and the translations products (proteins), was refined a new method, TSA-MISH, which could be very useful to analyze the whole regulation of gene expression as well as to analyze transcript accumulation in specific GFP-transgenic backgrounds. In the Diagram of work (Fig. 18.2) is combined both MISH and TSA-MISH method, in

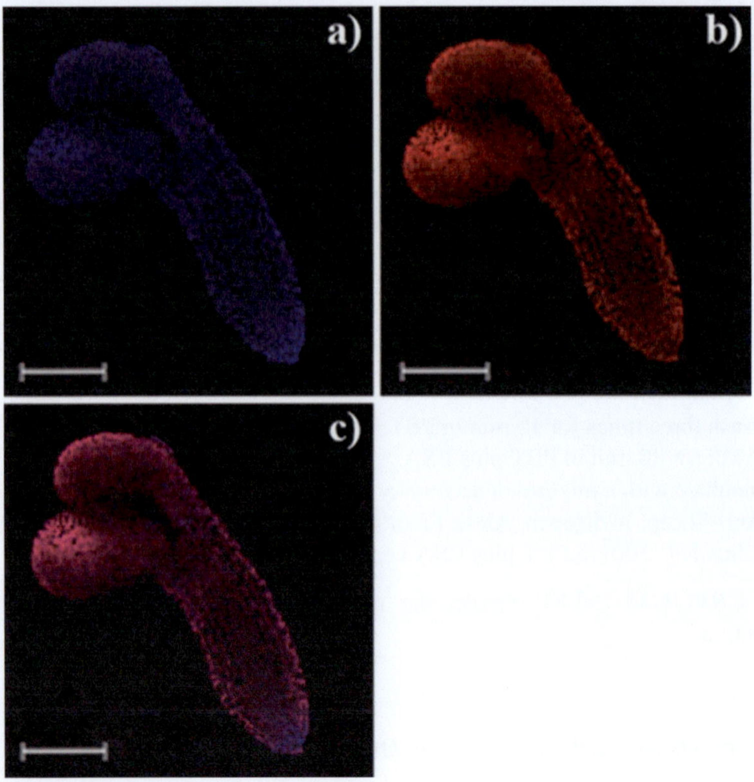

Fig. 18.1 Multiprobe in situ hybridization. Images were acquired using a SP8 confocal microscope (Leica, www.leica microsystems.com) with a 40 oil immersion objective: red, ELO3 digoxigenin riboprobe, sheep anti-digoxigenin and AF555 donkey anti-sheep; blue, HUB1 FITC riboprobe, rabbit anti-fluorescein isothiocyanate and AF647 chicken anti-rabbit; pink, merged HUB1 and ELO3 expression. Scale bar = 35 μm

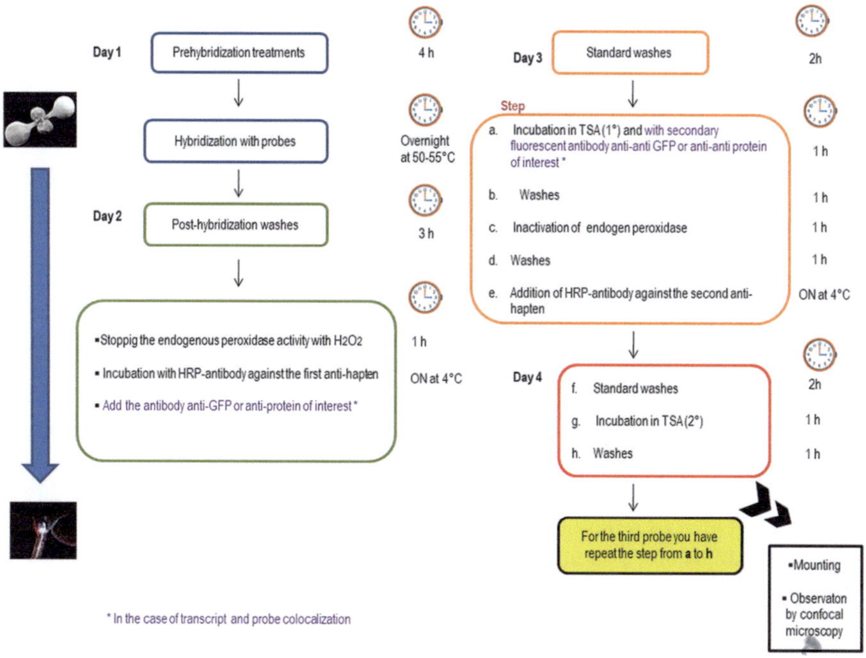

Fig. 18.2 Diagram of work in which is combined both MISH and TSA-MISH method. The steps for both techniques are distributed in days differently colored. In the boxes the red asterisk indicates highlight relevant points for the development of TSA-MISH procedure respect to MISH. Approximate times required for each step are indicated at the side of every single box. Day 1 is primarily dedicated to steps common to standard MISH. Days 2 and 3 are dedicated to endogenous peroxidase inactivation and detection of the first anti-hapten. Subsequent, a couple of days are dedicated to the detection of other anti-hapten. (This scheme is a reworking of the Workflow present in Bruno et al. 2015)

which for TSA-MISH method is necessary repeating the following steps: H_2O_2 treatment to inactivate the peroxidases treatment with the anti-antibody specific for the probe F-TSA (Tyramide associated to a fluorofor (F-TYR) peroxidase reaction. To realize the TSA-MISH method follow the previous protocol till step 37 and then the successive steps.

4.1 *Post-hybridization Washes*

1. Incubate in 0.2 × SSC with 1 mL 1% (v/v) H_2O_2 for 60 min at RT to inactivate the peroxidase activity relative to detection of the first ribo-probe.
2. Remove H_2O_2 and wash samples two times for 30 min in PBS, 0.1% (v/v) Tween-20 at RT.

3. Preincubate samples in PBS, 0.1% (v/v) Tween-20 with 1% (w/v) BSA for 30 min at RT.
4. Incubate sample with a selected primary antibody (Mouse anti-biotin) diluted (1:500) in PBT+BSA and anti-gfp (final concentration 100 ng mL^{-1}) diluted in PBT+BSA
5. Incubate samples ON in the above solution in the dark 4 °C under gentle shaking.

4.2 Washing and Fluorescent Detection

6. Wash samples four times for 20 min in PBS, 0.1% (v/v) Tween-20 at RT.
7. Wash samples in PBS, 0.1% (v/v) Tween-20 with 1% (w/v) BSA for 30 min at RT.
8. Incubate samples with an anti-hapten-HRP (488) antibody to second target (diluting the stock solution 1:100 in blocking) for 60 min at RT (no shaking).
9. Wash samples for 30 min in PBS 1×, 0.1% (v/v) Tween-20 at RT.
10. Incubate samples in Alexa Fluor dyes 488 Donkey Anti-Mouse-conjugated tyramide amplification reagent working solution (diluting the tyramide stock solution 1:100 in amplification buffer) for 60 min at RT.
11. Wash samples two times for 30 min in PBS, 0.1% (v/v) Tween-20 at RT.
12. Wash samples in PBS, 0.1% (v/v) Tween-20 with 1% (w/v) BSA for 30 min at RT.
13. Incubate sample in Alexa Fluor dyes 647 Donkey anti-rabbit diluted (1:500) in PBT+BSA for 2 h.
14. Wash samples twice for 15 min in darkness with PBS, 0.1% (v/v) Tween-20 at RT.
15. Incubate samples in a 4% hydrogen peroxide solution (v/v) H_2O for 60 min at RT (see step 17).
16. Wash samples for 20 min in PBS, 0.1% (v/v) Tween-20 at RT.
17. Wash samples in PBS, 0.1% (v/v) Tween-20 with 1% (w/v) BSA for 30 min at RT.
18. Incubate sample with a selected primary antibody (Mouse anti-dig) diluted (1:500) in PBT+BSA.
19. Incubate samples ON in the above solution in the dark 4 °C under gentle shaking.

4.3 Washes and Detection

20. Wash samples four times for 20 min each in PBS, 0.1% (v/v) Tween-20, at RT.
21. Wash samples in PBS, 0.1% (v/v) Tween-20 with 1% (w/v) BSA for 30 min at RT.

22. Incubate samples with an anti-hapten-HRP antibody to second target (dilute the stock solution 1:100 in blocking) for 60 min at RT (no shaking).
23. Wash samples for 30 min in PBS, 0.1% (v/v) Tween-20 at RT.
24. Incubate samples in Alexa 555-conjugated tyramide amplification reagent working solution for 60 min at RT.
25. Wash samples twice for 30 min in PBS, 0.1% (v/v) Tween-20 at RT.
26. Mount samples in a 2:1 mixture of antifading reagent (Fluka) and PBS, 0.1% (v/v) Tween-20 on a microscope slide for viewing.

- **TIMING MISH METHOD**

  Steps 1–8, Preparing DNA Template: 6–8 h
  Steps 9–16, Synthesizing labeled RNA probe: 5 h – 1 day
  Steps 17–20, Plant material fixation and dehydration: 1–5 day
  Steps 21–33, Prehybridization and hybridization: 1 day
  Steps 34–42, Post-hybridization: 1 day
  Steps 43–44, Washing and fluorescent detection: 1 day

- **TIMING TSA-MISH METHOD**

  Steps 1–5, Post-hybridization: 1 day
  Steps 6–18, Washing and fluorescent detection: 1 day
  Steps 19–25, Washes and detection: 1 day

References

Araniti F, Graña E, Krasuska U, Bogatek R, Reigosa MJ, Abenavoli MR, Sánchez-Moreiras AM (2016) Loss of gravitropism in farnesene-treated arabidopsis is due to microtubule malformations related to hormonal and ROS unbalance. PLoS One 11(8):e0160202

Araniti F, Bruno L, Sunseri F, Pacenza M, Forgione I, Bitonti MB, Abenavoli MR (2017) The allelochemical farnesene affects *Arabidopsis thaliana* root meristem altering auxin distribution. Plant Physiol Biochem 121:14–20

Birnbaum K, Jung JW, Wang JY, Lambert GM, Hirst JA, Galbraith DW, Benfey PN (2005) Cell type–specific expression profiling in plants via cell sorting of protoplasts from fluorescent reporter lines. Nat Methods 2(8):615–619

Bruno L, Chiappetta A, Muzzalupo I, Gagliardi C, Iaria D, Bruno A, Greco M, Giannino D, Perri E, Bitonti MB (2009) Role of geranylgeranyl reductase gene in organ development and stress response in olive (*Olea europaea*) plants. Funct Plant Biol 36(4):370–381

Bruno L, Muto A, Spadafora ND, Iaria D, Chiappetta A, Van Lijsebettens M, Bitonti MB (2011) Multi-probe in situ hybridization to whole mount Arabidopsis seedlings. Int J Dev Biol 55(2):197–203

Bruno L, Ronchini M, Gagliardi O, Corinti T, Chiappetta A, Gerola P, Bitonti MB (2015) Analysis of AtGUS1 and AtGUS2 in Arabidopsis root apex by a highly sensitive TSA-MISH method. Int J Dev Biol 59(4-5-6):221–228

Christine T, Bernard T (2008) High-resolution in situ hybridization to whole-mount zebrafish embryo. Nat Protoc 3:59–69

Chuaqui RF, Bonner RF, Best CJ, Gillespie JW, Flaig MJ, Hewitt SM, Phillips JL, Krizman DB, Tangrea MA, Ahram M, Linehan WM, Knezevic V, Linehan WM (2002) Post-analysis follow-up and validation of microarray experiments. Nat Genet 32:509–514

Drews GN, Beals TP, Bui AQ, Goldberg RB (1992) Regional and cell specific gene expression patterns during petal development. Plant Cell 4:1383–1404

Giannino D, Condello E, Bruno L, Testone G, Tartarini A, Cozza R, Innocenti AM, Bitonti MB, Mariotti D (2004) The gene geranylgeranyl reductase of peach (*Prunus persica* [L.] Batsch) is regulated during leaf development and responds differentially to distinct stress factors. J Exp Bot 55(405):2063–2073

Harrison PR, Conkie D, Affara N, Paul J (1974) In situ localization of globin messenger RNA formation. J Cell Biol 63(2):402–413

Kim JS, Jung HJ, Lee HJ, Kim K, Goh CH, Woo Y, Seung HO, Yeon SH, Kang H (2008) Glycine-rich RNA-binding protein7 affects abiotic stress responses by regulating stomata opening and closing in *Arabidopsis thaliana*. Plant J 55(3):455–466

Larkin JC, Oppenheimer DG, Pollock S, Marks MD (1993) Arabidopsis GLABROUS 1 gene requires downstream sequences for function. Plant Cell 5:1739–1748

Lefebvre V, North H, Frey A, Sotta B, Seo M, Okamoto M, Nambara E, Marion-Poll A (2006) Functional analysis of Arabidopsis NCED6 and NCED9 genes indicates that ABA synthesized in the endosperm is involved in the induction of seed dormancy. Plant J 45(3):309–319

Mizukami Y, Huang H, Tudor M, Hu Y, Ma H (1996) Functional domains of the floral regulator AGAMOUS: characterization of the DNA binding domain and analysis of dominant negative mutations. Plant Cell 8:831–845

Taylor C (1997) Promoter fusion analysis: an insufficient measure of gene expression. Plant Cell 9(3):273–275

Wellmer F, Riechmann JL, Alves-Ferreira M, Meyerowitz EM (2004) Genome-wide analysis of spatial gene expression in Arabidopsis flowers. Plant Cell 16(5):1314–1326

Chapter 19
Proteomics Analysis of Plant Tissues Based on Two-Dimensional Gel Electrophoresis

Jesus V. Jorrín-Novo, Luis Valledor-González, Mari A. Castillejo-Sánchez, Rosa Sánchez-Lucas, Isabel M. Gómez-Gálvez, Cristina López-Hidalgo, Victor M. Guerrero-Sánchez, Mari C. Molina Gómez, Inmaculada C. Márquez Martin, Kamilla Carvalho, Ana P. Martínez González, Mari A. Morcillo, Maria E. Papa, and Jeffrey D. Vargas Perez

1 Introduction

Proteomics, a molecular biology discipline, studies living organisms from the perspective of their proteins, the biomolecules responsible for executing the genetic information coded in the genes, aiming at deciphering and interpreting their life cycle, dynamics, and interactions, and, lately, genotype to phenotype translation.

From a methodological point of view, it comprises in vitro techniques and, to a much lesser extent, either in vivo or in situ approaches. As an in vitro technique, molecules, in this case proteins, are obtained directly (by extraction) or indirectly (e.g. by in vitro translation) from biological sources for ulterior characterization at the physico-chemical and biological level, and are also employed for translational purposes (e.g. for food traceability analysis). As an –omics approach, in the holistic (from the Greek *holos,* meaning entire or all) sense of the term, and differently from classical biochemistry, it investigates proteins as a whole rather than as individual entities, without discarding its use as a targeted, hypothesis-driven, and "proteinomics" strategy (Picotti et al. 2013).

As an adaptation of the "genome" term, M. Wilkins introduced the word "proteome" for the first time. That was in 1994, at the first "Genome to Proteome" Siena meeting (2D Electrophoresis–From Protein Maps to Genomes, Siena, Italy, September 5–7, 1994). Later, it appeared and was formally defined as "the PROTEin complement of a genOME" in a paper published in Electrophoresis by 1995

J. V. Jorrín-Novo (✉) · L. Valledor-González · M. A. Castillejo-Sánchez · R. Sánchez-Lucas
I. M. Gómez-Gálvez · C. López-Hidalgo · V. M. Guerrero-Sánchez · M. C. Molina Gómez
I. C. Márquez Martin · K. Carvalho · A. P. Martínez González · M. A. Morcillo · M. E. Papa
J. D. Vargas Perez
Agroforestry and Plant Biochemistry, Proteomics, and Systems Biology, Department of Biochemistry and Molecular Biology, University of Cordoba, Cordoba, Spain
e-mail: bf1jonoj@uco.es

© Springer International Publishing AG, part of Springer Nature 2018

A. M. Sánchez-Moreiras, M. J. Reigosa (eds.), *Advances in Plant Ecophysiology Techniques*, https://doi.org/10.1007/978-3-319-93233-0_19

(Humphery-Smith 2015). Since then, a new vision and era have arrived on the biochemistry and molecular biology/scene, transforming the classical protein chemistry or biochemistry into a holistic approach that opens up new possibilities to understanding the function of genes and the genotype to phenotype translation by tracking the total protein content of the cell.

As scientific disciplines grow in parallel, hand in hand, with the developments in technology, it is worth mentioning the three main advances that have contributed most to the birth of proteomics. First, it was the introduction, during the late 1980s, of soft ionization methods that allowed the analysis of peptides and proteins by mass spectrometry: the MALDI (matrix-assisted laser desorption/ionization) and the ESI (electrospray ionization) (Aebersold 2003; Aebersold and Mann 2003). Second, an increasing number of genomes were sequenced and DNA or EST sequences were made available thanks to progress in NGS (next generation sequencing) technologies (Buermans and den Dunnen 2014). And third, bioinformatics tools and algorithms were developed to identify and quantify proteins from MS spectra and to manage the statistical analysis of the huge amount of data generated (Baldwin 2004; Schubert et al. 2017). In addition, proteomics is based on classical protein biochemistry and cell biology methods including protocols for protein extraction, fractionation, purification, depletion, and labeling, in which electrophoresis has played a pivotal role, giving rise to one of the platforms most employed in plant research, i.e. two-dimensional gel electrophoresis, (2-DE), the focus of this chapter (Gorg et al. 2004).

Proteomics can be defined as being a scientific discipline or methodological approach, whose objective is the study of the living organism proteome, understood as the total set of protein species[1] present in a biological entity (subcellular fraction, cell, tissue, organ, organisms, population, ecosystem) at a certain time (specific growth and developmental stage), and under specific environmental conditions. It can also refer to a structural or functional group of proteins (proteases, phosphoproteome, membrane proteins, etc.). This definition emphasizes the dynamic character of the proteome that, together with the chemical complexity of the proteins, the number of protein species coded by individual genes, and the different concentration range within the cell, makes the approach quite challenging. By using proteomics we aim to find out "how", "where", "when", and "what for" are the several hundred thousand of individual protein species produced in a living organism. We wish to know how they interact with one another and with other molecules to construct the cellular building, and how they work in order to fit in with programmed growth and development, and to interact with their biotic and abiotic environment (Jorrín Novo 2015).

The objectives of proteomics research will define different areas within the field, including the simple identification and cataloguing of the protein species at the whole cell, tissue, organ or sub-cellular levels (descriptive and sub-cellular

[1] The term protein species will be utilized instead of proteins alone throughout this chapter referring to the different gene products of a gene as a result of post-transcriptional and post-translational events (Jungblut and Schlüter 2011).

proteomics), the qualitative and quantitative comparison of two or more biological samples in order to infer differences and biological interpretations of the variations among genotypes, organs, tissues, developmental stages, and environmental conditions (comparative proteomics), the identification and characterization of post-translational variants of a protein (post-translational proteomics), and the molecular interaction with other proteins or biomolecules (interactomics). Proteomics can be used for basic (gaining of biological knowledge) or translational purposes (Cox et al. 2011). This chapter is mostly focused on the first two premises, descriptive and comparative proteomics.

2 Proteomics in Plant Biology Research

Proteomics has become a priority in biological investigation, and plants are not an exception to this rule; together with other –omic approaches it is at the heart of Systems Biology. The relevance of the discipline can be deduced by considering the number of papers published since 1994 (Fig. 19.1), when the term proteome was coined, and when the first two papers on plant proteomics appeared (Egorov et al. 1994; Klabunde et al. 1994). The first works reporting a global plant proteome analysis date back to 1999 (Kehr et al. 1999; Peltier et al. 2000) and the first comparative proteomics 1–2000 (Chang et al. 2000; Natera et al. 2000). Since then, and up to

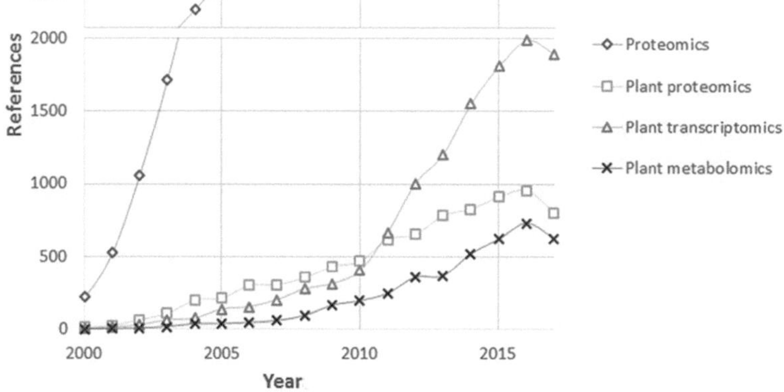

Fig. 19.1 Number of references reported at PubMed database during the 1994–2017 period when a search was performed with the words (all fields): proteomics, plant + proteomics, plant + transcriptomics, and plant + metabolomics

2017, the total number of citations listed in PubMed under the words "plant proteomics" was of 8011, representing 10% of the total items that came up when just "proteomics" was searched. For comparative purposes, the total number of plant transcriptomics items was of 11,776, and that of plant metabolomics 4113. The topic of plant proteomics has been extensively reviewed since 1999 (Thiellement et al. 1999), with some of them authored or co-authored by Prof. Jorrin-Novo (Jorrin-Novo et al. 2006, 2007, 2009; Jorrin-Novo and Valledor 2013; Jorrin-Novo 2014, 2015; Komatsu and Jorrin-Novo 2016; Sanchez Lucas et al. 2016).

Most of the original publications belonged to descriptive proteomics, including sub-cellular, and comparative categories, with two-dimensional gel electrophoresis-based proteomics being the dominant approach, although, for the last 5 years, label and gel-free label-free (shotgun) approaches have become dominant.

Proteomics papers have been published on close to one hundred plant species, including model systems (*Arabidopsis thaliana, Medicago truncatula, Lotus japonicus*), crops, including cereals, legumes, oil-bearing, vegetables, fruit and berries, sugar, and permanent, aromatics, weeds, and forest trees (reviewed in Jorrin Novo et al. 2015). As confident protein identification from mass spectra is only possible if the genome is sequenced, or there are enough well-annotated sequences available, proteomics with orphan species is highly challenging, as putative identified proteins corresponded, in the best of the cases, to orthologs rather than to gene products (Abril et al. 2011).

Proteomics experiments have been carried out with seeds, seedlings, and adult plants at the vegetative and flowering stages, either at the whole individual or organ, tissue, or cell level, including roots, hypocotyls, cotyledons, shoots, stems, leaves, buds, meristems, flowers, spikes, fruits, callus, cell suspensions, protoplasts, hairy roots, or somatic embryos (Jorrin Novo et al. 2015). The use of the different plant material is conditioned by the objectives of the research, but from a proteomics point of view the complexity and chemical composition determines, to a great extent, the final results in terms of the number of proteins that can be confidently identified and quantified. Plant organs comprise different type of tissues and cells, each one with its own protein signature, thus causing high biological variability. The presence of non-proteinaceous compounds in the tissue affects the amount and number of proteins extracted and solubilized prior resolution and mass spectrometry analysis. This is the case of salts, as for example in root tissues, polysaccharides, like in cereal seeds and fruits, lipids, in seed of oily plants, phenolics, in fruit and flowers proteases, in some fruit such as pineapple.

The protocol to be used for protein extraction, solubilization and resolution will depend on the chemical composition of the plant material, and the best one will capture the most protein species without being modified, eliminating, at the same time, non-protein compounds. Another important issue is the presence of major proteins that, like RubiSCO in leaves and reserve proteins in seeds, mask the visualization of minor proteins.

Proteomics has been used in plant studies for both basic research and translational purposes. In Table 19.1, a list of research objectives is summarized based on

Table 19.1 Basic and translational plant research objectives approached by using proteomics, as number of items references in PubMed. The list did not purport to be exhaustive

Objectives (plant proteomics + searching key words)	Number of items
Growth	2687
Development	2071
Hormones	448
Circadian responses	4
Symbioses	161
Mineral nutrition	25
Abiotic stress: temperature	228
Abiotic stress: light	184
Abiotic stress: water	334
Abiotic stress: salt	166
Abiotic stress: chemicals	148
Biotic stress: virus	193
Biotic stress: bacteria	167
Biotic stress: fungi	173
Biotic stress: insect	280
Biotic stress: parasitic plants	8
Translational: allergens	167
Translational: plant breeding	371
Translational: transgenic plants	315
Translational: food traceability	13

a search at PubMed on February 2nd, 2018. Plant development and responses to stresses are by far the topics most represented in the current literature.

3 Plant Proteomics Methods, Techniques and Protocols

In this section, the platforms employed in proteomics research will be mentioned and briefly discussed, with emphasis on 2-DE-MS, the one most used with plant species. It is not proposed to give many details or detailed protocols, but just a few guidelines that will help to approach a plant project using proteomics, to decide which protocol to use and to evaluate the results. A more detailed discussion will be found in the original publications, reviews or monographs by the author's group. Among them, Plant Proteomics Methods and Protocols, edited by Jorrin-Novo et al. (2014), is an excellent reference.

The workflow of a standard MS-based proteomics experiment includes the following steps, as illustrated in Fig. 19.2 for a 2-DE-based approach: experimental design, sampling material and storage, protein extraction, fractionation, purification, and/or depletion, protein electrophoresis (one- and two-dimensional), MS analysis, protein identification and quantification, and statistical analysis of the

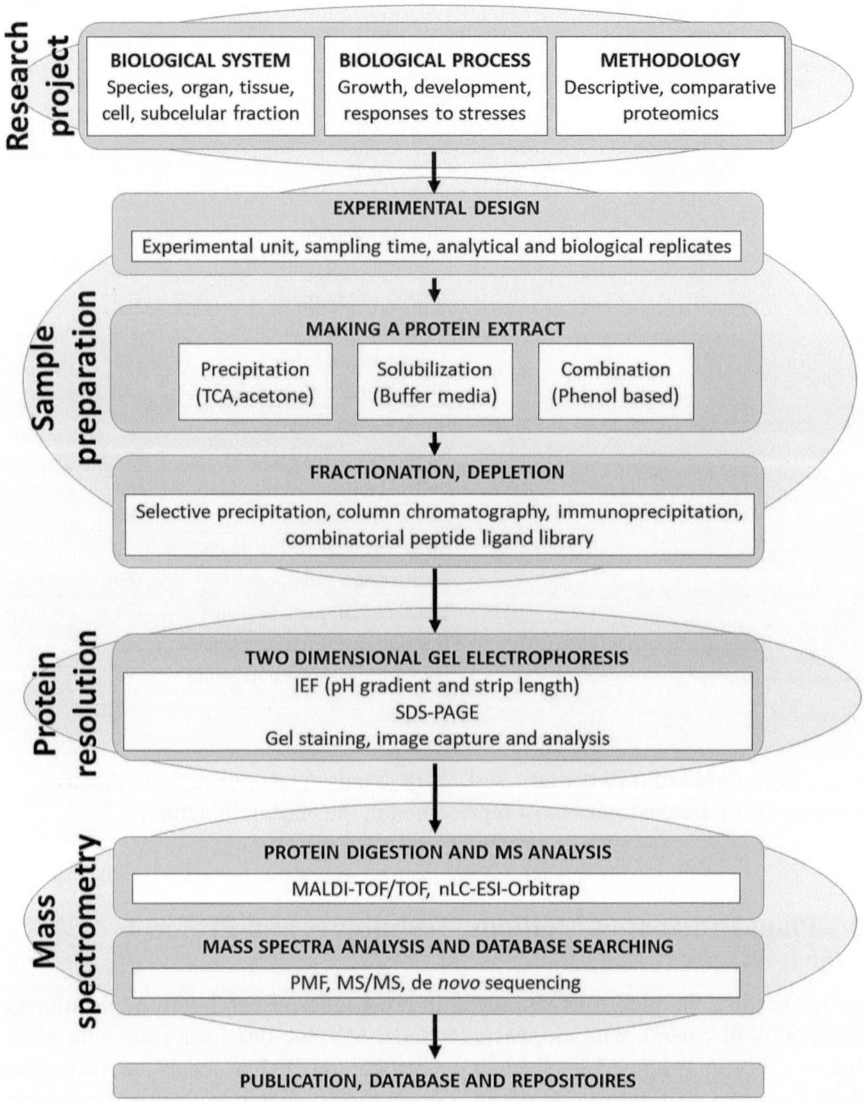

Fig. 19.2 Steps in a standard 2-DE based proteomics experiment workflow

data. For each stage, protocols have to be adapted to the experimental system and the objectives of the research (Jorrin-Novo et al. 2014).

3.1 Experimental Design

Although not always realized, this preliminary step is a key one, not only for proteomics but also for whatever approach is used in an investigation. The experimental unit must be clearly defined as well as the tissue to be sampled and the sampling time. Another important decision to be made is the number of analytical and biological replicates to be performed, which depends on the technique itself and the analytical and biological variability found. All these issues are discussed in depth in Jorrin-Novo et al. (2009), Valledor and Jorrin (2011), and Valledor et al. (2014), with clear examples from our work with Holm oak (*Quercus ilex*) (Jorge et al. 2005, 2006). Special attention should be paid to the statistical analysis of the data if we wish to confidently conclude from a biological point of view. The proteome should be analyzed as a whole so that a multivariate analysis of the data has to be performed. This test shows how homogeneous the replicates are and how different the samples, and also which spots contribute most to the biological variability.

3.2 Protein Extraction

Once plant tissue is sampled, it should be cleaned and sterilized in order to avoid contaminant proteins in the sample. If the proteins are not being extracted immediately after sampling, which is quite common, the tissue must be stored ensuring that it is not modified, and to avoid possible artifacts. In our hands freezing in liquid nitrogen and storing at −80 °C, or even better, lyophilizing before storing, has provided good results.

It is a maximum that, to detect and identify a protein, it has to be extracted and solubilized, so the proteomics experiment depends to a great extent on the extraction protocol. Two general methods can be used for protein extraction from plant tissue, either based on solubilization in a buffer medium, or precipitation by using organic solvents and acids; however both protocols can be combined. In our hands the precipitation protocols have also given the best results in terms of protein yield and number of bands or spots resolved by 1- or 2-D electrophoresis. The choice of the precipitation procedure is justified because of the low protein content in plant cells, and the chemical composition of the plant tissue, as most problems related to protein solubilization and resolution are associated with the co-extraction of non-protein compounds, such as salts, polysaccharides, polyphenols, lipids, and the presence of proteases (Jorrin Novo et al. 2009). The artifacts generated by all these compounds are minimized in precipitation protocols. The protocol must be optimized in each experimental system, as has been reported, for example, in Maldonado et al. (2008) for Arabidopsis leaf tissue, and carnation stem (Ardila et al. 2014). Hydrophobic proteins, as well as those with extreme pIs, are usually elusive to most of the standard protocols and their study requires specific methods, whose discussion is outside the objectives of the present chapter. Once extracted and solubilized in 1- or 2-D electrophoresis medium (Gorg et al. 2000), the protein content must be quantified by using colorimetric assays such as Bradford, Lowry or Biciconinic (BCA). From these data the extraction protocol has to be validated by comparing experimental yield data with the total protein content of the plant system under analysis as determined by Kjeldahl or NIRS technology (Romero Rodriguez et al. 2014). It is sometimes observed that the protein yield is low, which is often disappointing, but this is quite common. Thus, for example, and by using a protein sequential extraction of Holm oak seeds, it was not possible to solubilize more than 15% of the total protein content. But, even so, the number of spots resolved in a 2-DE gel was high enough to provide relevant information to the system, with more than 400 spots visualized (Sghaier-Hammami et al. 2016).

The proteome is, by definition, of a great complexity, with the number of protein species being the result of the number of genes and the post-transcriptional and post-translational events that make the total number much higher than that of the genes or transcripts. In order to obtain a deep proteome coverage, subcellular fractionation or sample pre-fractionation by using chromatographic or preparative electrophoresis techniques are two valid strategies, whose discussion is outside the scope of this chapter (Martínez-Maqueda et al. 2013).

Low-abundant proteins are another important issue in proteomics where they are usually masked by major ones. To overcome this limitation, depletion techniques have been utilized, with the most common one implicating the use of antibodies against abundant proteins such as RubisCO (Cellar et al. 2008) or the equalizer (combinatorial peptide ligand library) technology (Boschetti et al. 2009).

3.3 Two-Dimensional Gel Electrophoresis

Electrophoresis (the separation of ions under the influence of an electric field) is undoubtedly about the most powerful preparative and analysis technique most employed in protein research. Its origin dates back to the late 1920s, to Arne Thiselius, considered to be the father of the technique, pioneering the moving-boundary method. Since then, continuous improvements of the technique and different variants and applications have been developed, including zone electrophoresis, polyacrylamide gel electrophoresis (PAGE), disc electrophoresis, denaturing gel electrophoresis (sodium dodecyl sulphate, SDS-PAGE), isoelectrofocusing (IEF), and two-dimensional gel electrophoresis (2-DE), being among the most relevant. The 2-DE, with isoelectrofocusing as first and SDS-PAGE as second dimensions, was first reported by O'Farrel, Scheele and Klose in 1975 (Vesterberg 1989). Up to 2012, 2-DE, including the Differential Gel Electrophoresis (DIGE; Unlu et al. 1997) and the bidimensional variant Blue Native (BN)-SDS PAGE (Eubel et al. 2005) have been the dominant, almost unique, platforms in plant proteomics research. In the last 5 years plant proteomics has been moving towards second (labeling) and third (shotgun) approaches (Jorrin Novo et al. 2009, 2015).

2-DE is a consolidated technique with not much room for improvement (Gorg et al. 2000, 2004), so we did not claim to enter into the discussion of the technique details, but just to insist on the message of the need to optimize it for each experimental system. Detailed protocols can be found in our original publications in which we have employed 2-DE/MS in the proteomics analysis of different organs from Holm oak seedlings and plants, including fruit, seed embryo, leaves, root, and pollen (Jorrin Novo and Navarro Cerrillo 2014). The aim of that work was to characterize and catalogue provenances and to study development, growth and responses to biotic and abiotic stresses.

As shown in Ardila et al. (2014), some parameters of a general 2-DE protocol have to be fixed and optimized for each experimental system in order to obtain the maximum protein visualization (sensitivity), and resolution. The crucial ones are the amount of protein loaded, the IEF-strip pH gradient and length and the staining protocols (the classical visible Coomassie or silver and the fluorescent dyes), each one having different sensitivities and dynamic ranges.

Like any other technique, 2-DE has its own particular characteristics. It is a powerful one that, depending on the experimental conditions and the biological system, allows the detection of from a few hundred to up to a couple of thousand individual spots, each one corresponding to one or more protein species if comigration occurs, something quite common. This artifact can be avoided or minimized by using narrow pH gradients and long IEF strips, resulting in an increase in resolution of similar or closely related proteins, including different translation products of the same gene, allelic variants or isoforms. It gives precise information on the protein *Mr* and pI that will help in its identification. One great advantage is its multiplexing ability, allowing the combination of general or specific staining protocols, and its use in western analysis, activity-based profiling and labeling techniques. On the other hand, it has some limitations such as low reproducibility that is solved with the DIGE protocol, and the difficulty in analyzing recalcitrant, hydrophobic and extreme pI, proteins. Finally, and unlike liquid chromatography, the competitor technique, automation is not possible. All these technical and analytical issues are discussed in some of the excellent monographs edited by the companies selling equipment and reagents and some of the reviews published by Prof. Rabilloud (e.g. Rabilloud 2014; Rabilloud and Lelong 2011).

2-DE is a quantitative technique, at least in relative, comparative, terms. It is based on spot intensity that depends on the protein abundance and the staining or labeling protocol. The difference between two samples may be qualitative (spot presence or absence) or quantitative (a more or less abundant or intense spot). Protein species abundance should not necessarily be related to the level of the corresponding gene expression. So the absence of a spot does not necessarily mean that the coding gene is not being translated. This could be because it is below the detection limit of the staining procedure or has suffered some post translational modification that resulted in a change of *Mr* and/or pI, thus moving to a different coordinate within the gel.

As the number of spots in a gel is very high, its analysis is performed after gel image capturing by using algorithms, some of them commercial and others free. It is a laborious and time consuming step not exempt from artifacts, as discussed in Berth et al. (2007). As indicated above, the data on protein abundance when two or more samples are compared and the significance of the differences must be subjected to uni- and multivariate statistical tests. It is recommended to be restrictive and conservative when deciding whether or not a spot is variable among samples. It should be consistent (always present or absent in all the biological and analytical replicates), its variability lower than the average biological variability of the whole sample, and the differences statistically significant (e.g. uni ANOVA test). Multivariate analyses, such as the principal component analysis (PCA) will show how homogeneous the replicates are, how different the samples, and which spots contribute most to the variability (Righetti et al. 2004; Valledor and Jorrin Novo 2011). Once the 2-DE gel has been analyzed and the quantitative data subjected to statistics, the next step is the identification of the spots, either the variable ones or the whole set, by using mass spectrometry and, in some cases, EDMAN N-terminal sequencing.

3.4 Protein Identification Through Mass Spectrometry

Mass spectrometry is an analytical technique that measures the mass-to-charge ratio of electrically gas-phase particles (Calvete 2014). Mass spectrometry as an alternative or complementary approach to EDMAN sequencing appeared on the protein research scenario in the late 1980s, once soft ionization procedures, MALDI and ESI, had been developed. In a very simple scheme, a mass spectrometer contains three basic elements: the ionizer, the mass analyzer, and the ion detector. Different machines result from the combination of ionizers (MALDI or ESI) and analyzers (quadrupole, Q, ion trap, T, time of flight, TOF, Orbitrap), each one having its own characteristics and particularities (mass accuracy, resolution, sensitivity, dynamic range, speed, sequencing capabilities) that determine the number of peptides/proteins identified and quantified (Calvete 2014). The spectrometer may operate in single MS (m/z values for the ions, parental ions, proteins or peptides) or tandem MS or MS/MS modes (the parental ion is fragmented in the collision cell and the m/z values for the fragments determined) (Nesvizhskii et al. 2007). In most of the cases reported plant proteomics work is based on a bottom-up 2-DE MS strategy, in which the proteins (e.g. spots from a 2-DE gel) are subjected to digestion by trypsin and the tryptic fragments are directed towards MS analysis, most commonly by using the MALDI-TOF/TOF strategy. So, the protein data are inferred from the peptides that get the mass spectrometer.

The rationale of protein identification from mass (MS) or tandem mass (MS/MS or MS^n) data is the comparison between the experimental data (m/z ratios) and the theoretical ones deduced from the protein/peptide sequence as found in protein databases. The correct assignment of that spectrum to a peptide sequence is a first and central step in proteomic data processing (Nesvizhskii et al. 2007). That is why protein identification requires the availability of sequenced genomes (at the organism level) or DNA and EST sequences for individual genes. A confident identification results from both, the experimental, MS and MS/MS, data and the quality of protein database derived from in silico translation of DNA/RNA sequences. For orphan, unsequenced, organisms or those poorly represented in the database it is necessary to construct a specific protein database from species-specific DNA or EST sequences deposited and dispersed in different databases (Romero Rodriguez et al. 2014). This custom-built protein database improves the rate and quality of identifications. Alternatively, the employment of a single Viridiplantae database (NCBI, UniProt and TAIR) will provide matches to orthologs, this being a confident identification for conserved proteins. The dilemma lies in identifying orthologs or gene products. From a practical point of view, for example in plant breeding, the former are useless.

Some algorithms and bioinformatics packages are available for the analysis, identification and quantification of proteins. Some of the most frequent algorithms employed are discussed in Nesvizhskii et al. (2007). They use three main strategies (Baldwin 2004):

1. Peptide mass fingerprinting, PMF. These results from the direct comparison of the mass parental peptide peak with the predicted, theoretical, one deduced in silico. This is only valid when matching against species-specific protein databases.
2. De novo sequencing, where peptide sequences are explicitly read out directly from fragment ion spectra.
3. Hybrid approaches, such as MS-Tag. Based on comparisons between the experimental mass of the parental ion fragments produced in the collision cell, and all the predicted fragments for all the hypothetical peptides of the appropriate molecular mass, based on known fragmentation rules.

In current publications, the results of the database search are presented in a table in which the identified proteins are presented as being grouped according to their function and with columns corresponding to the name (function), species and acronyms in the database, cellular location, theoretical and experimental *Mr* and pI, together with the parameters of confidence, including score, number of peptides and percentage of sequence covered. How confident an identification is should be probabilistically understood, and is a frequent subject of discussion. A ranking of high and low probabilities should at least be established for all the matches or hits found, with an attempt to be very conservative when interpreting the data from a biological point of view.

4 Conclusions

2-DE-based proteomics is a powerful technique that has generated a huge amount of data and information on different aspects of plant biology, from growth and development to responses to biotic and abiotic environmental cues. All the information is disseminated throughout the current literature, databases and repositories. However, the full potential of the technique is far from being fully exploited and future research should move in this direction, especially in PTMs and interactomics areas. As things stand at this moment, plant proteomics remains mostly descriptive and speculative. In this regard it is important to validate proteomics data from a functional point of view. It also means integration with classic approaches of plant physiology and biochemistry, and the modern –omics, including transcriptomics and metabolomics, in the biology system direction. The interpretation of proteomics data from a biological point of view is not always possible and we may convert our publications into simple speculations. The proteome covered is, in most cases, just one frame of a very complex film, which is the life cycle of any organism. A frame in which a minimal fraction of the total proteome appears or is visualized, but that is big enough to make its analysis in a classic format impossible. These, and other issues related to standards in plant proteomics publications, are discussed in Jorrín Novo (2015).

References

Abril N, Gion JM, Kerner R, Müller-Starck G, Navarro-Cerrillo RM, Plomion C, Renaut J, Valledor L, Jorrín Novo JV (2011) Proteomics research on forest trees, the most recalcitrant and orphan plant species. Phytochemistry 72:1219–1242

Aebersold R (2003) A mass spectrometric journey into protein and proteome research. J Am Soc Mass Spectrom 14:685–695

Aebersold R, Mann M (2003) Mass spectrometry-based proteomics. Nature 422:198–207

Ardila HD, Gonzalez Fernandez R, Higuera BL, Redondo I, Martinez ST (2014) Protein extraction and gel-based separation methods to analyze responses to pathogens in Carnation (*Diantus caryophyllus*). Methods Mol Biol 1072:573–591

Baldwin MA (2004) Protein identification by mass spectrometry issues to be considered. Mol Cell Proteomics 3:1–9

Berth M, Moser FM, Kolbe M, Bernhardt J (2007) The state of the art in the analysis of two-dimensional gel electrophoresis images. Appl Microbiol Biotechnol 76:1223–1243

Boschetti E, Bindschedler LV, Fasoli E, Righetti PR (2009) Combinatorial peptide ligand libraries and plant proteomics: a winning strategy at a price. J Chromatogr A 1216:1215–1222

Buermans HPJ, den Dunnen JT (2014) Next generation sequencing technology: advances and applications. Biochim Biophys Acta 1842:1932–1940

Calvete JJ (2014) The expanding universe of mass analyzer configurations for biological analysis. Methods Mol Biol 1072:61–81

Cellar NA, Kuppannan K, Langhorst ML, Ni W, Xu P, Youn SA (2008) Cross species applicability of abundant depletion columns for ribulose-1,5-bisphosphate carboxylase/oxygenase. J Chromatogr B 861:29–39

Chang WW, Huang L, Shen M, Webster C, Burlingame AL, Roberts JK (2000) Patterns of protein synthesis and tolerance of anoxia in root tips of maize seedlings acclimated to a low-oxygen environment, and identification of proteins by mass spectrometry. Plant Physiol 122:295–318

Cox J, Heeren RM, James P, Jorrin-Novo JV et al (2011) Facing challenges in Proteomics today and in the coming decade: report of Roundtable Discussions at the 4th EuPA Scientific Meeting, Portugal, Estoril 2010. J Proteome 75:4–17

Egorov TA, Musolyamov AK, Andersen JS, Roepstorff P (1994) The complete amino acid sequence and disulphide bond arrangement of oat alcohol-soluble avenin-3. Eur J Biochem 224:631–638

Eubel H, Braun H-P, Millar AH (2005) Blue-native PAGE in plants: a tool in analysis of protein-protein interactions. Plant Methods 1:1–13

Gorg A, Obermaier C, Boguth G, Harder A, Scheibe B, Wildgruber R, Weiss W (2000) The current state of two-dimensional electrophoresis with immobilized pH gradients. Electrophoresis 21:1037–1053

Gorg A, Weiss W, Dunn MJ (2004) Current two-dimensional electrophoresis technology for proteomics. Proteomics 4:3665–3685

Humphery-Smith I (2015) The 20th anniversary of proteomics and some of its origins. Proteomics 15:1773–1776

Jorge I, Navarro M, Lenz C, Ariza D, Porras C, Jorrín J (2005) The Holm oak leaf proteome. Analytical and biological variability in the protein expression level assessed by 2-DE and protein identification by MS/MS de novo sequencing and sequence similarity searching. Proteomics 5:222–234

Jorge I, Navarro M, Lenz C, Ariza D, Jorrín J (2006) Variation in the holm oak leaf proteome at different plant developmental stages, between provenances and in response to drought stress. Proteomics 1:S207–S214

Jorrin JV, Rubiales D, Dumas-Gaudot E, Recorbet G, Maldonado A, Castillejo MA, Curto M (2006) Proteomics: a promising approach to study biotic interaction in legumes. A Rev Euphytica 147:37–47

Jorrin-Novo JV, Komatsu S, Weckwerth W, Wienkoop S (2014) Plant proteomics methods and protocols. Humana Press-Springer, New York

Jorrín-Novo JV (2014) Plant proteomics methods and protocols. Methods Mol Biol 1072:3–13

Jorrín-Novo JV (2015) Scientific standards and MIAPEs in Plant Proteomics research and publications: do we follow them? Front Plant Sci 6:473

Jorrín-Novo J, Navarro-Cerrillo RM (2014) Variability and response to different stresses in Andalusia Holm oak (*Quercus ilex* L.) populations by a proteomic approach. Ecosistemas 23:99–107

Jorrín-Novo JV, Valledor L (2013) Translational proteomics special issue. J Proteome 93:1–4

Jorrín-Novo JV, Maldonado M, Castillejo MA (2007) Plant proteome analysis: a 2006 update. Proteomics 7:2947–2962

Jorrín-Novo JV, Maldonado M, Echevarría-Zomeño S, Valledor L, Castillejo MA, Curto M, Valero-Galván J, Sghaier-Hammami B, Donoso G, Redondo I (2009) Plant proteomics update (2007–2008). Second generation proteomic techniques, an appropriate experimental design and data analysis to fulfill MIAPE standards, increase plant proteome coverage and biological knowledge. J Proteome 72:285–314

Jorrín-Novo JV, Pascual J, Sánchez-Lucas R, Romero-Rodríguez MA, Rodriguez-Ortega M, Lenz C, Valledor L (2015) Fourteen years of plant proteomics reflected in "Proteomics": moving from model species and 2-DE based approaches to orphan species and gel-free platforms. Proteomics 15:1089–1112

Jungblut PR, Schlüter H (2011) Towards the analysis of protein species: an overview about strategies and methods. Amino Acids 41:219–222

Kehr J, Haebel S, Blechschmidt-Schneider S, Willmitzer L, Steup M, Fisahn J (1999) Analysis of phloem protein patterns from different organs of *Cucurbita maxima* Duch. by matrix-assisted laser desorption/ionization time of flight mass spectroscopy combined with sodium dodecyl sulfate polyacrylamide gel electrophoresis. Planta 207:612–619

Klabunde T, Stahl B, Suerbaum H, Hahner S, Karas M, Hillenkamp F, Krebs B, Witzel H (1994) The amino acid sequence of the red kidney bean Fe(III)-Zn(II) purple acid phosphatase. Determination of the amino acid sequence by a combination of matrix-assisted laser desorption/ionization mass spectrometry and automated Edman sequencing. Eur J Biochem 226:369–375

Komatsu S, Jorrín-Novo JV (2016) Food and crop proteomics. J Proteome 143:1–2

Maldonado AM, Echevarría-Zomeño S, Jean-Baptiste S, Hernández M, Jorrín-Novo JV (2008) Evaluation of three different precipitation protocols of protein extraction for *Arabidopsis thaliana* leaf proteome analysis by two-dimensional electrophoresis. J Proteome 71:461–472

Martinez-Maqueda D, Hernandez-Ledesma B, Amigo L, Miralles B, Gomez-Ruiz JA (2013) Extraction/fractionation techniques for proteins and peptides and protein digestion. In: Toldra F, LML N (eds) Proteomics in foods. Principles and applications. Springer, Dordrecht

Natera SH, Guerreiro N, Djordjevic MA (2000) Proteome analysis of differentially displayed proteins as a tool for the investigation of symbiosis. Mol Plant-Microbe Interact 13:995–1009

Nesvizhskii AJ, Vitek O, Aebersold R (2007) Analysis and validation of proteomic data generated by tandem mass spectrometry. Nat Methods 4:787–797

Peltier JB, Friso G, Kalume DE, Roepstorff P, Nilsson F, Adamska I, van Wijk KJ (2000) Proteomics of the chloroplast: systematic identification and targeting analysis of lumenal and peripheral thylakoid proteins. Plant Cell 12:319–341

Picotti P, Bodenmiller B, Aebersold R (2013) Proteomics meets the scientific method. Nat Methods 1:24–27

Rabilloud T (2014) How to use 2D gel electrophoresis in plant proteomics. Methods Mol Biol 1072:43–50

Rabilloud T, Lelong C (2011) Two-dimensional gel electrophoresis in proteomics: a tutorial. J Proteome 74:1829–1841

Righetti PG, Castagna A, Antonucci F, Piubelli C, Cecconi D, Campostrini N, Antonioli P, Astner H, Hamdan M (2004) Critical survey of quantitative proteomics in two-dimensional electrophoresis approaches. J Chromatogr A 1051:3–17

Romero Rodriguez MC, Pascual J, Valledor L, Jorrin Novo JV (2014) Improving the quality of protein identification in non-model species. Characterization of *Quercus ilex* seed and *Pinus radiata* needle proteomes by using SEQUEST and custom databases. J Proteome 105:85–91

Romero-Rodríguez MA, Maldonado-Alconada A, Valledor L, Jorrín-Novo JV (2014) Back to Osborne. Sequential protein extraction and LC-MS analysis for the characterization of the Holm oak seed proteome. Methods Mol Biol 1072:379–389

Sánchez-Lucas R, Mehta A, Valledor L, Cabello-Hurtado F, Romero-Rodríguez MA, Simova-Stoilova LP, Demir S, Rodríguez de Francisco L, Maldonado A, Jorrín-Prieto A, Jorrín-Novo JV (2016) A year (2014–2015) of plants in Proteomics journal. Progress in wet and dry methodologies, moving from protein catalogs, and the view of classic plant biochemists. Proteomics 16:866–876

Schubert OT, Rost HL, Collins BC, Rosenberg G, Aebersold R (2017) Quantitative proteomics: challenges and opportunities in basic and applied research. Nat Protoc 12:1289–1294

Sghaier-Hammami B, Redondo-López I, Valero-Galván J, Jorrín Novo JV (2016) Protein profile of cotyledon, tegument, and embryonic axis of mature acorns from a non-orthodox plant species: *Quercus ilex*. Planta 243:369–396

Thiellement H, Bahrman N, Damerval C, Plomion C, Rossignol M, Santoni V, de Vienne D, Zivy M (1999) Proteomics for genetic and physiological studies in plants. Electrophoresis 20:2013–2026

Unlu M, Morgan ME, Minden JS (1997) Difference gel electrophoresis: a single gel method for detecting changes in protein extracts. Electrophoresis 18:2071–2077

Valledor L, Jorrín-Novo JV (2011) Back to the basics: maximizing the information obtained by quantitative two dimensional gel electrophoresis analyses by an appropriate experimental design and statistical analyses. J Proteomics 74:1–18

Valledor L, Romero-Rodríguez MA, Jorrín-Novo JV (2014) Standardization of data processing and statistical analysis in comparative plant proteomics experiment. Methods Mol Biol 1072:51–56

Vesterberg O (1989) History of electrophoretic methods. J Chromatogr 480:3–19

Chapter 20
Metabolomics and Metabolic Profiling: Investigation of Dynamic Plant-Environment Interactions at the Functional Level

Dominik Skoneczny, Paul A. Weston, and Leslie A. Weston

1 Introduction

Sessile plants routinely face challenges associated with environmental extremes or neighbouring competitors, and have therefore developed mechanisms that allow them to withstand constant exposure to these diverse abiotic and biotic stressors. In some cases, the response to plant stress can be manifested on demand by the plant (so-called inducible responses), while other responses are expressed constitutively and are available at all times to counter the stressor. Thus, it can be said that environment shapes a plant's physiology and, in turn, also impacts the functioning of ecosystems (Wittstock and Gershenzon 2002; Arbona et al. 2013; Weston et al. 2015). Interactions between plants, their competitors, and the environment are always dynamic and as a result often difficult to characterize. It is therefore not surprising that recent studies of such complex interactions have utilised a multitude of advanced techniques for experimentation, and have eventually led to an enhanced understanding of the physiological basis for these interactions.

Plant response to environmental stress or stimuli frequently occurs rapidly, sometimes within seconds to minutes, by triggering biochemical pathways that can be measured in this brief time frame. Such rapid and specific changes *in planta* can be challenging to detect and quantify (Ye et al. 2013; Schuman and Baldwin 2016). Other environmental stimuli trigger broader response patterns causing secondary changes *in planta* resulting in, for example, differential resource allocation. Such stress responses can be monitored at the gene level by studying the plant's transcriptome as assessed by gene expression. Following transcription, translation of gene

D. Skoneczny · P. A. Weston · L. A. Weston (✉)
Graham Centre for Agricultural Innovation, Charles Sturt University, Wagga Wagga, NSW, Australia

School of Agricultural and Wine Sciences, Charles Sturt University, Wagga Wagga, NSW, Australia
e-mail: leweston@csu.edu.au

© Springer International Publishing AG, part of Springer Nature 2018
A. M. Sánchez-Moreiras, M. J. Reigosa (eds.), *Advances in Plant Ecophysiology Techniques*, https://doi.org/10.1007/978-3-319-93233-0_20

products into functional or non-functional proteins occurs, producing what is referred to collectively as the plant's proteome. However, regulation of most environmental interactions in the plant system occurs at the metabolite level. Metabolites clearly play a major role in elicitation of plant response to the environment and can act as mediators, signal molecules, defence metabolites or regulators of certain cellular functions. This complete and functional set of metabolites within a living plant is referred to as the plant metabolome (Rochfort 2005; Weston et al. 2015).

Plants are generally successful in combating or adapting to moderate levels of environmental stress, herbivory and pathogen infestation (Akula and Ravishankar 2011; Ryalls et al. 2016). Stimuli known to trigger well-coordinated defence responses include, but are not limited to, temperature, photoperiod, drought, as well as herbivores, pathogens or neighbouring plant species (Lee 2002; Franceschi et al. 2005). Following herbivory, some plants exhibit increased production of defence metabolites that deter the specific herbivore. These host-specific responses require recognition of the herbivore at the chemical level and the ability to rapidly produce defence metabolites (Schuman and Baldwin 2016). Such complex biochemical interactions can now be studied over time both qualitatively and quantitatively using an analytical approach known as metabolomics.

When performed in a non-targeted manner, metabolomics refers to the systematic study of all metabolites in a living organism; the plant's metabolome consists of both primary and secondary metabolites. In the plant, metabolomics can be performed at various organismal levels including study of a single cell, a tissue, an organ or the entire organism. In addition, it can be performed on matrices associated with plants, including soil from the rhizosphere containing water-soluble metabolites, or the headspace surrounding a plant containing plant-produced volatiles (Dunn and Ellis 2005; Rochfort 2005; Weston et al. 2015). Metabolomics is often performed to monitor metabolite flux over time and aims to quantify those metabolites associated with a particular biochemical response at a specific point in time (Kim and Verpoorte 2010).

Although metabolomics is performed using multiple techniques, the steps in a metabolic profiling analysis should be performed with the highest level of care and precision to successfully assess variation in metabolite concentrations and presence over time, and also to detect those less stable metabolites often present in trace quantities. Since the biochemistry of living plants is dynamic and occurs in association with other living organisms including the plant's living microbiome, sample preparation for metabolomics must first involve the termination or quenching of all biochemical processes in the system under study, followed later by extraction, separation and detection of all key metabolites present. Once detection has been performed successfully, the complex data set is then analysed using selected software packages and chemometrics approaches.

Analytical techniques commonly used in metabolomics studies generally involve liquid or volatile samples prepared for separation using liquid or gas chromatography (LC or GC, respectively) coupled to various types of mass spectrometry (MS) (Roessner and Bacic 2009; Weston et al. 2015). In addition, nuclear magnetic resonance (NMR) spectroscopy instrumentation can also be used for detection

(Smolinska et al. 2012). Recently, MALDI TOF/MS (matrix assisted laser desorption and ionisation coupled to time of flight mass spectrometry) has been used directly on solid samples for intensive protein and fat metabolomics analyses (Fuchs et al. 2010).

This chapter explores the most common applications of metabolomics in plant ecophysiology research by providing an overview of typical instrumentation and workflows used by plant scientists along with a discussion of the experimental outcomes of such studies. It is important to remember that this field is rapidly advancing and new technological improvements to equipment, techniques and data processing occur on a yearly basis.

Metabolomics requires the knowledge of separation science and analytical techniques, thus scientists who do not specialise in these areas can find suitable collaborators in well-equipped regional or national research hubs. A deeper understanding of organic chemistry, biochemistry, chromatography, bioinformatics and statistics will allow for successful experimentation and data analysis.

2 Metabolomics Tools

A plant's metabolome typically consists of upwards of thousands of metabolites that vary with species and cultivar as well as with phenology in both composition and abundance. Due to enormous chemical diversity present in a typical plant extract, no single analytical method can detect all metabolites present. As a result, metabolomics is generally performed using a variety of platforms. In contrast, analysis of a subset of metabolites (metabolic profiling) can be accomplished with a single platform, which results in reduced number and diversity of metabolites detected in a sample, allowing for a more simplified analysis (Hill and Roessner 2014; Weston et al. 2015).

Integrated platforms, with a diversity of separation techniques and instruments, have frequently been used to provide the most comprehensive analysis of metabolomes. Chromatography [LC, GC, or CE (capillary electrophoresis)] coupled with MS or NMR analysis of the entire sample (Table 20.1) are the most common high-throughput techniques employed to study a suite of low molecular weight (<1500 Da) constituents. LC/MS is frequently used in the analysis of compounds ranging in polarity, whereas GC/MS is most commonly employed for the analysis of volatile and derivatized metabolites.

LC or GC/CE/MS usually require extensive sample preparation including filtration in contrast to NMR/MS, which can be carried out with minimal sample preparation. NMR is typically only used for the analysis of mixtures of metabolites in high abundance due to its limited sensitivity (Zhang et al. 2012). NMR can also be used in some cases for detection of compounds that are labile or less stable, in contrast to typical GC or LC techniques (Smolinska et al. 2012).

Sample preparation for LC or GC/CE/MS is highly specific. Samples must conform to meet the requirements of each instrument employed for metabolomics

Table 20.1 Comparison and characteristics of mass spectrometry (MS) and nuclear magnetic resonance (NMR) spectroscopy for metabolomics applications (Pan and Raftery 2007; El-Aneed et al. 2009; Yan et al. 2017)

Platform	Characteristics	Considerations
Mass spectrometry (MS)	Sensitive (LOD 0.5 nM)	Sample preparation and separation techniques are based on polarity of constituents
	Sample preparation required (purification)	
	Challenging quantification	Samples need to be derivatized and/or prepared in a solvent
	Limited structural information	
	Time of analysis depends on separation technique (15–40 min/sample)	High throughput but slower than NMR
	Destructive to samples	Untargeted analysis of large numbers of metabolites at low concentrations
	Detection of >500 metabolites per run	
	Small sample volume 1–100 μL	
	Less expensive instrumentation	Targeted analysis with prior knowledge of compounds of interest or analytical standard
	Matrix-Assisted Laser Desorption Ionization (MALDI) MS allows for analysis of sample co-crystalized with a solid matrix	MALDI can be used in analysis of solids and large biomolecules (up to 500,000 Da, including proteins and DNA). In metabolomics typically used to study lipids and glycoconjugates
Nuclear magnetic resonance spectroscopy (NMR)	Less sensitive than MS (LOD 0.5 μM)	Broad-based analyses
	Limited sample pre-preparation	Samples in various solvent matrices
	Quantification easy and precise	
	Limited or no sample preparation needed	High throughput technique for metabolites in high concentration
	Rich structural information	
	Rapid analysis (2–3 min/sample)	Useful for non-targeted analysis of metabolites present in high abundance
	Non-destructive analysis of precious samples	
	Detection of 40–200 metabolites per run	
	High reproducibility	
	Large sample volume required (0.1–0.5 mL)	
	Higher cost of equipment	

analyses (e.g. high acid content can damage HPLC systems; unfiltered samples can block HPLC columns) and should be suitable for the separation technique selected (e.g. GC samples may require derivatization if analysis is performed on less volatile mixtures of constituents).

Table 20.2 Comparison of separation techniques coupled to mass spectrometers

	Gas chromatography	Liquid chromatography	Capillary electrophoresis
Sample preparation	Extensive sample preparation	Sample purification required	Minimum sample preparation
	Derivatization of non-volatile compounds	Sample volume 1–15 μL	Sample volume 1–20 nL
	Sample volume ~1 μL	Sample in solvent mixture	Sample in aqueous phase
	Sample in solvent mixture/inert gas		
Type of analytes	Volatiles and compounds that volatilise after derivatization	Liquid samples (solubilized solids) compatible with mobile phase	Ionic and very polar metabolites
			Small sample volume
			Faster than GC/LC
	Thermally stabile compounds	Polar and non-polar compounds	Lower resolution than normal-phase LC but generally results in detection of different metabolites
	Non-polar low MW compounds	Mainly secondary metabolites	
	Mainly primary but some secondary metabolites associated with fragrance and flavour		
Application	Availability of multiple libraries, rapid identification, high reproducibility	Lower reproducibility, lack of large reproducible spectral databases	Lower reproducibility, lack of large reproducible spectral databases
	Carbohydrates, amino acids, organic acids, sugars, oils, terpenoids	Phenolics, alkaloids, glucosinolates, terpenoids, etc.	

In general, all samples should be relatively free from salts, residues, proteins and other contaminants that might cause interference during the analysis. Once prepared, certain liquid or gaseous samples can be injected directly into the MS interface or analysed using NMR; however, most complex sample matrices will need to be separated by a chromatographic method prior to analysis. The separation technique selected (Table 20.2) depends on various physical and chemical properties of analytes under study. Solids can be directly profiled using certain LC/MS interfaces and also by matrix-assisted laser desorption ionization (MALDI) MS. During MALDI/MS, samples are dissolved in a solution or frozen and applied onto a solid matrix and dried to form crystals that are bombarded with a laser beam to allow ion formation using matrix as a mediator (Fuchs et al. 2010). Additionally, desorption electrospray ionization (DESI) mass spectrometry allows for profiling of metabolites directly from solid surfaces. This ionization technique allows for profiling of metabolites in situ and creates a metabolic profile as an image of the solid under study (Claude et al. 2017).

2.1 Mass Spectrometry

Mass spectrometry is currently the most widely applied microanalytical technique in metabolomics as it provides high sensitivity and resolution while allowing for analysis of a wide range of metabolites (Table 20.1). Mass spectrometry also allows for quantitative and qualitative analyses as it has the capacity to generate structural information. It is generally highly reproducible for quantitative analysis and allows for a simultaneous determination of the relative abundance of hundreds of metabolites in a single run. Ideally it is linked to a chromatographic separation prior to analysis but direct-injection MS can also be used for some applications as well as for profiling of solids (Dunn and Ellis 2005; Zhang et al. 2012; Weston et al. 2015).

For all MS analyses, analytes must be ionized and present in the gaseous state in order to be detected. When performing MS, movement of ions within the ionisation chamber is regulated by modification of electromagnetic fields. Analytes are ionized in the MS interface using a variety of ionization sources including electrospray ionisation (ESI) or atmospheric-pressure chemical ionization (APCI) before introducing them into the *m/z* analyser (Fig. 20.1). To ensure free movement and lack of contamination, the MS is always operated under high vacuum. The MS platform generally consists of one (single MS) or two mass analysers (tandem MS or MS-MS), with the latter able to provide additional structural information and therefore more precise quantification.

In an MS-MS experiment, additional ion pre-selection and collision-induced disassociation (CID) take place for clear determination of the fragmentation pattern of a selected ion. The resulting mass spectra display abundance of fragments covering a range of mass-to-charge ratios *(m/z),* which can be used to infer the molecular structure of metabolites. Neutral molecules and molecules that have not ionized are generally not detected in the MS (Watson and Sparkman 2007).

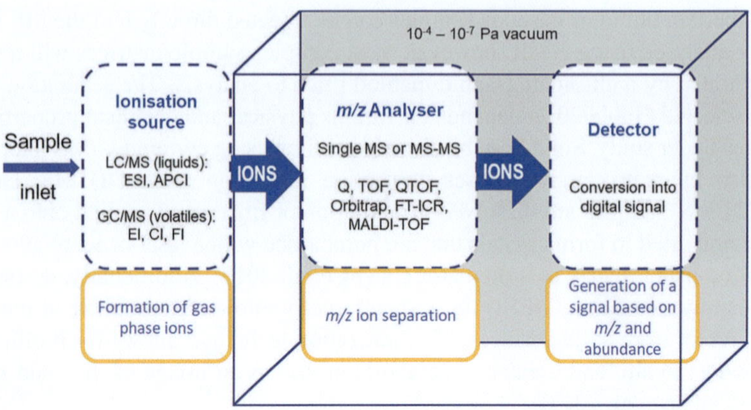

Fig. 20.1 General and simplified representation of a mass spectrometer platform typically used for metabolomics and metabolic profiling analysis. (Schematic representation based on Watson and Sparkman 2007)

2.2 Nuclear Magnetic Resonance

NMR spectroscopy is becoming of increasing importance in the field of metabolomics as sensitivity of NMR instrumentation increases. This analytical method is well suited for untargeted metabolomic analysis because it is non-selective and can often provide highly reproducible results and structural information, depending on concentration of analytes in the sample matrix. In contrast to LC/MS, hundreds of low molecular weight metabolites can be detected simultaneously in a single run with very limited sample preparation. NMR can also be used in vivo with samples without any sample preparation, such as a newly poured sample of wine studied for composition. NMR has been employed for metabolite fingerprinting as well as metabolic profiling, and since it is non-destructive, sample recovery is possible following analysis. Workflows for NMR are typically automated (Smolinska et al. 2012; Zhang et al. 2012), permitting the method to be used for larger sample sets depending on the platform; in some cases, up to 500 samples can be processed per day. The major drawback of NMR is its overall low sensitivity, limiting its utility to samples where profiling compounds of higher abundance is desirable (Pan and Raftery 2007).

During NMR experimentation, magnetic fields are applied to samples and nuclei of atoms with an odd atomic number such as ^{1}H or ^{13}C gain what is referred to as nuclear spin. At the same time, applied radio frequencies allow nuclei to reach high-spin energies, and the radiation generated during the relaxation of the magnetic field is then detected and compared to that of reference atoms. Commonly, ^{1}H or proton NMR is used in metabolomics studies as the majority of metabolites contain ^{1}H atoms or protons (Dunn and Ellis 2005).

2.3 Application of NMR and MS for Metabolomics Approaches for the Study of Plant Response to Stress

In a study using *Arabidopsis* as a model system, Jänkänpää et al. (2012) studied the impact of light intensity on dynamic changes in the *A. thaliana* metabolome using GC/QTOF MS. Results indicated that more than 70 compounds fluctuated in concentration with varying light intensity when plants were grown under controlled conditions. Over 30 metabolites, including amino acids, lipids and carbohydrates, were later identified to be responsive to light intensity. When plants were moved from controlled environment growth chambers (artificial light) to natural light conditions in the field, changes in the leaf metabolome were observed within a 3-day period, and pronounced shifts in metabolism were noted within 4 h following transition to the field.

A metabolomics approach was also used to evaluate the nutritional value of a fresh vegetable product (Maldini et al. 2015) under varying levels of light. This study used LC ion mobility-QTOF to investigate changes in the metabolome of broccoli (*Brassica oleracea*) sprouts under alternating light/dark conditions. In this study, increased light intensity was associated with increased chlorophyll biosynthesis as well as elevated levels of phytosterols, lipids, fatty acids and carotenoids.

Metabolomics can also prove useful for assessment of pharmaceutics and plant-produced medicinals. For example, alkaloid biosynthesis in the commercial poppy plant was assessed using of FT-ICR MS (Fourier-transform ion cyclotron resonance MS) in cell cultures. Metabolic differences between elicitor-treated and control cell cultures were observed, and targeted analysis of benzylisoquinoline alkaloid biosynthesis in opium poppy (*Papaver somniferum*) was performed with a focus on morphine, codeine and sanguinarine production. Approximately 1000 metabolites were annotated in the study, including compounds previously uncharacterized. ^{1}H NMR was also performed for comparative purposes and resulted in annotation of a greater number of metabolites in contrast to LC/MS QToF methods employed. Following exposure to an elicitor, changes in metabolism of cell suspension cultures were noted within hours following treatment (Hagel and Facchini 2008).

The impact of drought and climate change on the shoot and root metabolism in two invasive perennial grasses, *Holcus lanatus* (Yorkshire fog) and *Alopecurus pratensis* (meadow foxtail), were evaluated using LC/MS QToF with the LTQ Orbitrap XL and ^{1}H NMR. Over 850 metabolites were detected; however only 55 metabolites were positively identified. Shoot extracts were more chemically complex than root extracts, a typical observation in plant metabolomics. Simultaneous exposure to high temperature and drought resulted in differential responses in comparison to exposure to individual stressors, suggesting that the plant response to stress is complex and is mediated by multiple biochemical pathways (Gargallo-Garriga et al. 2015).

3 Metabolomics Pipeline: From Harvest to Data Analysis

As suggested previously, metabolomics workflows (the steps involved in processing and analysing samples and their resulting data files as part of a metabolomics analysis) are dependent on sample type, analytical instrumentation and desired outputs. However, metabolomics experiments should be designed to minimize sources of external variation while achieving sample-to-sample uniformity. When considering metabolomics and metabolic profiling analyses, four main steps—sample preparation, separation, detection and chemometrics analysis—are generally undertaken (Fig. 20.2). These approaches are explored in greater detail in Kim and Verpoorte (2010) and Hill and Roessner (2014).

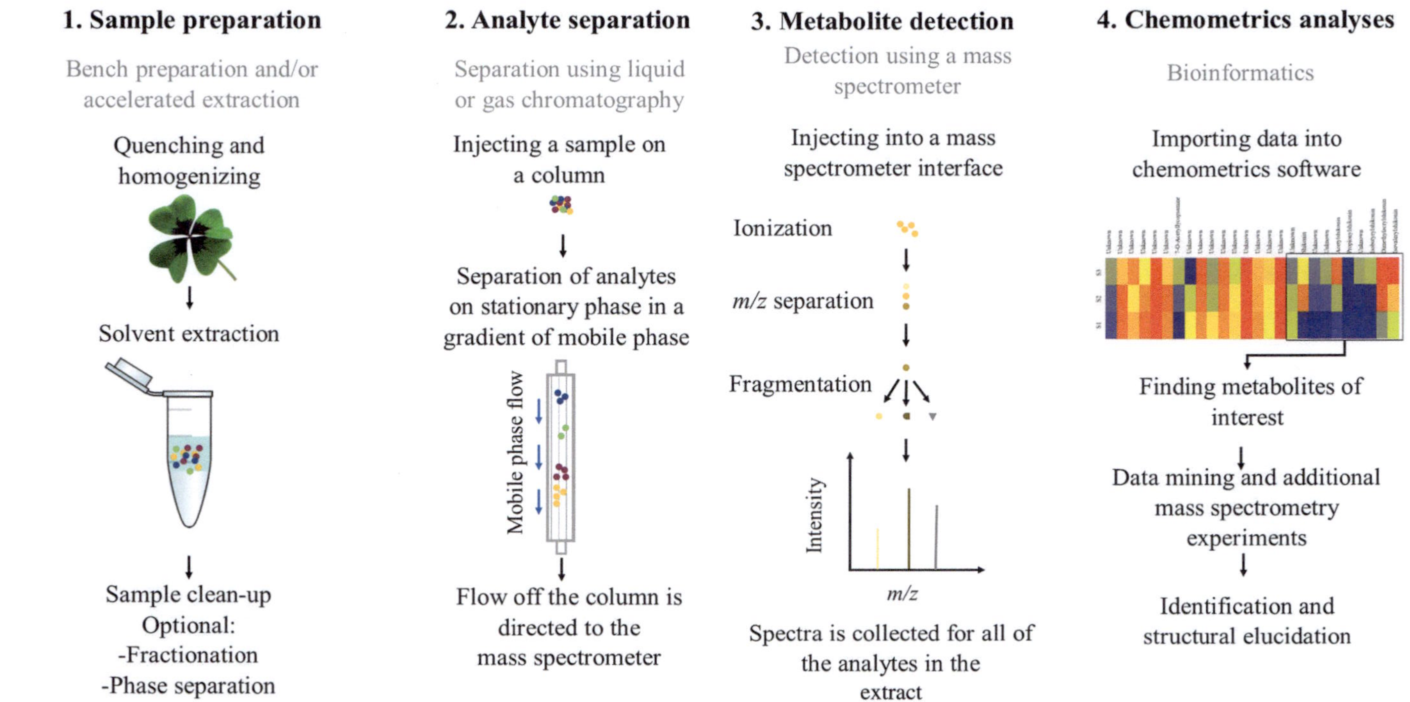

Fig. 20.2 Metabolomics pipeline used for identification and profiling of differentially regulated compounds. Although every step in this process can be modified, the general workflow remains the same

3.1 Sample Preparation

3.1.1 Harvest

It should be noted that plant-based experiments often require sampling at multiple time points as metabolite levels fluctuate due to circadian rhythms in higher plants, as well as with changes in plant phenology. The impacts of biotic and abiotic stress exposure on plants can therefore add complexity to a standard time-course experiment. Plant harvest should be performed similarly for all samples and processing of plant tissues should occur rapidly following harvest. Most protocols involve placing plants on ice, in an −80 °C freezer or freeze-drying tissues directly using liquid nitrogen to prevent biochemical changes post-harvest (Hill and Roessner 2014; Kim and Verpoorte 2010; Weston et al. 2015).

3.1.2 Quenching and Homogenizing

To conserve the sample and quench active biochemical processes in living tissues, plant tissue may be snap-frozen in liquid nitrogen and then stored at −80 °C until extraction. Several studies indicate the importance of drying using freeze-drying and homogenizing prior to extraction which can be performed using mortar and pestle or cryogenic mill (Hill and Roessner 2014). Protocols must be optimized for tissue type and/or compounds of interest in metabolic profiling to ensure stability of thermolabile and/or light-sensitive metabolites. In some cases, only fresh tissue should be processed immediately after harvest as freeze-drying can result in degradation of certain natural products such as porphyrins. For more stable metabolites, tissues can be processed to dryness at room temperature followed by grinding before extraction.

3.1.3 Extraction

Extraction is a critical step in sample preparation and varies with the choice of solvent, temperature and duration. Solvent extraction is most commonly employed but microwave-assisted extraction and supercritical fluid extraction (Kim and Verpoorte 2010) can also be used. The process of solvent extraction can be automated to increase uniformity and speed for high throughput (e.g. using BUCHI Speed Extractor) (Skoneczny et al. 2015; Weston et al. 2015).

3.1.4 Final Sample Preparation

Sample clean-up is the final step prior to analysis, and ensures compatibility of the sample matrix with selected instrumentation. For most NMR instruments, filtration and appropriate solvent choice will ensure sample homogeneity. Gas chromatography for non-volatile compounds such as amino acids, organic acids, fatty acids and sugars including sugar derivatives typically requires derivatization prior to chromatography to ensure volatility for separation. For liquid samples separated using liquid chromatography, samples are often prepared prior to separation by performing protein precipitation (Kim and Verpoorte 2010).

3.2 Separation of Analytes

Although NMR, MALDI/MS and direct infusion MS do not require separation preceding the analysis, most metabolic profiling experimentation involves a separation step for accurate identification and profiling. Analytes can be further separated using LC, GC or CE as summarized in Table 20.2 and previously described by numerous authors (Hill and Roessner 2014; Ramautar and de Jong 2014; Weston et al. 2015; García et al. 2017; Yan et al. 2017)

3.3 Metabolite Detection

Analytes are detected according to properties of the instrumentation used; thus, accuracy and resolution are instrument dependent. In general, QToF mass spectrometers used in metabolomics studies provide accurate mass (MS) and fragmentation pattern (MS-MS) for each molecule under study, which can later be compared to information available for known compounds in chemical libraries such as NIST or METLIN. In contrast to results obtained with LC and CE/MS, GC/MS fragmentation patterning is more robust and consistent for various instruments and therefore identification of constituents is generally more reliable.

3.4 Non-targeted Metabolomics

Non-targeted approaches aim to evaluate the presence and abundance of as many metabolites as possible in an often complex biological matrix or system. Unfortunately, many plant metabolites are uncharacterized which complicates identification of a large number of plant metabolites. In general, most holistic studies performed with plant extracts or tissues are not able to identify the majority of metabolites contained within samples and instead focus on metabolites that are significantly up- or down-regulated between treatment groups.

Various mass spectrometers and their application for use in non-targeted metabolomics and metabolic profiling studies have been previously described (Dunn and Ellis 2005; Watson and Sparkman 2007; El-Aneed et al. 2009; Viant and Sommer 2013). The most commonly used instrumentation for non-targeted studies include single and tandem MS along with quadrupole time-of flight (QTOF), time-of-flight

(TOF), Fourier-transformed ion cyclotron resonance (FT-ICR) and linear ion trap or orbitrap mass spectrometers. These platforms generally offer high mass accuracy and resolution required for metabolite identification and, more importantly, fast data acquisition and multiple fragmentation options required for complex metabolomics experiments.

Analytical standards are typically employed for confirmation of identification of metabolites. Existing databases can also be employed for annotation or tentative identification of metabolites in cases where standards are not available. Annotation should be performed according to minimum reporting standards such as those presented by the Metabolomics Standards Initiative (Sumner et al. 2007).

3.5 Data Analysis

Data is analysed following acquisition in situ using software packages that are provided by instrument vendors or are available on-line. Some useful software packages are available free of charge. Most of the data analysis workflows include data pre-processing, data pre-treatment and univariate and multivariate statistical analysis (Martínez-Arranz et al. 2015). Depending on the software, data obtained from samples separated via LC is typically deconvoluted; major peaks detected, integrated and aligned; and within-batch and between-batch normalization and baseline correction performed. Several software packages also allow complex metadata analyses across multiple datasets (Hill and Roessner 2014).

Large datasets from metabolomics studies generally require multivariate data analysis techniques, specifically the simultaneous analysis of more than one parameter or variable. Unsupervised classification methods such as principal component analysis (PCA), hierarchal cluster analysis (HCA) and supervised methods based on partial least square (PLS) regressions common when analysing metabolomics data. Such methods allow for identification of metabolites that are significantly up- or down-regulated between treatments, and may include comparisons among various cultivars, treatments or stressors. Selected metabolites are then compared to metabolites documented in databases and libraries, some of which are publicly available e.g. KEGG, METLIN, or NIST, or in-house libraries. The METLIN library as of 2018 currently contains over 100,000 metabolite entries, and most have a complete mass spectrum for referencing [31.10.2017] (Hill and Roessner 2014).

Final structural confirmation of investigated metabolites requires more detailed MS/MS experimentation on existing datasets and, ideally, comparison to analytical standards.

3.6 Bioassay-Driven Data Analysis

Results generated from chemometric analysis of metadatasets are often difficult to interpret. For those studying biological activity or toxicity, it is critical to employ bioassay-driven identification of key constituents associated with activity. In this case, one may use a statistical method to establish the relationship between abundance of molecular features and bioactivity, which is powerful and offers greater insights into biological activity of metabolic features. Such evaluations are now being utilised in advanced metabolomics and pathways analyses, but are currently not frequently undertaken due to the complexity of these analyses. However, in our opinion, these bioassay-driven associations have proven invaluable for discovery of families of compounds and biomarkers associated with biological activity.

3.7 Determination of the Metabolites

Identification of specific metabolites associated with bioactivity can often be accomplished using a number of multivariate statistical techniques. Partial least squares (PLS) regression is one technique suitable for the task of associating a large number of independent variables (in this case, metabolites) with bioactivity. The output from PLS regression, however, may be difficult to interpret because the relationship between all of the molecular features and bioactivity is revealed. Those entities that have a high degree of association with activity and are present in reasonable abundance are of greatest interest; however, the percentage of variance in activity that is associated with each of these entities is also of interest.

Stepwise linear regression can also be employed to deduce metabolite association with activity. Rather than trying to fit the abundance of all metabolites in the dataset with bioactivity, stepwise linear regression measures the degree of association between each metabolite and bioactivity, and incrementally adds metabolites with the highest degree of association to a predictive model. The output of this technique is a series of models that progressively account for much of the variation between metabolite abundance and bioactivity. In addition, the output indicates the degree to which variation is explained by a particular set of metabolites (r^2) and the statistical significance of the regression. Ideally, only a limited number of metabolites for a given study will result in a regression with a suitably high value of r^2.

4 Case Studies and Emergining Approaches in Plant Metabolomics

Rapid development of new strategies and platforms in metabolomics has recently triggered more insightful and novel research into plant metabolism. Below we present three case studies featuring various platforms to study plant metabolites produced in plants under stress, or those plants experiencing interactions with insects or other plants during plant/insect and plant/plant interactions. In addition, we have summarised a number of recent studies that exemplify a range of approaches to the study of plant metabolomics in Table 20.3.

Table 20.3 Recent research projects exemplifying current approaches to the study of plant metabolomics

Aim of study	Instrumentation	Citation
Comparison of suppressive and allelopathic metabolites in several wheat cultivars for selection of weed suppressive cultivars. Identification of 14 benzoxazinoids in plant tissues and after exudation into the rhizosphere	Targeted analysis using Applied Biosystems Q Trap LC/MS (Nærum, Denmark) and LC-QToF	Mwendwa et al. (2016)
Metabolomics of root exudates that foster the dialogue between belowground herbivores, nematodes, and microbial communities and other plants. Identification of up to 103 metabolites in root exudates. Understanding their role and degradation processes	Untargeted analysis of metabolites in root exudates using a variety of platforms including NMR and GC/LC-ToF	van Dam and Bouwmeester (2016) and Zhu et al. (2016)
Mass spectrometry imaging in plant tissue. Precise profiling of metabolites *in situ* with limited sample preparation. Mapping of the distribution of metabolites in cells or even organelles	Matrix assisted laser desorption ionisation (MALDI), desorption electrospray ionisation (DESI) and secondary ion mass spectrometry	Heyman and Dubery (2016)
Estimation and quantification of microbial metabolites in soils. Understanding their cycling in water-extractable organic matter fraction and decomposition	Untargeted analysis using Agilent GC/MS system (California, USA)	Swenson et al. (2015)
Metabolic profiling as a tool allowing for characterization of plant genotypes. This phenotyping approach can be applied to better understand genetically modified plant systems. Authors compared sucrose metabolism across four cultivars of potatoes tuber	Targeted GC/MS (Thermo-Quest, Manchester, UK)	Roessner et al. (2001)

4.1 Case study 1: Untargeted Metabolomics to Study Plant Response to Infestation by Weaver Ants Using GC/MS

4.1.1 Background

An interesting mutualistic interaction between plants and ants has been explored using a non-targeted metabolomics approach. It was previously noted that plants can provide nutrition or shelter for ants while ants aggressively defend the plant (Metlen et al. 2009).

More detailed insight into this mutualistic interaction was obtained from a study of weaver ants, *Oecophylla smaragdina,* and their association with cultivated coffee, *Coffea arabica*. Vidkjær et al. (2015) measured the nutritive effects of ant faeces on coffee plants; plants that hosted a colony of weaver ants were compared to control plants under glasshouse-controlled conditions. Previous reports had indicated that weaver ants could improve plant health and crop yield. Considering that some plants are able to take up nitrogen from urea through the leaves, the authors hypothesised that metabolic changes in *Coffea arabica* would occur due to nutrient transfer following infestation by ants.

4.1.2 Methodology

Leaves from control and treated plants were collected every 2 weeks and snap-frozen in liquid nitrogen (quenching) and stored at −80 °C until extraction (stable storage).

Samples were then extracted by shaking crushed leaves in a mixture of solvents. Dried extracts were derivatized within 24 h prior to GC analysis. Samples were analysed in duplicate in randomized order using gas chromatography (Agilent 7890A, Santa Clara, CA) coupled to Leco Pegasus HT 4010 time-of-flight mass spectrometer (St. Joseph, MI). In addition, total carbon and nitrogen were analysed.

4.1.3 Data Analysis

GC/MS data was deconvoluted and annotated using BinBase metabolomics database. Univariate and multivariate data analyses were conducted as well as principal component analysis (PCA) and partial least square discriminant analysis (PLS-DA) with rigorous validation of the models (Vidkjær et al. 2015).

4.1.4 Results and Discussion

This research provided insight into metabolic changes in *C. arabica*. Over 500 molecular features were identified, of which 96 were annotated across all samples. Plants that hosted ant colonies were observed to have increased levels of several fatty acids. In addition, an increase in total nitrogen levels was observed in colonized plants, and several nitrogen-containing metabolites originating from the phenylpropanoid biosynthetic pathway were also up-regulated, suggesting exposure to potential biotic stress. Results suggested that increased nitrogen levels were a result of assimilation of additional nitrogen through leaves due to the presence of ants and ant faeces. The up-regulation of the phenylpropanoid biosynthetic pathway in response to exposure to biotic stress may have led to increased production of secondary metabolites providing additional defence against potential predators (Vidkjær et al. 2015).

4.2 *Case study 2: Metabolic Profiling: Identification and Profiling of Naphthoquinones Under Environmental Stress in the Roots of* **Echium plantagineum** *Using LC/MS MS*

4.2.1 Background

Echium plantagineum is an invasive weed species in Australia that produces bioactive allelochemicals known as shikonins (naphthoquinones) in the periderm of living roots. Production of naphthoquinones was greater in field-grown plants collected from hotter and drier climates in Australia (Weston et al. 2013). To further understand the role of naphthoquinones in plant defence and their production in response to environmental factors, a platform for identification and profiling of shikonins was developed (Skoneczny et al. 2017).

4.2.2 Methodology

Echium plantagineum plants were exposed to drought, simulated herbivory and different temperature regimes in controlled condition experimentation. Roots were harvested and peeled periderm was extracted in ethanol. Composite sampling was performed initially to study the chemical diversity present in the root periderm. Rhizosphere soil was also sampled for shikonins using polydimethylsiloxane (PDMS) tubing.

Extracts were analyzed using UPLC (Agilent Infinity 1200) coupled with a QTOF mass detector (Agilent 6530). Negative ionization mode was selected for detection of shikonins in periderm extracts and MS-MS experimentation was performed to observe fragmentation patterns for individual and related shikonins present in the extracts (Fig. 20.3). The optimized method with tentatively identified nine compounds was then applied to samples collected from controlled condition experiments.

Data was processed using Mass Hunter Software and Mass Profiler Professional chemometrics software (Agilent Technologies, Santa Clara, CA, USA), (Skoneczny et al. 2017).

4.2.3 Results and Discussion

Nine red-pigmented naphthoquinones, also known as shikonins, were tentatively identified based on comparison of their MS-MS spectra to spectra of known standards and others previously reported (Fig. 20.3). Additionally, polymeric shikonins were detected but due to their structural complexity, complete elucidation of all structures was not performed (Skoneczny et al. 2017). Three shikonins were

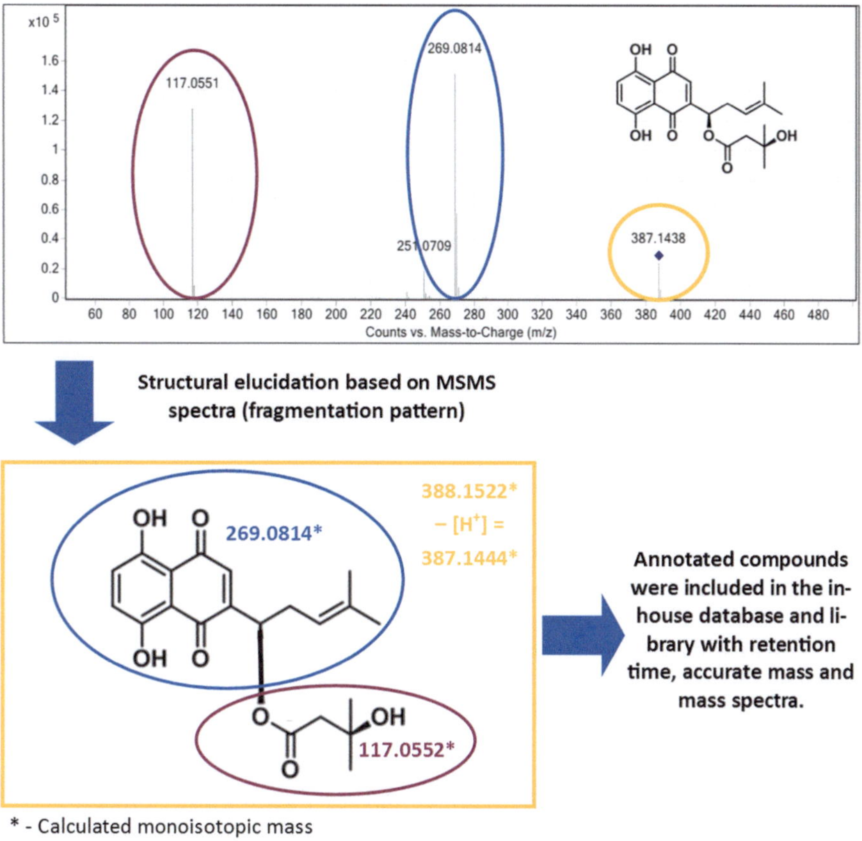

Fig. 20.3 Identification of metabolites by the analysis of the MS-MS spectra. In the case of shikonins, several analytical standards were obtained and permitted a comparison of generated spectra. Identified compounds mainly differed in side chain structure (purple) but all produced a characteristic 269.0814 ion, also present in analytical reference standards

identified from the soil extract suggesting that shikonins were exuded from living roots into the rhizosphere (Zhu et al. 2016).

Shikonins were also profiled in stress-treated plants. Up-regulation of the shikonin biosynthetic pathway was observed within 6 h after plant exposure to simulated herbivory and within days following exposure to increased temperature and water withholding. Temperature clearly impacted production of total shikonins under controlled environment conditions; production of the individual metabolite shikonin, a precursor to other shikonin metabolites, was significantly elevated following exposure to high temperature growth regimes (Fig. 20.4).

High abundance of shikonins was found to be correlated with red pigmentation of periderm and its extracts. Metabolites potentially associated with the shikonin biosynthetic pathway contributed to the clustering of differently coloured root

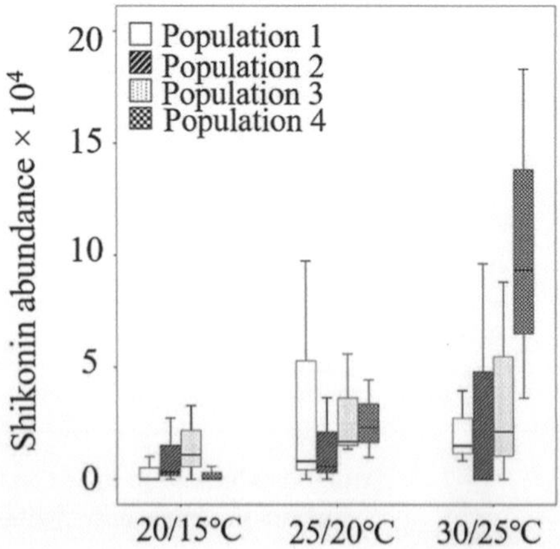

Fig. 20.4 Relative abundance of the individual metabolite, shikonin, in four *Echium plantagineum* populations grown in controlled conditions in different temperature regimes, measured 3 weeks after the beginning of the treatment

extracts in the principal component analysis. This study showed potential changes in secondary metabolism of differently coloured roots (Fig. 20.5).

In general, our studies showed that naphthoquinone production, as monitored through the production of specific shikonins, is rapidly elicited in the roots of healthy plants following exposure to a variety of biotic and abiotic stressors including neighbouring plants, pathogens, herbivory, water withholding and high temperatures.

4.3 *Case study 3: Identification of Thrips: Resistant* Senecio *Hybrids Using NMR*

4.3.1 Background

Thrips (*Frankliniella occidentalis*) are a global insect pest of food crops; feeding by thrips leads to reduced growth and yields. However, some plants are resistant to thrips due to production of toxic metabolites in their foliage. A study by Leiss et al. (2009) evaluated *Senecio* species F2 hybrids (susceptible and resistant) for their metabolite composition using an untargeted metabolomics approach.

4.3.2 Methodology

Over 30 different F2 hybrids of *S. jacobaea* and *S. aquaticus* were grown in tissue culture and then evaluated for their resistance to thrips by profiling both young and older leaves of individual plants. Thrips-resistant and -susceptible plants were further evaluated using a metabolomics approach performed with a 600 MHz Bruker DMX-600 NMR spectrometer (Bruker, Karlsruhe, Germany).

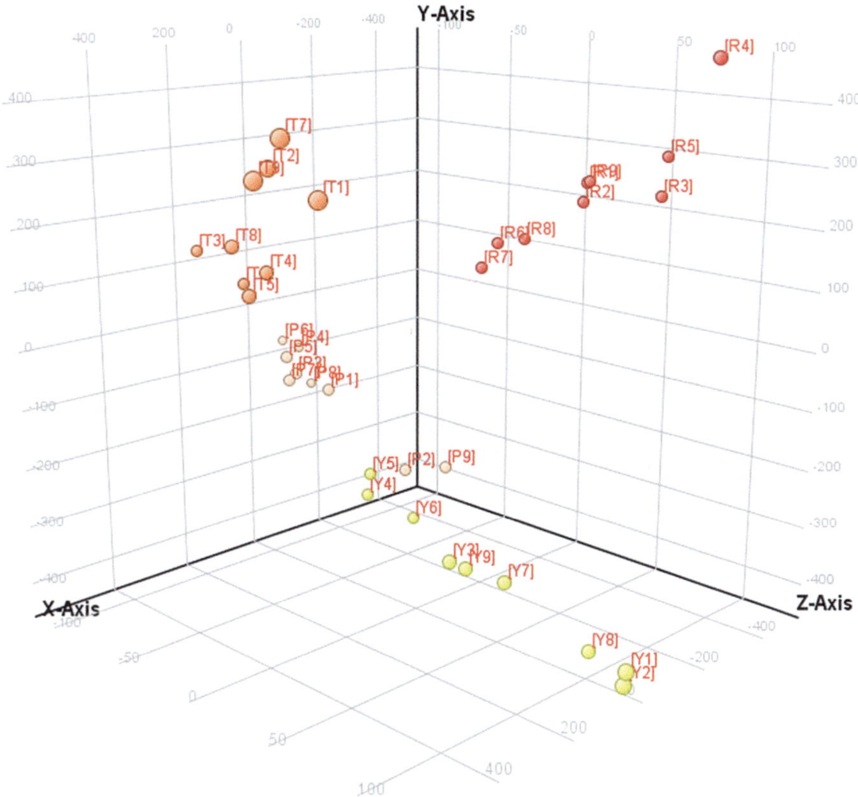

Fig. 20.5 Principal component analysis (PCA) of 36 root periderm extracts of *Echium plantagineum* with varying colouration from pale to red due to the variable presence of a group of naphthoquinones collectively referred to as the shikonins. A total of 162 entities (possible metabolites) were used in the analysis and contributed to the separation of coloured extracts as shown in the PCA plot above. This analysis clearly revealed that differential colouration of the samples is associated with variation in the shikonin biosynthetic pathway

4.3.3 Results and Discussion

NMR proved useful for the determination of secondary metabolite composition in *Senecio* and allowed the identification of variation in metabolic profiles between thrips-resistant and -susceptible hybrids. Resistant plants accumulated significantly greater quantities of toxic pyrrolizidine alkaloids (jaconine and jacobine *N*-oxide) and kaempferol glucoside. Interestingly, higher concentrations of defence metabolites were present in young leaves, resulting in less thrips damage. These findings are consistent with literature noting greater resistance in young leaves and the presence of constitutively expressed pyrrolizidine alkaloids in both *Senecio* species (Leiss et al. 2009).

5 Metabolomics in Systems Biology and Functional Genomics

Understanding the relationship between a living plant's functional phenotype (in this case the metabolites produced) and genotype is currently a hot topic in plant science. Focused studies are now providing strong evidence for the role of specific genes and their impacts on biological function and regulation of plant response to biotic and abiotic stresses (Bino et al. 2004; Saito and Matsuda 2010). In addition, a greater understanding of plant stress response and cellular physiology can be achieved by integrating transcriptomics, proteomics and metabolomics using various computation approaches. However, a key challenge in performing such studies is to create a unified network of genes, transcripts, proteins and metabolites through the integration and evaluation of sets of "-omics" data (Bunnik and Le Roch 2013).

Genes are now frequently annotated by correlating available transcripts with expressed metabolites (Schauer and Fernie 2006). Plants exposed to various stressors can be studied with respect to gene expression associated with metabolic response. Using this approach, gene function can be further predicted and associated metabolic responses can be elucidated (Saito and Matsuda 2010). Despite the complexity of such systems-based experimentation, global databases have been created which can be used to identify proteome, transcriptome and metabolome responses for specific plants under various environmental conditions (Hagel and Facchini 2008; Saito and Matsuda 2010). Such data sets can be especially useful for evaluation of biosynthetic pathways and metabolite flux.

6 Conclusions

A plant's metabolome is a complex and dynamic compilation of both identified and unknown metabolites. Metabolomics is a process that enables identification and quantification of key plant metabolites using a variety of methods and analytical instrumentation. Recent advances in analytical instrumentation and bioinformatics and the integration of multiple "omics" platforms now allows for advanced characterization of biosynthetic pathways and the study of their regulation. Thus, metabolic profiling and metabolomics approaches can be particularly useful for investigation of plant responses to stress.

Current challenges for those performing plant metabolomics include the paucity of metabolite databases containing identified plant metabolites, streamlining the processing of massive datasets generated in metabolomic studies, and bridging the gaps between laboratories employing various platforms for metabolomics on a global scale. Although metabolomics is of increasing importance in the plant sciences, costs associated with intensive metabolite analyses are often higher than those for other "omics" platforms. However, as we have shown in this review, metabolomics currently offers unparalleled opportunities to investigate the interactions between plants and their environment at a functional level.

Additional Information

The authors have attempted to provide an overview of common techniques, instrumentation and methods used in metabolomics studies in plant ecophysiology experimentation. However, as both metabolomics and chemometric analysis can be performed using a variety of platforms and approaches, not all were able to be described in detail in this review.

References

Akula R, Ravishankar GA (2011) Influence of abiotic stress signals on secondary metabolites in plants. Plant Signal Behav 6:1720–1731

Arbona V, Manzi M, Ollas C, Gómez-Cadenas A (2013) Metabolomics as a tool to investigate abiotic stress tolerance in plants. Int J Mol Sci 14:4885

Bino RJ, Hall RD, Fiehn O, Kopka J, Saito K, Draper J, Nikolau BJ, Mendes P, Roessner-Tunali U, Beale MH, Trethewey RN, Lange BM, Wurtele ES, Sumner LW (2004) Potential of metabolomics as a functional genomics tool. Trends Plant Sci 9:418–425

Bunnik EM, Le Roch KG (2013) An introduction to functional genomics and systems biology. Adv Wound Care 2:490–498

Claude E, Jones EA, Pringle SD (2017) DESI Mass Spectrometry Imaging (MSI). In: Cole LM (ed) Imaging mass spectrometry: methods and protocols. Springer New York, New York, pp 65–75

van Dam NM, Bouwmeester HJ (2016) Metabolomics in the rhizosphere: tapping into belowground chemical communication. Trends Plant Sci 21:256–265

Dunn WB, Ellis DI (2005) Metabolomics: current analytical platforms and methodologies. Trends Anal Chem 24:285–294

El-Aneed A, Cohen A, Banoub J (2009) Mass spectrometry. Review of the basics: electrospray, MALDI, and commonly used mass analyzers. Appl Spectrosc Rev 44:210–230

Franceschi VR, Krokene P, Christiansen E, Krekling T (2005) Anatomical and chemical defenses of conifer bark against bark beetles and other pests. New Phytol 167:353–376

Fuchs B, Süß R, Schiller J (2010) An update of MALDI-TOF mass spectrometry in lipid research. Prog Lipid Res 49:450–475

García A, Godzien J, López-Gonzálvez A, Barbas C (2017) Capillary electrophoresis mass spectrometry as a tool for untargeted metabolomics. Bioanalysis 9:99–130

Gargallo-Garriga A, Sardans J, Pérez-Trujillo M, Oravec M, Urban O, Jentsch A, Kreyling J, Beierkuhnlein C, Parella T, Peñuelas J (2015) Warming differentially influences the effects of drought on stoichiometry and metabolomics in shoots and roots. New Phytol 207:591–603

Hagel J, Facchini P (2008) Plant metabolomics: analytical platforms and integration with functional genomics. Phytochem Rev 7:479–497

Heyman HM, Dubery IA (2016) The potential of mass spectrometry imaging in plant metabolomics: a review. Phytochem Rev 15:297–316

Hill CB, Roessner U (2014) Advances in high-throughput untargeted LC-MS analysis for plant metabolomics. Future Science eBook Series "Advanced LC-MS applicafions for metabolomics". Future Science Group, London

Jänkänpää HJ, Mishra Y, Schröder WP, Jansson S (2012) Metabolic profiling reveals metabolic shifts in Arabidopsis plants grown under different light conditions. Plant Cell Environ 35:1824–1836

Kim HK, Verpoorte R (2010) Sample preparation for plant metabolomics. Phytochem Anal 21:4–13

Lee CE (2002) Evolutionary genetics of invasive species. Trends Ecol Evol 17:386–391

Leiss K, Choi Y, Abdel-Farid I, Verpoorte R, Klinkhamer PL (2009) NMR metabolomics of thrips (*Frankliniella occidentalis*) resistance in *Senecio* hybrids. J Chem Ecol 35:219–229

Maldini M, Natella F, Baima S, Morelli G, Scaccini C, Langridge J, Astarita G (2015) Untargeted metabolomics reveals predominant alterations in lipid metabolism following light exposure in broccoli sprouts. Int J Mol Sci 16:13678–13691

Martínez-Arranz I, Mayo R, Pérez-Cormenzana M, Mincholé I, Salazar L, Alonso C, Mato JM (2015) Data in support of enhancing metabolomics research through data mining. Data Brief 3:155–164

Metlen KL, Aschehoug ET, Callaway RM (2009) Plant behavioural ecology: dynamic plasticity in secondary metabolites. Plant Cell Environ 32:641–653

Mwendwa JM, Weston PA, Fomsgaard I, Laursen BB, Brown WB, Wu H, Rebetzke G, Quinn JC, Weston LA (2016) Metabolic profiling for benzoxazinoids in weed-suppressive and early vigour wheat genotypes. In: Conference proceedings of the 20th Australasian weeds conference, pp. 353–357. Weeds Society of Western Australia

Pan Z, Raftery D (2007) Comparing and combining NMR spectroscopy and mass spectrometry in metabolomics. Anal Bioanal Chem 387:525–527

Ramautar R, de Jong GJ (2014) Recent developments in liquid-phase separation techniques for metabolomics. Bioanalysis 6:1011–1026

Rochfort S (2005) Metabolomics reviewed: a new "omics" platform technology for systems biology and implications for natural products research. J Nat Prod 68:1813–1820

Roessner U, Bacic A (2009) Metabolomics in plant research. Aust Biochem 40:9–11

Roessner U, Luedemann A, Brust D, Fiehn O, Linke T, Willmitzer L, Fernie AR (2001) Metabolic profiling allows comprehensive phenotyping of genetically or environmentally modified plant systems. Plant Cell Online 13:11–29

Ryalls JMW, Moore BD, Riegler M, Johnson SN (2016) Above–belowground herbivore interactions in mixed plant communities are influenced by altered precipitation patterns. Front Plant Sci 7:345

Saito K, Matsuda F (2010) Metabolomics for functional genomics, systems biology, and biotechnology. Annu Rev Plant Biol 61:463–489

Schauer N, Fernie AR (2006) Plant metabolomics: towards biological function and mechanism. Trends Plant Sci 11:508–516

Schuman MC, Baldwin IT (2016) The layers of plant responses to insect herbivores. Annu Rev Entomol 61:373–394

Skoneczny D, Weston PA, Zhu X, Gurr GM, Callaway RM, Weston LA (2015) Metabolic profiling of pyrrolizidine alkaloids in foliage of two *Echium* spp. invaders in Australia—a case of novel weapons? Int J Mol Sci 16:26721–26737

Skoneczny D, Weston PA, Zhu X, Gurr GM, Callaway RM, Barrow RA, Weston LA (2017) Metabolic profiling and identification of shikonins in root periderm of two invasive *Echium* spp. Weeds in Australia Molecules 22:330

Smolinska A, Blanchet L, Buydens LMC, Wijmenga SS (2012) NMR and pattern recognition methods in metabolomics: from data acquisition to biomarker discovery: a review. Anal Chim Acta 750:82–97

Sumner LW, Amberg A, Barrett D, Beale MH, Beger R, Daykin CA, Fan TW-M, Fiehn O, Goodacre R, Griffin JL (2007) Proposed minimum reporting standards for chemical analysis. Metabolomics 3:211–221

Swenson TL, Jenkins S, Bowen BP, Northen TR (2015) Untargeted soil metabolomics methods for analysis of extractable organic matter. Soil Biol Biochem 80:189–198

Viant MR, Sommer U (2013) Mass spectrometry based environmental metabolomics: a primer and review. Metabolomics 9:144–158

Vidkjær NH, Wollenweber B, Gislum R, Jensen K-MV, Fomsgaard IS (2015) Are ant feces nutrients for plants? A metabolomics approach to elucidate the nutritional effects on plants hosting weaver ants. Metabolomics 11:1013–1028

Watson JT, Sparkman OD (2007) Introduction to mass spectrometry: instrumentation, applications, and strategies for data interpretation. Wiley, Chichester, p 862

Weston P, Weston L, Hildebrand S (2013) Metabolic profiling in *Echium plantagineum*: Presence of bioactive pyrrolizidine alkaloids and napthoquinones from accessions across southeastern Australia. Phytochem Rev 12(1–7):831e837

Weston LA, Skoneczny D, Weston PA, Weidenhamer JD (2015) Metabolic profiling: an overview – new approaches for the detection and functional anlysis of biologically active secondary plant products. J Allelochem Interact 2:15–27

Wittstock U, Gershenzon J (2002) Constitutive plant toxins and their role in defense against herbivores and pathogens. Curr Opin Plant Biol 5:300–307

Yan D, Afifi L, Jeon C, Trivedi M, Chang HW, Lee K, Liao W (2017) The metabolomics of psoriatic disease. Psoriasis 7:1–15

Ye M, Song Y, Long J, Wang R, Baerson SR, Pan Z, Zhu-Salzman K, Xie J, Cai K, Luo S (2013) Priming of jasmonate-mediated antiherbivore defense responses in rice by silicon. PNAS 110:3631–3639

Zhang A, Sun H, Wang P, Han Y, Wang X (2012) Modern analytical techniques in metabolomics analysis. Analyst 137:293–300

Zhu X, Skoneczny D, Weidenhamer JD, Mwendwa JM, Weston PA, Gurr GM, Callaway RM, Weston LA (2016) Identification and localization of bioactive naphthoquinones in the roots and rhizosphere of Paterson's curse (*Echium plantagineum*), a noxious invader. J Exp Bot 67:3777–3788

Chapter 21
SAR/QSAR

Marta Teijeira and María Celeiro

1 Introduction

(Q)Structure-Activity Relationships (QSAR and SAR) studies have been widely used in Medicinal Chemistry as a support in the drug's discovery and development process, as well as in the study of harmful and poisonous substances in Toxicological Chemistry (McKinney 2000; Anderson 2003). They have also been applied in other areas of the natural sciences as a tool for learning the behavior of biological systems, supporting the idea that the physiological effect of a compound is a function of its chemical structure (Avram et al. 2014; Shanmugam and Jeon 2017).

SAR studies aim to extract relevant chemical information in series of chemical compounds that share a similar biological activity. These studies can be used to determine which fragments of the chemical structure are responsible for the biological activity. Any change in the chemical structure that can modify its solubility in water, transport through the membranes, binding to the receptor and other kinetic properties of the compound, should be considered in a SAR study.

One of the fundamental assumptions of SAR analysis is that active and similar compounds interact with a biological target through similar mechanism/s of action. However, it is common to find compounds that do not meet these requirements, thus producing atypical values (Fig. 21.1).

Different types of structural changes in a compound, in shape, size or in type or number of functional groups, generally result in differences in power or even in a different activity. Usually, the presence of alkyl, aromatic or halogen groups increases the lipophilic character of a molecule. Lipophilia is a measure of the

M. Teijeira (✉)
Departamento de Química Orgánica. Facultade de Química, Universidade de Vigo, Vigo, Spain
e-mail: qomaca@uvigo.es

M. Celeiro
Instituto de Investigación Sanitaria Galicia Sur (IISGS), Universidade de Vigo, Vigo, Spain

© Springer International Publishing AG, part of Springer Nature 2018

A. M. Sánchez-Moreiras, M. J. Reigosa (eds.), *Advances in Plant Ecophysiology Techniques*, https://doi.org/10.1007/978-3-319-93233-0_21

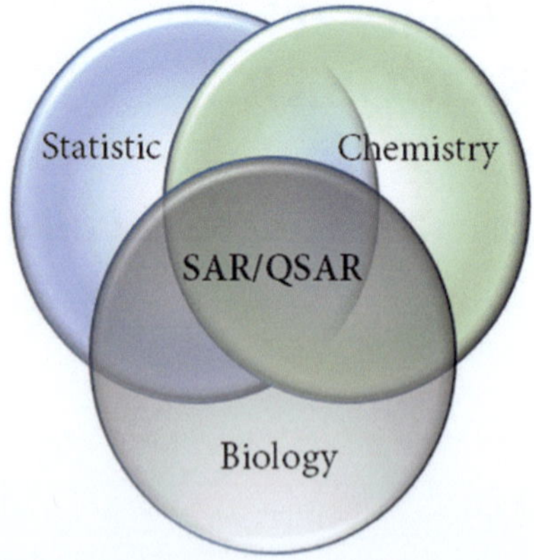

Fig. 21.1 Graphical representation of qualitative (SAR) and quantitative (QSAR) studies of structure-activity relationships

tendency of a compound to prefer a non-aqueous environment rather than an aqueous one, conditioning the absorption, distribution and elimination of the compound in the biological environment. It is also a measure of the solubility of a compound in cellular lipid membranes. These functional groups allow hydrophobic bonds with biological targets before establishing other polar bonds. Furthermore, the solubility in water can be related to the presence of hydrophilic functional groups such as carboxylic acids, amines or hydroxyl groups. In addition, these functional groups provide a hydrogen-bonding site in the interaction of the compound with biological targets. Some of these structural requirements are reflected in the SAR study of cis-cinnamic acid. The *cis* geometry of cinnamic acid is absolutely necessary for the inhibition of the growth of lettuce roots. The carboxylate group is important to establish hydrogen bonding to the target protein. The spatial arrangement of the aromatic ring is likewise important for its bioactivity (Abe et al. 2012).

Other key requirement in the structure of a bioactive compound is the three-dimensional arrangement of the atoms in the molecules (stereochemical properties). The enzymatic stereospecificity can restrict the activity of bioactive compounds. In a thorough review, relationships between the biological activity of some fungal phytotoxins and their stereochemistry were studied (Evidente et al. 2011). SAR analyses for some phytotoxins were also reported to be used in the elucidation of the herbicide potential of fungal metabolites (Cimmino et al. 2015).

Some SAR approaches divide the molecules into a common molecular nucleus on which substituents are distributed, which is very useful when studying sets of chemically homogeneous compounds or congeneric series (Hu et al. 2011). However, other approaches establish the relationship between molecules by using a holistic comparison (Macías et al. 2006), which is more suitable for the study of structurally heterogeneous compounds (Nagarajan et al. 2013).

The term "scaffold" is often used to describe the molecular nucleus or common element of a chemical series (Fig. 21.2). For chemically related compounds, a scaffold can also be defined as a maximum common substructure (MCS; Armitage and Lynch 1967; Ruiz et al. 2012). In recent years, different algorithms have been

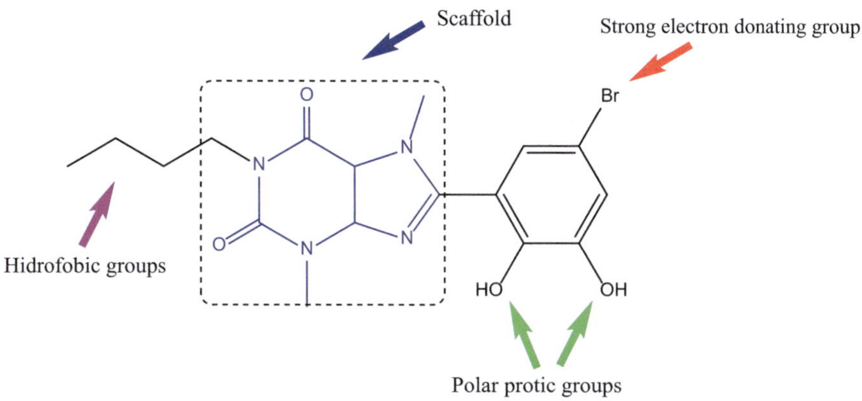

Fig. 21.2 Initial structure-activity relationships studies with a xanthine scaffold

described for the determination of MCS (Englert and Kovács 2015; Duesbury et al. 2017). Sometimes, scaffolds or MCS identification do not always result in significant substructures with intact functional groups or complete ring systems. This can be a problem when identifying chemically relevant fragments or substructures.

Recently, a SAR study of a series of umbelliferone-derived compounds was reported. The 7-hydroxycoumarin scaffold was chosen to evaluate the phytotoxic effect of hydroxyl and methyl substituents on positions 4 and 6 (Pan et al. 2017). Frequently, SAR studies of series of small and medium-sized compounds are performed routinely, without the need for computational approaches (Abe et al. 2012). However, in the study of sets of a large amount of data, or the simulation of biological environments where a compound interacts, the application of in silico methods is required (Geppert et al. 2010; Wassermann and Bajorath 2011; Wawer et al. 2014).

2 In Silico Methods in SAR and QSAR

In silico methods can have a predictive or descriptive objective. Predictive methods anticipate biological activity and include machine learning approaches and quantitative structure activity relationships (QSAR). On the other hand, the descriptive approaches help in understanding the biological and chemical behavior of compound sets (such as similarity analysis, molecular docking, dynamic molecular simulations, or QSAR).

SAR studies can also be classified into structure-based and ligand-based methods. Structure-based approaches use information of the structure of the biological target (structure of protein-targets), while ligand-based approaches use properties of the ligands (features of a set of chemical compounds).

In this chapter, the most important concepts of the ligand-based methods will be described, that is, those based on the biological or physico-chemical properties of the compounds, without taking into account their biological target.

In this sense, the understanding of many aspects of SARs of large series of chemical compounds can be established using metaphorically the term "activity landscape", since many of the characteristics of these are analogous to those present in ordinary geographical landscapes. The activity landscape is defined as a representation that integrates the analysis of structural similarity and power differences between compounds that share the same biological activity. The systematic characterization of the activity landscapes that describe certain SARs of a series of compounds is complex in nature (Stumpfe and Bajorath 2012). This activity landscape is often compared to a smooth region or continuous SAR (in which small changes in molecular structure are associated to small variations in activity) and a rugged landscape or discontinuous SAR (in which small changes in molecular structure lead to large variations in activity).

The existence of an activity landscape with wide variations in chemical structure but small variations in biological activity may suggest modes or binding sites to different biological targets or may reveal the effect of additional mechanisms of action.

2.1 Molecular Similarity Approaches

The concept of activity landscape is closely associated with the basic relationships between molecular structure and biological activity. It is based on the principle of property-similarity (Klopmand 1992), that is, similar structures should have similar biological properties, but there is no rigorous law that defines the notion of molecular similarity. Molecular representation is one of the most critical variables in SAR studies and can lead to inconsistencies when comparing and describing activity landscapes. Often, it is difficult to recognize which is the molecular property that best characterizes the activity landscape or allows the similarity between molecules to be evaluated (Bender et al. 2009; Willett 2014).

One way to measure molecular similarity between two compounds is setting a distance that varies between 0 and 1. Small distance values correspond to very similar compounds while large values correspond to dissimilar or very different compounds. It is often used as a criterion that distances of around 0.15 correspond to very similar compounds while values over 0.6 correspond to very different compounds.

Several binary fingerprints are commonly used for molecular similarity calculations, such as PubChem fingerprints, BCI (Barnard Chemical Information) fingerprints, SLN fingerprints (also called UNITY), TGT fingerprints, Daylight fingerprints, or MACCS (also called MDL) fingerprints, among others (Cereto-Massagué et al. 2015). These are strings of binary bits (0 or 1) where the absence or presence of functional groups, rings, fragments, etc. is represented and thus, a composite is represented numerically. Distance/similarity measures are used to describe the separation distance between the fingerprints of two compounds. One of the simplest formulas for this measurement is the Manhattan distance, which is a binary equivalent to taking a zigzag route following only ordinal axes. Other indices that have been used are Cosine, Dice, Euclidean, or Soergel (Todeschini et al. 2012). However, one of the most used and selected formulas for estimating a diagonal route in a straight line in the binary space of fingerprints is the Tanimoto index (Bajusz et al. 2015).

Tanimoto index (Tc) between two compounds is defined as:

$$Tc = \frac{c}{a+b-c}$$

where, given the fingerprints of two compounds A and B, a is the number of bits in A; b is the number of bits in B; and c is the number of bits in A and B.

Therefore, a Tanimoto index near 1 indicates a high degree of molecular similarity between those two compounds.

In recent years, the modeling of activity landscapes has been reported and reviewed in several publications (Guha and Van Drie 2008a, 2008b; Peltason and Bajorath 2008). Besides molecular similarity indices, landscapes modeling activity includes graphs (Chemical Neighborhood Graphs) and maps (Structure–Activity Similarity (SAS) maps) of landscape structure-activity (Stumpfe and Bajorath 2012). It should be noted that different measures of molecular similarity could give different values, depending on the sensitivity of the function of representation and similarity used. To solve this problem, a variety of methods has been developed based on combining the results of multiple similarity procedures (Stumpfe and Bajorath 2012).

It should not be overlooked that there are important exceptions to the similarity principle. Often, small changes in a structure cause a drastic change in biological activity. This phenomenon has recently been referred to as an "activity cliff" or "active zone". The importance of identifying the activity cliffs in the chemical series before their SAR study has been reported (Stumpfe et al. 2014; Bajorath 2017). In recent years, representations of landscapes of multitarget activity (High-Dimensional Activity Landscapes) have been published, designing a landscape based on maps that assemble structurally similar compounds and encode their relationships with the activity (Iyer et al. 2012).

2.2 *QSAR Approaches*

Those studies that seek to quantify the relationship between the chemical structure and the activity of a series of compounds, through the establishment of mathematical equations calculated by statistical approximations, are known by the acronym QSAR.

$$Biological\ respose = f\left(chemical\ structure, physicochemical\ property\right)$$

The variables involved in these studies can be discrete or continuous, which determines the usefulness of the QSAR model. Modeling of continuous variables allows obtaining classification or discriminant models, while the modeling of discrete variables allows correlation models. These approaches allow the interdisciplinary exploration of the activity of compounds in a wide range of aspects in chemistry, physics, biology or toxicology. Besides providing a reasonable basis to establish a predictive correlation model, QSAR studies allow the exploration of the chemical characteristics encoded in the descriptors.

The descriptors are numerical representation of attributes or properties of the chemical compound that define the characteristics of the analyzed molecules (Todeschini et al. 2009). The descriptors can be obtained from an experimental analysis or from a suitable theoretical algorithm, which diagnoses the chemical behavior of the molecules. Depending on the nature of the property studied (called endpoint), quantitative models can describe a chemical-physical property, a biological activity or a toxicity identified with the acronym QSPR/QSAR/QSTR, respectively. Additionally, they can use a response parameter of the activity/toxicity, toxicity-toxicity or property-property type as a predictor variable, in which case they are designated as QAAR, QTTR or QPPR, respectively. From a wider point of view, QSAR studies may include the analysis of the biochemical interactions involved in the mechanism/s of action of a chemical compound, providing relevant information on structural factors that determine the biological activity (Sukumar et al. 2012).

Three aspects must be addressed to establish a QSAR model: the preparation of the data, its computational processing and finally its interpretation. The data processed can either refer to the response or activity under study (variable dependent on the mathematical equation) or to the chemical attributes or molecular descriptors (independent variables of the mathematical equation). A classic QSAR model can be divided into the following main steps, as shown in Fig. 21.3. The data of the activity that is being studied or modeled must be of high quality, that means being reliable and consistent with the whole data set to be modeled. Any errors made in the data used in a QSAR study will result in models with low predictive power. It must be ensured that all these data have been obtained following a standard protocol, considering also their origin, i.e. if they come from a single laboratory or from several different laboratories. In addition, the chemical structures of the compounds from which the theoretical descriptors used for the QSAR models will be calculated, must also be correctly coded.

Many efforts have been made to provide a guide of good practices for the preparation of chemical data before the modeling process (Young et al. 2008; Fourches et al. 2010). Emphasis should be placed on the importance of creating and following a standardized data cleansing strategy, applicable to any set of compounds. In any case, the reliability of the data can be completely ensured if the publication of a QSAR study allows full access to all the data that have been used.

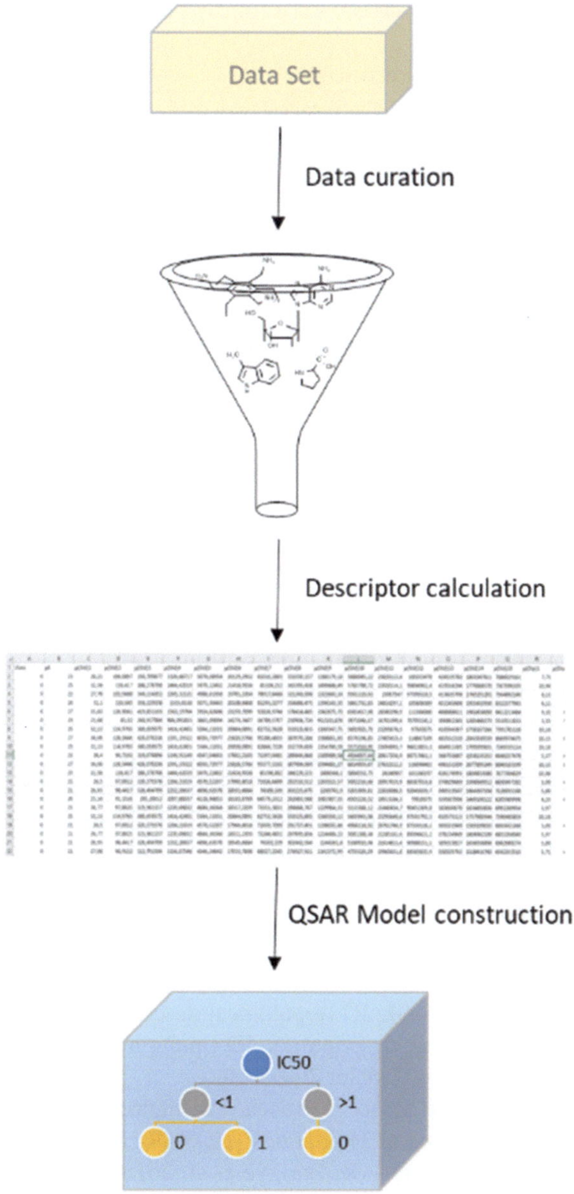

Fig. 21.3 General workflow of QSAR modeling

The selection of molecular descriptors can be done from a conservative perspective (Hansch's approach), where it is the modeler himself who selects the descriptors or, alternatively in a different approach, a large number of theoretical descriptors can be used and be then, in a second step, selected by chemometric methods. In the first case, *a priori* knowledge of the mechanism/s of action of the compound is required. Thus, in studies of biological modes of action, it is usual to use descriptors

Table 21.1 QSAR dimensional perspectives from different descriptors

QSAR dimension	Type of descriptor	Descriptor examples
0D	Descriptors obtained from the chemical formula	Atom number, molecular weight, atom-type count
1D	Constitutional descriptors representing a molecule or a list of structural fragments of a molecule	WLN, SMILES, SMARTS, CAST
2D	Topological descriptors what consider how the atoms are connected in the molecule in terms of presence and nature of chemical bonds	ISIDA fragments, spectral moments, Balaban index, Estrada index
3D	Geometrical descriptors that allow not only a representation of the nature and connectivity of the atoms, but also the overall spatial configuration of the molecule	WHIM, Getaway, RDF, 3D-Morse
4D	Stereoelectronic descriptors that consider molecular properties arising from electron distribution, molecule interactions with probes characterizing the space surrounding them	Molecular interaction fields

such as log Kw, a parameter that limits permeation in the cell membrane, or parameters such as bioaccumulation, sorption coefficient in soil, etc. In the second case, it is assumed that a QSAR modeler should not influence the selection of the descriptor, but that mathematical tools should be applied to select the descriptors that best correlate with the activity under study. This could help to maximize the predictive power of the model. A detailed review of a large number of more important molecular descriptors, from the origins to the present, is presented in the reference book "Handbook of Molecular Descriptors" (Todeschini et al. 2009), including a complete view of the different molecular representations and their chemical-physical meaning.

The molecular descriptors are usually classified according to the dimensionality of their molecular representation (Table 21.1; Terfloth 2003) in: 0D, 1D (constitutional descriptors), 2D (topological descriptors), 3D (geometrical descriptors), 4D (stereoelectronic descriptors), 5D (multiple representation of induced-fit hypotheses) (Vedani and Dobler 2002), etc.

Other more recent approaches to multidimensional QSAR use other types of descriptors such as Holograms (HQSAR; Jiao et al. 2016) or Multivariate Image Analysis (MIA-QSAR; Barigye et al. 2016), among others. Recently, Mia-QSAR modeling has been applied to study herbicide phytotoxicity on problematic weeds (Freitas et al. 2013).

Different types of mathematical and statistical methods can be used to carry out a QSAR study (Perez Gonzalez et al. 2008; Liu and Long 2009). In general, they can be divided into regression methods and classification methods. Some of the general methods that have been mostly used and published in QSAR studies are presented below.

- Multiple Linear Regression (MLR) is one of the first regression methods used. This method stands out for its simplicity of interpretation. It has been used to selectively identify polyphenols that exhibit their antioxidant activity by electron donation coupled proton transfer pathway (Chakraborty and Basu 2014).
- From MLR, QSAR model is expressed in a linear correlation, such as:

$$Y = a_0 + a_1X_1 + a_2X_2 + \ldots + a_nX_n$$

where Y is the dependent variable that represents the modeled activity; X_1, X_2, ..., X_n are the independent variables that denote different structural characteristics or molecular descriptors and a_1, a_2, ..., a_n, are the contributions of the descriptors (regression coefficients) to the Y response; a_0, is the error.
Following the aforementioned equation, a positive correlation coefficient suggests that the corresponding descriptor contributes positively to the property studied, while a negative descriptor suggests a negative contribution. The weight of the coefficient can reveal the degree of influence of the corresponding descriptor on the modeled property. However, the assumption of a linear relationship in the MLR does not always fit well with the modeling of complex problems where multiple mechanisms can interact to produce a biological response. In addition, this type of models should not be used when the variables of the equation are correlated, that is, when the descriptors are collinear.
- The method of partial least squares (PLS; Tobias 1995) has been widely used to solve the problem of collinearity between descriptors. This methodology is also very useful when the number of descriptors far exceeds the number of compounds studied. Thus, to reduce the number of descriptors, the PLS method extracts the "latent variables" that can explain the variations of all the molecular descriptors. A PLS regression method was applied to find relationships between various molecular properties and membrane transport observed in leaf pulvinar cells of *Mimosa pudica* (Rocher et al. 2017). The structural properties of plant flavonoids and their physiological properties were related to PLS regression approach, which assisted to understanding their mode of action as anti-oxidants (Mishra et al. 2016). Other related analysis techniques include Principal Component Regression (PCR), and/or Principal Component Analysis (PCA).
- Artificial Neural Networks (ANNs) are one of the most popular non-linear regression methods in QSAR (Cartwright 2015). This methodology belongs to the self-learning algorithms, in which the neural network learns the relationship between the studied biological activity and the molecular descriptors. Learning is achieved through iterative prediction and improvement cycles. The ANN attempts to imitate a biological neural network, inspired by the structure, processing and learning method of a human brain. Recently, ANN-QSAR models with quantum chemical descriptors, for predicting and analyzing antioxidant activities of carotenoids, were reported (Jhin and Hwang 2015). The successful application of ANN methods to QSAR analysis also has been confirmed in a study of analogs of curcumin, produced from the rhizomes of *Curcuma longa*

plants (Satpathy et al. 2010). Probabilistic Neural Network (PNN) is similar to the ANNs, but it is used in classification problems.

- Among the classification methods, Logistic Regression (LR) is similar to linear regression, but it is used to model the probability that a compound belongs to a given destination property. In the same way, regression coefficients with a large value show that the molecular descriptor strongly affects the probability of the predicted biological activity, whereas small values indicate that the molecular descriptor has no influence. Likewise, the sign of the coefficients indicates whether the probability is positive or negative for the studied biological activity.
- Within the classification models, Linear Discriminant Analysis (LDA) has been widely used. LDA models the biological activity defined as a categorical property through a linear discriminant function that is a linear combination of molecular descriptors. Multi-resistance risk LDA models, for classification and prediction of agrochemical fungicides, have been reported (Speck-Planche et al. 2011). PNNs models showed advantages over LDA approach for the classification of anticancer activities of active compounds in medicinal plants (Xue et al. 2005).
- The Decision Tree (DT) allows constructing many possible tree variations from a set of descriptors, systematically subdividing the information from a set of rules and relationships. It is a hierarchical structure, with branches and nodes. The classification of a compound is reached after going through a series of questions and answers. Thus, the compound will be classified with a biological activity if it meets a certain condition for a given molecular descriptor. Decision Trees usually have an easy interpretation, especially if they are not very large.
- Conversely, the Ramdom Forest (RF) uses the growth of many classification trees; in such a way, it makes a prediction based on the majority prediction of each of the trees. Generally, it is recommended to grow a large number of trees and to take a number of descriptors of the square root of the total descriptors.
- The nearest k-neighbor (kNN) uses the distance between an unknown compound and each the studied compounds, looking for compounds that are similar in characteristics to the unknown compound. Among the most used distance measurements are the Euclidean distance and the Manhattan distance (BB). Using kNN-MFA approach, QSAR models were successfully carried out to analyze the steric, electrostatic and hydrophobic effects on a series of pyrazinecarboxamide derivatives with herbicidal activity (Shaikh et al. 2015).
- Finally, Support Vector Machine (SVM) can be mentioned as a method based on the principle of structural risk minimization (SRM) of statistical learning theory (Barakat and Bradley 2010).

All these methods have been updated in recent years to improve their performance in QSAR studies (Puzyn et al. 2010). In any case, it is necessary to keep in mind that the combination of models that are statistically robust and biologically predictive, is not an easy task (Tropsha et al. 2003). The European Commission (REACH: Registration, Evaluation, Authorization and Restriction of Chemical

Products) (http://europa.eu.int/comm/environment/chemicals/reach.htm) has established the need to use SAR models to reduce experimental tests on animals. For this purpose and for regulatory purposes, some "OECD Principles for Model Validation (Q) SAR" have been established (http://www.oecd.org/chemicalsafety/risk-assessment/oecdquantitativestructureactivityrelationshipsprojectqsars.htm).

According to these principles, for a QSAR model to be reliable, it is necessary to take the following information in consideration:

1. PRINCIPLE 1. It must have a definite end point.
   The chosen end point can be related to processes that take place in the environment (such as hydrolysis reactions and oxidation and atmospheric reduction, biodegradation and bioaccumulation), ecological effects that can cause a series of compounds (such as aquatic or terrestrial toxicity), chemical-physical properties of the compounds (such as their boiling and melting points, Kw, water solubility or vapor pressure) or some properties related to human health (such as carcinogenicity, hepatotoxicity or cardiotoxicity).
2. PRINCIPLE 2. An unambiguous algorithm must be used.
   The algorithms used in the QSAR studies must be carefully described, in such a way that the user can understand exactly how the estimated value has been achieved and is able to reproduce it.
3. PRINCIPLE 3. It must possess a defined applicability domain.
   In general, QSAR models can only reliably predict molecules similar to the series of studied compounds. Even if a robust and validated model is available, the modeled property cannot be predicted for the entire universe of chemical substances. The applicability domain or the chemical space of a QSAR model is that spatial region in which the modeled response and the used molecular descriptors of the studied compounds are collected. Therefore, it is essential to define the applicability domain of a model to determine the chemical structural subspace that can be predicted reliably. Many approaches have been described for the estimation of the applicability domain, according to the methodology used: methods used for the characterization of the interpolation space of the descriptor, geometric methods, distance-based methods, methods based on the range of the response variable and methods based on the distribution of probability density, all of which have been reviewed in recent literature (Sahigara et al. 2012; Gajewicz 2018).
4. PRINCIPLE 4. It must include appropriate measures of goodness of fit, robustness and predictability.
   Every model must be validated before being used for the prediction of the endpoint of new compounds. If a model is not validated, it may be overfitted, i.e. too many variables have been selected. The relationship between the number of compounds and the descriptors included in a QSAR model must always be more than five if the data set is small. If the data set is large, the number of descriptors should be as small as possible. Finally, the preferred model is the one with the highest values of prediction parameters and the most balanced results among the validation parameters. In addition, it is necessary to carry out a validation with

compounds that have not been part of the data set in the construction of the model (external validation). Therefore, a last step must be to evaluate the predictive power of the model with external prediction chemicals (Tropsha 2010). Often, the limiting factor of the external validation of a QSAR model is the availability of the data, which should never be less than 5–20% of the data set.

5. PRINCIPLE 5. It should allow an interpretation of the mechanism of action of the compounds under study, if possible.
   The multifactorial nature of the mechanisms involved in a biological response makes it difficult, in most cases, to search for an interpretation of the selected descriptors in a QSAR model. It must be borne in mind that in addition, a predictive application in modeling is not always sought. In any case, for the QSAR models to be beneficial for the experimental chemist or biologist, the properties they describe must be unambiguous and easily interpretable, and the model must be robust, reliable and reproducible. The SAR and QSAR studies can facilitate the understanding of the nature of the regions of the chemical space that are relevant to biology and advance knowledge of biological processes.

QSAR studies have progressively evolved from their beginnings towards the analysis of large data sets with large amounts of molecular structures, assuming the existence of several mechanisms of action and the study of several compounds against different biological targets, using a wide variety of statistical techniques including machine learning (Wassermann et al. 2010) or multiple QSAR models (Roy et al. 2018).

Other classical computational methodologies for SAR studies use simulations of the coupling between ligands (small compounds) and biological targets (proteins, DNA, RNA, etc.). They are based on the hypothesis that a specific biological response is triggered by the favorable interaction between a molecule and a specific target into a particular binding site. These computational approaches are referred to as structure-based methods *versus* the ligand-based methods (QSAR, similarity analysis) mentioned above.

Due to their wide range of applications in the analysis of molecular recognition events the docking and scoring functions, as well as the molecular dynamic simulation, are among the most frequently used strategies. In recent years, there has been a trend towards integrating ligand-based and structure-based approaches in order to enhance the reliability and efficiency of both methodologies combining information from both the ligand and the protein.

The structure-based methods are another area of research of great interest and large scope, which have led to interesting reviews in the literature (Rognan 2011; Sliwoski et al. 2014; Ferreira et al. 2015), but this goes beyond the objectives of this chapter.

References

Abe M, Nishikawa K, Fukuda H, Nakanishi K, Tazawa Y, Taniguchi T, Park SY, Hiradate S, Fujii Y, Okuda K, Shindo M (2012) Key structural features of cis-cinnamic acid as an allelochemical. Phytochemistry 84:56–67

Anderson AC (2003) The process of structure-based drug design. Chem Biol 10:787–797

Armitage JE, Lynch MF (1967) Automatic detection of structural similarities among chemical compounds. J Chem Soc C Org:521–528

Avram S, Funar-Timofei S, Borota A, Chennamaneni SR, Manchala AK, Muresan S (2014) Quantitative estimation of pesticide-likeness for agrochemical discovery. J Chem Inform 6:1–11

Bajorath J (2017) Representation and identification of activity cliffs. Expert Opin Drug Discovery 12:879–883

Bajusz D, Rácz A, Héberger K (2015) Why is Tanimoto index an appropriate choice for fingerprint-based similarity calculations? J Chem Inform 7:1–13

Barakat N, Bradley AP (2010) Rule extraction from support vector machines: a review. Neurocomputing 74:178–190

Barigye SJ, Duarte MH, Nunes CA, Freitas MP (2016) MIA-plot: a graphical tool for viewing descriptor contributions in MIA-QSAR. RSC Adv 6:49604–49612

Bender A, Jenkins JL, Scheiber J, Sukuru SCK, Glick M, Davies JW (2009) How similar are similarity searching methods ? A principal component analysis of molecular descriptor space. J Chem Inf Model 49:108–119

Cartwright H (2015) Artificial neural networks. Springer, New York

Cereto-Massagué A, Ojeda MJ, Valls C, Mulero M, Garcia-Vallvé S, Pujadas G (2015) Molecular fingerprint similarity search in virtual screening. Methods 71:58–63

Chakraborty S, Basu S (2014) Mechanistic insight into the radical scavenging activity of polyphenols and its application in virtual screening of phytochemical library: an in silico approach. Eur Food Res Technol 239:885–893

Cimmino A, Masi M, Evidente M, Superchi S, Evidente A (2015) Fungal phytotoxins with potential herbicidal activity: chemical and biological characterization. Nat Prod Rep 32:1629–1653

Duesbury E, Holliday J, Willett P (2017) Comparison of maximum common subgraph isomorphism algorithms for the alignment of 2D chemical structures. Chem Med Chem. https://doi.org/10.1002/cmdc.201700482

Englert P, Kovács P (2015) Efficient heuristics for maximum common substructure search. J Chem Inf Model 55:941–955

Evidente A, Adolfi A, Cimmino A (2011) Relationships between the stereochemistry and biological activity of fungal phytotoxins. Chirality 23:674–693

Ferreira LG, Dos Santos RN, Oliva G, Andricopulo AD (2015) Molecular docking and structure-based drug design strategies. Molecules 20:13384–13421

Fourches D, Muratov E, Tropsha A (2010) Trust but verify: on the importance of chemical structure curation in chemoinformatics and QSAR modeling research. J Chem Inf Model 50:1189–1204

Freitas MR, Matias SVBG, Macedo RLG, Freitas MP, Venturin N (2013) Augmented multivariate image analysis applied to quantitative structure-activity relationship modeling of the phytotoxicities of benzoxazinone herbicides and related compounds on problematic weeds. J Agric Food Chem 61:8499–8503

Gajewicz A (2018) How to judge whether QSAR/read-across predictions can be trusted? Novel approach for establishing model's applicability domain. Environ Sci Nano 14. https://doi.org/10.1039/C7EN00774D

Geppert H, Vogt M, Bajorath J (2010) Current trends in ligand-based virtual screening: molecular representations, data mining methods, new application areas, and performance evaluation. J Chem Inf Model 50:205–216

Guha R, Van Drie JH (2008a) Structure – activity landscape index: identifying and quantifying activity cliffs. J Chem Inf Model 48:646–658

Guha R, Van Drie JH (2008b) Assessing how well a modeling protocol captures a structure-activity landscape. J Chem Inf Model 48:1716–1728

Hu Y, Stumpfe D, Bajorath J (2011) Lessons learned from molecular scaffold analysis. J Chem Inf Model 51:1742–1753

Iyer P, Dimova D, Vogt M, Bajorath J (2012) Navigating high-dimensional activity landscapes: design and application of the ligand-target differentiation map. J Chem Inf Model 52:1962–1969

Jhin C, Hwang KT (2015) Adaptive neuro-fuzzy inference system applied qsar with quantum chemical descriptors for predicting radical scavenging activities of carotenoids. PLoS One 10:1–13

Jiao L, Zhang X, Qin Y, Wang X, Li H (2016) Hologram QSAR study on the electrophoretic mobility of aromatic acids. Chemom Intell Lab Syst 157:202–207

Klopmand G (1992) In: Johnson MA, Maggiora GM (eds) Concepts and applications of molecular similarity. Wiley, New York 1990, J Comput Chem 13:539–540

Liu P, Long W (2009) Current mathematical methods used in QSAR/QSPR studies. Int J Mol Sci 10:1978–1998

Macías FA, Marín D, Oliveros-Bastidas A, Castellano D, Simonet AM, Molinillo JMG (2006) Structure-activity relationship (SAR) studies of benzoxazinones, their degradation products, and analogues. Phytotoxicity on problematic weeds *Avena fatua* L. and *Lolium rigidum* Gaud. J Agric Food Chem 54:1040–1048

McKinney JD (2000) The practice of structure activity relationships (SAR) in toxicology. Toxicol Sci 56:8–17

Mishra AK, Tyagi C, Pandey B, Chakraborty O, Kumar A, Jain AK (2016) Structural insights into the mode of action of plant flavonoids as anti-oxidants using regression analysis. Proc Natl Acad Sci 86:1023–1036

Nagarajan M, Maruthanayagam V, Sundararaman M (2013) SAR analysis and bioactive potentials of freshwater and terrestrial cyanobacterial compounds: a review. J Appl Toxicol 33:313–349

Pan L, Li X, Jin H, Yang X, Qin B (2017) Antifungal activity of umbelliferone derivatives: synthesis and structure-activity relationships. Microb Pathog 104:110–115

Peltason L, Bajorath J (2008) Molecular similarity analysis in virtual screening. In: Varnek A, Tropsha A (eds) Chemoinformatics approaches to virtual screening. The Royal Society of Chemistry Publishing, Cambridge, UK, pp 120–149

Perez Gonzalez M, Teran C, Saiz-Urra L, Teijeira M (2008) Variable selection methods in QSAR: an overview. Curr Top Med Chem 8:1606–1627

Puzyn T, Leszczynski J, Cronin MTD (2010) Recent advances in QSAR Studies. Methods and applications. Springer, New York 423 pp

Rocher F, Roblin G, Chollet JF (2017) Modifications of the chemical structure of phenolics differentially affect physiological activities in pulvinar cells of *Mimosa pudica* L. II. Influence of various molecular properties in relation to membrane transport. Environ Sci Pollut Res 24:6910–6922

Rognan D (2011) Docking methods for virtual screening: principles and recent advances. In: Sotriffer C, Mannhold R, Kubinyi H, Folkers G (eds) Virtual screening: principles, challenges, and practical guidelines. Wiley-VCH, Weinheim, pp 153–176

Roy K, Ambure P, Kar S, Ojha PK (2018) Is it possible to improve the quality of predictions from an "intelligent" use of multiple QSAR/QSPR/QSTR models? J Chemom e2992. https://doi.org/10.1002/cem.2992

Ruiz IL, García GC, Angel M (2012) Structural-similarity-based approaches for the development of clustering and QSPR / QSAR Models in chemical databases. In: Dehmer M, Varmuza K, Bonchev D, Emmert-Streib F (eds) Statistical modelling of molecular descriptors in QSAR/QSPR. Wiley-VCH Verlag GmbH & Co. KGaA, UK

Sahigara F, Mansouri K, Ballabio D, Mauri A, Consonni V, Todeschini R (2012) Comparison of different approaches to define the applicability domain of QSAR models. Molecules 17:4791–4810

Satpathy R, Guru RK, Behera R (2010) Computational QSAR analysis of some physiochemical and topological descriptors of curcumin derivatives by using different statistical methods. J Chem Pharm Res 2:344–350

Shaikh AR, Gonsalves SI, Nikam A, Kshirsagar SJ, Thombare Y (2015) Predicting pyrazinecarboxamides derivatives as an herbicidal agent: 3d Qsar by kNN-MFA and multiple linear regression approach. World Appl Sci J 33:980–989

Shanmugam G, Jeon J (2017) Aided drug discovery in plant pathology. Plant Pathol J 33:529–542

Sliwoski G, Kothiwale S, Meiler J, Lowe EWE (2014) Computational methods in drug discovery. Pharmacol Rev 66:334–395

Speck-Planche A, Kleandrova VV, Rojas-Vargas JA (2011) QSAR model toward the rational design of new agrochemical fungicides with a defined resistance risk using substructural descriptors. Mol Divers 15:901–909

Stumpfe D, Bajorath J (2012) Methods for SAR visualization. RSC Adv 2:369–378

Stumpfe D, Hu Y, Dimova D, Bajorath J (2014) Recent progress in understanding activity cliffs and their utility in medicinal chemistry. J Med Chem 57:18–28

Sukumar N, Das S, Krein M, Godawat R, Vitol I, Garde S, Bennett K, Breneman CM (2012) Molecular descriptors for biological systems. In: Guha R, Bender A (eds) Computational approaches in cheminformatics and bioinformatics. Wiley-VCH, Weinheim, pp 107–143

Terfloth L (2003) Calculation of structure descriptors. In: Engel JG (ed) Chemo-informatics. Wyley-VCH, Weinheim, pp 401–437

Tobias RD (1995) An introduction to partial least squares regression. SAS Conf Proc SAS Users Gr Int 20 (SUGI 20) 2–5

Todeschini R, Consonni V, Mannhold R, Kubinyi H, Folkers G (2009) Molecular descriptors for chemoinformatics, vol I & II. Wiley-VCH, Weinheim

Todeschini R, Consonni V, Xiang H, Holliday J, Buscema M, Willett P (2012) Similarity coefficients for binary chemoinformatics data: overview and extended comparison using simulated and real data sets. J Chem Inf Model 52:2884–2901

Tropsha A (2010) Best practices for QSAR model development, validation, and exploitation. Mol Inform 29:476–488

Tropsha A, Gramatica P, Gombar V (2003) The importance of being earnest: validation is the absolute essential for successful application and interpretation of QSPR models. QSAR Comb Sci 22:69–77

Vedani A, Dobler M (2002) 5D-QSAR: the key for simulating induced fit? J Med Chem 45:2139–2149

Wassermann AM, Bajorath J (2011) A data mining method to facilitate SAR transfer. J Chem Inf Model 51:1857–1866

Wassermann AM, Peltason L, Bajorath J (2010) Computational analysis of multi-target structure-activity relationships to derive preference orders for chemical modifications toward target selectivity. ChemMedChem 5:847–858

Wawer MJ, Jaramillo DE, Dancik V, Fass DM, Stephen J, Shamji AF, Wagner BK, Schreiber SL, Paul A (2014) Automated structure–activity relationship mining: connecting chemical structure to biological profiles. J Biomol Screen 19:738–748

Willett P (2014) The calculation of molecular structural similarity: principles and practice. Mol Inform 33:403–413

Xue CX, Zhang XY, Liu MC, Hu ZD, Fan BT (2005) Study of probabilistic neural networks to classify the active compounds in medicinal plants. J Pharm Biomed Anal 38:497–507

Young D, Martin T, Venkatapathy R, Harten P (2008) Are the chemical structures in your QSAR correct? QSAR Comb Sci 27:1337–1345

Chapter 22
Elucidating the Phytotoxic Potential of Natural Compounds

Adela M. Sánchez-Moreiras, Elisa Graña, Carla Díaz-Tielas, David López-González, Fabrizio Araniti, María Celeiro, Marta Teijeira, Mercedes Verdeguer, and Manuel J. Reigosa

1 Introduction

Crop protection in agroecosystems is still mainly focused on the use of synthetic herbicides, which although increase yield and the quality and quantity of crops short-term, are a major source of toxicity in the ecosystem and favor the presence and development of undesired resistant weeds (Concenço et al. 2009; De Boer et al. 2011). The low diversity in the modes of action of the commercially available synthetic herbicides and their indiscriminate use over the past 50 years has led to an increasing rise in the number of resistant weeds and harmful effects, which resulted even in the withdrawal of certain herbicides in different countries. In this context, the role of phytochemicals in regulating plant growth can be crucial for sustainably controlling weeds because of their short half-life in soil, their high specificity and their potentially benign toxicological profile (Dayan et al. 2000; Duke et al. 2014). These natural compounds often have highly diverse chemical structures with novel and multiple modes of action different from synthetic herbicides, so that they represent a

A. M. Sánchez-Moreiras (✉) · M. J. Reigosa
Department of Plant Biology and Soil Science, University of Vigo, Vigo, Spain
e-mail: adela@uvigo.es

E. Graña · C. Díaz-Tielas · D. López-González
Department of Plant Biology and Soil Science, Faculty of Biology, and Agri-Food Research and Transfer Centre of the Water Campus (CITACA), University of Vigo, Vigo, Spain

F. Araniti
Dipartimento di Agraria, Facoltà di Agraria, Università Mediterranea di Reggio Calabria, Reggio Calabria, Italy

M. Celeiro · M. Teijeira
Department of Organic Chemistry, Faculty of Chemistry, University of Vigo, Vigo, Spain

M. Verdeguer
Instituto Agroforestal Mediterráneo, Universitat Politècnica de València, Valencia, Spain

© Springer International Publishing AG, part of Springer Nature 2018

A. M. Sánchez-Moreiras, M. J. Reigosa (eds.), *Advances in Plant Ecophysiology Techniques*, https://doi.org/10.1007/978-3-319-93233-0_22

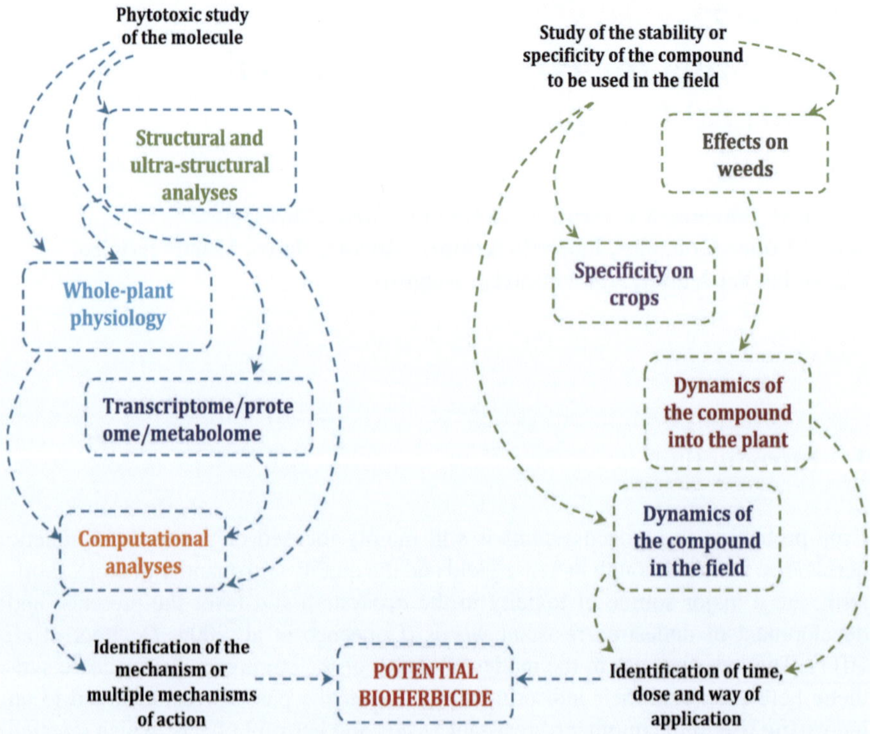

Fig. 22.1 Phases for approaching the phytotoxic study of a secondary metabolite with bioherbicide potential

real alternative to the massive use of conventional herbicides. Although there are numerous insecticides and bactericides from natural compounds that have been marketed, development and marketing of herbicides from plant secondary metabolites is still in an initial phase. A holistic approach is recommended when starting to study the herbicidal potential of a secondary metabolite. In the study of the potential use of natural compounds as phytotoxins is necessary to address the phytotoxic capacity of the compound as well as its specificity against crops and its stability in the ecosystem. A strongly phytotoxic metabolite with a very short half-life or a very limited translocation and transport capacity may be very unhelpful in the management of weeds under natural conditions. That is why a triple approach is proposed for the testing of natural compounds (Fig. 22.1).

2 Approaching the Herbicide Potential of a Natural Phytotoxin

When a new compound (previously isolated from a plant extract, leachate or exudate; selected due to its biological activity on other organisms; or due to the activity of other chemically-related compounds) arrives to our hands, its phytotoxic potential is addressed following the next consecutive phases (i) the study of the phytotoxic potential and the mode of action of the compound on model species (ii) the study of the phytotoxic potential of the natural compound on agriculturally interesting crop species and their associated adventitious flora, and (iii) the study of the stability and transport of the natural compound into the plant and its persistence in the soil (Fig. 22.1).

2.1 *Knowing the Mode of Action of Secondary Metabolites*

Knowing the exact mode/s of action of different plant secondary metabolites will allow establishing how, when and where these compounds must be used for weed management. This study is carried out from a morphological, physiological, biochemical and molecular perspective with special emphasis on the specific site of action and on the characterization of the primary and secondary effects of the studied compounds (Fig. 22.2).

2.1.1 Structural and Ultra-Structural Studies

The study of the potential phytotoxic effects of a natural compound, and the in detail study of the physiological and biochemical parameters induced on plant metabolism, must be developed in different consecutive phases, starting with the establishment of dose-response curves (see Chap. 1) and the careful study of structural and ultra-structural changes. Germination and radicle length curves will allow establishing the IC_{50} and IC_{80} values (concentrations of the tested compound that induce a 50% or 80% inhibition, respectively), which will be used for the morphological and structural measurements. Morphological alterations, such as root structure and thickness, number of lateral roots, presence or absence of root hairs, and growth direction must be studied by petri dishes scanning and magnifying of seedlings prior to tissue structural studies by light microscopy and ultra-structural studies of cells and organelles by Transmission Electron Microscopy (TEM). The combination of these analyses recorded at different times and concentrations will give us a picture of the effects of the compound/s on plant metabolism and will point the way for the following physiological or biochemical measurements.

Once that *de visu* morphological analysis has been done, structural (by light microscopy) and ultra-structural (by transmission electron microscopy, TEM) studies (see Chap. 15 for protocol details) must be performed in order to see the effects on cell organelles, cell walls, cell division and differentiation, etc. Most of the information recorded at this level will be the basis for later physiological and biochemical measurements and will be essential for understanding the full amount of data obtained by the – omics approach.

In this way, as shown in Fig. 22.2, structural and ultra-structural analyses can reveal root torsion, left-handed growth of roots, multinucleated cells and zigzag cell walls, suggesting microtubule alteration (Fig. 22.2a–c; for further information see

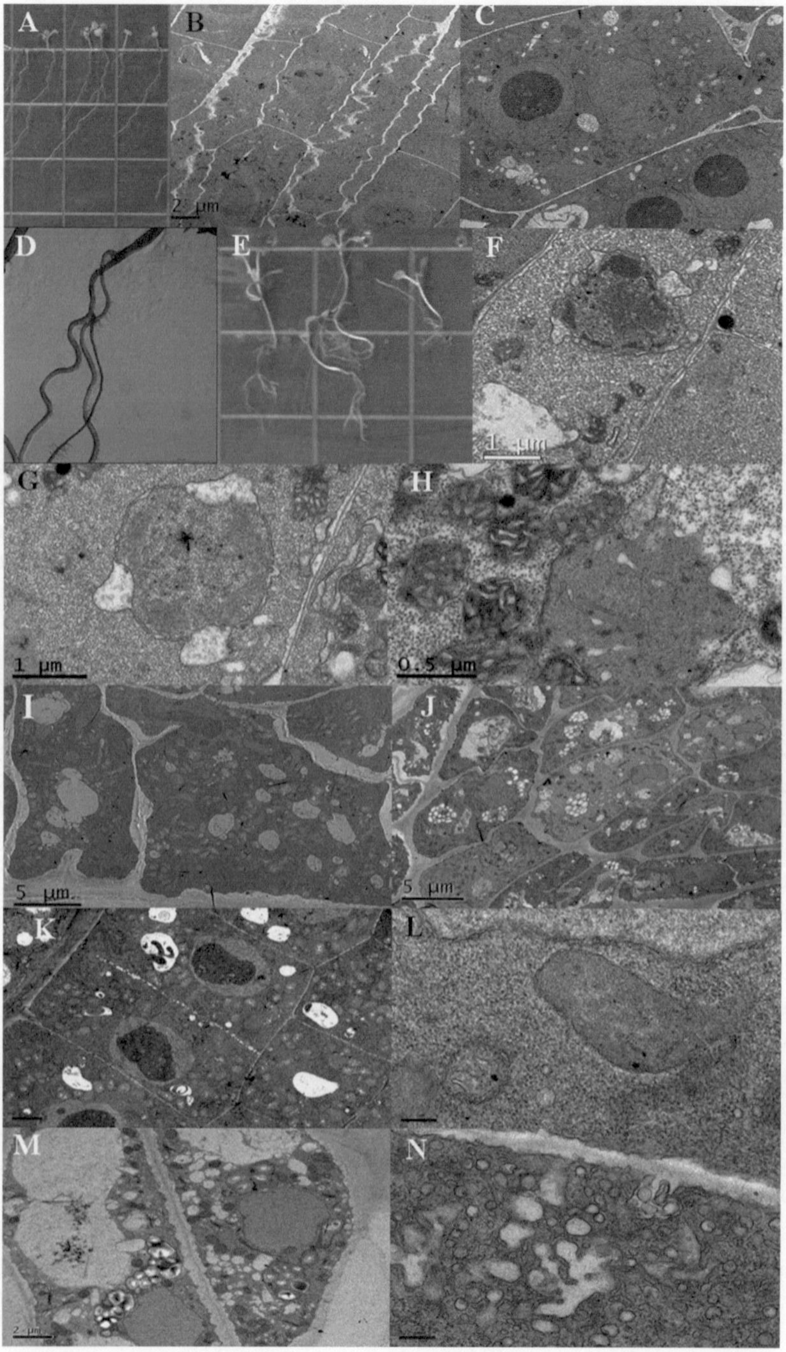

Fig. 22.2 Scanner (**a**, **e**), magnifier (**d**), and Transmission Electron Microscopy, TEM, (**b**, **c**, **f–n**) images of Arabidopsis radicles treated with myrtenal (**a**), citral (**b**, **i**, **j**), farnesene (**c**), norharmane (**d**), scopoletin (**e**), chalcone (**f**, **g**, **h**), rosmarinic acid (**k**, **l**), and eugenol (**m**, **n**)
(**a**) Left-handed roots; (**b**) Zigzag cell walls; (**c**) multinucleated cells; (**d**, **e**) Multiple adventitious roots and altered root growth; (**f**) Fragmented chromatin and nuclear hernia; (**g**) Nucleus with hernia; (**h**) Multiple condensed mitochondria; (**i**) Incomplete cell walls; (**j**) Disorganized cell division; (**k**) Increased number of mitochondria; (**l**) Broken mitochondria; (**m**) Strong vacuolization; (**n**) Increased Golgi activity.

Graña et al. 2013a; Araniti et al. 2016); multiple adventitious roots, altered growth direction and disorganized quiescent center, suggesting hormonal unbalance (Fig. 22.2d, e; unpublished data; for further information see Araniti et al. 2016); condensed mitochondria, fragmented chromatin and nuclear hernia, suggesting apoptotic-like programmed cell death (Fig. 22.2f–h; for further information see Díaz-Tielas et al. 2012); altered cell planes and unfinished cell walls, suggesting an effect on cell division (Fig. 22.2i, j; for further information, see Graña et al. 2013a, 2017); increasing number of mitochondria, or broken and with light matrix mitochondria, suggesting a deficit of ATP production (Fig. 22.2k, l; for further information, see Graña et al. 2017); or increased Golgi activity and vacuolization, suggesting detoxification processes (Fig. 22.2m, n; unpublished data), among others.

All together, dose-response curves and magnifier, light microscopy and TEM images, give a nice first picture of how our compound could be acting into the plant metabolism. Once morphological and ultra-structural studies have been performed, physiological and biochemical analyses can be established to go on with the study of the mode of action.

2.1.2 Physiological and Biochemical Studies

Structural and ultra-structural study of the tested compounds gives the necessary information to approach the mode of action and to determine the effect of the compounds. Further physiological and biochemical measurements can include measurements on microtubules (immunostaining with Alexa Fluor 488; Fig. 22.3a, b; for further information, see Araniti et al. 2016; Graña et al. 2017; and Chap. 17); potential of mitochondrial membrane (staining with JC-1 and confocal microscopy; Fig. 22.3c, d; for further information see Díaz-Tielas et al. 2012); apoptotic-like cell death (staining with acridine orange and ethidium bromide; Fig. 22.3e, f, g; for further information see Díaz-Tielas et al. 2012); increase of pectin in the cell walls (staining with ruthenium red; Fig. 22.3h, i; for further information see Graña et al. 2013a, b); blockage or delay of cell division (by flow cytometry and mitotic index count; Fig. 22.3l; for further information see Coba de la Peña and Sánchez-Moreiras 2001; Sánchez-Moreiras et al. 2001, 2008; Graña et al. 2013a, b and Chaps. 12 and 13); balance of auxins and ethylene (by HPLC; for further information see Araniti et al. 2016), etc.

Measurements on oxidative metabolism: *in situ* staining and quantification of the content of Reactive Oxygen Species, such as hydrogen peroxide, H_2O_2, or superoxide, O_2^- (ROS; Fig. 22.3j, k; for further information see Martínez-Peñalver et al. 2012a, b; Araniti et al. 2016), nitric oxide by electron microscopy, peroxidase, catalase, superoxide dismutase and NADPH oxidase activities by colorimetric methods, measurement of the concentration of carotenoids, ascorbate, glutathione and polyamines and the measurement of lipid peroxidation, can also be made (for further information see Pedrol and Tiburcio 2001; Sánchez-Moreiras and Reigosa 2005; Sánchez-Moreiras et al. 2009), as well as chlorophyll *a* fluorescence measurements

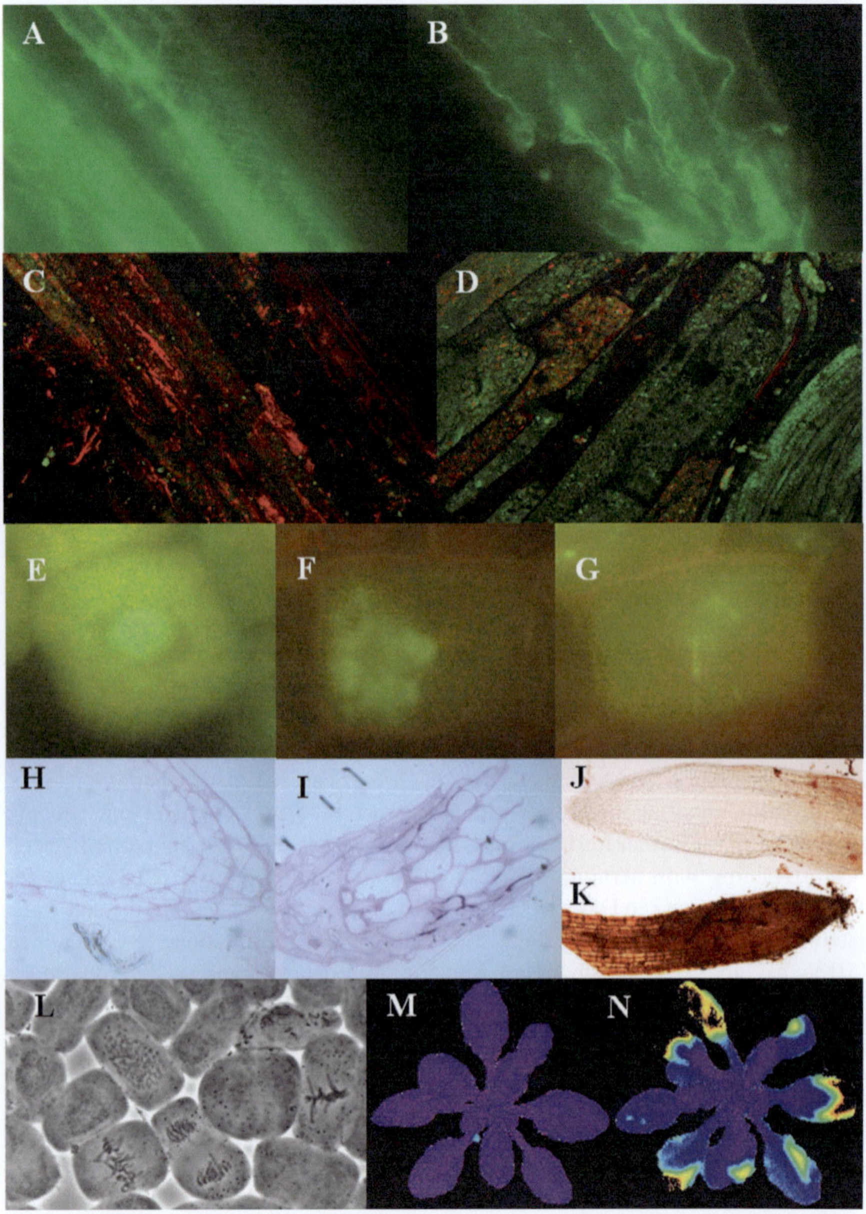

Fig. 22.3 (**a**, **b**) Fluorescence immunostaining of microtubules in control and eugenol-treated *Arabidopsis* cells, respectively; (**c**, **d**) Confocal image of mitochondrial membrane potential (JC-1 staining) in control and chalcone-treated cells; (**e**, **f**, **g**) Healthy, early-apoptotic and late-apoptotic nuclei, respectively, after chalcone treatment on *Arabidopsis* cells; (**h**, **i**) Pectin staining on control and citral-treated *Arabidopsis* roots, respectively; (**j**, **k**) *in situ* H_2O_2 localization (DAB staining) on control and citral-treated *Arabidopsis* roots, respectively; (**l**) Dividing cells in lettuce roots; (**m**, **n**) Maximum photosynthetic efficiency (Maxi-Imaging PAM) in control and BOA-treated *Arabidopsis* plants.

(maximum – F_v/F_m – and apparent – ϕ_{II} –efficiency of photosystem II; heat – ϕ_{NPQ} – and fluorescence – ϕ_{NO} – emission; and apparent electron rate – ETR – by Maxi-Imaging PAM fluorometer; Fig. 22.3m, n; for further information see Sánchez-Moreiras et al. 2011; Martínez-Peñalver et al. 2011, 2012a, b), and water status measurements (relative water content, water potential, osmotic potential, membrane permeability and membrane transport, stomatal density, etc.; for further information see Sánchez-Moreiras and Reigosa 2005; Sánchez-Moreiras et al. 2009) of the plants treated with the different natural compounds.

In this phase, whole-plant physiological measurements are focused on the roots or shoots of treated plants depending on the previously obtained morphological and structural information. These measures are intended to obtain the maximum amount of information that allows to correctly interpret the results obtained in later –omics approach.

2.1.3 Omics Analyses

The molecular mechanisms underlying the perception and transduction of the stress signal to the target genes are mostly unknown. The stress signal generated after treatment with a natural compound will be perceived as a chemical signal bound to receptors on the plant plasma membrane or in the cytosol (e.g. ABA), which can activate a large number of genes and induce post-transcriptional and post-translational modifications that can result in a different set of proteins and metabolites present in the plant. The last generation technologies, as transcriptomics, proteomics and metabolomics, are used in the search for the mode of action in order to generate a database of interconnected networks of genes, proteins and metabolites that give an integrated image of the response of the plant to the studied compounds, with different doses and at different application times.

Transcriptomic, proteomic and metabolomic profiles are monitored on seedlings of the model species *Arabidopsis thaliana*, with the IC_0, IC_{50} e IC_{80} values obtained after the dose-response curves and at different times established after the analyses of the previous ultra-structural, physiological and biochemical studies.

- Transcriptomic profile. Transcriptome analysis makes possible obtaining a snapshot of all the genes expressed simultaneously in the plant at a specific time, resulting in very precise comparisons of controls and samples. Looking simultaneously at the whole plant genome we can characterize the differences and similarities in gene expression profiles, understanding gene expression as a link among the genotype of the organism and its corresponding phenotype (Fig. 22.4). When the study of the effects of natural products on plant metabolism is being performed, DNA should be an increasing focus of attention, even when the interest is focused on small molecules, as DNA is by itself a target site for many natural products (furanocoumarins from angiosperm plants or duacarmycin, doxorubicin, diastamycin A all from Streptomyces, among others), but even more because by this way we are able to identify the stress-induced genes and the

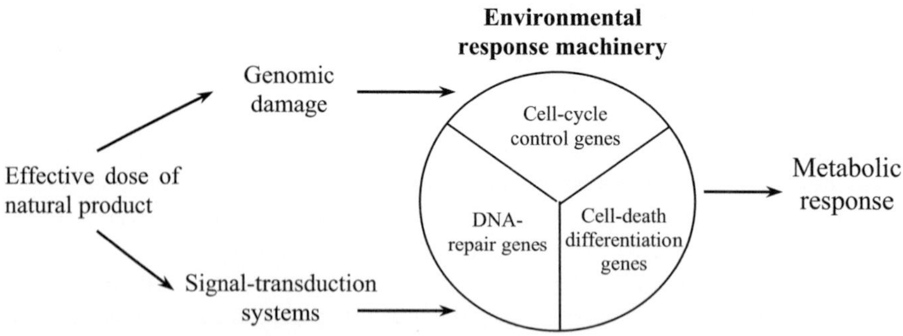

Fig. 22.4 The activation of different enzymatic systems and the interaction between some of them will regulate the final metabolic response under stress conditions. (Redrawn from Albers 1997)

so called "environmental response genes" into the gene sequence and getting then a better understanding about the stress response and detoxification mechanisms into the cell (Berenbaum 2002). Among the stress-induced genes there are signaling genes, transcription factors, and effector genes, which include enzymes that alter the cells structure and properties in response to stress.

A sequential analysis is done at different times to define the primary response of the plant to the tested metabolites. Once the treatment has started, the seedlings are harvested at 0, 0.25, 0.5, 1, 3, 6, 12 and 24 h. Gene expression is considered to be over-expressed, or under-expressed, when the expression of the gene is at least twice as large or smaller than the control. The data obtained by microarrays is contrasted with the RT-PCR real time quantification assays. The profile of genes overexpressed or deleted after treatment with each of the compounds is used for later comparisons with other natural compounds or synthetic herbicides (for further information see Baerson et al. 2005).

- Proteomic profile. Schulze and Downward (2001) pointed that one of the problems limiting the use of microarrays to elucidate phenotypic changes and establish gene networks is that the regulation of mRNA levels is one aspect of the biological control, but not the only one. Sometimes the alteration in mRNA levels is not traduced finally in altered protein levels, which are also controlled by post-translational modifications, and in fact the correlation among mRNA content and protein has proven to be lacking in different experimental systems (Gygi et al. 1999). Protein post-translational modifications and scaffold-mediated protein-protein interactions are important for biological processes, particularly in the propagation of cellular signals (like in the stress signal transduction). There are more than 300 reported different post-translational modifications such as removal of signal peptides, phosphorylation, glycosylation, etc., which confer diverse properties with physiological relevance to the modified proteins. (Xiong and Zhu 2001). The constitutively presence of H^+-ATPases and aquaporins in plasma membrane could be a good example of this because it is regulated by Ca^{2+}-dependent phosphorylation, a post-translational modification.

Proteomics can clarify the structural predictions of genome sequence information and assess the protein modifications and protein-ligand interactions that are relevant to stress-response (Cushman and Bohnert 2000). Diverse up- or down-regulated proteins have been identified to determine the final phenotype of an organism under different types of stress, as well late embryogenesis-abundant proteins (LEA), enzymes involved in the protection against oxidative stress, in the ABA signalling transduction, enzymes with chaperon function (Heat shock proteins), involved in lignin synthesis or in glycolytic reactions (Zivy and de Vienne 2000).

The study of the proteome is carried out in Arabidopsis seedlings treated with the compound for 0, 1, 3, 6, 12, 24 and 48 h. This lag with respect to the transcriptomic analysis allows the expressed genes to be translated into proteins and be able to compare them statistically. The protein profile is obtained by two-dimensional electrophoresis with a pre-purification of TCA-acetone. Both, first and second dimension, are done on a horizontal system. Once the most important spots have been identified by silver staining, they are cut and digested with a Bruker Daltonics chopper and a digester and are analyzed with a Bruker Microflex mass spectrometer (MALDI-TOF) and a DIONEX 3000 two-dimensional Nano-HPLC. The profile of the altered proteins after the treatment with each of the compounds is compared with the profile of untreated plants (Fig. 22.5, unpublished data; for further information see Sánchez-Moreiras and Pedrol Bonjoch 2001 and Sánchez-Moreiras and Reigosa 2018).

- Metabolomic profile. Metabolomic is the large-scale study of all the small molecules, with a mass range of 50–1500 daltons, whose interactions within a biological system are known as metabolome (Kim et al. 2010). The power of this technique is due to the fact that the fluctuation of metabolites concentration, unlike other -omics measures, directly reflect the molecular phenotype of cells / tissues/organs (Fiehn 2002). Metabolites identification is carried out through Nuclear Magnetic Resonance (NMR) as well as with hyphenated techniques such as gas (GC), liquid (LC, HPLC, UPLC) and other affinity chromatography coupled to mass spectrometry (MS) (Zhang et al. 2012). In metabolomics samples are subjected to a multi-step process and an example on GC-MS based metabolomic pipeline is reported in Fig. 22.6.

  Since metabolites concentration is extremely fluctuating along the organs and during the day, the experimental design and sample collection should be consistent. Immediately after collection, samples metabolism is immediately quenched and the samples are deep frozen in liquid nitrogen. Successively, we extract the samples, eventually we derivatize them (depending on the technique that will be used since NMR do not need sample derivatization), and we prepare the extracts for the analysis (Lisec et al. 2006). To follow early and late events, due to stressors factors, is useful to plan samples collection at different times. For example, in experiments carried on *Arabidopsis* seedlings, treated with natural phytotoxins, we collect the samples after 0, 3, 6, 12, 24, 48 and 72 h from the beginning of the treatment and then we analyze them. Once again, there is a lag regarding

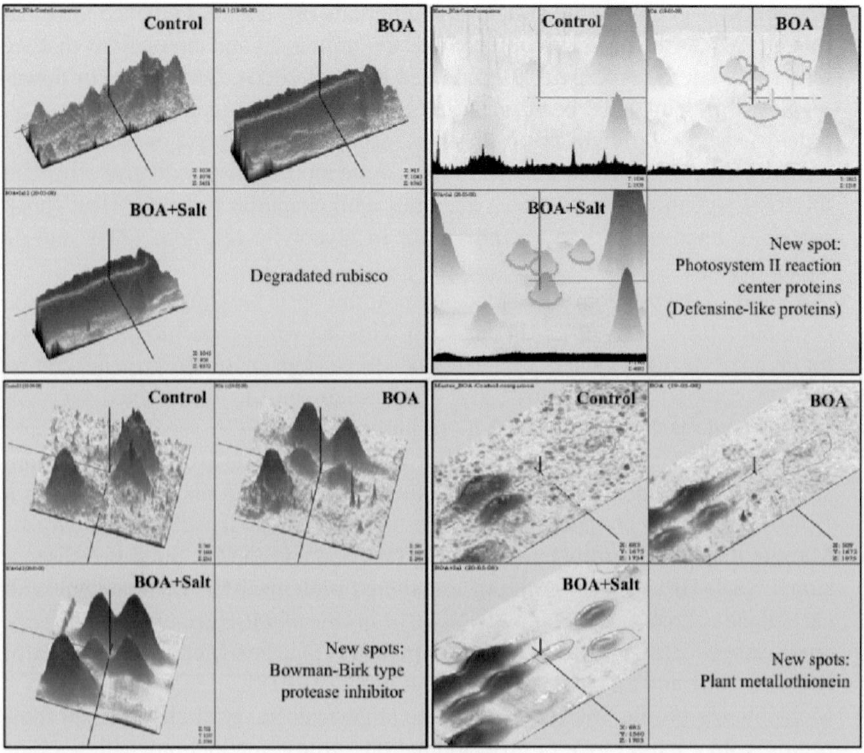

Fig. 22.5 Comparison of subsets (3-D plot graphs) from representative gels with variable spots of control, BOA and BOA+Salt treated plants. Lettuce were totally untreated (control) or treated with 1 mM BOA (BOA), or 1 mM BOA +60 mM ClNa (BOA+Salt)
First dimension gels were Immobilized pH Gradient gel strips (IPG strips, GE Healthcare) of 7 cm length with a pH range of 3–10 NL. Second dimension gels were precast polyacrylamide gels, ExcelGel 2-D homogeneous 12.5% from GE Healthcare (Sánchez-Moreiras and Reigosa 2018). Spots were silver stained for identification.

transcriptomic and proteomic analyses with the intention of being able to compare them statistically.

In metabolomic studies data analysis includes two key steps. We focus the first step in acquiring a readable dataset from the analytical analysis and in identifying and quantifying the metabolites using commercial libraries and/or public databases. During, the second step we statistically analyze the data using both multivariate and univariate techniques. The last step that we carry out during metabolomic studies pipeline consists in the pathway analysis, which combines the Pathway Enrichment Analysis (PEA), which allow us to quantify the pathway impact exerted by the treatment (Xia and Wishart 2010). Recently, new software based on Machine-Learning techniques (neural networks) have been developed in order to compare -omics data, but once again, there is a lag regarding transcriptomic and proteomic analyses with the intention of being able to compare them statistically (Huan et al. 2017). Therefore, the experimental design newly confirms its pivotal

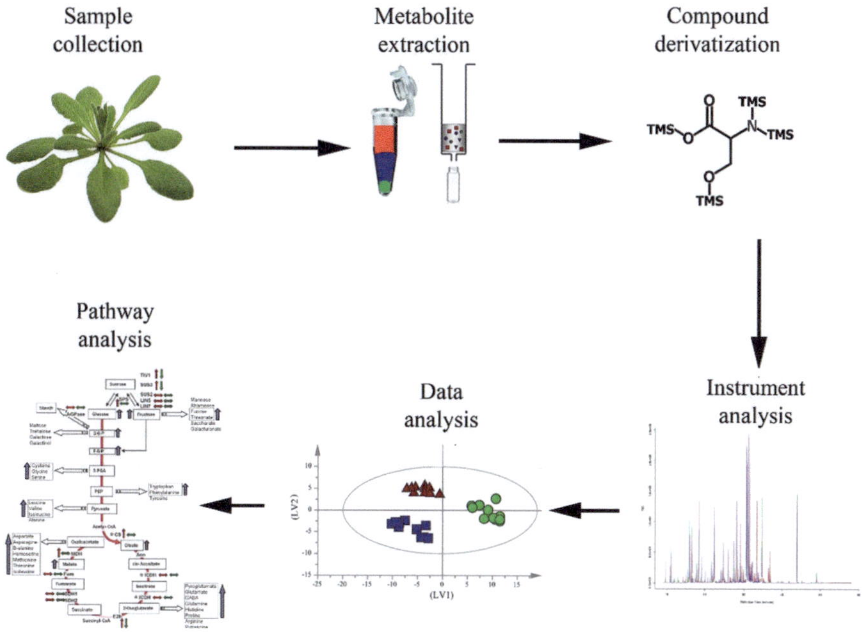

Fig. 22.6 Procedure generally adopted in GC-MS-based metabolomics

role in -omics studies. The profile of the altered metabolites after the treatment with each of the compounds is established to obtain the altered metabolic pathways (for further information see Araniti et al. 2017a).

The transcriptomic, proteomic and metabolomic profiles are statistically treated to approach to the mode of action from a complete plant perspective.

2.1.4 Structure-Activity Relationship (SAR) and Molecular Docking

Molecular docking and SAR studies complement all the data previously obtained of the effects on plant metabolism, and give more information about the behavior of the tested molecules into the plant. Two approaches are proposed to establish the structure activity relations and the mechanism of molecular action. One based on the structure of the ligands (QSAR) and another based on the structure of the biological receptor (Molecular Docking). For further information see González et al. 2008a, b and Chap. 20.

- QSAR methodologies. A quantitative study of structure-activity relationships from series of tested is carried out, following the steps shown below: (1) Construction of herbicide data sets. Phytotoxic compounds and plants growth promoters from scientific publications and patents are exported using the CAS (Chemical Abstract Service database). The chemical structures are ensured through their standardization and cleanliness using "QSARDW" application; (2) Classification of families of compounds and molecular fragmentation. In a first division, homogeneity is sought in the biological test protocol and in a second sub-classification, common structural areas (scaffolds) are defined, which will be explored from a toxicological point of view; (3) Calculation of molecular descriptors. Topological and fragment descriptors are calculated making a selection of the most appropriate according to the property to study. If required, those containing the same information or having a correlation higher than 85%, can be eliminated by Pairwise Correlation (González et al. 2005) in order to avoid multicollinearity among the independent variables of the future linear models. (4) Construction and validation of QSAR models. In order to guarantee the reliability of the models obtained, a validation with an external prediction series is carried out. Therefore, the dataset is divided into training, prediction and external validation sets, using Cluster analysis; The OECD principles are prioritized (Guidance Document On The Validation Of (Quantitative) Structure-Activity Relationships [(Q)Sar] Models). Different statistical methodologies are explored, looking for robust models, with good predictability and selection of descriptors easily interpretable.

- Molecular docking modeling of the ligand-receptor complex is carried out to study the binding sites between herbicide compounds and their biological target while making a comparison of the herbicidal potency between different compounds on the same receptor. The three-dimensional structure of the receptor (X-ray or NMR structures) is obtained from the Protein Data Bank (PDB) database and, when it is not available, homology models are constructed with a reference protein that has a certain similarity in the sequence and whose crystallized structure is known (for further information see Graña et al. 2017).
- Comparison of the results obtained with the already known modes of action of chemical herbicides. The results obtained in the previous analyses are included into a database with modes of action of herbicides, xenobiotics and other known molecules and a comparison of the altered genes, proteins and metabolites is carried out for each of the compounds in order to identify their mode/s of action. The non-correlation can be indicative of a novel mode of action that needs to be established through the integrated analysis of all known data for the tested compound.

2.2 Phytotoxic Potential of Secondary Metabolites on Crops and Associated Weeds

The phytotoxic potential on different species of crops and associated weeds is investigated *in vitro* by laboratory bioassays (see Chap. 1).

The phytotoxic capacity of the selected compounds is assessed first by germination and growth bioassays on crops and weeds, performing dose-response curves from which key parameters related to the phytotoxic potency of the tested compound are obtained. The commercially interesting crops that we usually use in these studies are *Hordeum vulgare* L., *Triticum aestivum* L., *Zea mays* L., *Oryza sativa* L., *Glycine max* L. and *Lactuca sativa* L., and their tested associated weeds are *Amaranthus retroflexus* L., *Echinochloa crus-galli* L., *Portulaca oleracea* L., *Plantago lanceolata* L., *Convolvulus arvensis* L., *Avena fatua* L., *Solanum nigrum* L. and *Thlaspi arvense* L (for further information see Graña et al. 2013b; Díaz-Tielas et al. 2014; Araniti et al. 2017b).

The number of germinated seeds, the weight of the seedlings and the length of the roots are recorded. At least 5 replications are used per concentration and compound in these bioassays. The most active secondary metabolites in the laboratory are subsequently applied in field and greenhouse bioassays in order to develop their potential natural herbicidal actions. For the realization of these bioassays, we usually select the crops and weeds that have shown a greater specificity and sensitivity, respectively. The objective is to find the most interesting crop–weed systems for further studies in integrated weed management.

Greenhouse and field experiments are carried out on loamy textured soil, homogenized by mechanical work on 25 cm depth. An experimental design with randomized tables (4 repetitions per treatment) of 1 × 1 m is established. The molecules are applied by spraying or irrigation, according to the proposed objectives. Germination is periodically evaluated and viable germinated weeds identified. Weeds are sampled at the time of flowering and their yield in fresh and dry weight is determined. Also, soil samples are taken to determine the persistence of the treatments applied. Prior to carrying out the tests, the weeds present in the test field and surroundings are inventoried. The tested concentrations of the compound and the selected crops and/or weeds species depend on the data obtained in the greenhouse bioassays for each molecule (Fig. 22.7).

2.3 Stability and Transport of Secondary Metabolites on Receptor Plants

The analysis of the amount of compound or its derivatives present in the plants in the different organs (root, stem and leaves) is done by GC-MS or HPLC-MS and is carried out on *Arabidopsis* or lettuce seedlings treated with IC_0, IC_{50} and IC_{80} concentrations for each compound (for further information see Sánchez-Moreiras et al. 2010. Once the treatment has started, the seedlings are harvested at different times. Gas-mass chromatography (GC-MS) is the most powerful method for the identification and quantification of relatively apolar organic molecules. For low volatility compounds that are not suitable for gas-mass chromatography, liquid chromatography is used. For this study, the compounds are applied by spraying or irrigation in order to elucidate whether their transport into the plant occurs via xylem, phloem, or both. The harvest times are determined based on the effect found in the previous analyses, in order to be able to correlate the presence of the compound into the plant with the effects measured in phase 1.

Knowing the interactions between plants, and the positive or negative effects that molecules (mainly secondary metabolites) released to the environment exert on other plants, will allow the advance of integrated agriculture, since the combination

Fig. 22.7 Effects of 0.62 mM citral on plants of *Avena fatua* grown in soil under greenhouse conditions. (Unpublished data)

of crops or the mulching of plants is a common practice in this type of agriculture that can induce changes in yield of future crops. Moreover, knowing the mode of action of plant growth regulating phytochemical substances will allow their use in the control of weeds in the framework of sustainable agriculture and at the same time will provide the industry with new modes of action for the preparation of organic substances more respectful with the environment to substitute as much as possible the traditional herbicides and to help to solve with their new target sites the problem of weed resistance.

References

Albers JW (1997) Understanding gene-environment. Environ Health Persp 105:578–580

Araniti F, Graña E, Krasuska U, Bogatek R, Reigosa MJ, Abenavoli MR, Sánchez-Moreiras AM (2016) Loss of gravitropism in farnesene-treated arabidopsis is due to microtubule malformations related to hormonal and ROS unbalance. PLoS One 11(8):e0160568

Araniti F, Lupini A, Sunseri F, Abenavoli MR (2017a) Allelopatic potential of Dittrichia viscosa (L.) W. Greuter mediated by VOCs: a physiological and metabolic approach. PLoS One 12:e0170161

Araniti F, Sánchez-Moreiras AM, Graña E, Reigosa MJ, Abenavoli MR (2017b) Terpenoid *trans-caryophyllene* inhibits weeds germination and induces plant water-status alteration and oxidative damage in *Arabidopsis* adult plants. Plant Biol 19:79–89

Baerson S, Sánchez-Moreiras AM, Pedrol N, Schulz M, Kagan IA, Agarwal AK, Reigosa MJ, Duke SO (2005) Detoxification and transcriptome response in *Arabidopsis* seedlings exposed to the allelochemical benzoxazolin-2(3*H*)-one. J Biol Chem 280:21867–21881

Berenbaum MR (2002) Postgenomic chemical ecology: from genetic code to ecological interactions. J Chem Ecol 28:873–896

Coba de la Peña T, Sánchez-Moreiras AM (2001) Cell cycle. In: Reigosa MJ (ed) Handbook of plant ecophysiology techniques. Kluwer Academic Publishers, Dordrecht, pp 65–80

Concenço G, Silva AF, Ferreira EA, Galon L, Noldin JA, Aspiazu I, Ferreira FA, Silva AA (2009) Effect of dose and application site on quinclorac absorption by barnyardgrass biotypes. Planta Daninha 27:541–548

Cushman JC, Bohnert HJ (2000) Genomic approaches to plant stress tolerance. Curr Opin Plant Biol 3:117–124

Dayan FE, Romagni JG, Duke S (2000) Investigating the mode of action of natural phytotoxins. J Chem Ecol 26:2079–2094

De Boer GJ, Thornburgh S, Gilbert J, Gast RE (2011) The impact of uptake, translocation and metabolism on the differential selectivity between blackgrass and wheat for the herbicide pyroxsulam. Pest Manag Sci 67:279–286

Díaz-Tielas C, Graña E, Sotelo T, Reigosa MJ, Sánchez-Moreiras AM (2012) The natural compound trans-chalcone induces programmed cell death in *Arabidopsis thaliana* roots. Plant Cell Environ 35:1500–1517

Díaz-Tielas C, Sotelo T, Graña E, Reigosa MJ, Sánchez-Moreiras AM (2014) Phytotoxic potential of trans-chalcone on crop plants and model species. J Plant Growth Regul 33:181–194

Duke SO, Owens DK, Dayan FE (2014) The growing need for biochemical bioherbicides. In: Biopesticides: state of the art and future opportunities, ACS symposium series, vol 1172. American Chemical Society, Washington, DC, pp 31–43

Fiehn O (2002) Metabolomics—the link between genotypes and phenotypes. In: Functional genomics. Springer, Dordrecht, pp 155–171

González MP, Terán C, Fall Y, Diaz LC, Helguera AM (2005) A topological sub-structural approach to the mutagenic activity in dental monomers. 3. Heterogeneous set of compounds. Polymer 46:2783–2790

González MP, Terán C, Teijeira M (2008a) Search for new antagonist ligands for adenosine receptors from QSAR point of view. How close are we? Med Res Rev 28:329–371

González MP, Terán C, Saíz-Urra L, Teijeira M (2008b) Variable selection methods in QSAR: an overview. Curr Topics Med Chem 8:1606–1627

Graña E, Sotelo T, Díaz-Tielas C, Araniti F, Krasuska U, Bogatek R, Reigosa MJ, Sánchez-Moreiras AM (2013a) Citral induces auxin-mediated malformations and arrests cell division in *Arabidopsis thaliana* roots. J Chem Ecol 39:271–282

Graña E, Sotelo T, Díaz-Tielas C, Reigosa MJ, Sánchez-Moreiras AM (2013b) The phytotoxic potential of the terpenoid citral on seedlings and adult plants. Weed Sci 61:469–481

Graña E, Costas-Gil A, Longueira S, Celeiro M, Teijeira M, Reigosa MJ, Sánchez-Moreiras AM (2017) Auxin-like effects of the natural coumarin scopoletin on *Arabidopsis* cell structure and morphology. J Plant Physiol 218:45–55

Gygi SP, Rochon Y, Franza BR, Aebersold R (1999) Correlation between protein and mRNA abundance in yeast. Mol Cell Biol 19:1720–1730

Huan T, Forsberg EM, Rinehart D, Johnson CH, Ivanisevic J, Benton HP, Fang M, Aisporna A, Hilmers B, Poole FL, Thorgersen MP, Adams MWW, Krantz G, Fields MW, Robbins PD, Niedernhofer LJ, Ideker T, Majumder EL, Wall JD, Rattray NJW, Goodacre R, Lairson LL, Thorgersen MP (2017) Systems biology guided by XCMS Online metabolomics. Nat Meth 14:461–462

Kim HK, Choi YH, Verpoorte R (2010) NMR-based metabolomic analysis of plants. Nat Protoc 5(3):536–549

Lisec J, Schauer N, Kopka J, Willmitzer L, Fernie AR (2006) Gas chromatography mass spectrometry–based metabolite profiling in plants. Nat Protoc 1(1):387–396

Martínez-Peñalver A, Reigosa MJ, Sánchez-Moreiras AM (2011) Imaging chlorophyll *a* fluorescence reveals specific spatial distributions under different stress conditions. Flora 206:836–844

Martínez-Peñalver A, Graña E, Reigosa MJ, Sánchez-Moreiras AM (2012a) The early response of *Arabidopsis thaliana* to cadmium- and copper-induced stress. Environ Exp Bot 78:1–9

Martínez-Peñalver A, Pedrol N, Reigosa MJ, Sánchez-Moreiras AM (2012b) The tolerance of *Arabidopsis thaliana* to the allelochemical protocatechualdehyde. J Plant Growth Regul 31:406–415

Pedrol N, Tiburcio AF (2001) Polyamines determination by TLC and HPLC. In: Reigosa MJ (ed) Handbook of plant ecophysiology techniques. Kluwer Academic Publishers, Dordrecht, pp 335–364

Sánchez-Moreiras AM, Pedrol Bonjoch N (2001) Two-dimensional electrophoresis. Stress proteins. In: Reigosa MJ (ed) Handbook of plant ecophysiology techniques. Kluwer Academic Publishers, Dordrecht, pp 297–333

Sánchez-Moreiras AM, Reigosa MJ (2005) Whole plant response of lettuce after root exposure to BOA (2(3*H*)-benzoxazolinone). J Chem Ecol 31:2689–2703

Sánchez-Moreiras AM, Reigosa MJ (2018) Proteomic analyses of BOA effects on lettuce leaves. J Allelochem Interact 4(2):in press

Sánchez-Moreiras AM, Coba de la Peña T, Martínez Otero A, Santos Costa XX (2001) Mitotic index. In: Reigosa MJ (ed) Handbook of plant ecophysiology techniques. Kluwer Academic Publishers, Dordrecht, pp 81–94

Sánchez-Moreiras AM, Coba de la Peña T, Reigosa MJ (2008) The natural compound benzoxazolin-2(3*H*)-one selectively retards cell cycle in lettuce root meristems. Phytochemistry 69:2172–2179

Sánchez-Moreiras AM, Pedrol N, González L, Reigosa MJ (2009) 2-3*H*-Benzoxazolinone (BOA) induces loss of salt tolerance in salt adapted plants. Plant Biol 11:582–590

Sánchez-Moreiras AM, Oliveros-Bastidas A, Reigosa MJ (2010) Reduced photosynthetic activity is directly correlated with 2-(3*H*)-benzoxazolinone accumulation in lettuce leaves. J Chem Ecol 36:205–209

Sánchez-Moreiras AM, Martínez-Peñalver A, Reigosa MJ (2011) Early senescence induced by 2-3*H*-benzoxazolinone (BOA) in *Arabidopsis thaliana*. J Plant Physiol 168:863–870

Schulze A, Downward J (2001) Navigating gene expression using microarrays – a technology review. Nat Cell Biol 3:E190–E195

Xia J, Wishart DS (2010) MetPA: a web-based metabolomics tool for pathway analysis and visualization. Bioinformatics 26(18):2342–2344

Xiong L, Zhu JK (2001) Abiotic stress signal transduction in plants: molecular and genetic perspectives. Physiol Plant 112:152–166

Zhang A, Sun H, Wang P, Han Y, Wang X (2012) Modern analytical techniques in metabolomics analysis. Analyst 137(2):293–300

Zivy M, de Vienne D (2000) Proteomics: a link between genomics, genetics and physiology. Plant Mol Biol 44:575–580

Chapter 23
Exploring Plants Strategies for Allelochemical Detoxification

Margot Schulz, Meike Siebers, and Nico Anders

1 Introduction

Plants must defense their habitat against individuals of other or sometimes of the own species. During evolution they developed methods to utilize suitable molecules (allelochemicals) which are released into the air or the soil by active or passive mechanisms and which suppress other plants.

In the soil, the rhizoplane and the rhizosphere can be temporarily enriched with such compounds, mostly secondary metabolites, which influence the composition of the microbiome in the narrow neighborhood (Ikeda et al. 2008; Wang et al. 2013). Alteration of the microbiome can be important, as microorganisms are often involved in allelopathic interactions. When the level of toxicity is high enough, which depends firstly on the chemical nature and the amounts of the accumulating compounds and secondly on the biomass of the releaser, germination of seeds, seedlings growth and even older plants can be directly or indirectly inhibited or injured, sometimes to death, by the original or converted compounds. These effects are most noted as they augur alternative practices for weed control. However, the affected plants start defense reactions, which can be well developed and imply possible recovery. These mechanisms include detoxification and degradation reactions. Such reactions are also of particular importance to avoid autotoxicity.

The knowledge about how plants cope with phytotoxins is a prerequisite for the design of alternative agricultural methods that make use of allelopathic interactions. It is essential to know which plants, cultivars or ecotypes of weeds and crops are

M. Schulz (✉) · M. Siebers
IMBIO Institut für Biotechnologie der Pflanzen, Universität Bonn, Bonn, Germany
e-mail: ulp509@uni-bonn.de

N. Anders
AVT-Enzyme Process Technology, RWTH Aachen University, Aachen, Germany

© Springer International Publishing AG, part of Springer Nature 2018

A. M. Sánchez-Moreiras, M. J. Reigosa (eds.), *Advances in Plant Ecophysiology Techniques*, https://doi.org/10.1007/978-3-319-93233-0_23

sensitive to an allelochemical and which ones are not, in particular as allelochemicals are seldom only phytotoxic but bioactive compounds directed to a broad spectrum of organisms. For instance, many allelochemicals are also focused as biological fungicides or insecticides. The molecular targets of receiver organisms are multiple, but the broad mode of action is doubtlessly of evolutionary advantage.

In this chapter, methods to recognize allelopathic effects on the soil microbiome, and the elucidation of plant defense and detoxification reactions are focused by means of examples. Experimental procedures are described which allow the use of innovative methods for analyses.

2 Recognition of Allelopathy Caused Problems in Agricultural Ecosystems

2.1 *Analyses of Allelochemical Induced Shifts in Microbial Biodiversity by High-Throughput Sequencing Methods*

Growth inhibition by plant-released allelochemicals is difficult to verify, because microorganisms are more and more recognized as mediators of the effects. Either they convert compounds to the functional phytotoxins or converted compounds extinguish beneficial microorganisms which impacts plant growth negatively. Thus, allelopathic effects can be indirectly triggered by the microbiome composition. On the other hand, detoxification properties of plants may substantially weaken by the loss of the microbial diversity since microorganisms often support detoxification processes.

High-throughput sequencing methods are presently the best way to study effects of allelochemicals on the microbial diversity since more than 99% of the microbiome presents uncultured organisms. For probing, bacterial 16S (U3441F-U806R) and, for instance, fungal ITS (ITS7F-ITS4R) regions are sequenced. Aside from the characterization of the soil collection site, the analyses of soil parameter, such as nitrogen, phosphate, sulfur and metal contents, pH and others are necessary. The cultivation history of the particular soil must be known, for instance which crops were previously cultured or which agrochemicals were used. After collection of the soil, the samples must repose for several days until the microbiome has recovered from disturbances caused by the collection (Siebers et al. submitted). For a standard DNA extraction 250–500 mg of fresh soil are used for DNA extraction with suitable DNA extraction kits according to the manufacturer's instruction. Concentrations of DNA are adjusted to a minimum of 5 ng μL^{-1}. Samples must be stored at −80 °C.

2.2 Soil Signature Lipid Biomarker Analysis

Signature lipid biomarker (SLB) analysis can be employed to give detailed insights into soil microbial community structure (Kruse et al. 2015). All living cells are surrounded by a plasma membrane, which is a barrier against the environment. The plasma membrane is constituted by a lipid bilayer, containing a diverse array of lipids, which are essential structural components of the cell membranes, comprising, among others, glycerolipids, sphingolipids, and sterols (Fig. 23.1).

Lipids can also serve as storage compounds (triacylglycerols) and participate in many biochemical processes such as signaling (jasmonic acid, lyso-phosphatidylcholine), protein modification (dolichol) and plant-microbe interactions (Siebers et al. 2016). They are defined according to their solubility properties, being insoluble in water, but soluble in non-aqueous solvents such as chloroform or alcohols. Lipids are characterized by a high structural diversity and therefore, suitable biomarkers to monitor adaptations to altered environmental conditions. Phospholipids are composed of two fatty acids esterified at *sn*-1 and *sn*-2 position to a glycerol backbone and a polar headgroup attached at *sn*-3 position.

After cell death, phospholipids are characterized by a rapid degradation into neutral lipids such as diacylglycerol (DAG) or into phosphatidic acid (PA) due to enzymatic hydrolysis. Hence, the amount of intact phospholipids in soil is highly correlated with the living microbial biomass in soil and can be employed as a quantitative measure of the viable biomass (White et al. 1993). The diacylglycerol moiety

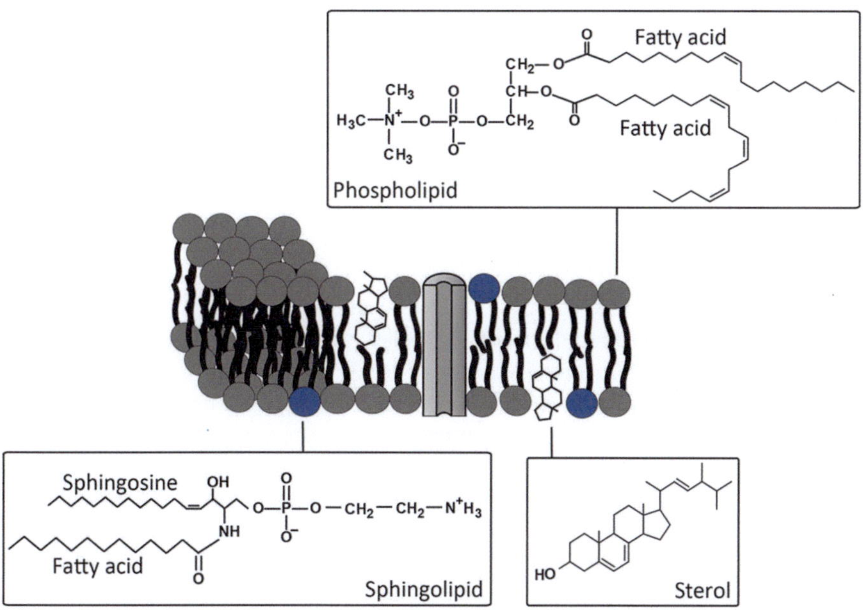

Fig. 23.1 Simplified presentation of the lipid bilayer and the different membrane lipids included

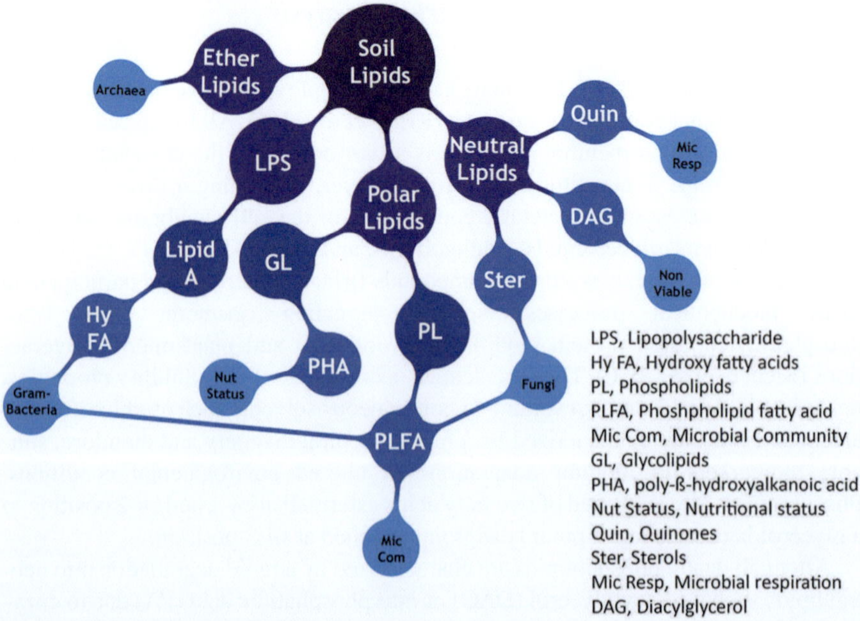

Fig. 23.2 Signature lipid biomarker in soil

of PA is incorporated into different phospholipids by phosphate ester condensation. Distinct amino-alcohols [e.g. phosphatidylserine (PS), phosphatidylcholine (PC), phosphatidylethanolamine (PE), phosphatidylglycerol (PG) and phosphatidylinositol (PI)] can be attached to the phosphate moiety of the headgroups. The fact that the different headgroups can be combined with a large number of fatty acids, varying in chain length and degree of desaturation leads to a vast number of different phospholipid molecular species. In addition there are also bacteria that ubiquitously occur in soil having ether-linked glycerophospholipids as membrane lipids which are characterized by an ether-linked acyl chain at the *sn*-1 and *sn*-2 position of the glycerol backbone. These ether bond phospholipids are most prominent in Archaea (Albers et al. 2000) and are more resistant to oxidation and high temperatures than ester bonds. Since the fatty acid composition in phospholipids varies widely among different microorganisms, their distribution profiles can be employed reflecting the soil microbial community structure (White et al. 1993) (Fig. 23.2).

Phospholipid-derived fatty acid (PLFA) analysis is a widely used method to measure the microbial biomass and the community structure composition (Zelles 1999; Frostegård et al. 2011; Buyer and Sasser 2012). A tremendous increase in our knowledge of the role of lipids in plant-microbe-soil interactions has been achieved since the availability of highly sensitive and quantitative analytical technologies utilizing mass-spectrometry (MS), gas chromatography (GC), and high-pressure liquid chromatography (HPLC) (Welti and Wang 2004; Wewer et al. 2011). To give a reliable community fingerprint of soil microorganisms, several analytical methods has to be assessed.

2.3 Lipid Extraction from Soil Samples

The described procedure for phospholipid extraction (Fig. 23.3) from soil is based on Bligh and Dyer (1959) with slight modification. To minimize contamination only glass- and teflon-ware is used. All solvents are HPLC or GC grade and contain 0.01% (w/v) butylated hydroxytoluene as antioxidant (Welti et al. 2002). Glassware is washed with chloroform prior to usage. For lipid extraction from soil, samples (fine earth <2 mm, ~5 g wet weight) are placed into a 50 mL glass vial with a teflon-lined screw cap.

Total lipids are extracted in two steps. Lipid-extract A: 10 mL of chloroform/methanol/formic acid [1:1:0.1] is added to the sample and incubated under continuous shaking at 150 rpm for 1 h at room temperature. The presence of formic acid in the solvent prevents lipid degradation by lipases during lipid isolation (Browse et al. 1986). After centrifugation at 1500 x g for 10 min, the supernatant is transferred to another 50 mL glass vial with teflon-lined screw cap and evaporated under N_2. Lipid-extract B: the residue of extraction A is re-extracted with 8 mL of chloroform/methanol [2:1]. After incubation for 30 min at RT and subsequent centrifugation at 1500 x g for 10 min, the supernatant are combined with lipid extract A. Then, 2 mL of 1 M KCl/0.2 M H_3PO_4 is added to the lipid extract and the mixture is shaken vigorously.

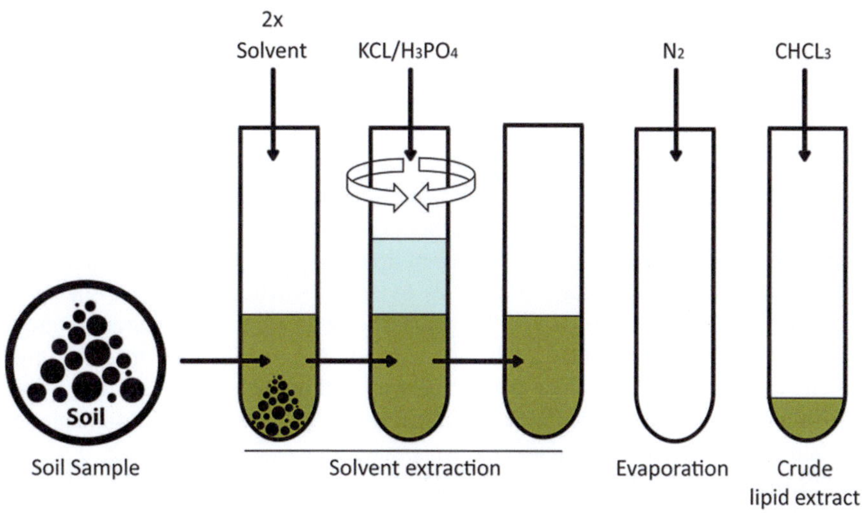

Fig. 23.3 Flowchart for the extraction of crude lipids from soil samples

Phase separation is achieved by centrifugation at 1500 x g for 10 min. The lower organic phase containing the total lipids is harvested, completely evaporated, and eluted in chloroform. Samples can be stored at −20 °C for later preparation or directly used. Finally, samples are dried for 48 h at 105 °C to determine the dry weight.

2.4 Lipid Fractionation by Normal-Phase SPE and FAME Analysis

A commonly used method to separate phospholipids from other lipids is solid phase extraction (SPE). The normal phase SPE typically involves a polar analyte, a mid- to nonpolar matrix (e.g. acetone, chloroform, or hexane), and a polar stationary phase. Silica is the sorbent of choice for lipid fractionation, since lipids are first recovered in nonpolar solvents such as chloroform. The retention of an analyte under normal phase conditions is primarily due to the interactions between the polar functional groups of the analyte (e.g. lipids) and the polar groups on the sorbent surface (e.g. silica), for example due to hydrogen bonding. A compound which is adsorped by these mechanisms is eluted by passing a solvent that disrupts the binding – usually a solvent that is more polar than the sample's original matrix. Hence, the polar silanol groups at the surface of the silica sorbent will more strongly adsorb polar compounds such as phospholipids, rather than neutral lipids such as triacylglycerol or sterols (Fig. 23.4). Several procedures to isolate distinct lipid classed by this method have been described (Hamilton and Comai 1988; Kim and Salem 1990).

For lipid fractionation by normal phase SPE, silica columns (e.g. Strata Si, 100 mg), are equilibrated with 2 x column volume of chloroform. The lipid-extracts are subsequently loaded onto the column and the non-polar lipids are eluted with 4 mL of chloroform. In a second step the more polar glycolipids are eluted with 4 mL of acetone/isopropanol [1:1]. Finally, the polar lipid fraction, including the phospholipids, is eluted by applying 4 mL of methanol. The methanol phase, which is highly enriched in phospholipids is then evaporated under N_2 and dissolved in the respective solvent (e.g. chloroform) prior to further analysis.

Direct analysis of phospholipids and free fatty acids via gas chromatography (GC) is impaired due to their high polarity. Therefore, derivatisation to the more volatile fatty acids methyl esters (FAMEs) of lipids and fatty acids by acid-catalyzed methylation is necessary for the analysis by GC. FAME synthesis can be done according to Browse et al. (1986). On this account ~150 μL lipid extract is incubated at 80 °C for 1 h in 1 mL 1 N HCl in methanol. After cooling down to RT, 1 mL hexane and 1 mL 0.9% NaCl is added and vortexed thoroughly. To achieve phase separation samples are centrifuged for 3 min at 1000 x g. The upper hexane phase, containing the FAMEs, is completely evaporated under N_2. Subsequently, FAMEs are dissolved in a defined volume of hexane. The quantification of fatty acids by GC depends on the availability of suitable internal standards, which must be absent in the sample. Therefore, a defined amount of an internal standard (e.g. pentadecanoid acid, 15:0) is added to the sample.

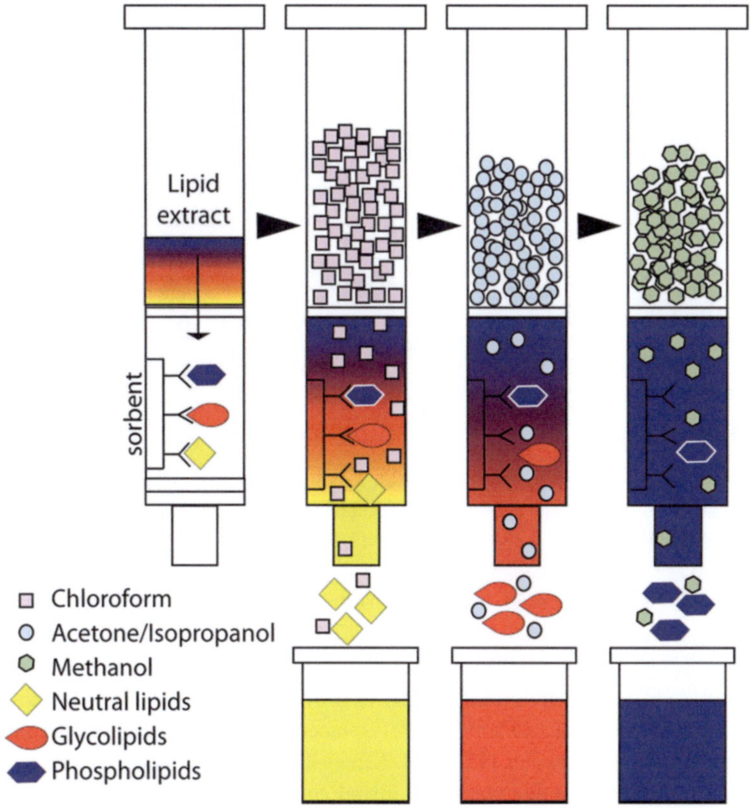

Fig. 23.4 Scheme of normal phase solid phase extraction (SPE). Lipids are fractionated according to their polarity by the use of a three solvent system

2.5 TLC of Phospholipids

For the PLFA analysis by GC the extracted and separated phospholipids are methanolyzed and the acyl groups are converted into their methyl esters after cleavage. In this approach only the acyl groups but not the headgroups can be analyzed. The GC-based methods are unable to provide information about the different molecular species of phospholipids. To this end the separation by thin layer chromatography (TLC) prior to FAME synthesis is required (Wu et al. 1994). On this account silica gel is most commonly used combined with different solvent systems (e.g. acetone/toluene/water [91:30:8]) (Kahn and Williams 1977). Once separated, the acyl groups can be quantified within the different phospholipid classes by isolating individual lipids from the silica material (Benning and Somerville 1992).

For TLC, silica gel plates and a suitable mobile phase can be used as a simple method for the separation of the different phospholipid classes. The one-dimensional TLC (Fig. 23.5a) system described here is a commonly used method to separate the

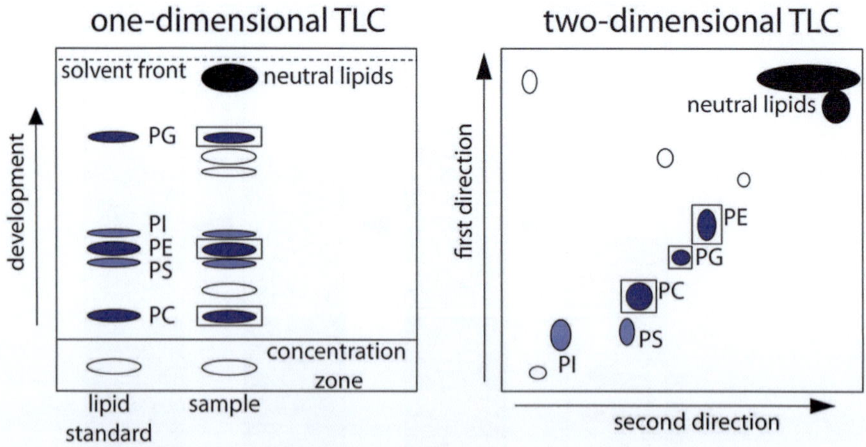

Fig. 23.5 Schematic separation of phospholipids on silica gel by one-dimensional TLC and two-dimensional TLC. PC, phosphatidylcholine; PS, phosphatidylserine; PE, phosphatidylethanolamine; PI, phosphatidylinositol; PG, phosphatidylglycerol

different phospholipid classes extracted from soil. First, TLC plates are submerged in 0.15 M ammonium sulfate and dried at RT for several days before activation at 120 °C for 2.5 h prior to use. Lipids, eluted in chloroform/methanol [2:1] are loaded onto the plate as 1 or 2 cm parallel streaks in the concentration zone. Plates are short air-dried in a fume hood and subsequently placed in the chromatography tank containing acetone/toluene/water [91:30:8]. After ~45–60 min migration time the plates are dried in a fume hood and bands are visualized using different staining methods. The lipids are identified with the help of co-migrating standards.

Two-dimensional (2D) TLC (Fig. 23.5b) using two different solvent systems is performed to further separate lipid extracts with complex lipid compositions. On this account, the lipid extract is applied to the lower left corner of the silica plate, without concentration zone. After developing the first dimension the plate is air dried in a fume hood and placed in a second chromatography chamber containing a different solvent for the development of the second direction. The most efficient solvent system is to use a neutral or basic solvent in the first direction followed by an acidic solvent in the second direction, e.g. chloroform/methanol/conc ammonia [65:35:4] for the first dimension and butanol/acetic acid/water [60:20:20] for the second dimension. The main disadvantages of this method are the time-consuming double development and that no standard sample can be used to appreciate the variations, which may occur from one run to another. Lipids are visualized by spraying with *p*-anisaldehyde (ANS, 0.2% (w/v) 8-anilino-1-naphtalene sulfonic acid in Methanol), which is a general-purpose stain particularly good with groups having nucleophilic properties. ANS is also particularly good for preparative TLC, because it does not interfere with subsequent analyses by GC or mass spectrometry (MS). Quadrupole Time-of-Flight Mass Spectrometry (Q-TOF MS) represents, for instance, a method for the non-destructive identification and quantification of intact lipids such as phospholipids (Kruse et al. 2015).

3 Detoxification and Degradation Reactions

3.1 *Microorganisms as Helpers in Detoxification Processes-Concept of Synergism*

The consortium of root colonizing bacteria and fungi can be involved in detoxification networks by acting in concert with the plant (Fig. 23.6; Schulz et al. 2013; Haghi Kia et al. 2014; Pagé et al. 2015). This field is almost not investigated but implies intriguing possibilities to design novel agricultural practices. To gain more insights in the root associated cultivable microorganisms, roots of weeds and crops can be harvested, cut in pieces and placed on different agar media suitable for the growth of fungi and bacteria. Emerging colonies have to be purified by subsequent culture steps and the finally obtained isolates must be identified. Extracted DNA is used for PCR with ITS/additional or 16S RNA primers and PCR products must be purified (Haghi Kia et al. 2014). Kits are available for all of these steps. Isolation of fungal DNA can be problematic because cell disruption is more difficult than with bacteria and plant tissues. Many microorganisms excrete nuclease, which has to be taken in account. Sequencing for the identification of the microorganisms is mostly done by companies.

When pure isolates are obtained, microorganisms can be cultured in liquid media (Fig. 23.7). Culture conditions and media must be adapted to the requirements of a given microorganism. The organisms can be tested for their ability to convert an allelochemical. For the tests, aliquots of the growing culture are supplemented with increasing amounts of the allelochemical of interest. Controls are run without supplements and with the organic solvent used for the solution of the allelochemical. Aliquots (100–200 μL) of the media are taken every 6–12 h and the samples are monitored for compound degradation, transformation products and detoxification intermediates by HPLC and MS-methods. If new compounds related to the applied allelochemical occur and they are stable enough, they can be isolated from the centrifuged and concentrated media by preparative HPLC or TLC. Compound identification by common spectrometric methods requires a high degree of purity. For most NMR-analyses 1–5 mg of the highly purified substance are necessary. The culture

Fig. 23.6 Microorganisms can convert allelochemicals to compounds with a reduced or enhanced phytotoxicity. Modulation of the molecules to initiate complete degradation is a common process. The intermediates are often unstable and only temporarily present, but may possess nevertheless bioactivity. Here, a *Pantoea ananatis* strain was incubated with different amounts of BOA-6-OH, yielding an orange colored transformation product, presenting the starter molecule for a hitherto unknown degradation pathway for BOA/BOA-6-OH (Schulz et al. 2017a)

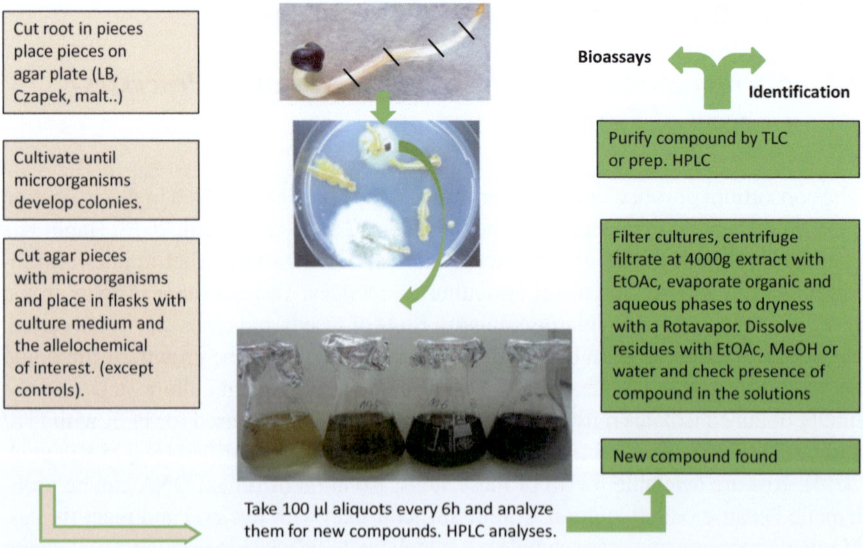

Fig. 23.7 Scheme of experimental steps to isolate a cultivable microorganism from the root surface, which may be able to degrade or to convert allelochemicals

medium sometimes influences the conversion of compounds, which has to be considered. Microbial participations in detoxification processes may be investigated by incubation of isolates or of microbial communities with allelochemicals. Isolation of degradation/detoxification products and bioassays with the isolated compounds are prerequisites to elucidate pathway sequences that result in conversion and degradation of allelochemicals. The following scheme explains the experimental steps.

3.2 *Peroxidases and Glucosyltransferases – Two Key Classes of Enzymes Involved in Different Types of Detoxification*

Plants can collaborate with microorganisms to overcome the toxicity of a phytotoxin. When the plant gets in contact with the compound specific defense reactions are started, indicated by genome wide alterations of gene expression to adapt the metabolism to stress conditions and to mobilize detoxification/rescue processes. The changed gene expression after application of allelochemicals, demonstrated by several researchers (Baerson et al. 2005; Venturelli et al. 2015), is not focused here. Instead, events subsequently occurring to gene expression and protein biosynthesis are emphasized.

There seem to be common reactions of plants, which are started almost immediately after exposure to many different compounds. H_2O_2 production and enhanced activities of peroxidases in the apoplast and on the root surface belong to the unspecific responses. Metabolic sequences for detoxification are very similar to the ones

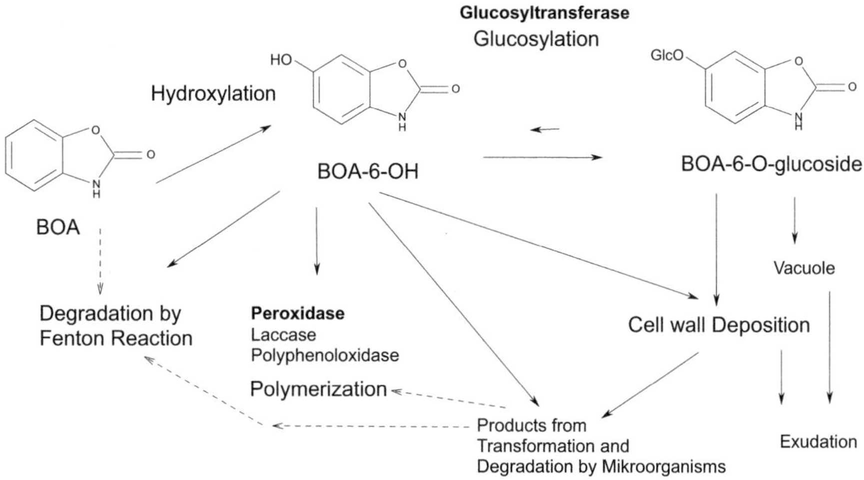

Fig. 23.8 Scheme of the possible fates of the allelochemical benzoxazolinone except for reaction sequences yielding phenoxazinones. (Haghi Kia et al. 2014)

known from synthetic herbicide detoxification. The process starts with the conversion yielding metabolites with increased susceptibility to further processing. Hydroxylations involving sometimes cyt P450s are common steps. In the second phase, conjugation with primary metabolites occur, such as glucosylations and/or malonylation, increasing water solubility of the metabolite, concomitant with a reduction of toxicity by masking reactive groups. In a third step, the detoxification products are fixed to cell wall material or they are transiently stored in the vacuole and exuded later. These steps require transporters. Step 1 detoxification products can also undergo polymerization reactions initiated by peroxidases, laccases or even by non-enzymatic reactions. Generated radicals of phenolic allelochemicals can be incorporated into cell wall polymers without further conversions. When the cell wall/root surface is the major detoxification site, detoxification products do not accumulate in the vacuole. Two important enzyme classes, peroxidases and glucosyltransferases have key functions in the different pathways, for instance of benzoxazolinone BOA detoxification (Fig. 23.8): transient detoxification (glucosylation/malonylation) and irreversible detoxification (polymerization and degradation).

Simple activity determinations of the two important enzyme classes are described more detailed.

3.2.1 Root Surface Peroxidase

ABTS (2,2-azinobis(3-ethylbenzthiazoline)-6-sulfonic acid) is a common substrate for peroxidase activity determination. ABTS is oxidized by the enzyme in presence of H_2O_2 which lead to the generation of a relatively stable, green colored radical. The plant seedlings are preincubated with an allelochemical, control seedlings with tap

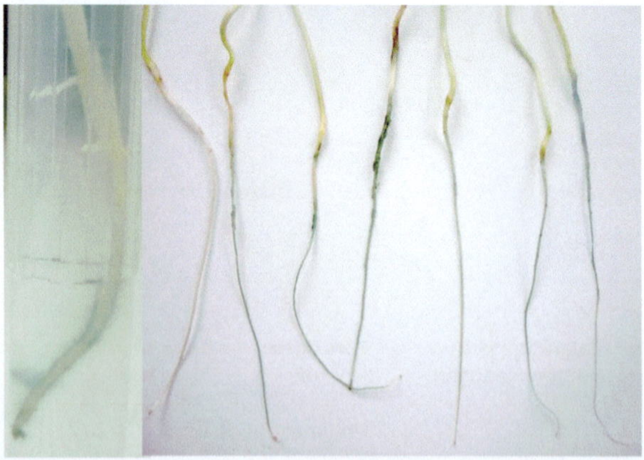

Fig. 23.9 Release of the green colored ABTS radical from *Abutilon* root surfaces placed in a cuvette into the assay mixture (Schulz et al. 2017b). Also the roots are green colored after the assay. (First root left: control, not incubated)

water (Courty et al. 2011). After washing the roots with water, the seedlings are placed into cuvettes (1 seedling/microcuvette) containing 1 mL tap water, 0.2 mL 100 mM phosphate buffer pH 7 supplemented with 80 μL H_2O_2 (1:60 dilution of 50% H_2O_2). The enzymatic reaction is started by adding of 200 μL 2 mM ABTS solution. The seedlings are incubated for 30 min. Then 200 μL of the assay solution is transferred into a in a microtiter plate and the color intensity immediately measured at 405 nm as an end determination. The enzyme activity can also be determined by monitoring the oxidation of ABTS at 405 nm. Aliquots of the reaction solution are taken every 5 min and the absorbance is determined with the plate reader. The enzyme activity is expressed as units per assay volume, where 1 U is defined as 1 μmol of substrate oxidized per min. The development of the green color also visualizes the presence of peroxidase at the root surface (Fig. 23.9). It is necessary to correct the measurements for laccase activity, another enzyme that can be involved in polymerization reactions in absence of H_2O_2 and to add catalase inhibitors (e.g., 3-amino-1,2,4-triazole). The total enzyme activity is composed of peroxidases from microorganisms and exuded plant peroxidase(s). It is necessary to establish standard curves with purchased enzymes.

3.2.2 Measurement of Soluble Glucosyltransferases

A common strategy to elucidate detoxification capacities of a plant for hydroxylated phenolics is to incubate seedlings with the compound of interest and to analyze the methanolic extract for stable detoxification products by HPLC/DAD or HPLC/MS methods. Assumed detoxification products are identified by common spectrometric methods (MS, NMR). When the compound is identified, it is possible to develop concepts which enzymes might be involved in the detoxification pathway (Fig. 23.10).

When glucosides are produced as detoxification products, glucosyltransferases are often the responsible enzymes for the conjugation of sugar moieties. Constitutive cytosolic glucosyltransferases involved in detoxification steps are relatively easy to measure in crude protein extracts. Allelochemical-incubated and control plants are used for soluble protein extraction (Schulz et al. 2016b). MES or Tris buffer pH 6–7,

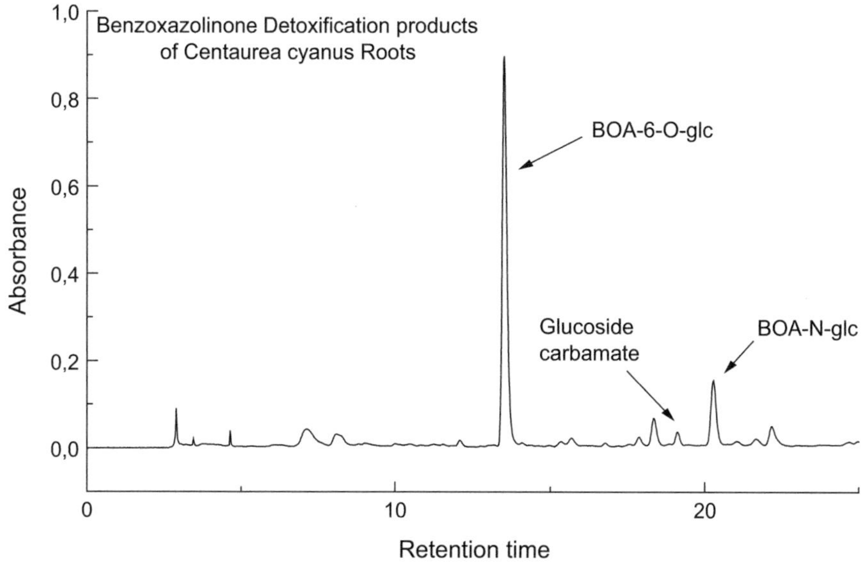

Fig. 23.10 Detoxification products can finally present major compound(s) in methanolic extracts. The chromatogram of the methanolic extract prepared from benzoxazolinone incubated cornflower roots shows glucosylated BOA-6-OH as the dominant compound. The two other arrows mark BOA-*N*-glucoside and glucoside carbamate as minor detoxification products in *Centaurea cyanus*. (Schulz and Wieland 1999)

supplemented with DTT, protease inhibitor cocktail (ready to use mixtures can be purchased from companies), is mixed 3:1, w/v with the plant material (roots or shoots) and homogenized in presence of 1–10% /FW PVPP (PVPP: polyvinylpyrrolidone, to bind extracted phenolics) on ice. Optimization of the extraction procedure is recommended. The slurry is centrifuged at 4 °C until the cell debris and PVPP are pelleted, the supernatant transferred into a fresh tube and stored on ice. The protein content is determined with the common Bradford method. Glucosyltransferase assays are composed of the extraction buffer, UDPG as sugar donor, the sugar acceptor (allelochemical) and protein solution.

The optimal assay composition (concentrations of UDPG and the sugar acceptor, pH, buffer concentration) depends on the properties of the enzyme and must be elucidated for any given glucosyltransferase. The amount of buffer and protein solution depends on the protein content of the extract. Concentrations and volumes to start with can be: 70 μL buffer, 10 μL UDPG (10 mM), 5 μL sugar acceptor (5 mM) and 30 μL (15–20 μg) protein solution. Activity determination should be within the linear phase of the reaction. Starting conditions for optimization of the incubation could be 30 min at 25 °C. The reaction is stopped by boiling, adding of acid or methanol or by extraction with ethyl acetate and use of the phases for product determinations. The optimal way to stop the reaction depends on the nature of the detoxified molecule. The assays or the assay extracts are centrifuged and analyzed by HPLC-DAD or HPLC-MS. The determination of the product concentration is necessary to calculate the enzyme activity (U).

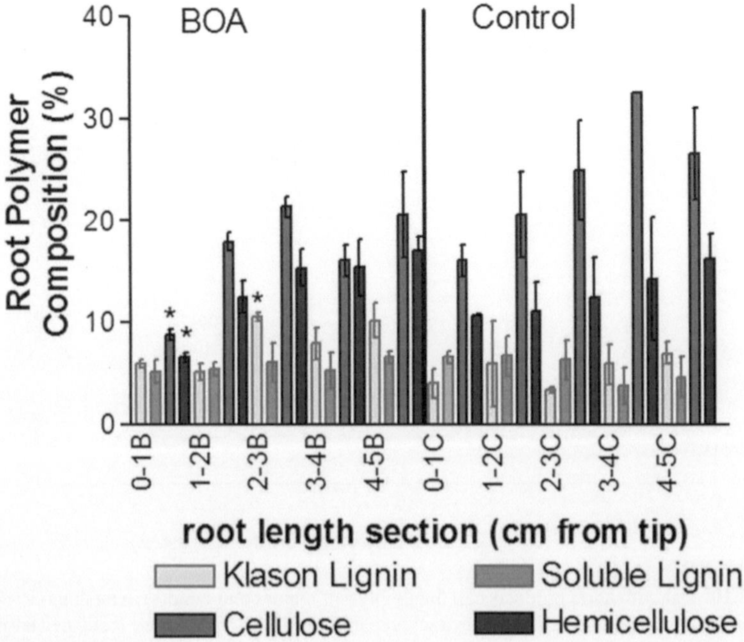

Fig. 23.11 Alteration of cell wall polymer upon exposure of maize seedlings to benzoxazolinone (BOA), (Schulz et al. 2016a)

The identification of the enzyme exhibiting activity against the allelochemical of interest is often possible by proteome analysis. A purification of the enzyme is time consuming and often difficult, but sometimes necessary because recombinant enzymes are not always active.

The cell wall is another site where glucosylations may occur, catalyzed by cell wall located glucosyltransferase. The hydrolysis of previously fixed radicals of allelochemicals to cell wall polysaccharides presents another possibility to generate glucosylated detoxification products. In maize, the amount of cell wall polysaccharides decreases during glucoside carbamate production, which a major detoxification product of benzoxazolinone in this plant (Schulz et al. 2016a).

The determination of cell wall polymer contents gives evidence of hydrolytic processes during detoxification (Fig. 23.11).

4 Cell Wall Analysis

The analysis of the plant cell walls for the natural polymers cellulose, hemicellulose, lignin and pectin was described first at the beginning of the twentieth century (reviewed in Sluiter et al. 2010). Here, the interwoven natural polymers (Fig. 23.12)

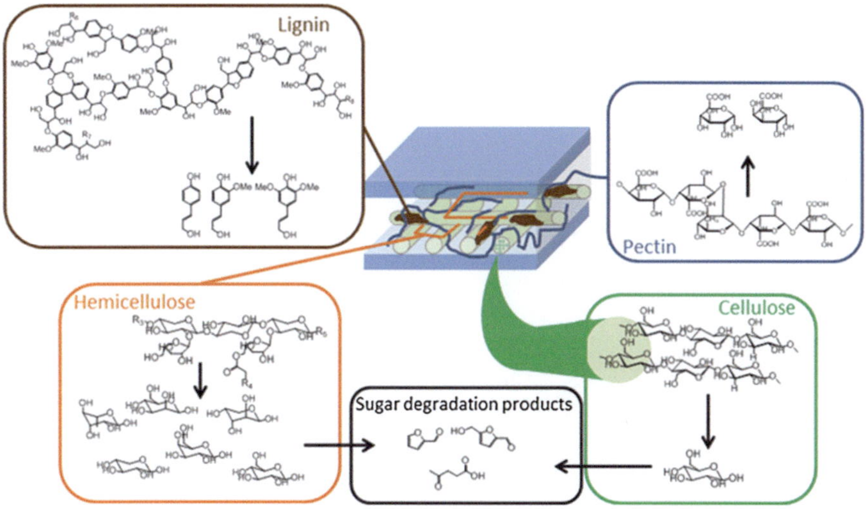

Fig. 23.12 Simplified presentation of the plant cell wall, the natural polymers that can be isolated from the plant cell wall and the degradation products of the monosaccharides isolated from both cellulose and hemicellulose

are pulled apart by a two-step acid hydrolysis of the polysaccharide fractions into the corresponding monomers. These monomers can be classified as the hexose sugar glucose derived from the natural polymer cellulose and the pentose sugars arabinose and xylose as well as the hexose sugars galactose, glucose and mannose from the natural polymer hemicellulose. The remaining solid resulting from the acid hydrolysis is declared as Klason-Lignin. Additionally, several publications mention the presence of soluble lignin compounds in the hydrolyzate (Sluiter et al. 2012).

Several new techniques and fractionation steps such as biomass dissolution in ionic liquids and nuclear magnetic resonance spectroscopy (NMR) as well as cell wall isolation and subsequent targeted hydrolysis have been published since the beginning of the twentieth century and have been improved and extensively used since (Foster et al. 2010a, b Part I and Part II; Chen et al. 2013; Picart et al. 2017; Viell et al. 2016). Nonetheless, the acid hydrolysis of the whole plant remains one of the fastest and easiest to realize methods.

4.1 Hydrolysis Mechanism

The two-step acid hydrolysis is divided into the break-up of the crystalline structure and the hydrolysis of the mainly accrued amorphous polysaccharides into the corresponding mono-saccharides. The first step also referred to as liquefaction is nowadays realized using sulfuric acid (72%) and moderate temperatures (30 °C) (Sluiter et al. 2012). Next to the desired cleavage of the hydrogen bond a small proportion

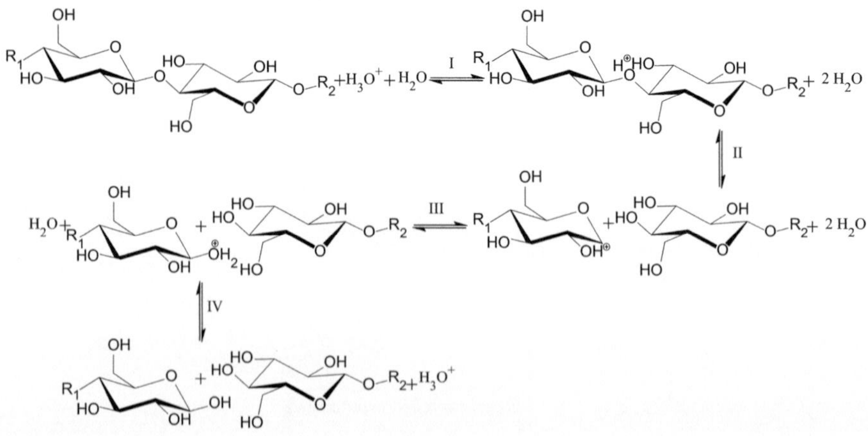

Fig. 23.13 Mechanism of the cleavage of the β-1,4 bonds using an acid catalyst according to Nevel and Upton 1976

of β-1,4-glycosidic bonds is cleaved. However, the lack of free water due to the high acid concentration prevents the complete hydrolysis of the polymer (see Fig. 23.13 – step III). In particular under moderate temperatures this phenomenon is known as recombination. The polysaccharide is cleaved after the protonation of the oxygen in the β-1,4-bond, however, under almost water-free conditions the chemical equilibrium tends to work against this compulsion by closing the bond again.

In order to hydrolyze the cell wall polysaccharides to the corresponding monosaccharides the second step is conducted under a moderate acid concentration (sulfuric acid, 4%) and high temperatures (121 °C) (Sluiter et al. 2012). Here, the acid attacks the β-1,4-glycosidic bond and after the addition of a proton the bond is cleaved. As the water to acid ratio is much higher for the second step the bond is cleaved by addition of a molecule of the free water (see. Fig. 23.13 – step III). The dilute acid and high temperatures as well as the low amount of biomass compared to the aqueous solvent guarantee a minimum of monosaccharide degradation products such as 5-hydroxymethylfurfural and furfural. The formed monosaccharides can subsequently be measured using chromatographic techniques allowing for a quantification of the hydrolysis reaction and a characterization of the starting material.

4.2 *Hydrolyzate Analysis*

Starting from compound identification via selective chemical reactions in the early nineteenth century the analytical efforts have become more and more complex as the hydrolysis of plant cell walls leads to compounds of diverse polarity and molecular size. Thus, various analytical techniques such as spectroscopy and chromatography are required.

As the analysis of monosaccharides is crucial for the determination of the cellulose and hemicellulose content of hydrolyzed lignocellulosic biomass the separation can be obtained using both liquid and gas chromatography (Cowie and Hedges 1984; Agblevor et al. 2004; Anders et al. 2015). However, gas chromatography is time-consuming due to the additional derivatization step, which is required for the transport of the monosaccharides into the gas phase (Medeiros and Simoneit 2007). In contrast, for liquid chromatography (LC), the liquid hydrolyzates can directly be measured after liquid-solid separation. In particular, the advantage of providing additional information about degradation products of monosaccharides and potential fermentation inhibitors leads to the wide distribution of LC in hydrolyzate analysis. High performance anion exchange chromatography coupled to pulsed amperometric detection (HPAEC-PAD) became one of the most utilized devices for hydrolyzate analysis. The opportunity to separate soluble degradation products from all natural polymers is exemplarily shown in Fig. 23.3 for maize roots (Anders et al. 2015; Anders et al. 2017; Cuerten et al. 2018; Schulz et al. 2016a). Soluble maize derived compounds are separated in the HPAEC-PAD using an anion-exchange mechanism (Fig. 23.14). The anions are produced due to the acidity of the monosaccharides and alcoholic as well as phenolic compounds. Subsequently, after the separation on the anion-exchange column material, the anions are oxidized in the detector leading to the peaks in the chromatogram. In order to draw a conclusion of the amount of the natural polymer cellulose or hemicellulose the amount of the monosaccharides needs to be multiplied with the factor 0.9 for hexose and 0.88 for pentose monosaccharides. This calculation is based on the amount of water which is added for the cleavage of the β-1,4-bonds.

Separation due to liquid-solid separation occurs for cell wall elements such as Klason-lignin or salts which are not soluble or degradable to smaller compounds in 4% sulfuric acid.

Subsequently, the solid is washed and dried in order to determine the non-acid-soluble residue content (Sluiter et al. 2012). After this determination the solid is burned at 550 °C in order to determine the ash and Klason-lignin content of the plant (Fig. 23.15).

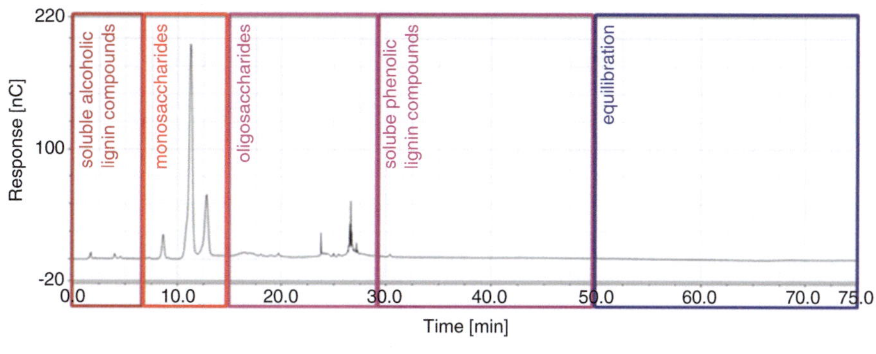

Fig. 23.14 HPAEC-PAD Chromatogram of an acidic liquefied maize root sample

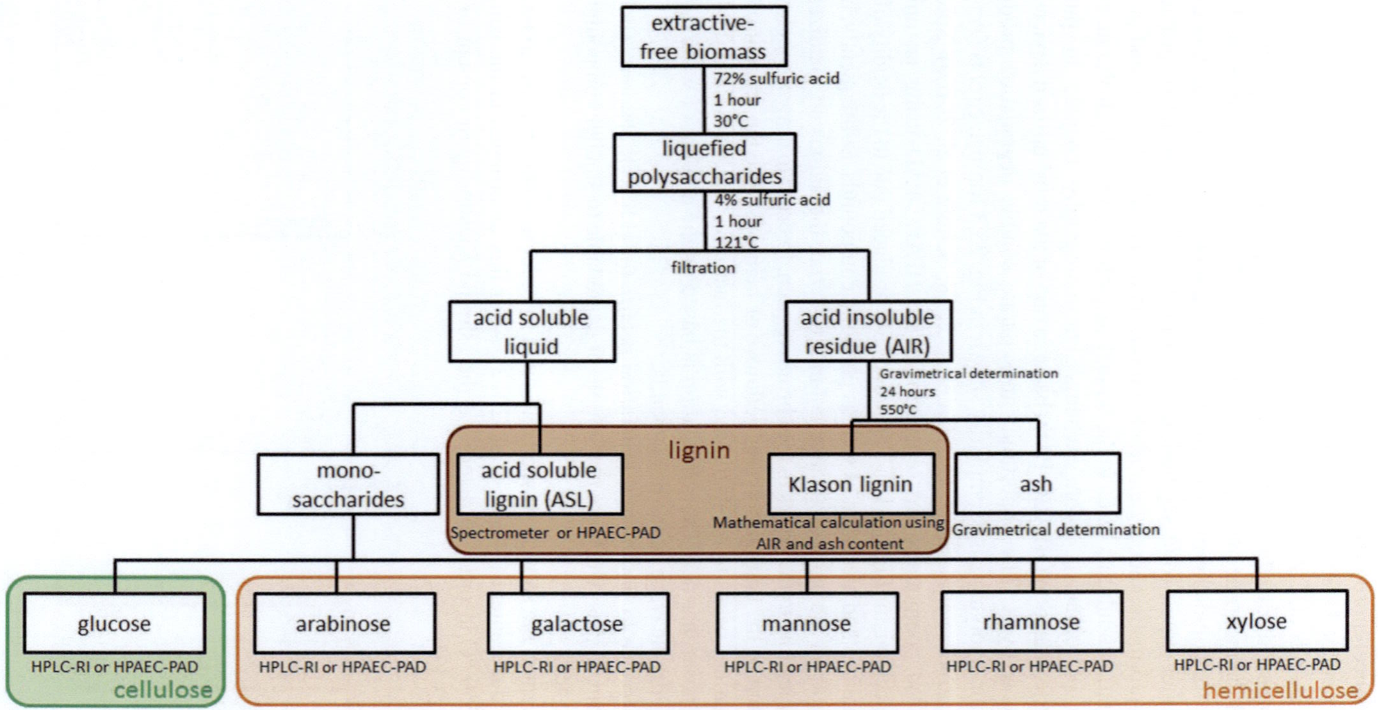

Fig. 23.15 The scheme depictures several experimental steps necessary for cell wall analysis and presents a flowchart for the characterization of extractive-free biomass

References

Agblevor FA, Murden A, Hames BR (2004) Improved method of analysis of biomass sugars using high-performance liquid chromatography. Biotechnol Lett 26:1207–1210

Albers SV, Van de Vossenberg JL, Driessen AJ, Konings WN (2000) Adaptations of the archaeal cell membrane to heat stress. Front Biosci 5:813–820

Anders N, Humann H, Langhans B, Spiess AC (2015) Simultaneous determination of acid-soluble biomass-derived compounds using high performance anion exchange chromatography coupled with pulsed amperometric detection. Anal Methods 7:7866–7873. https://doi.org/10.1039/C5AY01371B

Anders N, Schelden M, Roth S, Spiess AC (2017) Automated chromatographic laccase-mediator-system activity assay. Anal Bioanal Chem 409(20):4801–4809

Baerson SR, Sánchez-Moreiras AM, Pedrol-Bonjoch N, Schulz M, Kagan IN, Agarwal AK, Reigosa MJ, Duke SO (2005) Detoxification and transcriptome response in Arabidopsis seedlings exposed to the allelochemical benzoxazolinone. J Biol Chem 280:21867–21881

Benning C, Somerville CR (1992) Isolation and genetic complementation of a sulfolipid-deficient mutant of *Rhodobacter sphaeroides*. J Bacteriol Parasitol 174:2352–2360

Bligh EG, Dyer WJ (1959) A rapid method of total lipid extraction and purification. Can J Biochem Physiol 37:911–917

Browse J, McCourt PJ, Somerville CR (1986) Fatty acid composition of leaf lipids determined after combined digestion and fatty acid methyl ester formation from fresh tissue. Anal Biochem 152:141–145

Buyer JS, Sasser M (2012) High throughput phospholipid fatty acid analysis of soils. Appl Soil Ecol 61:127–130

Chen K, Sorek H, Zimmermann H, Wemmer DE, Pauly M (2013) Solution-state 2D NMR spectroscopy of plant cell walls enabled by a dimethylsulfoxide-d6/1-ethyl-3-methylimidazolium acetate solvent. Anal Chem 85:3213–3221

Courty PE, Labbe´ J, Kohler A, Marcxais B, Bastien CJ, Churin JL, Garbaye J, Le Tacon F (2011) Effect of poplar genotypes on mycorrhizal infection and secreted enzyme activities in mycorrhizal and non-mycorrhizal roots. J Exp Bot 62:249–260

Cowie GL, Hedges JI (1984) Determination of neutral sugars in plankton, sediments, and wood by capillary gas chromatography of equilibrated isomeric mixtures. Anal Chem 56:497–504

Cuerten C, Anders N, Juchem N, Ihling N, Volkenborn K, Knapp A, Jaeger KE, Buechs J, Spiess AC (2018) Fast automated online xylanase activity assay using HPAEC-PAD. Anal Bioanal Chem 410(1):57–69

Foster CE, Martin TM, Pauly M (2010a) Comprehensive compositional analysis of plant cell walls (lignocellulosic biomass) part I: lignin. J Vis Exp 37:e1837. https://doi.org/10.3791/1745

Foster CE, Martin TM, Pauly M (2010b) Comprehensive compositional analysis of plant cell walls (lignocellulosic biomass) part II: carbohydrates. J Vis Exp 37:e1837. https://doi.org/10.3791/1837

Frostegård Å, Tunlid A, Bååth E (2011) Use and misuse of PLFA measurements in soils. Soil Biol Biochem 43:1621–1625

Haghi Kia S, Schulz M, Ayah E, Schouten A, Müllenborn C, Paetz C, Schneider B, Hofmann D, Disko U, Tabaglio V, Marocco A (2014) *Abutilon theophrasti*'s defense against the allelochemical benzoxazolin-2(3h)-one: support by *Actinomucor elegans*. J Chem Ecol 40:1286–1298

Hamilton JG, Comai K (1988) Separation of neutral lipid, free fatty acid and phospholipid classes by normal phase HPLC. Lipids 23:1150–1153

Ikeda S, Rallos LEE, Okubo T, Eda S, Inaba S, Mitsui H, Minamisawa K (2008) Microbial community analysis of field-grown soybeans with different nodulation phenotypes. Appl Environ Microbiol 74:5704–5709

Kahn MU, Williams JP (1977) Improved thin-layer chromatographic method for the separation of major phospholipids and glycolipids from plant lipid extracts and phosphatidyl glycerol and bis(monoacylglyceryl) phosphate from animal lipid extracts. J Chromatogr 140:179–185

Kim HY, Salem N (1990) Separation of lipid classes by solid phase extraction. J Lipid Res 31:2285–2289

Kruse J, Abraham M, Amelung W, Baum C, Bol R, Kühn O, Lewandowski H, Niederberger J, Oelmann Y, Rüger C, Santner J, Siebers M, Siebers N, Spohn M, Vestergren J, Vogts A, Santner J (2015) Innovative methods in soil phosphorus research: a review. J Plant Nutr Soil Sci 178:43–88

Medeiros PM, Simoneit BRT (2007) Analysis of sugars in environmental samples by gas chromatography – mass spectrometry. J Chrom A 1141:271–278

Pagé AP, Yergeau É, Greer CW (2015) *Salix purpurea* stimulates the expression of specific bacterial xenobiotic degradation genes in a soil contaminated with hydrocarbons. PLoS One 10(7):e0132062. https://doi.org/10.1371/journal.pone.0132062

Picart P, Liu H, Grande PM, Anders N, Zhu L, Klankermayer J, Leitner W, Domínguez de María P, Schwaneberg U, Schallmey A (2017) Multi-step biocatalytic depolymerization of lignin. Appl Microbiol Biotechnol 101(15):6277–6287

Schulz M, Wieland I (1999) Variation in metabolism of BOA among species in various field communities – biochemical evidence for co-evolutionary processes in plant communities? Chemoecology 9:133–141

Schulz M, Knop M, Muellenborn C, Steiner U (2013) Root-associated microorganisms prevent caffeine accumulation in shoots of *Salvia officinalis* L. Int J Agric Forest 3:152–158

Schulz M, Filary B, Kühn S, Colby T, Harzen A, Schmidt J, Sicker D, Hennig D, Hofmann D, Disko U, Anders N (2016a) Benzoxazolinone detoxification by N-glucosylation: the multi-compartment-network of *Zea mays* L. Plant Sign Behav 11:e1119962. https://doi.org/10.1080/15592324.2015.1119962

Schulz M, Kant S, Colby T, Harzen A, Schmidt J, Sicker D, Pourmoayyed P (2016b) *Zea mays* glucosyltransferase BX9 - an essential enzyme for benzoxazolinone detoxification. JAI 2:25–38

Schulz M, Sicker D, Schackow O, Hennig L, Hofmann D, Disko U, Ventura M, Basyuk K (2017a) 6-Hydroxy-5-nitrobenzo[d]oxazol-2(3H)-one—A degradable derivative of natural 6-Hydroxybenzoxazolin-2(3H)-one produced by Pantoea ananatis. Commun Integr Biol 10:e1302633. https://doi.org/10.1080/19420889.2017.1302633

Schulz M, Sicker D, Schackow O, Hennig L, Yurkov A, Siebers M, Hofmann D, Disko U, Ganimede C, Mondani L, Tabaglio V, Marocco A (2017b) Cross-cooperations of Abutilon theophrasti Medik. and root surface colonizing microorganisms disarm phytotoxic hydroxybenzoxazolin- 2(3H)-ones. Plant Signal Behav 12(8):e1358843. https://doi.org/10.1080/15592324.2017.1358843

Siebers M, Brands M, Wewer V, Duan HG, Dörmann P (2016) Lipids in plant–microbe interactions. Biochim Biophys Acta 1861:1379–1395

Sluiter JB, Ruiz RO, Scarlata CJ, Sluiter AD, Templeton DW (2010) Compositional analysis of lignocellulosic feedstocks. 1. Review and description of methods. J Agr Food Chem 58:9043–9053

Sluiter AD, Hames B, Ruiz RO, Scarlata CJ, Sluiter JB, Templeton DW, Crocker D (2012) Determination of structural carbohydrates and lignin in biomass, laboratory analytical procedure. In: NREL/TP-510-42618

Venturelli S, Belz RG, Kämper A, Berger A, von Horn K, Wegner A, Böcker A, Zabulon G, Langenecker T, Kohlbacher O, Barneche F, Weigel D, Lauer UM, Bitzer M, Becker C (2015) Plants release precursors of histone deacetylase inhibitors to suppress growth of competitors. Plant Cell 27:3175–3189

Viell J, Inouye H, Szekely NK, Frielinghaus H, Marks C, Wang Y, Anders N, Spiess AC, Makowski L (2016) Multi-scale processes of beech wood disintegration and pretreatment with 1-ethyl-3-methylimidazolium acetate/water mixtures. Biotechnol Biofuels 9(1):7

Wang C-M, Li T-C, Jhan Y-L, Weng J-H, Chou C-H (2013) The impact of microbial biotransformation of catechin in enhancing the allelopathic effects of *Rhododendron formosanum*. PLoS One 8(12):e85162. https://doi.org/10.1371/journal.pone.0085162

Welti R, Wang X (2004) Lipid species profiling: a high-throughput approach to identify lipid compositional changes and determine the function of genes involved in lipid metabolism and signaling. Curr Opin Plant Biol 7:337–344

Welti R, Li W, Li M, Sang Y, Biesiada H, Zhou HE, Rajashekar CB, Williams TD, Wang X (2002) Profiling membrane lipids in plant stress responses role of phospholipase Dα in freezing-induced lipid changes in Arabidopsis. J Biol Chem 277:31994–32002

Wewer V, Dombrink I, vom Dorp K, Dörmann P (2011) Quantification of sterol lipids in plants by quadrupole time-of-flight mass spectrometry. J Lipid Res 52:1039–1054

White DC, Meadows P, Eglinton G, Coleman ML (1993) In situ measurement of microbial biomass, community structure and nutritional status and discussion. Phil Trans R Soc Lond Ser A 344:59–67

Wu JR, James DW, Dooner HK, Browse J (1994) A mutant of *Arabidopsis* deficient in the elongation of palmitic acid. Plant Physiol 106:143–150

Zelles L (1999) Fatty acid patterns of phospholipids and lipopolysaccharides in the characterisation of microbial communities in soil: a review. Biol Fert Soils 29:111–129

Chapter 24
Chemical Characterization of Volatile Organic Compounds (VOCs) Through Headspace Solid Phase Micro Extraction (SPME)

Fabrizio Araniti, Sebastiano Pantò, Antonio Lupini, Francesco Sunseri, and Maria Rosa Abenavoli

1 Biosynthesis of Volatile Organic Compounds

Plants are sessile, dumb and deaf living organisms that produce a wide array of chemicals ranging from ethylene to complex nitrogen-containing alkaloids (Wink 2010).

They produce more than 200,000 chemicals and each species, depending on its complexity, contains among 5000 and 30,000 compounds and more than 1700 of them are volatile organic compounds (VOCs), identified in several families belonging to both gymnosperms and angiosperms (Knudsen et al. 2006; Knudsen and Gershenzon 2006). It has been estimated that plants could yearly release in the atmosphere about 500 tera-grams of carbon as isoprene, and probably a similar amount of carbon as monoterpenes (Dicke and Loreto 2010). Moreover, VOCs emission is not only restricted to the aerials parts but also to root system participates in their release (Wenke et al. 2010). VOCs can freely cross membranes and successively released, from both aerial parts (fruits, stems, leaves and flowers) and roots, into the atmosphere and into the soil (Pichersky et al. 2006). They cover a wide range of ecological roles such as attract pollinators, seed dispensers, signal in plant-plant communication, attractants or repellents of herbivores and pathogens, and finally they could regulate mycorrhizal growth and development (Dudareva et al. 2006; Dudareva and Pichersky 2008).

The biochemical pathways, from which VOCs originate, include four primary metabolic pathways (Fig. 24.1): green leaf volatiles are produced by the lipoxygenase

Authors Fabrizio Araniti and Sebastiano Pantò have equally contributed to this chapter.

F. Araniti (✉) · A. Lupini · F. Sunseri · M. R. Abenavoli
Department Agraria, Mediterranea University, Reggio Calabria, Italy
e-mail: fabrizio.araniti@unirc.it

S. Pantò
LECO European Application and Technology Center Biotechpark, Berlin, Germany

© Springer International Publishing AG, part of Springer Nature 2018

A. M. Sánchez-Moreiras, M. J. Reigosa (eds.), *Advances in Plant Ecophysiology Techniques*, https://doi.org/10.1007/978-3-319-93233-0_24

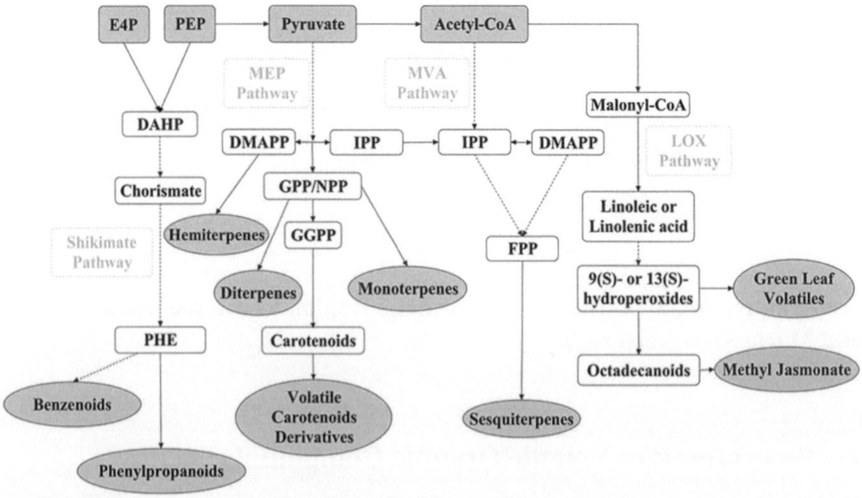

Fig. 24.1 Biosynthetic pathways involved in the synthesis and emission of plant volatile organic compounds (VOCs). Erythrose 4-phosphate (E4P); phosphoenolpyruvate (PEP); pyruvate; and acetyl-CoA are the main precursors of the majority of plant VOCs, which originate from primary metabolism. In dashed squares are reported the four major VOC biosynthetic pathways. In particular, the shikimate/phenylalanine, the mevalonic acid (MVA), the methylerythritol phosphate (MEP) and lipoxygenase (LOX) pathways, which products (VOCs) are reported in the light grey ellipses (benzenoids/phenylpropanoids, terpenoids, methyl jasmonate etc.). Solid and dashed arrows indicate single and multiple step enzymatic reactions, respectively. (DAHP) 3-deoxy-D-arabinoheptulosonate-7 phosphate; (DMAPP) dimethylallyl pyrophosphate; (FPP) farnesyl pyrophosphate; (GGPP) geranylgeranyl pyrophosphate; (GPP) geranyl pyrophosphate; (IPP) isopentenyl pyrophosphate; (NPP) neryl pyrophosphate; (PHE) phenylalanine. (Adapted from Dudareva et al. 2013)

pathway (GLV-s) (Hatanaka 1993); aromatic volatiles by the shikimic acid pathway (Paré and Tumlinson 1996); isoprene and monoterpenoids by the methylerythritol pathway (MEP) (Pichersky et al. 2006; Memari et al. 2013), whereas volatile sesquiterpenoids are produced through the mevalonic acid pathway (MVA) (Taveira et al. 2009; Memari et al. 2013; Rosenkranz and Schnitzler 2013). Thus, according to their origins, VOCs could be divided into several groups, including terpenoids, fatty acid derivatives, aromatic compounds and volatiles derived from amino acids other than L-phenylalanine (Fig. 24.2) reflecting both the complexity and the diversity of their metabolic origins (Dudareva et al. 2004).

1.1 Terpenoids: Biosynthetic Pathways

The largest group of plant volatiles is the group of terpenoids, which are primarily synthetized from dimethylallyl pyrophosphate (DMAPP) and its isomer isopentenyl pyrophosphate (IPP) (McGarvey and Croteau 1995) (Fig. 24.1). These two

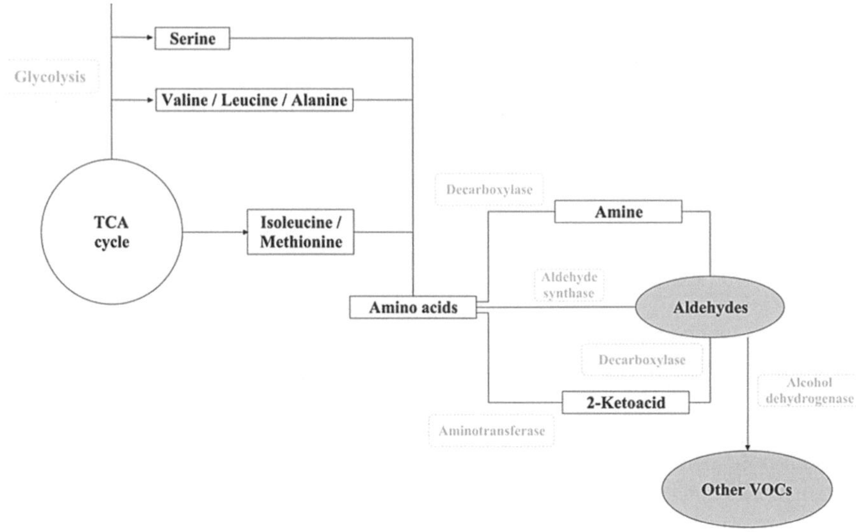

Fig. 24.2 Schematic representation of the biosynthesis of branched-chain amino acids-derived VOCs. It should be noted that VOCs production proceeds through different biosynthetic routes starting from derived- and non-derived-TCA (tricarboxylic acid) amino acids. (Adapted from Dudareva et al. 2013)

C5-isoprene building units are biosynthesized through two different pathways: the methylerythritol phosphate (MEP) and the mevalonic acid (MVA) (Fig. 24.1). Through the MEP pathway the precursors for the biosynthesis of the volatile hemiterpenes (C5), monoterpenes (C10) and diterpenes (C20) are provided, whereas the volatile sesquiterpenes (C15) originates from the MVA pathway (McGarvey and Croteau 1995) (Fig. 24.1).

Concerning the spatial localization of these two pathways, the full set of the enzymes involved in the MEP pathway is localized only in the plastids (Hsieh et al. 2008). On the other hand, the subcellular localization of the MVA pathway is still unclear but it seems to be spread among the peroxisomes, cytosol and endoplasmic reticulum (Simkin et al. 2011; Pulido et al. 2012).

Both, monoterpenes and sesquiterpenes, as well as the hemiterpene isoprene, are only a relatively small part of the different groups of isoprenoids emitted by plants (Owen and Penuelas 2005). Isoprene, which is the terpenoid (C5) emitted by plants with the simplest chemical structure, is synthesized from DMAPP by the isoprene synthase (Sharkey et al. 2008). As proposed by Ruzicka (1959) volatile monoterpenes (C10), mainly produced by leaves, have two isoprene units, whereas sesquiterpenes, which are mainly floral fragrances, have three isoprene units (C15) (Dudareva et al. 2006; Maffei 2010). Anyway, both mono and sesquiterpenes production is also stimulated, in leaf glandular trichomes, by herbivore-induced damages.

The family of terpene synthases (TPS), which uses as substrates the geranyl diphosphate (GPP) and the farnesyl diphosphate (FPP), produces a huge structurally diverse number of terpenoids. Moreover, several TPSs that synthetize both mono and sesquiterpenes have been characterized in wide number of plant species (Owen et al. 1997, 2001; Lin et al. 2007; Arimura et al. 2008a, b, 2009; Wu and Baldwin 2009). Finally, VOCs such as β-ionone are derived by the action of carotenoid cleavage dioxygenases, from the cleavage of carotenoids (D'Auria et al. 2002; Maffei 2010).

1.2 Phenylpropanoids and Benzenoids: Biosynthetic Pathways

Benzenoids and phenylpropanoids, which precursor is the aromatic amino acid phenylalanine (Phe), are also one of the largest group of plant VOCs (Knudsen et al. 2006) (Fig. 24.1). These chemicals play a pivotal role in protecting plants from herbivores, ultraviolet light and pathogens. Moreover, they are extremely important in plant-pollinator interactions (Dudareva et al. 2013). Interestingly, it has been demonstrated that the Phe biosynthesis is localized in plastids (Maeda and Dudareva 2012), and it is successively converted to VOCs outside from this organelle.

Phe biosynthesis takes place through the shikimate pathway, which connects the primary metabolism to Phe, through seven enzymatic reactions (Tzin and Galili 2010; Maeda and Dudareva 2012). The primary metabolites that serve as precursors for the shikimate pathway are the phosphoenolpyruvic acid (PEP) and erythrose 4-phosphate (E4P), which originate from the glycolysis and the pentose phosphate pathways (Fig. 24.1). From the same pathways originates the pyruvate (Pyr), putative precursor for the MEP pathway, which compete with the shikimate/phenylpropanoid pathway for carbon allocation (Razal et al. 1996) (Fig. 24.1). This means that under stress conditions the allocation of the photosynthetically fixed carbon is preferentially fated to the Phe biosynthesis to produce lignin.

1.3 Fatty Acid Derivatives: Biosynthetic Pathway

The third class of plant VOCs is represented by fatty acids derivatives. They include chemicals originating from linolenic or linoleic unsaturated fatty acids (C18). Among fatty acids VOCs is worthy of note the methyl jasmonate, which play a pivotal role in plant defense. This compound can induce plants in producing multiple different types of defense chemicals such as phytoalexins, nicotine or proteinase inhibitors (Cheong and Do Choi 2003).

The biosynthesis of these fatty acid derivatives is confined in the plastids and the acetyl-CoA Pyr-derived is its putative precursor (Fig. 24.1).

The unsaturated fatty acids, linoleic and linolenic, once entered in the lipoxygenase pathway (LOX) are stereo specifically oxygenated in 9-hydroperoxy and 13-hydroperoxy intermediates (Feussner and Wasternack 2002) (Fig. 24.1). Successively, these two compounds are metabolized through the LOX pathway generating VOCs. In particular, the branch of the LOX pathway driven by the enzyme hydroperoxide lyase converts both 9-hydroperoxy and 13-hydroperoxy intermediates into C6 and C9 aldehydes. Since these volatiles are produced in the green parts of the plants they are also known as green leaf volatiles (Fig. 24.1). These aldehydes, through the alcohol dehydrogenases, could be further reduced to alcohols, which are successively converted to their esters (D'Auria et al. 2007; Gigot et al. 2010). Conversely, the biosynthetic pathway involving the enzyme allene oxide synthase leads to the production of jasmonic acid using the 13-hydroperoxy intermediate only. Successively, jasmonate is transformed in methyl jasmonate through the enzyme jasmonate carboxyl methyl transferase (Song et al. 2005) (Fig. 24.1).

1.4 Branched-Chain Amino Acid-Derived Volatiles: Biosynthetic Pathways

The precursors of branched-chain amino acids-derived VOCs are alanine, leucine, isoleucine, valine and methionine, or their biosynthetic intermediates containing nitrogen and sulfur (Knudsen et al. 2006) (Fig. 24.2).

The main VOCs derived from this pathway are aldehydes, which can be produced through three different ways: (i) from amino acids through the enzymatic activity of the aldehyde synthase; (ii) through the decarboxylation of amino acids followed by deamination and/or (iii) by transamination of amino acids followed by carboxylation (Gonda et al. 2010; Tzin and Galili 2010) (Fig. 24.2). Successively, the aldehyde could be further reduced, esterified and/or oxidized forming acids, alcohols and esters (Reineccius 2016) (Fig. 24.2).

These molecules play a wide range of ecological roles in plant-plant and plant-herbivore interactions acting as signaling molecules. When the plant tissues are intact, their production is tightly suppressed but upon herbivore attack, biotic and mechanical wounding as well as during abiotic stresses, they are immediately produced (Hassan et al. 2015).

2 Ecological Role of Vocs

Volatile and semi-volatile terpenoids (monoterpenes, diterpenes, sesquiterpenes, etc.) are strictly involved in several ecological roles such as above- and belowground biotic interactions and plant defense against both abiotic and biotic stresses (Pierik et al. 2014; Yazaki et al. 2017).

VOCs play a pivotal role in pollinator attraction and in repelling florivores. It has been demonstrated that several species can switch the blend of their volatiles emission after pollination, repelling the pollinators and directing them towards unpollinated flowers (Schiestl and Ayasse 2001). In carnivorous plants, several volatiles are produced to attract not only their pollinators but also their prey. As well, several species bloom in a different period from trap production in order to avoid devouring their pollinators (Jürgens et al. 2009). Moreover, floral scents have anti-fungal and antimicrobial activities protecting the reproductive organs that are pivotal for species survival (De Moraes et al. 2001; Raguso 2004).

The self-protection ability of plants against herbivores, through the release of plant volatiles, has been well largely documented. Plants can directly interact with herbivorous producing *ex novo* repelling volatiles in response to wound or they can produce several blends that attract herbivores enemies (Pichersky and Gershenzon 2002). The phytotoxic compound farnesene (Araniti et al. 2013a, 2016, 2017a), a sesquiterpene largely found in the Labiateae family, is an alarm pheromone produced by aphids (Francis et al. 2005). It has been demonstrated that, in response to insect attack, plants increase the production of these volatiles discouraging the aphids from plant colonization (Wohlers 1981; Kunert et al. 2005). Similarly, the phytotoxic sesquiterpene insect-induced *trans*-caryophyllene was actively produced from maize roots to attract nematodes that kill the larvae of the attacking herbivores (Rasmann et al. 2005; Araniti et al. 2017c).

Plant VOCs play also an important role in above- and below-ground plant-plant communication. It has been demonstrated, through both laboratory and field experiments, that plants can actively communicate with neighboring plants (inter- and intra-specific interactions). Dolch and Tscharntke (2000) demonstrated that the partial defoliation of the tree species *Alnus glutinosa* resulted in an induced resistance in its conspecific neighbors, showing that the potential chemicals involved in this phenomenon were ethylene, 4,8-dimethylnona-1,3,7-triene, 4,8,12-trimethyltrideca -1,3,7,11-tetraene and β-ocimene (Tscharntke et al. 2001). It has been also demonstrated that during the interspecific plant-plant chemical communication, volatiles might not necessarily penetrate the leaf tissue of the neighboring plant but they can exert their activity through the leaf surface. Himanen et al. (2010) observed that the sesquiterpenoids produced by rhododendron leaves were adsorbed on the leaves of *Betula* sp. trees exerting direct herbivore-repellent activities.

Furthermore, Araniti et al. (2017b) demonstrated, in laboratory experiments that tried to mimic field conditions, that the pioneer species *Dittrichia viscosa* was able to affect growth and development of other species through the release of VOCs. Similarly, they observed that foliar volatiles produced by *Calamintha nepeta* were able to affect seedling establishment of several species (Araniti et al. 2013b).

Several volatiles produced by plants can induce the *de novo* production of defense VOCs. A similar interaction was observed on healthy maize plants that once exposed to the volatile (Z)-3-hexenol stared to emit a volatile blend, which was typically released after herbivorous feeding attracting entomoparasitic insects (Ruther and Kleier 2005).

Beside to their involvement in reproductive processes, defense against herbivores and plant communication, VOCs are also strongly implicated in plant resistance to abiotic stresses. For example, volatile isoprenoids were involved in plant thermotolerance. In fact, they protected leaves from heat damages and high temperatures allowing the plant to maintain a good photosynthetic rate (Sharkey et al. 2001; Velikova and Loreto 2005). Indeed, the inhibition of the isoprenoid biosynthetic pathway resulted in a reduced thermotolerance of the photosynthetic machinery (Copolovici et al. 2005). Interestingly, when fumigated with low atmospheric monoterpenes concentrations, whose synthesis was previously blocked, plants were able to restore their ability to resist heat stress (Copolovici et al. 2005).

Monoterpenes emission is also implicated to ozone defense. In fact, in several species if monoterpene production was artificially blocked, they became susceptible to ozone, reducing their photosynthetic efficiency. This alteration of the photosynthetic machinery was mainly due to the accumulation of reactive oxygen species followed by lipid peroxidation (Loreto and Velikova 2001; Loreto et al. 2004). All these findings highlighted the pivotal role of terpenoids in protecting plants from oxidative stress (Loreto and Velikova 2001).

Moreover, also relatively low volatile terpenoids play several roles in plant physiology and biochemistry. For example, the tetraterpenoids carotenoids are photosynthetic pigments that unroll primary functions in photosynthesis. Plastoquinone plays a pivotal role in the light-dependent reactions of photosynthesis. Compounds such as phytosterols are key elements of membrane structures, whereas compounds such as cytokinins, abscisic acid, gibberellins, brassinosteroids etc. play a crucial role as plant growth regulators (Lichtenthaler 2007; Ramel et al. 2012; Davies 2013; Nisar et al. 2015).

3 Analytical Methods for Determination of Vocs in Plants

The interest in VOCs biosynthetic pathways and their implication on plant biochemistry and physiology, under various environmental/experimental conditions has led to improve a large number of analytical methodologies for the isolation and analysis of these VOCs (Tholl et al. 2006; Deasy et al. 2016). In particular, in the last two decades, the introduction of more sensitive and relatively low-price instrumentation such as gas chromatography-mass spectrometry technique (GC-MS) along with more precise sampling techniques than traditional solvent extraction and/or steam distillations (Tholl et al. 2006) have improved the qualitative and quantitative VOCs analysis (McCormick et al. 2014; Soto et al. 2015; Silva et al. 2017).

In this respect, Head Space-Solid Phase Micro Extraction (HS-SPME) plays an important role in the analysis of VOCs compounds (Pawliszyn 2011) as will be discussed in the next paragraph.

3.1 HS-SPME for VOCs Samples in Plants

Analyses of VOCs, organic environmental pollutants and flavors or fragrances generally start with the concentration of the analytes of interest through headspace, purge-and-trap, liquid-liquid extraction, and other methods. However, these procedures need rather always complicated equipment, excessive time and/or usage of high volumes of organic solvents (Stashenko et al. 2004; Pourmortazavi and Hajimirsadeghi 2007).

Conversely, HS-SPME, an adsorption/desorption technique removes the need to use solvents or complex apparatus for concentrating both, non-volatile or volatile chemicals in headspace, solid or liquid samples or (Pawliszyn 2011). Furthermore, this technique provides linear results over a wide analytes' concentration range and is compatible with all the instruments (e.g. GC-FID, GC-MS) and columns (Pawliszyn 2011).

Overall, by choosing an appropriate polarity and thickness of the coating phase, the analyst can attain consistent results, also for trace components (McCormick et al. 2014; Yazaki et al. 2017).

3.2 Fiber Selection

The choice of the coating phase for an SPME fiber is strictly dependent by the nature of the analytes that should be extracted/analyzed. The coatings available on market are polydimethylsiloxane (PDMS), polyacylate (PA), Carboxen (CAR), divinylbenzene (DVB), and Carbowax (CW). In this respect, the most widespread are PA and PDMS, this latter tolerates high temperatures, up to 300 °C and it is characterized by a high stability (Balasubramanian and Panigrahi 2011).

There are several fibers available, which differ in coating combinations as well as in assemblies and film thicknesses. Furthermore, mixed coatings with a primary porous solid extraction phase are available (Górecki et al. 1999).

Generally, the best extraction efficiencies for a wide range of analytes with different polarities and molecular weights are provided by the combinations DVB/CAR/PDMS and CAR/PDMS (Balasubramanian and Panigrahi 2011). Nevertheless, PA fibers are mainly used for polar analytes, whereas the PDMS for nonpolar (Balasubramanian and Panigrahi 2011). Also, DVB is a polar coating fiber that extracts polar compounds such as sulfides (Cai et al. 2001). The extraction of bipolar compounds like aldehydes, ketones, alcohols, carboxylic acids and ethers is generally performed using fibers that are a combination of different e.g., polar material (DVB) and non-polar material (PDMS) (Balasubramanian and Panigrahi 2011).

An exhaustive description of coating materials and affinities, in terms of polarity and molecular weight, was reported by Pawliszyn (2011).

4 Materials and Methods for Vocs Collection and Identification

4.1 Materials

1. Sample: *Ormenis multicaulis* (Chamomile oil, Moroccan), collected at the flowering stage
2. Scissor and/or scalpel
3. 20 mL screw cap glass vials with PTFE/silicone septum
4. Gas chromatograph (e.g. Thermo Fisher G-Trace 1310) coupled with a single quadrupole mass spectrometer (e.g. ISQ LT, Thermo Fisher) and an autosampler (e.g. Tri-Plus, Thermo Fisher)
5. Capillary column (e.g. TG-5MS 30 m × 0.25 mm ID × 0.25 μm film thickness)
6. GC carrier gas: helium with high degree of purity (6.0 or 99.9999%)
7. Splitless direct injection liners for SPME applications (e.g. Topaz 1.8 mm ID Straight/SPME Inlet Liner)
8. SPME Fiber Holder for use with manual sampling
9. SPME fiber (55/30 μm DVB/CAR on polydimethylsiloxane coating [PDMS/DVB/CAR] for use with manual holder, needle size 24 gauge (e.g. Supelco)
10. Heated orbital shaker
11. Laboratory support stand, clamp holder and burette clamp
12. *n*-Alkanes standard solution C7-C30 (e.g. Supelco) for Linear Retention Index (LRI) calculation.

4.2 Methods

4.2.1 SPME Static Headspace: Sample Preparation and Fiber Conditioning and Desorbing

Before going to the method description, is worth to remember that the fiber has to be conditioned before analysis according to the recommendations supplied by the producer.

For routine analysis, after each extraction, 10 min reconditioning of the fiber at 250 °C is enough to avoid any carryover effect and thus to allow a good repeatability between consecutive injections. When used a 55/30 μm divinylbenzene/Carboxen on polydimethylsiloxane coating (PDMS/DVB/CAR), as in this case, a temperature of 270 °C has to be applied for 60 min.

The first step of the method is to cut, with scissors, a branch of *Ormenis multicaulis*, including both leaves and flowers. Then, using a razor blade, excise leaves and flowers (1 g for each organ and for each replicate) and transfer the organs into 20 mL glass vial, which has to be immediately closed with the PFTE cap.

The vial is placed onto the orbital shaker at room temperature to allow the volatiles to accumulate into the vial headspace for 20 min (incubation time). Then, expose the fiber for 20 min (optimized extraction time) to extract a suitable and representative number of volatiles from the sample. Once the extraction time is over, withdraw the fiber into the fiber holder and expose for 2 min into the GC injector temperature (250 °C) in order to release all the components adsorbed into the injector.

Finally, leave the fiber exposed in the same injector for additionally 10 min (reconditioning time); alternatively, backflush in a different injection port at the same temperature. After this period, the fiber will be ready for another sampling process.

Using the same methodology of the extraction procedure and GC conditions, perform the analysis of the C7–C30 alkane mixture in order to extrapolate the LRIs. Please, note that alkanes mixture is likely to contain the dilution solvent, which must be excluded from the acquisition time.

4.2.2 GC and MS Settings

The GC oven column temperature ranged from 45 °C (2 min) to 250 °C at 3 °C min^{-1}, with a final hold time of 10 min. Helium (He) was used as carrier gas at 1.0 mL min^{-1} constant flow (*ca.* 36 cm/s linear velocity). Injector port temperature was 250 °C while the injector operated in the splitless mode in order to transfer quantitatively all the components onto the column. The interface temperature was set to 280 °C, while the quadrupole and ion source temperatures were settled at 260 °C and 280 °C, respectively. Mass spectra were recorded in Electronic Impact (EI) mode at 70 eV within the mass range 40–400 m/z.

5 Data Analysis

The qualitative information extracted from the HS-SPME-GC-MS analysis of a sample such as the *Ormenis multicaulis* has to be evaluated cautiously. In fact, this sample often contains multiple components belonging to the same chemical class, thus having a similar fragmentation pattern in GC-MS operating in EI mode. Nowadays, the availability of various mass spectral libraries (e.g. Wiley, NIST MS Database) could help researchers in selecting the authentic standards for comparison. However, as above-mentioned, the use of this criterion only can lead to misleading results and as consequence, multiple identification confirmations are

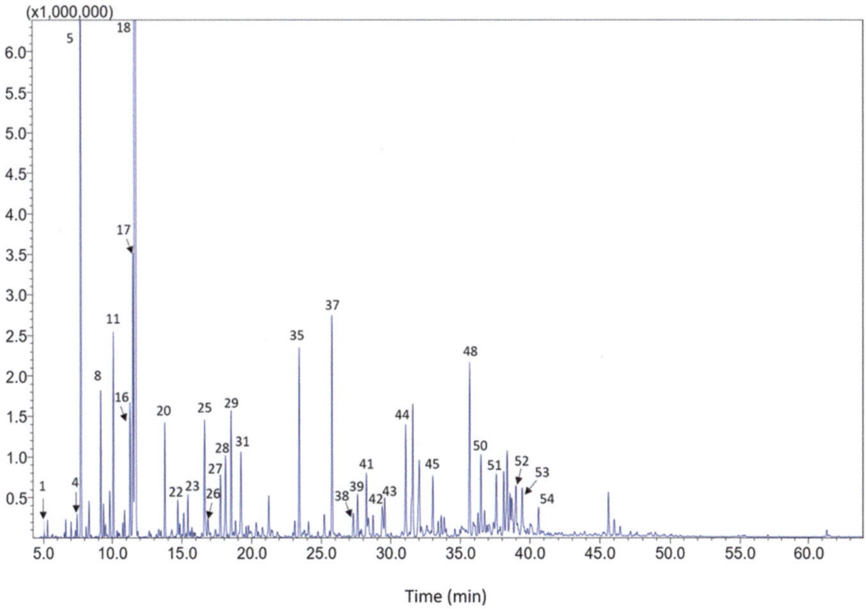

Figure 24.3 A typical GC-MS chromatogram obtained from the analysis of the volatile fraction of *Ormenis multicaulis* using HS-SPME extraction method

needed. In this respect, LRIs play an important role for the confirmation of structurally related molecules within the same sample.

The most thoroughly accepted retention index calculation in temperature programming conditions was proposed by Van den Dool and Kratz (1963). These values are known in bibliography as Retention Index (RI) or Linear Retention Index (LRI). Their calculation is based on the following equation:

$$I^{T} = 100\mathrm{n}\left(\mathrm{t}_{\mathrm{Ri}}^{\mathrm{T}} - \mathrm{t}_{\mathrm{Rz}}^{\mathrm{T}} / t_{\mathrm{R}(z+1)}^{\mathrm{T}} - t_{\mathrm{Rz}}^{\mathrm{T}}\right) + 100\mathrm{z}$$

where *I* represents the isothermal retention index at temperature *T*, $\mathrm{t}^{\mathrm{T}}_{\mathrm{Ri}}$ the retention time of the unknown component and $z + 1$ are *n*-alkanes with z and $z + 1$ number of carbons, respectively.

For the current experiments, the identification of constituents within the chromatogram of *Ormenis multicaulis* was based on:

(a) computer matching against NIST commercial mass spectral library (NIST 2014)
(b) comparison of the LRIs with those of authentic compounds and literature data
(c) comparison of spectra with literature data

Figure 24.3 shows the GC-MS chromatogram of the volatile fraction of *Ormenis multicaulis*. Additionally, Table 24.1 reports the tentatively identified components

Table 24.1 Identified peaks in the GC-MS chromatogram of *Ormenis multicaulis*

No	Name	RTs (min)	LRI^{exp}	LRI^{Theor}	Similarity (%)
1	*n*-Hexanol	5.06	863	860	98
2	Santolinatriene	6.62	902	908	98
3	Isobutyl-Isobutyrate	7.01	913	910	95
4	α-Thujene	7.44	927	929	98
5	α-Pinene	7.75	933	937	99
6	2-Methylpropyl Butyrate	8.09	942	947	98
7	Camphene	8.30	953	952	97
8	Sabinene	9.16	976	974	98
9	β-Pinene	9.35	981	979	98
10	Myrcene	9.81	994	991	95
11	(E)-2,5,5-Trimethylhepta-3,6-dien-2-ol	10.08	1001	1000	99
12	Isobutyl-2-methyl-Butyrate	10.34	1007	1004	98
13	Hex-(3Z)-enyl acetate	10.46	1010	1005	93
14	Isobutyl Isovalerate	10.54	1012	1005	96
15	Isopentyl Isobutyrate	10.76	1017	1015	95
16	*p*-Cymene	11.30	1029	1025	99
17	Limonene	11.51	1033	1030	98
18	8-hydroxy-Menthol	11.69	1037	1031[a]	90
19	5-ethenyldihydro-5-methyl-2(3H)-Furanone	11.85	1041	1043	95
20	Artemisia alcohol	13.79	1085	1084	95
21	Isobutyl-Tiglate	14.29	1096	1093	94
22	Linalool	14.71	1105	1099	93
23	3-Methyl pentyl-Isobutyrate	15.44	1121	1115[a]	95
24	cis-, para-Menth-2-en-1-ol	15.85	1124	1122	93
25	*trans*-Pinocarveol	16.63	1146	1138	96
26	*trans*-Verbenol	16.86	1151	1144	97
27	β-Artemisia acetate	17.75	1170	1173[a]	99
28	Borneol	18.12	1177	1167	96
29	Terpinen-4-ol	18.53	1186	1177	93
30	*p*-Cymen-8-ol	18.84	1193	1183	91
31	α-Terpineol	19.23	1201	1189	94
32	Verbenone	19.77	1213	1205	96
33	*trans*-Carveol	20.35	1225	1217	95
34	Cuminaldehyde	21.44	1248	1239	94
35	Bornyl acetate	23.44	1290	1285[a]	98
36	3-Methylpentyl, tiglate 2-	24.08	1304	1300[a]	93
37	Δ-Elemene	25.79	1342	1344	99
38	Cyclosativene	27.30	1375	1367[a]	94
39	α-Copaene	27.61	1382	1376	94
40	*trans*-Geranyl acetate	27.76	1385	1382	94
41	β-Elemene	28.24	1395	1390[a]	94

(continued)

Table 24.1 (continued)

No	Name	RTs (min)	LRI^{exp}	LRI^{Theor}	Similarity (%)
42	Isobornyl isobutyrate	29.38	1421	1418	93
43	(E)-Caryophyllene	29.56	1426	1424[a]	97
44	β-Farnesene	31.06	1460	1457	98
45	α-Muurolene	33.02	1506	1499	98
46	γ-Cadinene	33.62	1520	1512[a]	96
47	Δ-Cadinene	33.84	1525	1518[a]	93
48	(E)-Nerolidol	35.67	1570	1564	97
49	Spathulenol	36.28	1585	1576[a]	93
50	Caryophyllene oxide	36.49	1590	1581	97
51	Copaborneol	37.60	1617	1613[a]	92
52	T-Muurolol	39.00	1653	1642	92
53	Cadin-4-en-10-ol	39.45	1664	1653	95
54	α-Bisabolol	40.62	1694	1684	92

[a]Literature LRI
LRI^{exp} Experimentally calculated LRI
LRI^{Theor} NIST LRI for Semi-standard non-polar column

having a library march against the NIST database higher than 90% and an LRI tolerance window of ±10 points.

The combination of these two filters greatly reduces the number of false positive candidates in the peak table.

References

Araniti F, Graña E, Reigosa MJ, Sánchez-Moreiras AM, Abenavoli MR (2013a) Individual and joint activity of terpenoids, isolated from *Calamintha nepeta* extract, on *Arabidopsis thaliana*. Nat Prod Res 27(24):2297–2303

Araniti F, Lupini A, Sorgonà A, Statti GA, Abenavoli MR (2013b) Phytotoxic activity of foliar volatiles and essential oils of *Calamintha nepeta* (L.) Savi. Nat Prod Res 27:1651–1656

Araniti F, Graña E, Krasuska U, Bogatek R, Reigosa MJ, Abenavoli MR, Sánchez-Moreiras AM (2016) Loss of gravitropism in farnesene-treated arabidopsis is due to microtubule malformations related to hormonal and ROS unbalance. PLoS One 11(8):e0160202

Araniti F, Bruno L, Sunseri F, Pacenza M, Forgione I, Bitonti MB, Abenavoli MR (2017a) The allelochemical farnesene affects *Arabidopsis thaliana* root meristem altering auxin distribution. Plant Physiol Biochem 121:14–20

Araniti F, Lupini A, Sunseri F, Abenavoli MR (2017b) Allelopatic potential of *Dittrichia viscosa* (L.) W. Greuter mediated by VOCs: a physiological and metabolomic approach. PLoS One 12:e0170161

Araniti F, Sánchez-Moreiras AM, Graña E, Reigosa MJ, Abenavoli MR (2017c) Terpenoid *trans*-caryophyllene inhibits weed germination and induces plant water status alteration and oxidative damage in adult Arabidopsis. Plant Biol 19(1):79–89

Arimura GI, Garms S, Maffei M, Bossi S, Schulze B, Leitner M, Mithoefer A, Boland W (2008a) Herbivore-induced terpenoid emission in *Medicago truncatula*: concerted action of jasmonate, ethylene and calcium signaling. Planta 227:453–464

Arimura GI, Kopke S, Kunert M, Volpe V, David A, Brand P, Dabrowska P, Maffei ME, Boland W (2008b) Effects of feeding *Spodoptera littoralis* on lima bean leaves: IV. Diurnal and nocturnal damage differentially initiate plant volatile emission. Plant Physiol 146:965–973

Arimura G, Matsui K, Takabayashi J (2009) Chemical and molecular ecology of herbivore-induced plant volatiles: proximate factors and their ultimate functions. Plant Cell Physiol 50:911–923

Balasubramanian S, Panigrahi S (2011) Solid-phase microextraction (SPME) techniques for quality characterization of food products: a review. Food Bio Tech 4:1–26

Cai J, Liu B, Su Q (2001) Comparison of simultaneous distillation extraction and solid-phase microextraction for the determination of volatile flavor components. J Chromatogr A 930:1–7

Cheong J-J, Do Choi Y (2003) Methyl jasmonate as a vital substance in plants. Trends Gen 19:409–413

Copolovici LO, Filella I, Llusia J, Niinemets Ü, Peñuelas J (2005) The capacity for thermal protection of photosynthetic electron transport varies for different monoterpenes in *Quercus ilex*. Plant Physiol 139:485–496

D'Auria JC, Chen F, Pichersky E (2002) Characterization of an acyltransferase capable of synthesizing benzylbenzoate and other volatile esters in flowers and damaged leaves of *Clarkia breweri*. Plant Physiol 130:466–476

D'Auria JC, Pichersky E, Schaub A, Hansel A, Gershenzon J (2007) Characterization of a BAHD acyltransferase responsible for producing the green leaf volatile (Z)-3-hexen-1-yl acetate in *Arabidopsis thaliana*. Plant J 49:194–207

Davies P (2013) Plant hormones: physiology, biochemistry and molecular biology. Springer Science & Business Media, Berlin

De Moraes CM, Mescher MC, Tumlinson JH (2001) Caterpillar-induced nocturnal plant volatiles repel conspecific females. Nature 410:577

Deasy W, Shepherd T, Alexander CJ, Birch ANE, Evans KA (2016) Development and validation of a SPME-GC-MS method for in situ passive sampling of root volatiles from glasshouse-grown broccoli plants undergoing below-ground herbivory by larvae of cabbage root fly, *Delia radicum* L. Phytochem Anal 27:375–393

Dicke M, Loreto F (2010) Induced plant volatiles: from genes to climate change. Trends Plant Sci 15:115–117

Dolch R, Tscharntke T (2000) Defoliation of alders (*Alnus glutinosa*) affects herbivory by leaf beetles on undamaged neighbours. Oecologia 125:504–511

Dudareva N, Pichersky E (2008) Metabolic engineering of plant volatiles. Curr Opin Biotech 19:1–9

Dudareva N, Pichersky E, Gershenzon J (2004) Biochemistry of plant volatiles. Plant Physiol 135:1893–1902

Dudareva N, Negre F, Nagegowda DA, Orlova I (2006) Plant volatiles: recent advances and future perspectives. Critical Rev Plant Sci 25:417–440

Dudareva N, Klempien A, Muhlemann JK, Kaplan I (2013) Biosynthesis, function and metabolic engineering of plant volatile organic compounds. New Phytol 198:16–32

Feussner I, Wasternack C (2002) The lipoxygenase pathway. Annu Rev Plant Biol 53:275–297

Francis F, Vandermoten S, Verheggen F, Lognay G, Haubruge E (2005) Is the (E)-β-farnesene only volatile terpenoid in aphids? J Appl Entomol 129:6–11

Gigot C, Ongena M, Fauconnier M-L, Wathelet J-P, Du Jardin P, Thonart P (2010) The lipoxygenase metabolic pathway in plants: potential for industrial production of natural green leaf volatiles. BASE 14:451

Gonda I, Bar E, Portnoy V, Lev S, Burger J, Schaffer AA, Tadmor Y, Gepstein S, Giovannoni JJ, Katzir N, Lewinsohn E (2010) Branched-chain and aromatic amino acid catabolism into aroma volatiles in *Cucumis melo* L. fruit. J Exp Bot 61:1111–1123

Górecki T, Yu X, Pawliszyn J (1999) Theory of analyte extraction by selected porous polymer SPME fibres. Analyst 124:643–649

Hassan MN, Zainal Z, Ismail I (2015) Green leaf volatiles: biosynthesis, biological functions and their applications in biotechnology. Plant Biotech J 13:727–739

Hatanaka A (1993) The biogeneration of green odour by green leaves. Phytochemistry 34(5):1201–1218

Himanen SJ, Blande JD, Klemola T, Pulkkinen J, Heijari J, Holopainen JK (2010) Birch (*Betula* spp.) leaves adsorb and re-release volatiles specific to neighbouring plants–a mechanism for associational herbivore resistance? New Phytol 186:722–732

Hsieh M-H, Chang C-Y, Hsu S-J, Chen J-J (2008) Chloroplast localization of methylerythritol 4-phosphate pathway enzymes and regulation of mitochondrial genes in ispD and ispE albino mutants in Arabidopsis. Plant Mol Biol 66:663–673

Jürgens A, El-Sayed AM, Suckling DM (2009) Do carnivorous plants use volatiles for attracting prey insects? Funct Ecol 23:875–887

Knudsen JT, Gershenzon J (2006) The chemical diversity of floral scent. In: Dudareva N, Pichersky E (eds) Biology of floral scent. CRC Press, London, pp 27–52

Knudsen JT, Eriksson R, Gershenzon J, Ståhl B (2006) Diversity and distribution of floral scent. Bot Rev 72:1–120

Kunert G, Otto S, Röse US, Gershenzon J, Weisser WW (2005) Alarm pheromone mediates production of winged dispersal morphs in aphids. Ecol Lett 8:596–603

Lichtenthaler HK (2007) Biosynthesis, accumulation and emission of carotenoids, α-tocopherol, plastoquinone, and isoprene in leaves under high photosynthetic irradiance. Photos Res 92:163–179

Lin C, Owen SM, Penuelas J (2007) Volatile organic compounds in the roots and rhizosphere of *Pinus* spp. Soil Biol Biochem 39:951–960

Loreto F, Velikova V (2001) Isoprene produced by leaves protects the photosynthetic apparatus against ozone damage, quenches ozone products, and reduces lipid peroxidation of cellular membranes. Plant Physiol 127:1781–1787

Loreto F, Pinelli P, Manes F, Kollist H (2004) Impact of ozone on monoterpene emissions and evidence for an isoprene-like antioxidant action of monoterpenes emitted by *Quercus ilex* leaves. Tree Physiol 24:361–367

Maeda H, Dudareva N (2012) The shikimate pathway and aromatic amino acid biosynthesis in plants. Annu Rev Plant Biol 63:73–105

Maffei ME (2010) Sites of synthesis, biochemistry and functional role of plant volatiles. South Afr J Bot 76(4):612–631

McCormick AC, Gershenzon J, Unsicker SB (2014) Little peaks with big effects: the role of minor plant volatiles in plant insect interactions. Plant Cell Environ 37(8):1836–1844

McGarvey DJ, Croteau R (1995) Terpenoid metabolism. Plant Cell 7:1015

Memari HR, Pazouki L, Niinemets Ü (2013) The biochemistry and molecular biology of volatile messengers in trees. In: Niinemets U, Monson RK (eds) Biology, controls and models of tree volatile organic compound emissions. Springer, Dordrecht, pp 47–93

National Institute of Standards and Technology (2014) PC version 2.2 of the NIST/EPA/NIH Mass Spectral Database

Nisar N, Li L, Lu S, Khin NC, Pogson BJ (2015) Carotenoid metabolism in plants. Mol Plant 8:68–82

Owen SM, Penuelas J (2005) Opportunistic emissions of volatile isoprenoids. Trends Plant Sci 10:420–426

Owen S, Boissard C, Street RA, Duckham SC, Csiky O, Hewitt CN (1997) Screening of 18 Mediterranean plant species for volatile organic compound emissions. Atmosph Environ 31:101–117

Owen SM, Boissard C, Hewitt CN (2001) Volatile organic compounds (VOCs) emitted from 40 Mediterranean plant species: VOC speciation and extrapolation to habitat scale. Atmos Environ 35:5393–5409

Paré PW, Tumlinson JH (1996) Plant volatile signals in response to herbivore feeding. Fla Entomol 95:93–103

Pawliszyn J (2011) Handbook of solid phase microextraction. Elsevier, London

Pichersky E, Gershenzon J (2002) The formation and function of plant volatiles: perfumes for pollinator attraction and defense. Curr Opin Plant Biol 5:237–243
Pichersky E, Noel JP, Dudareva N (2006) Biosynthesis of plant volatiles: nature's diversity and ingenuity. Science 311:808–811
Pierik R, Ballare CL, Dicke M (2014) Ecology of plant volatiles: taking a plant community perspective. Plant Cell Environ 37:1845–1853
Pourmortazavi SM, Hajimirsadeghi SS (2007) Supercritical fluid extraction in plant essential and volatile oil analysis. J Chromatogr A 1163:2–24
Pulido P, Perello C, Rodriguez-Concepcion M (2012) New insights into plant isoprenoid metabolism. Mol Plant 5:964–967
Raguso RA (2004) Why are some floral nectars scented? Ecology 85:1486–1494
Ramel F, Birtic S, Cuiné S, Triantaphylidès C, Ravanat J-L, Havaux M (2012) Chemical quenching of singlet oxygen by carotenoids in plants. Plant Physiol 158:1267–1278
Rasmann S, Kollner TG, Degenhardt J, Hiltpold I (2005) Recruitment of entomopathogenic nematodes by insect-damaged maize roots. Nature 434:732
Razal RA, Ellis S, Singh S, Lewis NG, Towers GN (1996) Nitrogen recycling in phenylpropanoid metabolism. Phytochemistry 41:31–35
Reineccius G (2016) Flavor chemistry and technology. CRC Press, Boca Raton
Rosenkranz M, Schnitzler JP (2013) Genetic engineering of BVOC emissions from trees. In: Biology, controls and models of tree volatile organic compound emissions. Springer, Dordrecht, pp 95–118
Ruther J, Kleier S (2005) Plant–plant signaling: ethylene synergizes volatile emission in *Zea mays* induced by exposure to (Z)-3-hexen-1-ol. J Chem Ecol 31:2217–2222
Ruzicka L (1959) Isoprene rule and the biogenesis of terpenic compounds. Experientia 9:357–367
Schiestl FP, Ayasse M (2001) Post-pollination emission of a repellent compound in a sexually deceptive orchid: a new mechanism for maximising reproductive success? Oecologia 126:531–534
Sharkey TD, Chen X, Yeh S (2001) Isoprene increases thermotolerance of fosmidomycin-fed leaves. Plant Physiol 125:2001–2006
Sharkey TD, Wiberley AE, Donohue AR (2008) Isoprene emission from plants: why and how. Ann Bot 101:5–18
Silva ÉAS, Saboia G, Jorge NC, Hoffmann C, dos Santos Isaias RM, Soares GL, Zini CA (2017) Development of a HS-SPME-GC/MS protocol assisted by chemometric tools to study herbivore-induced volatiles in *Myrcia splendens*. Talanta 175:9–20
Simkin AJ, Guirimand G, Papon N, Courdavault V, Thabet I, Ginis O, Bouzid S, Giglioli-Guivarc'h N, Clastre M (2011) Peroxisomal localisation of the final steps of the mevalonic acid pathway in planta. Planta 234:903
Song MS, Kim DG, Lee SH (2005) Isolation and characterization of a jasmonic acid carboxyl methyltransferase gene from hot pepper (*Capsicum annuum* L.). J Plant Biol 48:292–297
Soto VC, Maldonado IB, Jofré VP, Galmarini CR, Silva MF (2015) Direct analysis of nectar and floral volatile organic compounds in hybrid onions by HS-SPME/GC–MS: relationship with pollination and seed production. Microchem J 122:110–118
Stashenko EE, Jaramillo BE, Martínez JR (2004) Comparison of different extraction methods for the analysis of volatile secondary metabolites of *Lippia alba* (Mill.) NE Brown, grown in Colombia, and evaluation of its in vitro antioxidant activity. J Chromatogr A 1025:93–103
Taveira M, Fernandes F, Guedes de Pinho P, Andrade PB, Pereira JA, Valentão P (2009) Evolution of *Brassica rapa* var. *rapa* L. volatile composition by HS-SPME and GC/IT–MS. Microchem J 93(2):140–146
Tholl D, Boland W, Hansel A, Loreto F, Röse US, Schnitzler JP (2006) Practical approaches to plant volatile analysis. Plant J 45:540–560
Tscharntke T, Thiessen S, Dolch R, Boland W (2001) Herbivory, induced resistance, and interplant signal transfer in *Alnus glutinosa*. Biochem Syst Ecol 29:1025–1047

Tzin V, Galili G (2010) New insights into the shikimate and aromatic amino acids biosynthesis pathways in plants. Mol Plant 3:956–972

Van den Dool H, Kratz PD (1963) A generalization of the retention index system including linear temperature programmed gas-liquid partition chromatography. J Chromatogr A 11:463–471

Velikova V, Loreto F (2005) On the relationship between isoprene emission and thermotolerance in *Phragmites australis* leaves exposed to high temperatures and during the recovery from a heat stress. Plant Cell Environ 28:318–327

Wenke K, Kai M, Piechulla B (2010) Belowground volatiles facilitate interactions between plant roots and soil organisms. Planta 231:499–506

Wink M (2010) Introduction: biochemistry, physiology and ecological functions of secondary metabolites. Annual plant reviews vol 40: biochemistry of plant secondary metabolism, 2nd ed, pp 1–19

Wohlers P (1981) Effects of the alarm pheromone (e)-β-farnesene on dispersal behaviour of the pea aphid *Acyrthosiphon pisum*. Entomol Exp Appl 29:117–124

Wu JQ, Baldwin IT (2009) Herbivory-induced signalling in plants: perception and action. Plant Cell Environ 32:1161–1174

Yazaki K, Arimura G-i, Ohnishi T (2017) “Hidden” terpenoids in plants: their biosynthesis, localisation and ecological roles. Plant Cell Physiol 58:1615–1621

Chapter 25
Carbon Radiochemicals (^{14}C) and Stable Isotopes (^{13}C): Crucial Tools to Study Plant-Soil Interactions in Ecosystems

Geneviève Chiapusio, Dorine Desalme, Philippe Binet, and François Pellissier

1 Introduction

The study of plant-environment interactions has grown steadily during the past two decades. This trend will continue as many environmental changes impact the functioning of ecosystems. One aspect of studying the plant-environment interactions is to focus on the way plants react to abiotic or biotic stresses including chemical mediation between plants or plants-microorganisms (allelopathy) and plant reaction to pollutants. This chapter proposes to focus on carbon radiochemicals (^{14}C) and stable carbon isotope tracers (^{13}C). Indeed, they are powerful techniques in plant ecophysiological researches to describe the transfer and effects of allelochemicals and pollutants in plants and their environment.

^{14}C radiochemicals are widely used in agronomy to understand pesticide's translocation into plants and so they offer promising research perspectives in ecology. Because ^{14}C is not present in natural environment, this is a powerful technique that offers the clue of studying where, when, how many target

G. Chiapusio (✉)
Université Bourgogne Franche Comté, CNRS, Chrono-Environnement UMR 6249, Montbéliard cedex, France

Université Savoie Mont Blanc, INRA, CARRTEL, Thonon-les-Bains, France
e-mail: genevieve.chiapusio@univ-fcomte.fr

D. Desalme
Université Lorraine, INRA, Ecologie et Ecophysiologie Forestières, Vandoeuvre-lès-Nancy, France

P. Binet
Université Bourgogne Franche Comté, CNRS, Chrono-Environnement UMR 6249, Montbéliard cedex, France

F. Pellissier
Université Savoie Mont Blanc, CNRS, LECA, Le Bourget-du-Lac Cedex, France

© Springer International Publishing AG, part of Springer Nature 2018
A. M. Sánchez-Moreiras, M. J. Reigosa (eds.), *Advances in Plant Ecophysiology Techniques*, https://doi.org/10.1007/978-3-319-93233-0_25

radiochemicals and their related metabolites are transferred into plants and in the surrounding soils. As example, the study of plant secondary metabolites (PSMs) represents an important area of research in plant ecophysiology. PSMs served to defend plants against herbivores, pathogens and abiotic stress but also as chemical mediators of interactions with competitors (Chiapusio et al. 2005). A single compound can influence multiple components within an ecological system and can have effects at different scales, from physiological to structural community roles. ^{14}C radiochemical techniques offer then a good way to discover PSM fate in plants (Chiapusio and Pellissier 2001; Chiapusio et al. 2004).

Natural abundance of ^{13}C is currently used for a diverse range of applications in environmental and plant sciences since this isotope is naturally occurring in all compartments of ecosystems. As reviewed by Brüggemann et al. (2011), isotopic signature of C varies according to plant photosynthesis strategy (C3 *vs* C4 *vs* CAM), plant water use efficiency, plant organs and its constituting compounds (e.g. lipids are ^{13}C-depleted whereas cellulose and carbohydrates are ^{13}C-enriched). Pulse labeling plants with stable carbon isotope $^{13}CO_2$ has a great efficiency to trace *in situ* recently assimilated C in plant organs, soil, and microorganisms including its release through respiration (Kuzyakov and Gavrichkova 2010; Epron et al. 2012). Indeed, leaves incorporate CO_2 into carbohydrates to then "feed" physiological processes of all organs such as growth, respiration, maintenance, storage, osmotic regulation and defense. C allocation among plant organs and rhizospheric microorganisms is then related to the activity of C sources (leaves) and C sinks (roots, trunk, fruits, symbiotic microorganisms). C allocation is furthermore affected by environmental variations, such as shading (Warren et al. 2012; Bahn et al. 2013), drought (Ruehr et al. 2009; Zang et al. 2014; Hartmann et al. 2015), nitrogen supply (Högberg et al. 2010), atmospheric carbon dioxide enrichment (Johnson et al. 2013) and atmospheric pollutants such as Polycyclic Aromatic Hydrocarbons (PAHs) (Desalme et al. 2011a) or ozone (Andersen 2003; Kasurinen et al. 2012).

To explore various applied examples of ^{14}C and ^{13}C techniques in an ecological context, we propose to focus on the transfer and effects in the environment of naturally occurring plant secondary metabolites (PSMs) and organic pollutants (PAHs) on plants. Among the numerous PSMs produced by plants, we focused on allelochemicals such as phenolic compounds, involved in plant-plant and plant-microorganisms interactions. PAHs are ubiquitous organic pollutants emitted by incomplete combustion of biomass or fossil fuel and recovered in all compartments of ecosystems (air, soil, water, plants). From an ecophysiological point of view, challenges still concern the way allelochemicals and organic pollutants are transferred in plants and affect their metabolism but also their surrounding microbial communities.

This chapter aims to present (1) basic aspects of extraction and quantification procedures of radiochemical (^{14}C) and stable isotopes (^{13}C) techniques in ecophysiological experiments and (2) some practical examples using labeled organic molecules (^{14}C-PSMs and ^{13}C-PAH) to trace their fate in plants and their surrounding soil, and labeled $^{13}CO_2$ to study the way carbon allocation is changed in plant-soil systems exposed to atmospheric pollution by PAHs.

2 Overview of Some ^{14}C Radiochemical and ^{13}C Stable Isotope Techniques

2.1 ^{14}C Radiochemical Techniques (Table 25.1)

These powerful techniques allow to follow the ^{14}C compound into seedlings, plants or in the soil in a quantitative and/or qualitative way to describe the transfer of radiolabeled molecules in mesocosms. Such techniques prove the absorption, the distribution, the metabolisation and/or degradation of the ^{14}C compound in plants and their surrounding soils. Time course of the^{14}C compound from the soil to the plant, including remanence in the soils is then well described. The identification of crucial environmental factors influencing its kinetic can also be evaluated. Once in plants, the absorbed ^{14}C compound can be chemically transformed by oxidation, glycosylation or polymerization to be completely or partially metabolized and stocked into the vacuoles or strongly linked into cell walls. The choice for using a molecule partially or fully marked with ^{14}C, its specific activity (mCi $mmol^{-1}$) and its radiochemical purity depends of the goal of the study (global signal *vs* precise chemical fate in inner tissues of plant or soil).

Table 25.1 Some comparisons between usual ^{14}C radiolabeled techniques

	Objectives	Advantages	Limits
Extraction by			
Grinding	To get incomplete extraction of ^{14}C samples	Simple	Chemical risk (inhalation) Need time for the extraction
Oxidizing (in an oxidizer)	To get complete extraction of ^{14}C samples	Simple and fast in manipulating Few radioactive waste	
Quantification by			
Liquid scintillation counting (in a liquid scintillation counter)	To give the global radiochemical amount of ^{14}C samples after grinding	Simple	Chemical risk Radioactive waste
TLC coupled with an imaging scanner	To give the distribution of radioactivity on the plant	Classes of compound involved: Chromatograms are analyzed with the imaging scanner	No quantitative data available
TLC coupled with UHPLC or GC-MS	To find out which molecule is radiolabeled (initial molecule, metabolites)	Higher resolution for the separation of closely related molecular species Quantitative/qualitative data	Time required is longer than other techniques

UHPLC Ultra High Performance Liquid Chromatography, *TLC* Thin Layer Chromatography, *GC-MS* Gas Chromatography Mass spectrometry

2.1.1 Extraction Procedure

Currently, ^{14}C tissues are extracted by grinding or by oxidizing. The grinding extraction gives part of the recovered plant radioactivity because it is function of the selected solvent (from polar to non-polar solvent). The oxidizer extraction (oxidative combustion of the plant) gives the total recovered radioactivity in the target organ or in the plant (^{14}C is transformed in $^{14}CO_2$). Comparisons between the two extractions within a same sample offer then complementary information.

2.1.2 Quantification Procedure

^{14}C extracts are diluted with a liquid of scintillation to be quantified by a Liquid Scintillation Counter (LSC). More specifically analyses can be performed to discover the ^{14}C metabolites by Thin Layer Chromatography (TLC) or by Ultra High Phase Liquid Chromatography (UHPLC- UV with radioactive counter detector with continuous flux) or by Gas Chromatography Mass Spectrometry. The use of such techniques will depend of the chemical structure of the selected ^{14}C compound. The autoradiography technique is also a qualitative technique, which gives the full image of the ^{14}C labeled plant with a specific imager (phosphor-imager). It allows visualizing the distribution of the radioactivity into the plant. The plant is usually contaminated in a solution into the soil or in the air through ^{14}C chemical deposition in the leaves.

2.1.3 Warning

Manipulating such extracts needs specific room and material devoted to this. Glass materials including beakers to grow plants are strongly recommended to recover the entire radioactivity including material rinsing. Therefore a trap of $^{14}CO_2$ (NaOH for example), coming from the degradation of the initial ^{14}C molecule, is often added in the microcosms. For any experimentation, radiochemical measurements are validate only if the total recovered ^{14}C in all compartments (including culture medium, plants, soil, rinsing wall beaker, etc.) is at minimum 80% of the initial radioactive solution. Under 80%, experiments are considered as non-valuable and must be redone.

2.2 *Stable ^{13}C Isotope Techniques (Table 25.2)*

Enriched isotope methods involve applying huge amounts of a labeled substance and permit one to follow the flows and fates of an element without altering its natural behavior. Because the substances are enriched relative to the background, tracer studies remove or minimize problems of interpretation brought about by fractionation among pools that mix. The signal (the label) is amplified relative to the noise (variation caused by fractionation). Manipulation of stable isotopes requires

Table 25.2 Some comparisons between usual ^{13}C tracers techniques

	Objectives	Advantages	Limits
Extraction by			
Grinding the plant or soil samples	To study the allocation of ^{13}C in the soil-plant system	Simple and fast	Grinding must be done until having a fine powder to obtain good results
Specific purification with solvents of soil or plant samples	To study the partitioning of ^{13}C in different biomolecule pools (e.g., non structural C compounds, starch, lipids)	Simple	Contamination with C must be avoided during the purification steps, using C-less chemicals and volatile solvents Time to obtain pure fractions can be long
Quantification of C content by			
EA	To get the total organic C content of a solid sample	Few amounts of sample (1 mg for plant organ, 25 mg for soil, 0.5 mg for starch) are needed	Liquid samples must be freeze-dried (water, acid or base) or vacuum-dried (organic solvent) before analysis ^{13}C label amounts incorporated must be in sufficient amount to be detected (according to the total initial amount of ^{12}C and ^{13}C in plant and soil samples)
Liquid or gas chromatography	To get the concentrations of specific molecules	Separation of specific biomolecules to get their isotopic signature	Time to analyze one sample is long. The coupling with IRMS is not so easy because the most abundant molecule is not necessarily the most concentrated in ^{13}C. If so, dilutions of the sample are difficult to determine and a fraction collector is often needed to separate minor/major compounds
Quantification of ^{13}C			
IRMS	To determine the $\partial^{13}C$ (ratio $^{13}CO_2/^{12}CO_2$)	High precision	Saturation and memory effect when ^{13}C amounts are too high
IRIS	To determine the concentrations of $^{13}CO_2$ and $^{12}CO_2$	Suitable in the field for analysis of CO_2 High temporal resolution	Interference with non-target gases (CH_4, H_20, H_2S, volatile organic compounds VOCs)

EA elemental analyzer, *IRMS* Isotopic Ratio Mass Spectrometer, *IRIS* Isotopic Ratio Infrared Spectroscopy special room and material devoted to this to prevent the contamination of samples mostly during their preparation.

2.2.1 Pulse-Labeling with CO_2

Pulse-labeling consists in submitting the foliage to ambient concentration of CO_2 having a different signature of ambient atmospheric CO_2 (average $\partial^{13}C = -8‰$) during few minutes or hours. Injected CO_2 is generally highly enriched in ^{13}C (up to 50–100 times background levels) but ^{13}C-depleted CO_2 can also be used ($\partial^{13}C = -47‰$) (Plain et al. 2009; Streit et al. 2013; Zang et al. 2014). The main objective of $^{13}CO_2$ labeling is to study how fast and where the assimilated C is allocated among competing organs and partitioned among pools of C compounds. Labeling can be useful to study C allocation and partitioning in any plant or organ collected. The continuous-labeling is also possible. The choice between pulse or continuous depends on the process(es) investigated and whether the organs and C compounds involved turnover rapidly (from hours to months; pulse-labeling) or slowly (over several months to years; continuous-labeling) (Epron et al. 2012; Hartmann and Trumbore 2016).

Warning Note that the injection of $^{13}CO_2$ can stress the plant and can alter its natural behavior. It is possible to perform labeling under controlled conditions as well as in the field even though it is more difficult to handle and requires several persons, especially in the case of adult trees.

2.2.2 Purification of Samples

In plant samples, total organic matter is analyzed directly but some purification with solvents can be performed to separate different target biomolecules. For example, soluble sugars, starch, lipids and structural C compounds can be purified from various organs in order to get the dynamics of C partitioning in these pools of molecules (Streit et al. 2013; Blessing et al. 2015; Hartmann et al. 2015; Heinrich et al. 2015; Desalme et al. 2017). The objective of such purification is then to get the purest fraction of target molecules rather than the biggest quantity.

Warning Caution must be taken to do not add more C in the different fractions during purification. It is recommended to use volatile solvent and carbon-less chemicals.

2.2.3 Quantification of $^{13}CO_2/^{12}CO_2$ Ratio

$^{13}CO_2/^{12}CO_2$ ratio in plant and soil samples (solid, liquid and gas) is analyzed by an Isotopic Ratio Mass Spectrometer (IRMS). The principle of IRMS is to ionize the CO_2 issued from the oxidation of the sample (mainly 2 isotopologues, $^{12}C^{16}O^{16}O$

and $^{13}C^{16}O^{16}O$), to separate the 2 isotopologues in an electromagnet according to their mass/charge ratio ($^{12}CO_2 = 44$ and $^{13}CO_2 = 45$), and to collect each one in separated faraday cups, allowing the calculation of a $^{13}CO_2/^{12}CO_2$ ratio. Because the IRMS gives a ratio, total C content of the sample or concentration of the molecules must be determined by different techniques. Solid samples containing organic C are analyzed by an elemental analyzer (EA) in which all the organic C is oxidized into CO_2 in a combustion reactor at >1000 °C coupled with an IRMS (EA-IRMS). Liquid samples (e.g., phloem sap extract, latex, exudates) can be analyzed in EA-IRMS after being dried. Specific compound analysis can also be performed with a chromatography coupled with an IRMS. Molecules in the samples are first separated and quantified by chromatography, then oxidized to produce CO_2, and finally injected in the IRMS. Chromatographic separation of the specific molecules is performed by gas or liquid chromatography by following the same procedure as a simple chromatographic analysis. In liquid chromatography, the eluent phase must be inorganic and solutions have to be degassed to prevent CO_2 contamination.

2.2.4 In Situ Quantification of $^{13}CO_2/^{12}CO_2$ Ratio

High temporal resolution and near-continuous measurements of $^{13}CO_2/^{12}CO_2$ ratio are also possible *in situ* by using Isotope Ratio Infrared Spectrometers (IRIS) that are more portable than IRMS, which use remains limited to the lab. The principle of IRIS is to emit a radiation by a laser, and to measure the radiation absorbed or emitted by the target gas species with a spectrometer (Kerstel and Gianfrani 2008; Griffis 2013).

3 Measurement of ^{14}C-Labeled Compounds

The way to obtain target plants with incorporated radiolabeled allelochemicals will influence the selection of the technique to detect the radioactivity. Thus, this section will explain the way to prepare radioactive solution, will indicate how to express results and will give two practical examples of using ^{14}C allelochemicals illustrating grinding and oxidizing extractions. This part will be illustrated by the study of the effect of ^{14}C *p*-hydroxybenzoic acid (POH) from the sample preparation, its extraction by grinding and oxidizing to its quantification in seedlings (Chiapusio and Pellissier 2001).

3.1 Sample Preparation (Fig. 25.1)

Preparation of the Radiochemical: The « Mother Solution » (Step 1) Most of the time, radiochemicals are sold in a powder form. Adding solvent is then necessary to obtain a solution. The adequate quantity will depend on the radiochemical specific activity. In general, the solubilisation in the appropriate solvent is made to obtain a convenient dilution to prepare the further radiochemical solutions. For example,

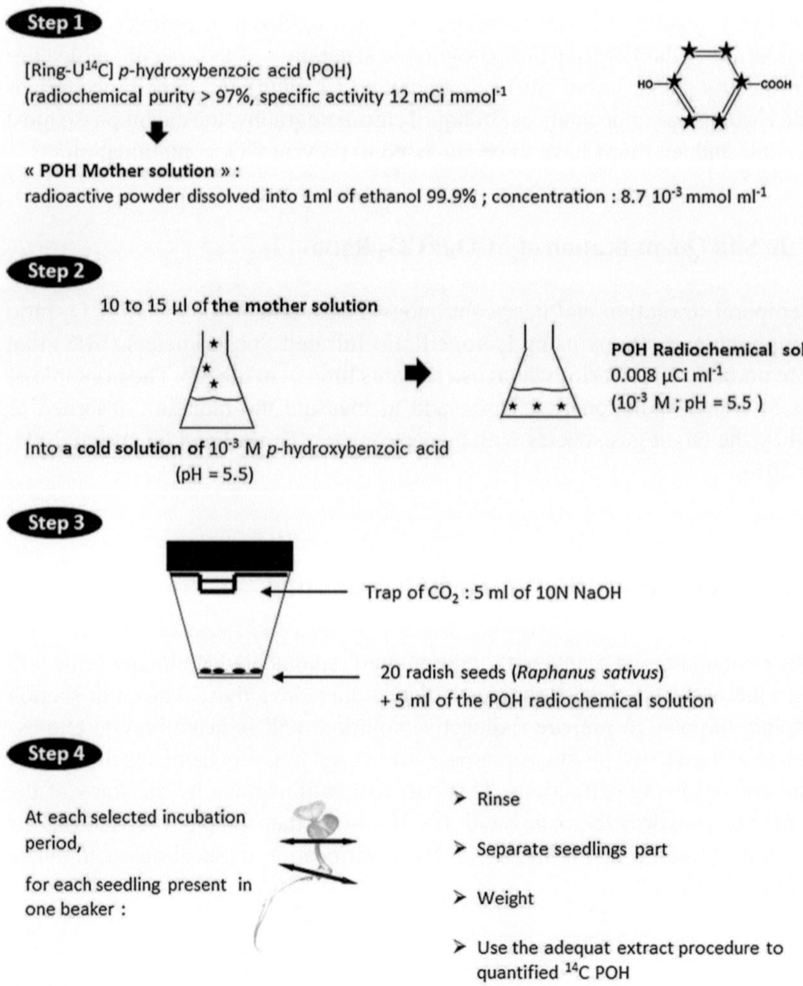

Fig. 25.1 Sample preparation for ^{14}C compounds: the example of ^{14}C *p*-hydroxybenzoic acid (POH) translocation in radish seedlings (non-sterile culture)

ethanol or methanol are commonly used with radiolabeled phenols. This so-called « mother solution», must be prepared carefully because it will be the base of preparation of all further radiochemical solutions. This solution must be kept in freezer until further use.

Radiochemical Solution (Step 2) This solution is a mix of (1) the accurate tested concentration of non-radioactive chemical, called «the cold solution» and (2) an adequate quantity of the «mother solution» to get sufficient radioactivity to remain upper the threshold measurement of the radioactivity detector.

Plant Growth Conditions (Step 3) No special culture conditions are required for the radiochemical experiments compared to usual cultures. The choice of sterile or non-sterile culture depends on the objective of the work but attention must be paid for the potential release of $^{14}CO_2$, which can be trapped in a NaOH solution if necessary.

Samples Preparation (Step 4) No special sample preparation is required. At the end of each selected incubation period, seedlings from each beaker are washed three times in adequate solvent (e.i. ethanol/water for phenolics) to remove adsorbed compounds. Then, seedling parts are separated. Organs of the same type removed from the same beaker are weighed and frozen together (−25 °C). These constitute one sample of one organ type.

3.2 *Expression of Results*

Results of the radiochemicals having penetrated into the seedlings can be either expressed as concentration (for example μmol g^{-1} FW) or as quantity (% of applied ^{14}C chemical) for one beaker. Quantity is the ratio between the Disintegration Per Minute (DPM) of the initial ^{14}C allelochemical solution, which was added to each beaker, and the recovery obtained from separated organs of seedlings. Allelochemical concentration is calculated at the equivalence of the cold solution moles, represented by x DPM detected in organs. Concentration pattern reflects the distribution of allelochemical in the plant organs. This unit is widely used to obtain a ratio between the allelochemical content and the plant biomass, which allows then comparisons between different studies. However, using only concentration could lead to erroneous conclusion because seedlings biomass *per se* is necessary to understand distribution of allelochemical in the organs of target plants (i.e. radish cotyledon biomass is about three times more than roots in 96 h seedlings). Thus, allelochemical content is used to detect preferential accumulation organ.

3.3 Practical Example of ^{14}C Allelochemical Quantification: Extraction by Grinding and Quantification by Liquid Scintillation (Table 25.3)

Principle Samples are ground with a mortar and pestle and soluble radiolabeled allelochemical is extracted with the adequate solvent. Homogenates are centrifuged (3000 rpm) for 3 min at 20 °C. The supernatant is set aside for further analysis after each extraction. The ground residue is extracted twice in the mortar and pestle, supernatants being combined.

Counting Supernatants and pellets are mixed with liquid scintillation cocktail. The choice and the quantity of the added liquid scintillation cocktail are made according to manufacturer recommendations. Radioactivity within samples is counted using a Liquid Scintillation Counter and express as Disintegration Per Minute (DPM).

Conclusion *p*-hydroxybenzoic acid (POH) is absorbed by radish seedlings. POH concentration recovered in each organ is depending of the time and the type of organ. Focusing on the % of applied ^{14}C POH, cotyledons are the POH sink organ at any time (Chiapusio and Pellissier 2001).

3.4 Practical Example of ^{14}C Allelochemical Quantification: Extraction by Oxidizing and Quantification by Liquid Scintillation (Table 25.4)

Principle Frozen radioactive samples are wrapped in a paper (Germaflor) for oven-drying at 80 °C for 48 h or lyophilized. Each sample is then combusted in a biological oxidizer at 900 °C for about 3 min. $^{14}CO_2$ released by samples is directly trapped into a vial containing liquid scintillation cocktail.

Table 25.3 *p*-hydroxybenzoic acid (POH) translocation in radish seedlings in non-sterile condition of incubation

Incubation period	POH quantification	Roots	Hypocotyls	Cotyledons	Total in seedlings
72 h	*µmol g⁻¹FW*	5.9 ± 1.4	5.8 ± 0.9	10.3 ± 1.8[a]	22.0 ± 4
	% of applied	8 ± 1	5 ± 0.4	35 ± 3[a]	47 ± 4
96 h	*µmol g⁻¹FW*	2.8 ± 1.4	2.7 ± 1.3	5.5 ± 1.9	10.9 ± 4.5
	% of applied	5 ± 2	4 ± 1	23 ± 8[a]	32 ± 11

POH concentration (expressed as µmol g^{-1} Fresh Weight) and quantity (expressed as % of applied ^{14}C POH for 20 seedlings ± standard deviation) were measured after grinding extraction method

Data represent the means ± SD (n = 3)

[a]Indicates inside a same incubation time, POH content of the organ is statistically different from the others according to Mann-Whitney *U* non-parametric test ($P < 0.05$)

Table 25.4 *p*-hydroxybenzoic acid (POH) translocation in radish seedlings in sterile condition of incubation

Incubation period	POH quantification	Roots	Hypocotyls	Cotyledons	Total in seedlings
72 h	*μmol g⁻¹FW*	15.7 ± 4.1	7.1 ± 0.9[a]	12.4 ± 2.4	35.1 ± 5.9
	% of applied	5 ± 1	5 ± 1	30 ± 2[a]	40 ± 4
96 h	*μmol g⁻¹FW*	14.8 ± 3.1	5.5 ± 0.9[a]	12.4 ± 1.0	33.2 ± 2.4
	% of applied	11 ± 5	6 ± 1	38 ± 3[a]	53 ± 7

POH concentration (expressed as μmol g^{-1} Fresh Weight) and quantity (expressed as % of applied ^{14}C POH for 20 seedlings ± standard deviation) were measured after oxidizer extraction method
Data represent the means ± SD (n = 3)
[a]Indicates inside a same incubation time, POH content of the organ is statistically different from the others according to Mann-Whitney *U* non-parametric test ($P < 0.05$)

Counting Radioactivity within samples is counted using a Liquid Scintillation Counter and results are expressed as DPM.

Conclusion *p*-hydroxybenzoic acid (POH) is absorbed by radish seedlings. Cotyledons and roots represent the radish organ having the highest POH concentrations. Focusing on the % of applied ^{14}C allelochemical, cotyledons are the POH sink organ. Comparing results of Tables 25.3 and 25.4, the observed decreased of POH in total seedlings (Table 25.3) is due to a fraction of non-extractable POH in plant tissue (Chiapusio and Pellissier 2001). Indeed, such decrease is not observed when using the oxidizer for POH quantification (Table 25.4).

4 Measurement of ^{13}C-Labeled Compounds

This section will explain the way to label plants with $^{13}CO_2$, to collect the samples and how to express results. Two practical examples are selected, both focusing on plants exposed to a phenanthrene (PHE) atmospheric pollution. The first example aims at considering the way carbon allocation is changed in plant-soil systems exposed to PHE by labeling plants with $^{13}CO_2$ and the second one the way PHE is transferred in plants-soil systems exposed to ^{13}C-labeled PHE.

4.1 Labeling and Sample Collection

Plant Growth Conditions No special culture conditions are required for plants prior to be labeled with $^{13}CO_2$.

Labeling of the Plants A transparent plastic chamber must be specifically designed to label the whole crown of the plants. It means that the design and size of the chamber is related to the dimensions of the target plant. As the chamber modifies the microclimate around the crown, attention must be paid to avoid disturbance. Air temperature and hygrometry inside the chamber are maintained at those of the outside thanks to an air conditioner. ^{13}C-enriched or ^{13}C-depleted CO_2 is injected through pipes from a bottle of gas into the chamber at a controlled flow rate. The adequate volume of $^{13}CO_2$ delivered to the plant depends on plant assimilation and the studied process (which molecule? in which compartment? which turnover of the target molecule?) and so is attained by adjusting the duration of the labeling. During the labeling, it is preferable to follow the $^{12}CO_2$ and $^{13}CO_2$ assimilation rate of the labeled plant using an infrared gas analyzer (e.g. S710, SICK/Maihac). If not available, $^{12}CO_2$ assimilation can be measured with a gas analyzer before the labeling and the amount of $^{13}CO_2$ needed to label correctly the plant can be calculated. At the end of labeling period, the bottle is closed, and after around 15 min (the time for plants to assimilate ^{13}C remaining in the chamber) the chamber is removed.

Samples Collection Samples are collected before the labeling to determine the natural abundance of ^{13}C in each studied compartment and at different selected dates after the labeling to follow the fate of ^{13}C in each studied compartments. It is important to know the total quantity of ^{13}C assimilated by the labeled plant. For trees, total assimilated ^{13}C is generally considered as the ^{13}C recovered in leaves collected just after the end of labeling. However, for herbaceous and small plants, ^{13}C can have been already exported out of leaves to other organs, and total assimilated ^{13}C corresponds rather to the total of ^{13}C recovered in all the organs of the plants collected just after labeling. If possible, it is helpful to make some preliminary tests to determine the time needed to recover the ^{13}C in the different parts of the plant (stems/trunks and roots) and in the soil.

Samples Preparation Samples are either frozen in liquid nitrogen or at −20 °C and then rapidly freeze-dried or dried with microwave (useful when only ^{13}C of total organ is needed). They are ground to obtain a fine powder like flour. According to the aim of the study, samples are directly injected in EA-IRMS, or treated with solvent for purification of biomolecules and dried before injection in EA-IRMS, or injected in liquid phase in LC or GC-IRMS.

4.2 *Expression of Results*

The formula of the ^{13}C isotope composition ($\partial^{13}C$), the relative abundance of ^{13}C (x(Ai)), the ^{13}C label recovered (LRi), the fraction of total assimilated ^{13}C ($FLR_{i,t}$) and the partitioning of LR ($PLR_{i,t,}$) are described below.

- EA-IRMS gives the signature of ^{13}C ($\partial^{13}C$, expressed in ‰) of each studied organ. By convention, ^{13}C isotope composition was expressed relative to Vienna Pee Dee Belemnite standard ($R_{PDB} = 0.011802$).

$$\delta^{13}C = \frac{\frac{^{13}CO_2}{^{12}CO_2}}{R_{PDB}} - 1$$

- Relative abundance of ^{13}C in any compartment i ($x(^{13}C)_i$, expressed in atom %):

$$x\left(^{13}C\right)_i = 100 \times \frac{^{13}C}{\left(^{13}C + {}^{12}C_i\right)} = 100 \times \frac{\left(\frac{\delta 13C_i}{1000} + 1\right) \times \mathrm{R_{VPDB}}}{\left[\left(\frac{\delta 13C_i}{1000} + 1\right) \times \mathrm{R_{VPDB}}\right] + 1}$$

- The ^{13}C label recovered in any compartment i (LR_i, mg ^{13}C):

$$LR_i = \frac{\left(x\left(^{13}C\right)_{lp} - x\left(^{13}C\right)_{up}\right)}{100} \times Q_i$$

Q_i the quantity of C in the compartment i (mg C); lp and up denote labeled and unlabeled pots.

- Because C assimilation differs between plants and treatments, data can be expressed in fraction of total assimilated ^{13}C ($FLR_{i,t}$) by taking into account the total assimilated ^{13}C (i.e., quantity of ^{13}C recovered in the total plant or in leaves at day 0).

$$FLR_{i,t} = \frac{LR_{i,t}}{\sum_i LR_{i,t=0}} \times 100$$

- The partitioning of LR in each compartment at a given harvest-time ($PLR_{i,t}$, %) is expressed in percentage of the total ^{13}C recovered in the system at the respective harvest-time *t* (i.e. the total ^{13}C recovered at T1, T2 or T3)

$$PLR_{i,t} = \frac{LR_{i,t}}{\sum_i LR_{i,t}} \times 100$$

4.3 Practical Example of ^{13}C Quantification in the Different Compartments of a Plant-Soil System Submitted to PHE Atmospheric Pollution: $^{13}CO_2$ Pulse Labeling and Quantification by EA-IRMS (Figs. 25.2 and 25.3)

Principle Plant-soil microsystems were subjected to a PHE atmospheric pollution during 28 days in specific chambers (Desalme et al. 2011a). Each chamber (3 dedicated to polluted atmosphere and 3 to control atmosphere) contains four red clover (*Trifolium pratense* L.) cultivated pots.

Labeling At the end of PHE exposure, pure $^{13}CO_2$ was injected in chambers containing the plant-soil systems during 1.5 h. The chambers were opened and vented for 10 min to remove all the $^{13}CO_2$ in excess then the light were turned off.

Sample Collection Samples of leaves, stems, roots and soil were sampled over a 4 days-period: 2 h, 14 h, 38 h and 86 h after the beginning of labeling. Samples of unlabeled pots were collected just before the labeling for each treatment (polluted and control). After being weighted, all the samples were frozen in liquid nitrogen, freeze dried and ground into fine powder. Microbial biomass was extracted from each fresh soil sample by chloroform fumigation of soils during 24 h to solubilize microflora, and extraction with K_2SO_4 of fumigated and non-fumigated samples.

^{13}C abundance Determination in the Different C Pools C content and C isotopic composition of plants, soil and microbial biomass were determined using an

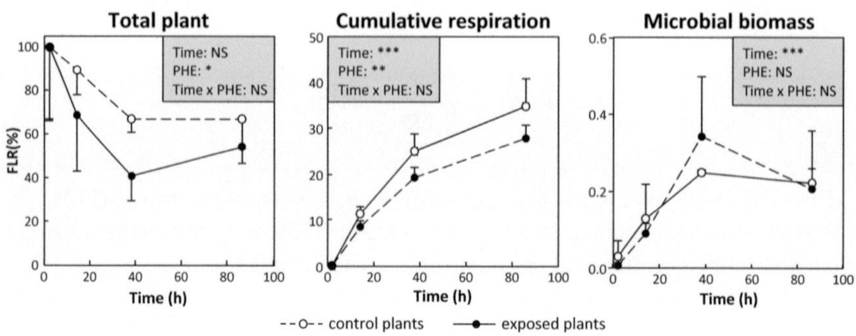

Fig. 25.2 Kinetics of the fraction of ^{13}C label recovered (FLR, %) in total plant, cumulative respiration, and microbial biomass after 1 month of red clover (*Trifolium pratense*) exposure to atmospheric PHE (exposed, closed circle) and to ambient air (control, open circle)
Data are expressed in percentage of ^{13}C label recovered in each compartment at each harvest-time according to the total ^{13}C recovery in the soil-plant system just after the end of labeling. Each point represents the mean ± standard deviation (n = 3). Overall differences across time, between treatments (PHE) and their interaction for the FLR in total plant, cumulative respiration and microbial biomass were tested by a linear mixed effect test ($P < 0.05$, *; $P < 0.01$, **; $P < 0.001$, ***; NS: non significant)

elemental analyzer coupled with an isotopic ratio mass spectrometer (EA-IRMS). Isotope composition of respired CO_2 was analyzed for 3 consecutive days on the 3 polluted and 3 control pots collected at 86 h by introducing them into a 2.4 L tight chamber and analyzing the accumulation of $^{13}CO_2$ and $^{12}CO_2$ during 30 min in the dark.

Expression of Results Results of ^{13}C amounts recovered in plant organs, soil, microbial biomass and cumulative respiration were expressed in fraction of total assimilated ^{13}C recovered (FLR, %) (Fig. 25.2). The quantity of label recovered in each compartment of the plant was also expressed as the partitioning of the total label recovered at each harvest time (PLR) (Fig. 25.3).

Conclusion The amount of ^{13}C recovered in plants and in cumulative respiration was lower in polluted than in control systems (Fig. 25.2). In terms of ^{13}C partitioning among plant organs at each harvest-time, ^{13}C was preferentially retained in the aboveground organs in both polluted and control systems (Fig. 25.3).

Exposure to PHE had a significant overall effect on the carbon partitioning among the clover organs. More ^{13}C are retained in the leaves (51% *versus* 39% on average in polluted and control plant soil systems, respectively) at the expense of the other plant sinks (roots).

4.4 *Practical Example of ^{13}C-PHE Quantification in Different Compartments of a Plant-Soil System Submitted to PHE Atmospheric Pollution (Table 25.5)*

Principle Recent studies showed that using ^{13}C-labeled chemical compounds is a relevant way to trace the fate of organic pollutants in plant-soil mesocosms (Cebron et al. 2011; Cennerazzo et al. 2017). We propose here to use the ^{13}C-phenanthrene/^{12}C-

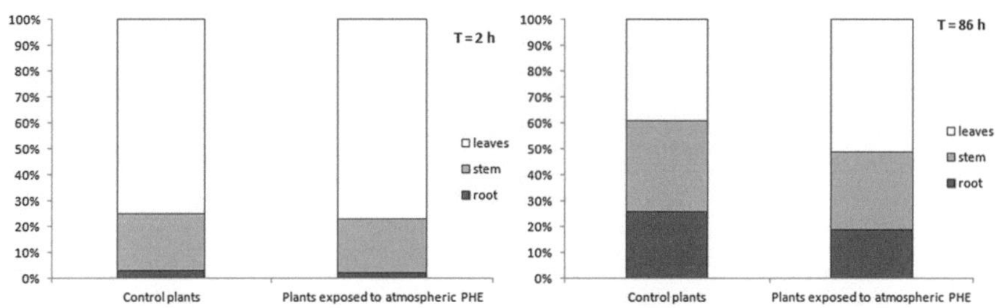

Fig. 25.3 Partitioning of ^{13}C (PLR, %) among red clover organs (leaves, stems and roots) after 1 month of exposure to atmospheric PHE (plants exposed to atmospheric PHE) and to ambient air (control plants). Each fraction of the bar represents the mean percentage of ^{13}C recovery (n = 3) in leaves, stems, and roots of red clover (*Trifolium pratense*) according to the total ^{13}C recovery in the plant at the respective harvesting time

Table 25.5 Phenanthrene (PHE) concentrations and ratio ^{13}C-PHE/^{12}C-PHE recovered in red clover shoot and soil after 8 days to exposure to atmospheric ^{12}C PHE or ^{13}C + ^{12}C PHE

Treatments	PHE quantification	PHE weight in the evaporator	Leaflets	Petioles	Soil
^{12}C PHE	PHE	50 mg	16.8 ± 2 mg kg^{-1}	4.2 ± 1 mg kg^{-1}	0.009 ± 0.001 mg kg^{-1}
	$^{13}C/^{12}C$ PHE (%)	0	0.1 ± 0.02	0.2 ± 01	0
^{12}C + ^{13}C PHE	PHE	50 mg	12.7 ± 1.5 mg kg^{-1}	7.5 ± 1.7 mg kg^{-1}	0.009 ± 0.001 mg kg^{-1}
	$^{13}C/^{12}C$ PHE (%)	25	26 ± 4	26 ± 5	2 ± 0.8

Data represent the means ± SD (n = 3)

phenanthrene ratio (^{13}C-PHE/^{12}C-PHE) as a tool to identify the respective contribution of atmospheric and soil exposure pathways for red clover (*Trifolium pratense*) contamination.

Labeling Red clovers were exposed to PHE atmospheric pollution during 8 days in specific chambers (Desalme et al. 2011b) (3 dedicated to polluted atmosphere and 3 to control atmosphere). Before reaching the chambers, air passed through an evaporator filled with PHE pills. In three exposure chambers, evaporators were filled with PHE pills (50 mg) from ^{13}C PHE (10 mg) and ^{12}C PHE (40 mg) powder (Sigma–Aldrich). A control exposure chamber was performed with an evaporator filled with PHE pills from 50 mg of ^{12}C PHE powder.

Sample Collection Leaves were collected after 8 days of exposure. Leaflets and petioles were separated, lyophilized and grinded with a ball crusher. Soils were collected and dried at room temperature.

^{13}C-PHE/^{12}C-PHE Extraction and Analysis Extractions from leaflet, petiole and soil samples were performed twice with 15 mL of hexane/acetone (v/v: 2/1) by accelerated solvent extraction(ASE100®; Dionex, France) during 8 min (120 °C, 100 bar). Leaflet and petiole extracts were concentrated to 0.8 mL under N_2 flux at 35 °C (TurboVap) and purified using a SPE column 10 mL (SILICA GEL, Sigma). PHE quantification was performed with gas chromatography coupled with a mass spectrometer (GCMS-QP5050A; Shimadzu, France).

Conclusion The ^{13}C-PHE/^{12}C-PHE ratios recovered in leaflets and in petioles were in accordance with the ^{13}C-PHE/^{12}C-PHE ratio of the pills filled in the evaporators but were different from ^{13}C-PHE/^{12}C-PHE ratio of the soil (Table 25.5). These results confirmed that the ^{13}C-PHE/^{12}C-PHE ratio can be used to identify the major contributing pathway for plant contamination.

5 General Conclusion

This chapter gave basis to set up experiments using ^{13}C labeling tracers and ^{14}C radiochemicals. The ^{13}C is now currently used for a diverse range of applications in environmental and plant sciences and the ^{14}C remains a powerful technique to trace specifically ^{14}C molecule translocation into plants and their environment. Specific methodology, equipment and a quite high budget (i.e. the cost of the ^{13}C or ^{14}C molecules) are required for such experiments but they remain crucial tools to understand plant-environment interactions.

References

Andersen CP (2003) Source-sink balance and carbon allocation below ground in plants exposed to ozone. New Phytol 157:213–228

Bahn M, Lattanzi FA, Hasibeder R, Wild B, Koranda M, Danese V, Brüggemann N, Schmitt M, Siegwolf R, Richter A (2013) Responses of belowground carbon allocation dynamics to extended shading in mountain grassland. New Phytol 198:116–126

Blessing CH, Werner RA, Siegwolf R, Buchmann N (2015) Allocation dynamics of recently fixed carbon in beech saplings in response to increased temperatures and drought. Tree Physiol 35:585–598

Brüggemann N, Gessler A, Kayler Z, Keel SG, Badeck F, Barthel M, Boeckx P, Buchmann N, Brugnoli E, Esperschütz J, Gavrichkova O, Ghashghaie J, Gomez-Casanovas N, Keitel C, Knohl A, Kuptz D, Palacio S, Salmon Y, Uchida Y, Bahn M (2011) Carbon allocation and carbon isotope fluxes in the plant-soil-atmosphere continuum: a review. Biogeosciences 8:3457–3489

Cebron A, Louvel B, Faure P, France-Lanord C, Chen Y, Murrell JC, Leyval C (2011) Root exudates modify bacterial diversity of phenanthrene degraders in PAH-polluted soil but not phenanthrene degradation rates. Environ Microbiol 13:722–736

Cennerazzo J, de Junet A, Audinot JN, Leyval C (2017) Dynamics of PAHs and derived organic compounds in a soil-plant mesocosm spiked with ^{13}C-phenanthrene. Chemosphere 168:1619–1627

Chiapusio G, Pellissier F (2001) Methodological setup to study allelochemical translocation in radish seedlings. J Chem Ecol 27:1701–1712

Chiapusio G, Pellissier F, Gallet C (2004) Uptake and translocation of phytochemical 2-benzoxazolinone (BOA) in radish seeds and seedlings. J Exp Bot 55:1587–1592

Chiapusio G, Gallet C, Dobremez JF, Pellissier F (2005) Allelochemicals: tomorrow's herbicides? In: Regnault-Roger C, Philogène B Jr, Vincent C (eds) Biopesticides of plant origin. Intercept Ltd, Curepipe, pp 139–151

Desalme D, Binet P, Epron D, Bernard N, Gilbert D, Toussaint ML, Plain C, Chiapusio G (2011a) Atmospheric phenanthrene pollution modulates carbon allocation in red clover (*Trifolium pratense* L.). Environ Pollut 159:2759–2765

Desalme D, Binet P, Bernard N, Gilbert D, Toussaint ML, Chiapusio G (2011b) Atmospheric phenanthrene transfer and effects on two grassland species and their root symbionts: a microcosm study. Environ Exp Bot 71:146–151

Desalme D, Priault P, Gerant D, Dannoura M, Maillard P, Plain C, Epron D (2017) Seasonal variations drive short-term dynamics and partitioning of recently assimilated carbon in the foliage of adult beech and pine. New Phytol 213:140–153

Epron D, Bahn M, Derrien D, Lattanzi FA, Pumpanen J, Gessler A, Högberg P, Maillard P, Dannoura M, Gérant D, Buchmann N (2012) Pulse-labelling trees to study carbon allocation dynamics: a review of methods, current knowledge and future prospects. Tree Physiol 32:776–798

Griffis TJ (2013) Tracing the flow of carbon dioxide and water vapor between the biosphere and atmosphere: a review of optical isotope techniques and their application. Agric For Meteorol 174:85–109

Hartmann H, Trumbore S (2016) Understanding the roles of nonstructural carbohydrates in forest trees - from what we can measure to what we want to know. New Phytol 211:386–403

Hartmann H, McDowell NG, Trumbore S (2015) Allocation to carbon storage pools in Norway spruce saplings under drought and low CO_2. Tree Physiol 35:243–252

Heinrich S, Dippold MA, Werner C, Wiesenberg GLB, Kuzyakov Y, Glaser B (2015) Allocation of freshly assimilated carbon into primary and secondary metabolites after in situ ^{13}C pulse labelling of Norway spruce (*Picea abies*). Tree Physiol 35:1176–1191

Högberg MN, Briones MJI, Keel SG, Metcalfe DB, Campbell C, Midwood AJ, Thornton B, Hurry V, Linder S, Näsholm T, Högberg P (2010) Quantification of effects of season and nitrogen supply on tree below-ground carbon transfer to ectomycorrhizal fungi and other soil organisms in a boreal pine forest. New Phytol 187:485–493

Kasurinen A, Biasi C, Holopainen T, Rousi M, Maenpaa M, Oksanen E (2012) Interactive effects of elevated ozone and temperature on carbon allocation of silver birch (*Betula pendula*) genotypes in an open-air field exposure. Tree Physiol 32:737–751

Kerstel E, Gianfrani L (2008) Advances in laser-based isotope ratio measurements: selected applications. Appl Phys B Lasers Opt 92:439–449

Kuzyakov Y, Gavrichkova O (2010) REVIEW: time lag between photosynthesis and carbon dioxide efflux from soil: a review of mechanisms and controls. Glob Chang Biol 16:3386–3406

Plain C, Gérant D, Maillard P, Dannoura M, Dong YW, Zeller B, Priault P, Parent F, Epron D (2009) Tracing of recently assimilated carbon in respiration at high temporal resolution in the field with a tuneable diode laser absorption spectrometer after in situ $^{13}CO_2$ pulse labelling of 20-year-old beech trees. Tree Physiol 29:1433–1445

Ruehr NK, Offermann CA, Gessler A, Winkler JB, Ferrio JP, Buchmann N, Barnard RL (2009) Drought effects on allocation of recent carbon: from beech leaves to soil CO_2 efflux. New Phytol 184:950–961

Streit K, Rinne KT, Hagedorn F, Dawes MA, Saurer M, Hoch G, Werner RA, Buchmann N, Siegwolf RTW (2013) Tracing fresh assimilates through *Larix decidua* exposed to elevated CO_2 and soil warming at the alpine treeline using compound-specific stable isotope analysis. New Phytol 197:838–849

Warren JM, Iversen CM, Garten CT, Norby RJ, Childs J, Brice D, Evans RM, Gu L, Thornton P, Weston DJ (2012) Timing and magnitude of C partitioning through a young loblolly pine (*Pinus taeda* L.) stand using ^{13}C labeling and shade treatments. Tree Physiol 32:799–813

Zang U, Goisser M, Grams TEE, Haberle KH, Matyssek R, Matzner E, Borken W (2014) Fate of recently fixed carbon in European beech (*Fagus sylvatica*) saplings during drought and subsequent recovery. Tree Physiol 34:29–38

Chapter 26
Stable-Isotope Techniques to Investigate Sources of Plant Water

Adrià Barbeta, Jérôme Ogée, and Josep Peñuelas

1 Fractionation of Water Stable Isotopes in the Earth's Critical Zone

1.1 Meteoric Waters

The number of protons in the atomic nucleus defines each chemical element of the periodic table. Each element generally has several stable or radioactive isotopes, defined by the number of neutrons in the atomic nucleus. Stable isotopes of hydrogen exist with one or two neutrons (^1H and ^2H) and those for oxygen have 16, 17 or 18 neutrons (^{16}O, ^{17}O and ^{18}O). The most abundant form of water molecules is $^1H_2{}^{16}O$ but other forms also exist in relatively high natural abundances, mainly $HD^{16}O$ and $^1H_2{}^{18}O$. The difference in mass of these different water isotopologues lead to differential partitioning of heavy and light isotopologues during diffusion or phase changes, called isotopic fractionation (Dawson et al. 2002). Isotopic fractionation during the transfer of water among the various compartments of the water cycle lead to distinct isotopic compositions of the different water pools that can be exploited to trace the origin of water in the landscape and/or within an ecosystem.

During evaporation of oceanic water, light isotopologues tend to evaporate preferentially resulting in a depletion of atmospheric water vapor compared to oceanic water. That is, atmospheric water vapor has lower isotopic ratios ($^{18}O/^{16}O$

A. Barbeta (✉) · J. Ogée
ISPA, Bordeaux Science Agro, INRA, Villenave d'Ornon, France
e-mail: adria.barbeta-margarit@inra.fr

J. Peñuelas
CREAF, Cerdanyola del Vallès, Catalonia, Spain

Global Ecology Unit, CREAF-CSIC, Cerdanyola del Vallès, Catalonia, Spain

© Springer International Publishing AG, part of Springer Nature 2018
A. M. Sánchez-Moreiras, M. J. Reigosa (eds.), *Advances in Plant Ecophysiology Techniques*, https://doi.org/10.1007/978-3-319-93233-0_26

and D/H) compared to ocean water. These isotopic ratios are often expressed as a deviation from the Vienna Standard Mean Ocean Water (VSMOW) and noted $\delta^{18}O$ and $\delta^{2}H$. Isotopic fractionations during evaporation and condensation are complex but mass-dependent processes so that the $\delta^{18}O$ and $\delta^{2}H$ atmospheric vapor and meteoric water (precipitation) are linearly related, with a slope of about 8, forming the Global Meteoric Water Line (GMWL) (Craig 1961). The dynamics of water vapor condensation and precipitation in the atmosphere generate temporal and spatial isotopic variability in meteoric waters, with variations along the GMWL, depending mainly on condensation temperature, latitude, altitude and continentality (Dansgaard 1964). Indeed atmospheric water masses moving inland from the ocean, or polewards from the tropics, become progressively more depleted as they lose condensates (Gat and Carmi 1970). For a given location the isotopic composition of meteoric waters varies seasonally along the Local Meteoric Water Line (LMWL) with more depleted values in winter and more enriched ones in summer (Fig. 26.1). The temperature effect also explains why snow is usually depleted relative to rain at a given site (Fig. 26.1). The amount of precipitation also influences the isotopic signal of meteoric water; wetter months have an overall more depleted isotopic signal (Kurita et al. 2009). This amount effect also explains why fog water usually falls on the upper part of the LMWL (Fig. 26.1), at least in regions where fog originates from the same water body as rain water (Scholl et al. 2011). Seasonal variations in the isotopic composition of meteoric waters lead to distinct isotopic composition of the water pools in the critical zone accessible to plants, and this difference can thus be exploited to trace the origin of plant water.

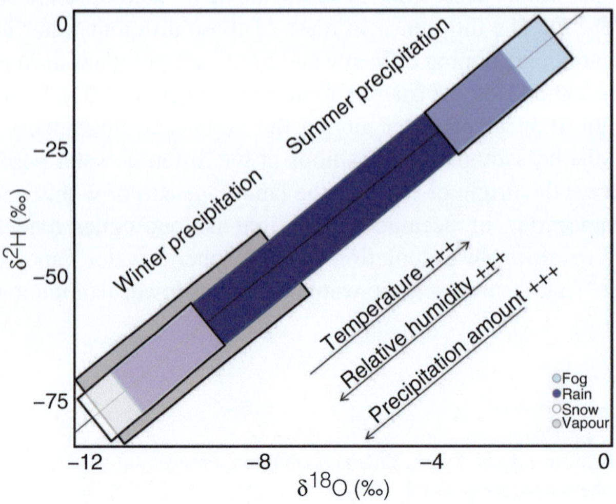

Fig. 26.1 Theoretical dual-isotope plot of meteoric water samples. Different colors indicate different types of precipitation, and the arrows reflect the effect of environmental factors on the isotopic composition. The black line corresponds to the Global Meteoric Water Line (GMWL)

1.2 Soil Waters

Plants take up water mostly through their roots, although foliar water uptake can also constitute a substantial water source in some fog-inundated ecosystems (Goldsmith et al. 2013). The signal of meteoric water is modified when the water is stored underground. Several processes contribute to the isotopic differentiation of underground water pools. Most of these processes lead to isotopic differences that are much larger than the precision of the techniques of isotopic determination, so they are measureable. The two main processes involved are percolation and evaporation. The isotopic signal of underground water pools will partly represent the signal of the precipitated water that percolates through the soil pores. The deviation from meteoric water depends mainly on depth.

Deep pools of groundwater are recharged during seasons in which the soil is close to field capacity, i.e. in wet seasons. Groundwater pools thus represent the average isotopic signal of precipitation during the wet season (Brooks et al. 2010) or in cold seasons with low evapotranspiration and high soil moisture. In addition, the isotopic signal of precipitation can also change during a rain event, with more depleted rain at the end of the event because of the *rainout effect* (Brooks et al. 2010). Deep groundwater pools in confined or perched aquifers created in past geological ages, however, can have a substantially different isotopic composition, indicating the signal of past precipitation (Darling et al. 2003). Signals of stream water can also be relevant in riparian ecosystems, because water accessed by plants and supplying streamflow can belong to the same water pool. Stream water will be more or less seasonally variable depending on its source (snowmelt, storms, old groundwater pools or seasonally recharged groundwater pools). In contrast to deep soil water, groundwater and stream water, the isotopic composition of water in surface soil is more likely to represent the isotopic composition of recent precipitation, regardless of the season, and is affected by soil evaporation.

Evaporative enrichment is another important process contributing to underground isotopic differentiation. Evaporation of soil water produces a progressive enrichment of the water at the soil surface, because lighter isotopes are more easily evaporated. The kinetic fractionation of soil water, however, differs from equilibrium fractionation in the atmosphere, so the slope of the relationship between $\delta^{18}O$ and $\delta^{2}H$ deviates from the GMWL (Horita et al. 1995; Sprenger et al. 2016). This evaporative enrichment creates a isotopic differentiation between enriched surface soil water and non-evaporated, depleted deep soil water or groundwater that can be used to estimate the soil evaporation line (SEL). The depth reached by evaporative enrichment may differ substantially between climatic zones, from 0.2 to 0.3 m in temperate zones to a maximum of 3 m in arid climates (Sprenger et al. 2016). An isotopic gradient with progressively more depleted water with depth should be measurable whenever evaporation occurs. Finally, the first few centimeters of the soil can be affected by atmospheric water

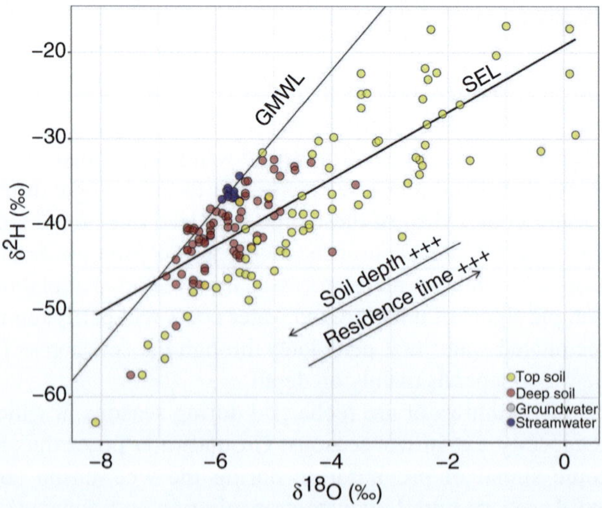

Fig. 26.2 Theoretical dual-isotope plot of belowground water pools. Different colors indicate different water pools, and the arrows reflect the effect of soil depth and residence time. The thin black line corresponds to the Global Meteoric Water Line (GMWL) and the thick black line represent the Soil Evaporation Line (SEL)

vapor, which depletes superficial soil and decreases with depth as evaporation dominates the signal. The isotopic signal of plant xylem water could be tracked if the isotopic signal of underground water pools is sufficiently distinct. Importantly, the characterization of potential plant-water sources will critically depend on the correct identification of each isotopically distinct underground water pool. A proper characterization of the isotopic profile with depth is thus recommended before the onset of any field study. A typical distribution of the isotopic composition of underground water pools is depicted in Fig. 26.2, representing an example of a temperate forest.

Early studies of the stable isotopes of plant and soil water concluded that no fractionation occurred during the uptake of water by roots (Allison et al. 1984; White et al. 1985; Ehleringer and Dawson 1992). Identifying the spatial and temporal patterns in plant-water sources is therefore possible by simultaneously sampling plant xylem water and its potential sources. The identification of plant-water sources, though, is only possible when at least two underground water pools are isotopically distinct. Xylem water is also very likely to indicate a mixture from different water pools, because many plants have a dimorphic root system that allows them to access more than one water pool at the same time (Dawson and Pate 1996). In addition, plant-water sources can differ between coexisting species that do not share the same ecohydrological niche (Silvertown et al. 2015), and root development implies that larger and/or older plants may access deeper water pools more than their smaller and/or younger conspecifics (Kerhoulas et al. 2013). These characteristics of plant-water sources require that the sources be investigated individually.

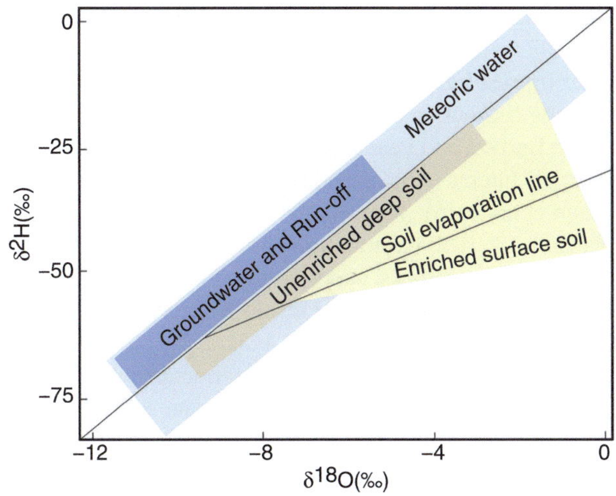

Fig. 26.3 Conceptual scheme depicting the most common distribution of the isotopic composition atmospheric and belowground water pools of an ecosystem

In summary, the sources of isotopic variability of underground water pools are reasonably well understood (Fig. 26.3). Together with the generally accepted absence of fractionation during the uptake of water by roots (Ehleringer and Dawson 1992), the theoretical framework for studies of plant-water sources has been straightforward and sound until very recently. The technological development in this discipline, however, has increased the number of studies applying stable isotopes to identify plant-water sources. Larger data sets with higher temporal and spatial resolution have been compiled, and some have challenged the main assumptions of prior studies, for both underground and plant-mediated isotopic fractionation. Even though these data sets do not invalidate previous work, researchers must understand the limitations and uncertainties of these techniques before conducting fieldwork or interpreting their data. We have synthesized the most relevant findings in the following section.

1.3 Soil Isotopic Heterogeneity and Plant-Mediated Fractionation

Soil-water isotopic signals are governed by precipitation inputs, evaporation and the mixing between new and old water pools in soil pores (Sprenger et al. 2016). Several fractionation processes in the soil, however, must be taken into account to properly track the movement of water to roots. Heavy and light isotopes may interact differently with soil minerals and organic matter. In soils with high contents of clay minerals the interactions of water with cations can entail isotopic differences between cation-absorbed water and the remaining free water (Oerter et al. 2014). A similar isotopic differentiation has been reported for water in contact with the surface of organic particles (Chen et al. 2016), which would also create isotopically different pools of water in soils with high organic content: one on the surface of organic particles and another formed by free water accessed by plants.

Roots explore the bedrock in search for water in porous rock and rock fractures (Schwinning 2010; Barbeta and Peñuelas 2017). Water infiltrating through rocky layers can be isotopically filtered (Coplen and Hanshaw 1973), resulting in more depleted rock water (Oshun et al. 2015). The residual solution, i.e. the free water potentially absorbed by plants, is likely to have a more enriched signal. Plants, however, would access structural water from rocks under some environmental conditions (Palacio et al. 2014). These processes can influence results if the methods used to analyze soil-water isotopes are based on the extraction of bulk soil water, whereas plants do not access all isotopically distinct pools. All these fractionation processes therefore require knowledge of the characteristics of the substrate at the field site. Soil texture and type and depth of bedrock or organic-matter content are particularly relevant characteristics potentially affecting the isotopic composition of underground water pools.

Recent studies have also found that xylem water may not necessarily indicate the source water. Early studies reported that xylem water did not change isotopically until it reached non-suberized twigs and leaves (Ehleringer and Dawson 1992). The evaporative enrichment of woody stems, however, can occur during periods of low water flow, such as in winter (Bowling et al. 2017) or droughts (Martín-Gómez et al. 2017). Analyzing water from the bottom of the plants (i.e. base of the trunk) less exposed to evaporation is thus preferable following or during winter or droughts, when possible. Finally, a recent pot experiment has reported discrimination during the uptake of water by roots and its dependence on soil texture and soil-water content (Vargas et al. 2017). The mechanisms were not fully resolved by the study, because the fractionation signal did not agree with any known fractionation process in the soil, described in the previous paragraph. The study, however, demonstrated that stable isotopes may not always identify plant-water sources with high precision. The simultaneous analysis of the isotopic composition of xylem and source water, though, is still a valuable tool that works well in most cases. Describing the temporal, spatial and/or species-specific patterns of plant-water uptake is possible following a cautious and informed interpretation of the data.

2 Existing Protocols for Sampling Water Pools in the Critical Zone

2.1 *Meteoric Waters*

Sampling meteoric data is highly recommended for any study aiming at describing the patterns of plant water uptake. These data provide the seasonal variations in the isotope composition input of soil water that will subsequently be modified by soil process dynamics. This allows to derive the local meteoric water line (LMWL) that can differ slightly from the GMWL. The LMWL is useful to assess possible isotopic deviations of ground and soil water pools. The International Atomic Energy Agency (IAEA), that coordinates the Global Network of Isotopes in Precipitation (GNIP), provide a protocol for installing rain collectors for the analysis of stable isotopes (IAEA/GNIP 2014). Ensuring a full exposure of the collector to rainwater and a placement away from any source of heat is necessary, and importantly, the design must avoid any evaporation during collection and storage of precipitated water. Rain collectors can be installed at meteorological stations for determining the relationship between precipitation amounts and isotope composition. Precipitation amounts will have consequences in input isotopic signals but also in water infiltration depths. If the establishment of a rain collector is not feasible, the GNIP database can be accessed and searched for the nearest station. Some areas, though, will not be well represented, as differences in altitude and distance to the sea from the nearest GNIP station can strongly affect the isotopic composition of precipitation (Gat and Carmi 1970). Thus, a site-specific isotopic data for precipitation is recommended, especially for mountainous or remote areas.

Precipitation in the form of snow should be sampled separately, because the accumulation of snow on the collectors and subsequent melting or evaporation could lead to incomplete sampling. A recent study reported an effective and simple method for sampling snow. Bowling et al. (2017) buried a plastic bucket leveled to the soil surface immediately before winter to collect melting snow. Snow can also be directly sampled from the snowpack cores (Bowling et al. 2017). A sampling of the entire snowpack profile and over different soil covers will ensure capturing the potential spatial variability in snow-water isotopes.

Plants in regions with frequent fog can also take this water up through leaves or roots from the condensate dripping from the canopy (Limm et al. 2009). Fog water should thus be sampled and its isotopic composition analyzed wherever it may represent a significant part of the plants water input. There are two main types of fog collectors: (1) passive fog collectors that rely on the wind to push air through the collection surface (either a mesh with vertical and horizontal strands, or a harp with only vertical strands), and (2) active fog collectors in which a fan pulls the air through the collection surface (Fisher et al. 2007). Both types of collectors take advantage of the difference in inertia between fog droplets and the surrounding air, leading to the impact of the droplets onto the strands during the deflection of the air stream when passing through the mesh or harp. Active fog collectors are preferable because they can be more easily protected from precipitation, minimizing potential mixing of fog and rainwater during windy rain events. On the other hand, passive collectors can be installed in remote areas with limited access to electric power. The bottom of the mesh or harp must be connected to a vessel similar to that used in rain collectors, ensuring no evaporation during storage. Alternatively, fog water can be sampled during individual field campaigns in a similar fashion as when sampling water vapor using a cold trap made of a glass coil submerged in a mixture of ethanol and dry ice and connected to a pump that will slowly pump air through the cold trap at a rate that ensures complete solidification (typically below 0.5 liter per minute).

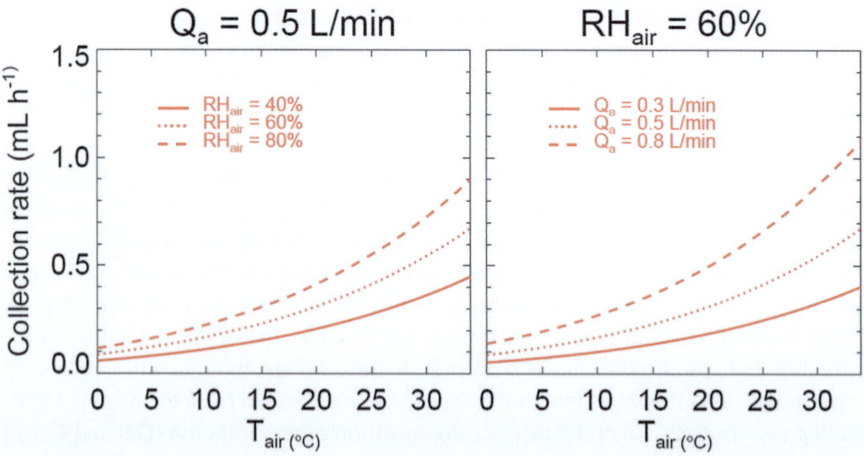

Fig. 26.4 Collection rates of water vapor using a cryogenic trap as a function of temperature, relative humidity and the flow rate of the pump. The left panel is for a set flow of 0.5 L min^{-1} and the right panel is for a constant relative humidity of 60%

This method will exclusively collect water vapor on days without fog. The time required to sample the required amount of water for isotopic analysis will depend on the airflow, temperature and relative humidity (Fig. 26.4). Upon collection from the field, all the meteoric water samples must be stored in a refrigerator in airtight glass bottles with a cap covered with Parafilm® to ensure that no vapor escapes during storage, and until stable isotope analysis.

2.2 *Soil Waters*

To make sure that all potential plant water sources are considered, sampling soil water along the entire plant root profile and even below the maximum rooting depth is desirable. Sampling soil water down to the water table (including groundwater) should be sufficient at sites with shallow water tables. The isotopic composition of the water in saprolite, weathered rock or bedrock should at least be obtained at rocky sites where roots do not reach the water table. Because the maximum rooting depth can be difficult, to assess experimentally, a modeling framework for rooting depth (e.g. Fan et al. 2017) can be used to infer the expected maximum rooting depth, as a function of climate, landscape position (valley bottom, hillslope, hilltop) and soil texture. Deep water pools may often not be accessible from the surface with a soil auger, either because the soil is too deep (i.e. > 2–3 m) or too rocky. In such cases, a natural spring or nearby well should provide an isotopic signal comparable to that accessed by deep roots.

The enriched surface soil and the soil below the evaporation front should at least be sampled. The number of depth increments will depend on the isotopic

Fig. 26.5 Theoretical isotopic profiles in depth. A corresponds to a steady isotopic depletion with depth, whereas B and C show different cases of evaporation fronts, with more stable isotopic compositions in depth

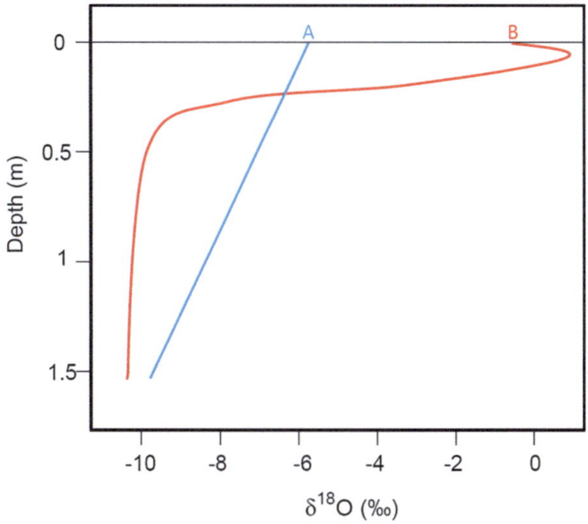

profile. The soil isotopic profiles that will require different sampling strategies are depicted in Fig. 26.5. In case A, typical of what should be observed after 1–2 days of days without rain, the isotopic change is approximately constant throughout the depth profile. All depths should thus be sampled at increments adapted to the sample processing capacity. In case B, typical of what should be observed after several (>5 days) of days without rain and with high evaporative enrichment, the surface soil and the soil below 0.5 m are clearly differentiated. Characterizing the surface and deep soil would thus be enough. The precision of the estimates of plant-water uptake will also depend on the horizontal variability of the isotopic composition of water in each soil layer. The evaporative enrichment of soil can vary substantially within short distances due to contrasting vegetation cover or orientation exposition to sunlight. The water isotope composition at the surface of the soil will consequently be spatially and temporally more variable than at depth. Obtaining as many replicates as possible is thus recommended, depending on the sample processing capacity. The vials for collecting and storing the soil samples should have caps with septa to avoid water vapor leaks and contamination and a volume of at least 10 mL. Collecting larger soil samples may be necessary at dry sites in order to have enough water to extract for isotopic analysis.

In most studies soil water samples are collected from the field and subsequently analyzed in the laboratory (Sect. 26.3), but new techniques have recently been developed enabling *in situ* measurements of soil water isotopes (Oerter and Bowen 2017). Briefly, the method is based on the assumption that soil pore water vapor is in isotopic equilibrium with liquid water. Water vapor probes are permanently inserted at different depths within the soil column and connected to a stable isotope water vapor analyzer (Oerter et al. 2017). The main advantage of this method is that it is non-destructive (once the probes are inserted), so that soil water isotopes can be recorded at the exact same location over time. The main disadvantage of the method is that it can only be used to survey a limited amount of closely related soil profiles. The method also requires main power, which is not always accessible in remote areas.

2.3 *Xylem Water*

Sampling xylem water to determine its isotopic composition in discrete campaigns is relatively straightforward and easy. Xylem water does not generally suffer isotopic fractionation in suberized plant stems and branches (Ehleringer and Dawson 1992). Evaporative isotopic enrichment of xylem water may occur in transpiring, non-suberized, young branches, especially when the water flow is low (Martín-Gómez et al. 2017). The back-diffusion of evaporatively enriched foliar water can also interfere with the signal from the source water in these terminal branches (Dawson and Ehleringer 1991). Removing the bark and phloem that may also enrich the signal from foliar water is also recommended when cutting a branch. Alternatively, wood cores could be collected with an increment borer from the base of the trunk or at breast height for large trees with high canopies that are difficult to access. The core would not need to be deep and should only contain sapwood (Ehleringer and Dawson 1992). The outermost part of the core, corresponding to bark and phloem, should also be removed. Wood cores or pealed branches must be placed immediately (within a couple of minutes maximum) in airtight glass bottles and kept in a cooler until being stored in a laboratory freezer. Plant water sources can change quickly, so each sampling campaign should be conducted as quickly as possible and preferentially at the same time of the day. Likewise, abrupt changes in plant water sources should be monitored after a rain or during extreme climatic events such as heatwaves or droughts.

Similar to what has been done in soils, a new method to continuously monitor *in situ* the isotopic composition of tree xylem water has recently been proposed (Volkmann et al. 2016). Xylem water is diffused through a porous membrane of the probe, pulled by the difference in partial pressure between a vacuum line and the wet xylem. The authors installed more than one probe per trunk and irrigated them with isotopically labeled water. The results were comparable to cryogenically extracted water, but different probes in the same trunk detected slightly different isotopic values. Damage to the xylem caused by probe insertion may therefore have altered the conductive capacity of the xylem, or different parts of the trunk's xylem may have been connected to different parts of the root system. The authors nonetheless recommended the installation of several probes per tree to obtain a more integrated isotopic signal. The main advantage of this method is the possibility of sampling at a high temporal resolution and thus determining daily cycles in plant water sources that are still poorly understood. This probe-based technique may be more difficult to apply in remote areas without access to power, similar to the soil *in situ* methods.

3 Water Isotope Analysis from Soil and Xylem Samples

3.1 Cryogenic Extraction of Water from Soils and Xylems

Compact equipment for monitoring xylem and soil-water isotopes *in situ* has been developed, such as laser-based isotopic analyzers, but most studies of plant-water sources still rely on cryogenic water extraction. This technique consists of a vacuum distillation in which the water contained in a solid sample is evaporated and condensed in a collection tube (West et al. 2006, and citations therein). Detailed descriptions of the design and functioning of cryogenic extraction lines have been provided in previous studies (West et al. 2006; Orlowski et al. 2013). Briefly, (1) the xylem or soil samples and the collection tubes are first frozen with liquid nitrogen and connected to the vacuum line, (2) the xylem or soil samples are then heated (with boiling water, mineral oil or heating blocks) and (3) the water is progressively evaporated from the sample and is trapped by the collection tubes submerged in liquid nitrogen. This extraction technique is a Rayleigh fractionation, because lighter isotopes will evaporate first. A long extraction time, however, will effectively remove all water from a solid sample, thus obtaining a similar isotopic composition of water trapped by the collection tube. The extraction times will depend on the soil and plant types; water in sandy soils will be completely extracted after 30 min, and water in woody stems will be completely extracted after 60–75 min (West et al. 2006). Most of the isotope laboratories extract for 2 h to standardize the time. This technique is widely applied and is consistent for xylem water, but tightly bound water in some types of soils may not be effectively removed during cryogenic extraction (see Sect. 26.3.2).

3.2 Issues with Cryogenic Extraction of Soil Water

Early methodological tests of extraction times of soil water have reported that nearly complete extraction from clay soils requires more time (West et al. 2006), because of a higher fraction of tightly bound water in clayey soils compared to coarser soils. Large effects of the physiochemical properties of soils on the isotopic signal of extracted water have recently been demonstrated (Meißner et al. 2013; Orlowski et al. 2013). In particular, cryogenically extracted water tends to be isotopically depleted relative to input water in re-wetting experiments (Meißner et al. 2013; Orlowski et al. 2013; Orlowski et al. 2016a, b; Newberry et al. 2017). Clay, calcium carbonate and soil-water contents can amplify the isotopic differences between input and extracted water (Meißner et al. 2013; Orlowski et al. 2016a, b; Newberry et al. 2017). The fractionation of water in soil, however, does not imply that tightly bound water in soil micropores or clay or organic particles does not mix with free water (Newberry et al. 2017; Vargas et al. 2017). This isotopic stratification within the soil matrix could be misleading for the study of plant-water sources if water extracted from soils did not represent the total water accessed by plants. Alternatively plants may not access these pools of tightly bound water, even with complete extraction. Testing the effects of various settings of the cryogenic extraction line is recommended for avoiding mismatches between the water extracted from plants and soils. For example, testing various extraction times and temperatures to extract water contents up to the permanent wilting point. Pre-weighing the samples, extracting the water, oven-drying the samples at high temperatures and then re-weighing them to quantify the amount of residual water after the extraction would be helpful. These considerations are more relevant for studies of soils with high clay and/or low soil-moisture contents.

Issues associated with the cryogenic extraction of soil-pore water may be overcome using the recently developed vapor-probe method (Oerter and Bowen 2017), as mentioned above. This method, however, would be difficult to apply in some cases. Many alternative techniques to cryogenic extraction are available. A recent comparison of techniques, though, found that all produced inconsistent results because the isotopic composition of extracted water always differed from the rehydration water added after oven-drying the soils, particularly for clayey soils with low soil-water contents (Orlowski et al. 2016a, b). Centrifugation (White et al. 1985; Böttcher et al. 1997) and mechanical squeezing (Peters and Yakir 2008) of the samples performed best (lower deviation from added reference water) and should be the best options for the precision required in studies of plant-water sources (Orlowski et al. 2016a, b). The time needed for sample processing, however, is longer than for most cryogenic extractions (10 samples per day for mechanical squeezing), and a larger quantity of soil is required. Cryogenic extractions highlight the need for a progressive shift towards alternative extraction techniques, such as centrifugation and mechanical squeezing, for extracting soil-pore water in studies of plant-water sources.

4 Isotopic Determination of Extracted Water

Isotopic determination is the last step in the laboratory after the extraction of water from solid samples. Early and many recent studies of plant water sources have used isotopic ratio mass spectrometry (IRMS), which requires the conversion of water to H_2 or CO or its isotopic equilibration with CO_2 (Gehre et al. 1996; West et al. 2010). Laser-based isotopic analyzers have recently become more popular, because they allow the simultaneous measurement of both the oxygen and hydrogen isotopes of water, reducing the time and cost of the analyses (Lis et al. 2008; Gupta et al. 2010). Off-axis integrated-cavity output spectroscopy (OA-ICOS) (Lis et al. 2008) and wavelength-scanned cavity ring-down spectroscopy (WS-CRDS) (Gupta et al. 2010) are two rather common laser-based techniques. Some studies have found that laser-based instruments can detect non-significantly different isotopic ratios for soil samples compared to IRMS (West et al. 2010; Orlowski et al. 2016a, b), although another study reported discrepancies (Martín-Gómez et al. 2015). Discrepancies for plant water samples (leaves and xylems) have been more consistent with OA-ICOS (Schultz et al. 2011) and WS-CDRS (West et al. 2010) than IRMS.

Organic compounds in plant materials can be distilled together with water during cryogenic extraction and volatilized during analysis into the instrument cavity. These compounds can produce spectral interferences, that can lead to large discrepancies with IRMS results (West et al. 2010; Schultz et al. 2011; Zhao et al. 2011; Martín-Gómez et al. 2015). Because soil and xylem samples may contain amounts of volatile organic compounds large enough to cause spectral interferences, a subset of samples should preferentially be analyzed also by IRMS when using a laser-based instrument. Post-processing softwares for these laser-based instruments also provide diagnostic tools that indicate if spectral interferences have been detected. These diagnostics can be used to decide how to proceed with the isotopic determination. In addition, post-processing correction methods have been published (Schultz et al. 2011) or are provided by the manufacturers of the laser-based instruments. A step-by-step decision-making scheme is provided in Fig. 26.6.

For OA-ICOS instruments, developing instrument-specific correction curves by analyzing series of dilutions with ethanol and methanol and producing correlations of the broadband and narrowband metrics with the isotopic offset from non-contaminated samples are necessary (Schultz et al. 2011).

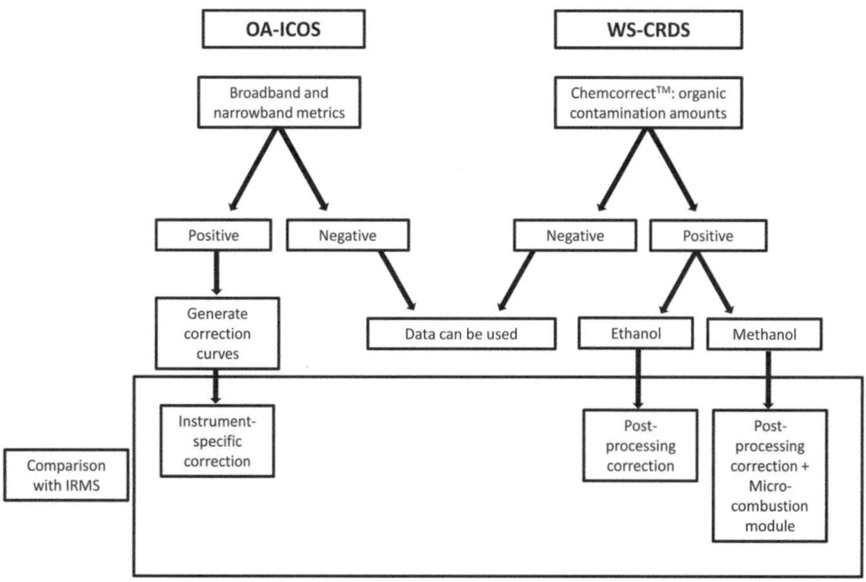

Fig. 26.6 Decision-making scheme for the isotopic determination of plant and soil water samples. For the two main laser-based instruments. After the positive detection of organic contamination by the corresponding post-processing softwares, the comparison of a subset of samples with IRMS results is always mandatory

5 Identifying Plant-Water Sources and Quantifying Their Relative Contributions

The ultimate aim of the application of stable isotopes described here is to identify the sources of water for plants and how they change spatially, temporally and/or between species or between life forms, plant size and other factors. This identification is achieved by determining the isotopic composition of the source water and xylem samples collected simultaneously. Plant water sources can be identified graphically by plotting isotopic compositions in the dual-isotope space (Fig. 26.2) and finding the most similar source water for each xylem sample. This method, however, is usually not simple, nor quantitative, because plants may be simultaneously tapping several sources, and the isotopic composition of the xylem water will indicate a mixture of these sources. Xylem water samples would then not match any of the water sources but would lie at the barycenter of their relative contributions. Early studies that used only one water isotope based their identification of plant-water sources on the statistical differences between the isotopic composition of xylem water and the various sources tested (Dawson and Ehleringer 1991).

Conclusions drawn from studies using a dual-isotope approach, however, can differ slightly (Bowling et al. 2017), because samples with the same δ^2H signal may not share the same $\delta^{18}O$ signal. Statistical differences may thus not be adequate if plants are assumed to take up water from more than two sources, which is very likely for deep rooting systems. Various numerical approaches have been developed for these (rather common) cases to estimate the relative contribution of each potential source of plant water. Exhaustive descriptions and comparisons have been provided (Phillips and Gregg 2001, 2003; Parnell et al. 2013; Ogle et al. 2014; Rothfuss and Javaux 2016), so we provide here only a synthesis of the most used approaches.

End-member mixing analyses were the first models to be applied to identify plant-water sources (White et al. 1985). These models assume that roots extract water from two sources without isotopic fractionation and that the water mixes completely within the xylem (Rothfuss and Javaux 2016). Likewise, the contributions of sources *a* and *b* to xylem isotopic composition (δ_p) is expressed as:

$$\delta_p = f_a\delta_a + f_b\delta_b$$

where δ_a and δ_b are the isotopic compositions of sources *a* and *b*, and *f* is the proportion in volume of water extracted from each source ($f_a + f_b = 1$). This equation can be applied to isotopic compositions with only one isotopologue, and the error associated with the estimation of *x* can also be calculated (Phillips and Gregg 2001; Rothfuss and Javaux 2016). The equation, however, can be expanded by adding the other isotopologue (Brunel et al. 1995).

Phillips and Gregg (2003) developed the program *IsoSource* based on a standard linear mixing model for obtaining a combination of source contributions that conserves mass balances for a system of two isotopes (in studies of plant-water sources). Equation (a) will not have a unique solution if a third water source is added, so the program iteratively creates each possible combination of solutions, the predicted isotopic signatures for each combination is then calculated and the predicted mixture signatures are compared with the observed mixing signatures (Phillips and Gregg 2003). *IsoSource* then provides all feasible source contributions with histograms for each source with the most likely source contribution. *IsoSource* has been extensively used in studies of plant-water sources, but more recent Bayesian mixing models have gained popularity in recent years.

SIAR (Parnell et al. 2010) and *MixSIR* (Moore and Semmens 2008) *Bayesian isotope mixing models* have been developed more recently, providing statistical uncertainties associated with the estimates of source contribution and an optimal solution rather than a range of feasible solutions (Rothfuss and Javaux 2016). These two models have been merged into *MixSIAR* (Parnell et al. 2013), which is implemented by an R package (R Core Development Team 2012). The user prepares three data sets, one with the mean $\delta^{18}O$ and $\delta^{2}H$ signals and corresponding standard deviations of all potential sources, another with $\delta^{18}O$ and $\delta^{2}H$ signals of the plants (grouped or not) and the third specifying an "enrichment factor", which should be set to 0 assuming no fractionation during the uptake of water by roots (see Sect. 26.2). The model uses Monte-Carlo Markov chains to produce a posterior distribution of the relative contribution of each water source. Histograms of the posterior distributions are plotted, allowing a visual representation of the results. The median of the posterior distribution, i.e. the most frequent result of the model, and the corresponding confidence intervals can then be calculated at individual, plot, species or site levels.

End-member and Bayesian mixing models have different routines but require similar data inputs (isotopic compositions of sources and plants). *Bayesian process-based mixing models* (Ogle et al. 2014) also allow the incorporation of prior information about the system (rooting depth, profiles of root density, soil moisture or plant-water potentials) and represent a promising method to improve the precision of plant-sourcing studies. The implementation of process-based models has the advantage of avoiding false positives under some conditions. For example, if the upper soil layers contain too little water to be extracted by roots, a simple linear model would still try to calculate the water's source contribution if it is set as a potential source. In contrast, process-based models can specify that water is not extractable below a certain level of soil moisture, or where roots are not present if information for root profiles was added. The specification of these models, however, is more complex than for *MixSIAR* and *IsoSource*. The additional measurements that should be carried out in the field also require correctly incorporating and specifying this information into a model before it is run, which may require some training in hierarchical Bayesian modeling.

A recent comparative study of simulated data for the uptake of water by roots (Rothfuss and Javaux 2016) has demonstrated that Bayesian mixing models (Parnell et al. 2010, 2013) perform better than *IsoSource* and previous models (Brunel et al. 1995), but only when the potential sources matched the actual plant-water sources. As mentioned above, the inclusion of sources not accessed by plants and the exclusion of some accessed sources can produce misleading results. In summary, we recommend applying Bayesian mixing models (such as *MixSIAR*) for estimating source contributions for plants. Bayesian process-based models (Ogle et al. 2004, 2014) are a better option if higher precision is required, because they can integrate other ecological data, but are more time-consuming.

References

Allison GB, Barnes CJ, Hugues MW, Leaney FWJ (1984) Effect of climate and vegetation on oxygen-18 and deuterium profiles in soils. Isotope hydrology, 1983: Proceedings of an International Symposium on Isotope Hydrology in Water Resources Development, pp 105–123

Barbeta A, Peñuelas J (2017) Relative contribution of groundwater to plant transpiration estimated with stable isotopes. Sci Rep 7(1):10580

Böttcher G, Brumsack HJ, Heinrichs H, Pohlmann M (1997) A new high-pressure squeezing technique for pore fluid extraction from terrestrial soils. Water Air Soil Pollut 94:289–296

Bowling DR, Schulze ES, Hall SJ (2017) Revisiting streamside trees that do not use stream water: can the two water worlds hypothesis and snowpack isotopic effects explain a missing water source. Ecohydrology 10:1–12

Brooks JR, Barnard HR, Coulombe R, McDonnell JJ (2010) Ecohydrologic separation of water between trees and streams in a Mediterranean climate. Nat Geosci 3:100–104

Brunel JP, Walker GR, Kennett-Smith AK (1995) Field validation of isotopic procedures for determining sources of water used by plants in a semi-arid environment. J Hydrol 167:351–368

Chen G, Auerswald K, Schnyder H (2016) Isotopic offset between unconfined water and water adsorbed to organic matter in equilibrium. Biogeosci Discuss 2016(March):1–19

Coplen TB, Hanshaw BB (1973) Ultratitration by a compacted clay membrane-I. Oxygen and hydrogen isotopic fractionation. Geochim Cosmochim Acta 37:2295–2310

Craig H (1961) Isotopic variation in meteoric water. Science 133:1702–1703

Dansgaard W (1964) Stable isotopes in precipitation. Tellus 16(4):436–468

Darling WG, Bath AH, Talbot JC (2003) The O & H stable isotopic composition of fresh waters in the British isles. 2. Surface waters and groundwater. Hydrol Earth Syst Sci 7:183–195

Dawson TE, Ehleringer JR (1991) Streamside trees that do not use stream water. Nature 350(6316):335–337

Dawson TE, Pate JS (1996) Seasonal water uptake and movement in root systems of Australian phraeatophytic plants of dimorphic root morphology: a stable isotope investigation. Oecologia 107:13–20

Dawson TE, Mambelli S, Plamboeck AH, Templer PH, Tu KP (2002) Stable isotopes in plant ecology. Annu Rev Ecol Syst 33:507–559

Ehleringer JR, Dawson TE (1992) Water uptake by plants: perspectives from stable isotope composition. Plant Cell Environ 15:1073–1082

Fan Y, Miguez-Macho G, Jobbágy EG, Jackson RB, Otero-Casal C (2017) Hydrologic regulation of plant rooting depth. Proc Nat Acad Sci 114:10572–10577

Fisher JB, Baldocchi DD, Misson L, Dawson TE, Goldstein AH (2007) What the towers don't see at night: nocturnal sap flow in trees and shrubs at two AmeriFlux sites in California. Tree Physiol 27:597–610

Gat JR, Carmi I (1970) Evolution of the isotopic composition of atmospheric waters in the Mediterranean Sea area. J Geophys Res 75:3039–3048

Gehre M, Hoefling R, Kowski P, Strauch G (1996) Sample preparation device for quantitative hydrogen isotope analysis using chromium metal. Anal Chem 68:4414–4417

Goldsmith GR, Matzke NJ, Dawson TE (2013) The incidence and implications of clouds for cloud forest plant water relations. Ecol Lett 16:307–314

Gupta P, Noon D, Galewsky J, Sweeney C, Vaughn B (2010) Demonstration of high-precision continuous measurements of water vapor isotopologues in laboratory and remote field deployments using wavelength-scanned cavity ring-down spectroscopy (WS-CRDS) technology. Rapid Commun Mass Spectrom 23:2534–2542

Horita J, Cole DR, Wesolowski DJ (1995) The activity-composition relationship of oxygen and hydrogen isotopes in aqueous salt solutions: III. Vapor-liquid water equilibration of NaCl solutions to 350 °C. Geochim Cosmochim Acta 59:1139–1151

IAEA/GNIP (2014) IAEA/GNIP precipitation sampling guide, v2.02, 20

Kerhoulas LP, Kolb TE, Koch GW (2013) Tree size, stand density, and the source of water used across seasons by ponderosa pine in northern Arizona. Forest Ecol Manag 289:425–433

Kurita N, Ichiyanagi K, Matsumoto J, Yamanaka MD, Ohata T (2009) The relationship between the isotopic content of precipitation and the precipitation amount in tropical regions. J Geochem Expl 102:113–122

Limm EB, Simonin KA, Bothman AG, Dawson TE (2009) Foliar water uptake: a common water acquisition strategy for plants of the redwood forest. Oecologia 161(3):449–459

Lis G, Wassenaar LI, Hendry MJ (2008) High-precision laser spectroscopy D/H and ^{18}O measurements of microliter natural water samples. Anal Chem 80:287–293

Martín-Gómez P, Barbeta A, Voltas J, Peñuelas J, Dennis K, Palacio S, Dawson TE, Ferrio JP (2015) Isotope-ratio infrared spectroscopy: a reliable tool for the investigation of plant-water sources? New Phytol 207:914–927

Martín-Gómez P, Serrano L, Ferrio JP (2017) Short-term dynamics of evaporative enrichment of xylem water in woody stems: implications for ecohydrology. Tree Physiol 37:511–522

Meißner M, Köhler M, Schwendenmann L, Hölscher D, Dyckmans J (2013) Soil water uptake by trees using water stable isotopes (δ^2H and $\delta^{18}O$)–a method test regarding soil moisture, texture and carbonate. Plant Soil 376:327–335

Moore JW, Semmens BX (2008) Incorporating uncertainty and prior information into stable isotope mixing models. Ecol Lett 11:470–480

Newberry SL, Prechsl UE, Pace M, Kahmen A (2017) Tightly bound soil water introduces isotopic memory effects on mobile and extractable soil water pools. Isot Environ Health Stud 53:368–381

Oerter EJ, Bowen G (2017) In situ monitoring of H and O stable isotopes in soil water reveals ecohydrologic dynamics in managed soil systems. Ecohydrology 10:e1841

Oerter E, Finstad K, Schaefer J, Goldsmith GR, Dawson T, Amundson R (2014) Oxygen isotope fractionation effects in soil water via interaction with cations (Mg, Ca, K, Na) adsorbed to phyllosilicate clay minerals. J Hydrol 515:1–9

Oerter EJ, Perelet A, Pardyjak E, Bowen G (2017) Membrane inlet laser spectroscopy to measure H and O stable isotope compositions of soil and sediment pore water with high sample throughput. Rapid Commun Mass Spectrom 31:75–84

Ogle K, Wolpert RL, Reynolds JF (2004) Reconstructing plant root area and water uptake profiles. Ecology 85:1967–1978

Ogle K, Tucker C, Cable JM (2014) Beyond simple linear mixing models: process-based isotope partitioning of ecological processes. Ecol Appl 24:181–195

Orlowski N, Frede H-G, Brüggemann N, Breuer L (2013) Validation and application of a cryogenic vacuum extraction system for soil and plant water extraction for isotope analysis. JSSS 2:179–193

Orlowski N, Breuer L, Mcdonnell JJ (2016a) Critical issues with cryogenic extraction of soil water for stable isotope analysis. Ecohydrology 9:3–10

Orlowski N, Pratt DL, McDonnell JJ (2016b) Intercomparison of soil pore water extraction methods for stable isotope analysis. Hydrol Process 30:3434–3449

Oshun J, Dietrich WE, Dawson TE, Fung I (2015) Dynamic, structured heterogeneity of water isotopes inside hillslopes. Water Resour Res 52:164–189

Palacio S, Azorín J, Montserrat-Martí G, Ferrio JP (2014) The crystallization water of gypsum rocks is a relevant water source for plants. Nat Commun 5:4660

Parnell AC, Inger R, Bearhop S, Jackson AL (2010) Source partitioning using stable isotopes: coping with too much variation. PLoS One 5(3):e9672

Parnell AC, Phillips DL, Bearhop S, Semmens BX, Ward EJ, Moore JW, Jackson AL, Grey J, Kelly DJ, Inger R (2013) Bayesian stable isotope mixing models. Environmetrics 24:387–399

Peters L, Yakir D (2008) A direct and rapid leaf water extraction method for isotopic analysis. Rapid Commun Mass Spectrom 22:2929–2936

Phillips DL, Gregg JW (2001) Uncertainty in source partitioning using stable isotopes. Oecologia 127:171–179

Phillips DL, Gregg JW (2003) Source partitioning using stable isotopes: coping with too many sources. Oecologia 136:261–269

R Core Development Team (2012) R: a language and environment for statistical computing. R foundation for statistical computing. Vienna, Austria. Retrieved from http://www.r-project.org/

Rothfuss Y, Javaux M (2016) Isotopic approaches to quantifying root water uptake and redistribution: a review and comparison of methods. Biogeosci Discuss 14:2199–2224

Scholl M, Eugster W, Burkard R (2011) Understanding the role of fog in forest hydrology: stableisotopes as tools for determining input and partitioningof cloud water in montane forests. Hydrol Process 25:353–366

Schultz NM, Griffis TJ, Lee X, Baker JM (2011) Identification and correction of spectral contamination in $^{2}H/^{1}H$ and $^{18}O/^{16}O$ measured in leaf, stem, and soil water. Rapid Commun Mass Spectrom 25:3360–3368

Schwinning S (2010) The ecohydrology of roots in rocks. Ecohydrology 130:126–130

Silvertown J, Araya Y, Gowing D (2015) Hydrological niches in terrestrial plant communities: a review. J Ecol 103:93–108

Sprenger M, Leistert H, Gimbel K, Weiler M (2016) Illuminating hydrological processes at the soil-vegetation-atmosphere interface with water stable isotopes. Rev Geophys 54:674–704

Vargas AI, Schaffer B, Yuhong L, Sternberg L d SL (2017) Testing plant use of mobile vs immobile soil water sources using stable isotope experiments. New Phytol 215:582–594

Volkmann THM, Kühnhammer K, Herbstritt B, Gessler A, Weiler M (2016) A method for in situ monitoring of the isotope composition of tree xylem water using laser spectroscopy. Plant Cell Environ 39:2055–2063

West AG, Patrickson SJ, Ehleringer JR (2006) Water extraction times for plant and soil materials used in stable isotope analysis. Rapid Commun Mass Spectom 20:1317–1321

West AG, Goldsmith GR, Brooks PD, Dawson TE (2010) Discrepancies between isotope ratio infrared spectroscopy and isotope ratio mass spectrometry for the stable isotope analysis of plant and soil waters. Rapid Commun Mass Spectrom 24:1948–1954

White J, Cook E, Lawrence J, Broecker W (1985) The D/H ratios of sap in trees - implications for water sources and tree-ring D/H ratios. Geochim Cosmochim Acta 49:237–246

Zhao L, Xiao H, Zhou J, Wang L, Cheng G, Zhou M, Yin L, McCabe MF (2011) Detailed assessment of isotope ratio infrared spectroscopy and isotope ratio mass spectrometry for the stable isotope analysis of plant and soil waters. Rapid Commun Mass Spectrom 25:3071–3082

Chapter 27
Soil Microorganisms

Joana Costa, Rui S. Oliveira, Igor Tiago, Ying Ma, Cristina Galhano, Helena Freitas, and Paula Castro

1 General Introduction

Soil can be defined has the upper layer of the Earth's crust altered by weathering, physical/chemical and biological processes, composed of mineral particles, organic substance, water, air, and living organisms organized in genetic soil horizons (ISO 2015). This concept includes soil organisms as an integral part of soil including plants, microorganisms, animals and their interactions, providing important ecosystem functions and services. Soil organisms are responsible for supporting plant productivity, nutrient and water cycling and soil formation, for regulating soil erosion and water purification, or simply for providing a pool of biodiversity (Beck et al. 2005; European Commission 2006; Brussaard 2012; Mendes et al. 2016).

Soil is a reservoir holding a great part of the world's biodiversity, dominated by microorganisms, i.e., bacteria and fungi, in terms of numbers and biomass. Besides these organisms, soil ecosystems generally contain a large variety of animals, nematodes, micro-arthropods, enchytraeids and earthworms. In addition, a large number of macrofauna species are living in the uppermost soil layers, the soil surface and

J. Costa · I. Tiago · Y. Ma · H. Freitas · P. Castro (✉)
Centre for Functional Ecology – Science for People & the Planet, Department of Life Sciences, University of Coimbra, Coimbra, Portugal

R. S. Oliveira
Centre for Functional Ecology – Science for People & the Planet, Department of Life Sciences, University of Coimbra, Coimbra, Portugal

Department of Environmental Health, Research Centre on Health and Environment, School of Health Sciences, Polytechnic Institute of Porto, Porto, Portugal

C. Galhano
Centre for Functional Ecology – Science for People & the Planet, Department of Life Sciences, University of Coimbra, Coimbra, Portugal

Department of Environmental Sciences, College of Agriculture – Polytechnic Institute of Coimbra, Coimbra, Portugal

© Springer International Publishing AG, part of Springer Nature 2018

A. M. Sánchez-Moreiras, M. J. Reigosa (eds.), *Advances in Plant Ecophysiology Techniques*, https://doi.org/10.1007/978-3-319-93233-0_27

the litter layer (Beck et al. 2005). Soil biota is involved in the global cycling of organic matter, energy, and nutrients. In this sense, soil microorganisms play important roles, namely has chemical engineers, through the decomposition of organic matter and carbon cycling; fixation of nitrogen and phosphorus solubilization making it available for plants; influencing soil pH by nitrification and denitrification; degradation of anthropogenic compounds and influencing heavy metal mobility. Soil microorganisms have direct importance for humans since they are the source of most of the antibiotic used to fight diseases (Beck et al. 2005; Mendes et al. 2016; Pajares and Bohannan, 2016; Alori et al. 2017; Masson-Boivin and Sachs 2017).

This chapter presents a detailed overview on the methods and techniques to assess the structural and functional diversity of soil microorganisms. It then focuses on the description of optimized methods and procedures to isolate and characterize plant growth promoting bacteria and to stain roots of legumes and detect arbuscular mycorrhizas. The chapter finalizes with some considerations regarding sampling and extraction methods of nematodes and presents optimized procedures to stain and detect these organisms (*Globodera* and *Meloidogyne*) in soil and plant roots.

2 Assessing the Structural and Functional Diversity of Soil Microorganisms

2.1 Introduction

This section pretends to provide a bibliography starting point to those that are eager to jump into microbial ecology studies for the first time. The main idea that one should keep is that the success of the project will depend/benefit greatly by carefully planning the approaches to be used. Sampling methodology and sample manipulation are one of the most important parts of the workflow (whether culture dependent and/or independent methodologies will be used), that will grant or not good quality working material and provide downstream valuable data for analyses.

All techniques have their vantages and disadvantages that have to be known and pounder before engaging in any unrealistic analyses that won't provide the data that one was aiming to. Patience and resilience are also requisites to have in such work, since some (a lot) brainstorming and repetition may be/are need to troubleshooting along the way.

Before engaging in Soil Microbiology, or in any microbiology study for that matter, one has to have in mind that it is very easy to disperse energy and lose focus. So it is paramount to establish clear goals from the start, which may suffer small adjustments on the go (Nemergut et al. 2014). Correct sampling methodologies are crucial for the downstream success of a soil microbial survey. One has to consider the characteristics of the soil in which the study will be performed (i.e. composition in organic matter, particle size, humidity, pH), on the main goal(s) of the microbial study, and if culture dependent and/or independent methodologies are to be employed.

Nevertheless, taking several samples from the same site/place (different depths, horizontal landscape, soil horizons, etc.) is the best approach to achieve a representative sampling of a particular site, providing robustness to the results (Pepper and Gerba 2009; Delmont et al. 2011). In some cases, when the amount of sampling sites is big, and the downstream treatment of large quantities of samples is a problem, combining samples (constructing a composite sample of specific locations) will reduce the amount of effort in the analyses while providing good and valuable results.

2.2 Culture Dependent Versus Culture Independent Analyses

"THE GREAT PLATE COUNT ANOMALY" Perhaps the first report of the plate count anomaly was that of Razumov, who noted a large discrepancy between the viable plate count and total direct microscopic count of bacteria from oligotrophic to mesotrophic aquatic habitats. He found higher numbers (by several orders of magnitude) by direct microscopic counting than by the plating procedure. He further noted that this same trend held in eutrophic canal water, but the differences between the two procedures were not nearly so great. A similar disparity between culturable and uncultarable methods was reported for marine habitats (Staley and Konopka 1985).

Based on the actual knowledge provided by culture-independent microbial surveys, it's estimated that the recovered microorganisms rarely represent the most abundant population existent in the environment (Blaser et al. 2016; Solden et al. 2016). Instead, they most likely represent fast-growers on artificial growth media and constitute less than 2% of all microbial species present (Wu et al. 2009; Rinke et al. 2013). This reality has led to an increased use of culture-independent techniques to conduct microbial surveys in extreme and/or high microbial diversity environments, such as soils, where an optimum approach of the cultivation methodologies is even harder to establish (Ferrer et al. 2007; Müller and Ruppel 2014; Xu et al. 2014; Lok 2015).

Whole metagenome sequencing from complex microbial communities makes possible high-throughput genome analyses, resulting in metagenomes that supply information on individual genes or organisms, determining the genomic potential and microbial diversity of a particular ecosystem (Berlec 2012; Müller and Ruppel 2014; Delmont et al. 2015). Equivalent techniques in metaproteomics and metatranscriptomics enable the study of all proteins and their expression profile by microbial communities recovered directly from environmental samples (for a review see Siggins et al. (2012). The combination of these analysis permits functional genes and metabolic pathways to be elucidated (metabolomic and metafluxes) (Jones et al. 2014; Saia et al. 2015; Heaven and Benheim 2016). This "omic" approach, based on the new generation sequencing (NGS) (Mardis 2013), yield huge amounts of data and is excellent to performed a thorough and holistic (or ecosystems biology approach) analysis, combining diversity (taxonomic and functional) (de Castro et al. 2014; Mendes et al. 2015) with adaptation (phenotypic plasticity and evolution) (Violle et al. 2014; Stubbendieck et al. 2016). NGS platforms applied to environmental microbiology owns two major advantages: (i) the circumvention of cultivation bias (technically more challenging when dealing with high diversity ecosystems hosting a panoply of different microorganisms with diverse, and most of the times intangible, growth requirements); (ii) the possibility of uncovering novel structure–function relationships due to the reliance on functional screening and the high and generally unknown diversity of the genetic source material, results that are very difficult to ascertain with isolation surveys due to its low diversity coverage results (Morris et al. 2002; Fakruddin et al. 2013; Krause et al. 2014).

So, does this mean that culture-dependent methodologies are useless? No, in fact it is quite the opposite (Nemergut et al. 2014; Delmont et al. 2015). Only by growing the microorganisms we can determine their phenotypic characteristics in specific conditions and perform a proper description and evaluation of their "true" role in the ecosystem. Questions regarding specific genomic characteristics determined by molecular methodologies for a given microorganism in a given ecosystem can only be answered when researchers work directly with the organism. So, it is not strange that nowadays research groups continue to invest time and resources to develop new methodologies to isolate and study this uncultured microbial majority, the so-called "microbial dark matter" (Rinke et al. 2013; Lok 2015; Solden et al. 2016). These new methodologies were (and continue to be) developed by combining the knowledge that huge amounts of *in-silico* data made available, with environmental data (where organisms can be found, survive and establish bounds with other microorganisms and with the abiotic elements). Currently, microbiologists have a holistic view of the ecosystem, were quorum sensing - microbial cell to cell communication quorum (Hawver et al. 2016), biofilm formation (Kim et al. 2015) and microbial interaction with eukaryotes (Ghannoum 2016; Gkorezis et al. 2016) are crucial to the microbial development, considerations that the "old school" classical isolation methodologies did not considered. This lead to the development of new multi-disciplinary strategies that allowed the isolation and/or study of specific microbial populations inaccessible until know (Ling et al. 2015; Lok 2015; Kang 2016).

2.3 Culture-Dependent Methodologies

Considering what was described above, and since the majority of the laboratories are not equipped with top-nosh equipment, in this part of the chapter we will discuss how can we still perform a good culture-dependent microbial survey in a given environment. First, compile a list of the resources available in ones lab, i.e. incubators (which temperature range can we use to incubate our cultures) or determine if one has the ability to perform aerobic and/or anaerobic incubations. Then, define the main objective(s) of the research. Do we want to determine the microbial diversity of a given environment, or just isolate a specific phylogenetic group of microorganisms? Am I considering isolating those that degrade specific substrates and/or produce specific molecules and/or grow in specific conditions (i.e. extreme temperature, atmosphere compositions, salinity, pH, etc.)? These questions will aid on what comes next: media composition and incubation conditions. Depending on your goal, the medium composition will vary from: nutrient to minimal and/or selective medium.

2.4 *Media*

Nutrient media are not selective and are constituted by generic elements that most bacteria need for growth, i.e. Luria-Bertani agar, plate count agar, nutrient agar, trypticase soy agar (all these have liquid version without agar). Minimal media contain the minimum nutrients possible for colony growth (in very low concentration), as such generally contains various salts, which provide essential elements such as magnesium, nitrogen, phosphorus, and sulfur to allow the bacteria to synthesize protein and nucleic acids. Selective media are used to selectively grow microorganisms with specific growth requirements and/or specific phenotypic characteristics, such as the ability to use a designated substrate, synthesize a certain metabolite or have resistance to specific antibiotics. To improve survey's success, no matter which final outcome you seek, one should always use more than one medium, and change media composition by testing different salinity concentration, pH values and use different incubating temperatures. This will increase the conditions of incubation, and for certain downstream laboratory work, but will also increase the chance to isolate more diverse microorganisms (leading to the achievement of diversity coverage providing robustness to the outcome), or even previously uncharacterized microorganisms.

2.5 *Sample Manipulation*

Soil microorganisms have to be detached from the soil particles and aggregates that constitute it. The best methodology is to make a suspension of the sample in a designated buffer or minimal media (water is not recommended since it can induce osmotic shock and cell lyses), by shaking the sample with or without mechanic help, followed by serial dilution of the suspension. Diluted suspensions can be used for spread plating and/or liquid enrichments in specific isolation media. Serial dilution methodology allows obtaining cell suspensions in concentrations that when spread and grown in solid media one should obtain isolated colonies. This permits to do CFU (colony forming unit) quantification, and makes the process of isolation of pure colony easier.

2.6 Isolation

Isolation process is time consuming but it has to be performed correctly so that the majority of the recovered diversity is isolated. One should try to sub-culture, if not all, the majority of the colonies that are obtain in each different incubation condition. This increases drastically the number of isolates to screen and identify. In order to reduce these numbers there are several molecular fingerprinting techniques that are helpful to group the isolates for downstream identification, i.e. DGGE/TTGE, ARDRA, T-RFLP, LH-PCR, RISA, and RAPD based on direct analysis of PCR products amplified from environmental DNA; fatty acid methyl esters - FAME analyses, matrix assisted laser desorption/ionization time of flight mass spectrometry - MALDI-TOF/MS and whole cell protein analysis (Ahmad et al. 2011; Fakruddin et al. 2013). This allows reducing the amount of isolates to work with since we can pick one or two representative from each of the obtained groups to do phylogenetic identification. One should have in mind, that the more exhaustive the work is, the closest one becomes to determine the true microbial diversity of the environment in define conditions.

Obtaining isolates, besides the ecologic point of view, provide us the possibility of determining the existence of novel organisms to science, but above all, grant us with isolates banks that can be screened for specific activities and that can be used for multiple applications in the future.

2.7 Community Activity

The microbial activity of a given site can be determined by testing the enzymatic activity and/or the metabolism towards single carbon sources. These tests are useful to determine shifts on the microbial composition of a given environment, and may be of importance in ecological studies where the determination of specific metabolic activity in a given site and/or comparison of such characteristics between sites, that may be under different stress conditions, is desired (Lipiec et al. 2016; Pajak et al. 2016). For that purpose, cell suspensions (discussed above in sample manipulation) can be used as inoculums to test a set of designated substrates. One can design the battery of substrates to use, but there are also commercial kits/systems that provide a more controlled platform to perform such analyses i.e. Biolog by Biolog-Inc and API® by bioMérieux.

2.8 Culture-Independent Methodology

Nowadays the use of environmental DNA/RNA has much more utilities than "a simple" 16S rRNA gene microbial diversity determination, that ~16 years ago was the paramount of environmental molecular biology. In a paper of 1998, Handelsman and colleagues (Handelsman et al. 1998) used the term "Metagenomics" for the first time to address the advantages of cloning fragments of environmental DNA in order to access genes and enzymes previously inaccessible due to cultivation bias (Handelsman et al. 1998). Since then, sequencing platforms (Mardis 2013) and meta-omics applications (Prosser 2015) have evolved in a way that one can only imagine (Hiraoka et al. 2016; Papadopoulou et al. 2015) and with that an all new perspective of microbial ecology studies arise (Prosser 2015; Dolinšek et al. 2016; Hahn et al. 2016; Kang 2016; Krause et al. 2014). But be aware that culture-independent methodology approach has its bias and the methodologies used will reflect greatly our results and final outcomes (Krakat et al. 2016).

2.9 Environmental DNA/RNA

Before engaging on molecular biology analyses of your soil samples, first you have to extract environmental DNA/RNA. This could be problematic when dealing with environmental samples, and gets even worse when the samples are soil, since these samples are very complex and diverse in its composition (as discussed in introduction) and may have inhibitory substances (generally referred to as "humic and fulvic compounds, and/or polyphenolic compounds") that if not properly eliminated during extraction steps can interfere in downstream processes, such as DNA-transforming processes including hybridization, quantification, amplification, and may have direct effect on DNA polymerases activity (Frostegård et al. 1999; Lim et al. 2016). For all those reasons, several protocols are published every year by research groups that work with the most different soil sample origins. One can search for two or three protocols (key words like "DNA extraction", "troubleshooting", "bias" will help on search), compare them and make your own protocol, taking in account all advices and tips (Delmont et al. 2011). On the other hand, there are several commercial kits that can be used for DNA/RNA extraction from soil samples, one should read the protocols available, compare them and choose the one more suitable to your samples (Mahmoudi et al. 2011; Young et al. 2014).

2.10 Molecular Biology Methodology

Well, the question that one might have when engaging in molecular biology based ecology studies for the first time is: from all available techniques and platforms which is the most appropriated to achieving the main objectives of the work? So, what do you intend to do? Specific genes surveys like (PCR-dependent amplification of genes): PCR, qPCR (quantitative PCR), clone library, NGS, community fingerprinting [DGGE/TTGE, ARDRA, T-RFLP, LH-PCR, RISA], or PCR-independent amplification of genes methodologies: NGS shotgun meta-genomics and -transcriptomics.

First of all, no technique is flawless and all have their inherent bias. Those based on PCR-dependent amplification of genes, besides several PCR biases (like inhibition) and artifacts, are constricted to the actual knowledge of the gene in study, that will impact the construction of primers (making them more or less "universal") thus influencing greatly the final result and the determined diversity (van Elsas and Boersma 2011). Additionally, the problem of working with complete gene sequences *vs* partial sequences which will depend significantly on the platform that will be used to obtain results (Birtel et al. 2015; Keisam et al. 2016). While Sanger platform allows performing primer walking (for instance from a clone library) making possible to work with complete sequences, the majority of the other platforms do not, namely DGGE or the NGS platforms. Thus the majority of the actual environmental research (PCR-dependent amplification of genes) relays on partial sequences, with all the associated bias (Tremblay et al. 2015). Nevertheless, the recent NGS platform PacBio sequencing promises the generation of high-quality near full-length sequence fragments, that could be a game changer, but not without some drawbacks (Schloss et al. 2015; Levy and Myers 2016).

Despite the fact that PCR-independent amplification of genes methodologies (shotgun metagenomics and metatranscriptomics) is not dependent on primer design, they are reliant on all the other abovementioned contingencies (sampling and/or DNA extraction efficiency) and also on budget (although each year the sequencing services is getting more affordable). Additionally, the common/nowadays platforms, i.e. Illumina, 454-Pyrosequencing, IonTorrent, and more recently PacBio, have their intrinsic bias (Goodwin et al. 2016), that have to be taken into consideration while choosing one over the other. Nevertheless, shotgun meta-analyses allow us to do taxonomic assignment, functional assignment, specific gene analyses, genome binning, gene and protein prediction and several other applications (Sharpton 2014), which will grant a holistic knowledge about the environment. But it is much more time consuming than single gene PCR-dependent surveys, requires higher computational power and deeper bioinformatics knowledge for result analyses.

3 Isolation and Characterization of Plant Growth Promoting Bacteria

3.1 *Introduction*

Plant growth promoting bacteria (PGPB) are a special class of bacteria that are involved directly or indirectly in enhancing plant growth and protecting plants from biotic and abiotic stresses via various mechanisms (Ma et al. 2011; Glick 2012). The PGPB include those that reside in the rhizosphere/rhizoplane (plant growth promoting rhizobacteria; PGPR), or those that inhabit in living plant tissue interior and establish close associations with host plants (plant growth promoting endophytic bacteria; PGPE). The mechanisms by which PGPB promote plant growth at different stages of the host plant life cycle can differ among species and strains. Several important bacterial biochemical characteristics, such as 1-aminocyclopropane-1-carboxylate (ACC) deaminase activity, phosphate solubilization, as well as production of phytohormone indole acetic acid (IAA), siderophores, ammonia (NH_3) and hydrogen cyanide (HCN) can be assessed as plant growth promotion traits. In this chapter, the methods of isolation and biochemical characterization of PGPB are addressed in detail.

3.2 *Isolation of Plant Growth Promoting Bacteria*

To isolate rhizobacteria, about 1 g of wet soil sample is serially diluted using 25 mM phosphate buffer and spread on Luria-Bartani (LB) agar medium. After incubating plates at 27 °C for 48 h, morphologically distinct bacterial colonies are randomly picked, purified and restreaked on LB media until the colonies of each isolate are morphologically homogeneous (Ma et al. 2009). To isolate endophytic bacteria, plant samples are washed with tap water followed by several rinses with sterile distilled water (SDW) and then separated into different organs (e.g. roots, stems and leaves). Healthy plant tissues are sterilized by sequential immersion in 70% (v/v) ethanol for 1 min, 3% sodium hypochlorite for 3 min and then washed several times with SDW to remove surface sterilization agents. Sterility should be checked by plating 0.1 L of final rinsed water on LB agar, in order to confirm that the surface disinfection process is successful. After that, plant tissues (e.g. roots, stems and leaves) are cut into small pieces, crushed in mortar and pestle, and titrated in SDW; appropriate dilutions are plated onto sucrose-minimal salts low-phosphate agar medium [1% sucrose; 0.1% $(NH_4)_2SO_4$; 0.05% K_2HPO_4; 0.05% $MgSO_4$; 0.01% NaCl; 0.05% yeast extract; 0.05% $CaCO_3$; pH 7.2] and incubated at 27 °C for 3 days.

To isolate plant growth promoting bacteria (PGPB), the rhizo- or endophytic bacterial strains are grown on Dworkin and Foster salts minimal (DFSM) medium (Dworkin and Foster 1958) amended without (blank) or with 3 mM ACC as a nitrogen (N) source at 27 °C for a week at 175 rpm. The bacterial growth is monitored by measuring the optical density (OD) at 600 nm after 5 days of incubation. The strains that can grow only in DFSM medium with ACC is known as ACC utilizing bacteria. Further, the ACC utilizing strains are assessed for their plant growth promoting activity by roll towel assay (ISTA 1966). Briefly, seeds of the model plants (e.g. Brassicaceae species) are surface sterilized in 2% $Ca(OCl)_2$ for 2 h and rinsed several times with SDW. The seeds are inoculated by soaking in a bacterial suspension after adjusting OD600 to 1 for 2 h, then placed in wet blotters and incubated in a growth chamber for 3 weeks.

The germination percentage of seeds is recorded and the vigor index is calculated with the following formula (Abdul-Baki and Anderson 1973).

$$\text{Vigour index} = \text{germination}\left(\%\right) \times \text{seedling length}\left(\text{shoot length} + \text{root length}\right)$$

The procedures described above are designed to screen isolate and screen PGPR and PGPE, which can be used as biological agents for sustainable agriculture and environmental decontamination.

3.3 *Biochemical Characterization of Plant Growth Promoting Bacteria*

3.3.1 1-Aminocyclopropane-1-Carboxylate (ACC) Deaminase Activity

Bacterial ACC deaminase activity is determined by monitoring the amount of α-ketobutyrate (αKB) generated enzymatically via hydrolysis of ACC (Belimov et al. 2009; Saleh and Glick, 2001). Bacterial strain is grown in LB medium for 24 h at 27 °C and harvested by centrifugation at 7000 rpm for 10 min at room temperature. Cell pellets are washed three times with 5 mL of 0.1 M Tris-HCl buffer (pH 7.5), resuspended in 1 mL of salts minimal (SM) medium (0.4 g KH_2PO_4; 2 g K_2HPO_4; 0.2 g $MgSO_4$; 0.1 g $CaCl_2$; 5 mg $FeSO_4$; 2 mg H_3BO_3; 5 mg $ZnSO_4$; 1 mg Na_2MoO_4; 3 mg $MnSO_4$; 1 mg $CoSO_4$; 1 mg $CuSO_4$; 1 mg $NiSO_4$; per liter; pH 6.4) and then 0.5 mL of each suspension is added to 2.5 mL of liquid SMN medium containing 3 mM ACC as a sole N source (0.5 g ACC; 1 g glucose; 1 g sucrose; 1 g Na-acetate; 1 g Na-citrate; 1 g malic acid; 1 g mannitol; per liter). Bacterial strain is incubated for 24 h at 27 °C, centrifuged and resuspended in 1 mL of 0.1 M Tris-HCl buffer (pH 7.5), and centrifuged at 7000 rpm for 10 min. The pellets are resuspended in 600 mL of 0.1 M Tris-HCl buffer (pH 8.5) and cells are disrupted by adding 30 mL of toluene and vigorous vortexing. After reaction of mixtures containing 100 mL of cell suspension, 10 mL of 0.5 M ACC and 100 mL of 0.1 M Tris-HCl buffer (pH 8.5) for 30 min at 27 °C, 1 mL of 0.56 N HCl is added, and the mixtures are centrifuged at 10,000 rpm for 5 min. The mixtures containing no cell suspension or no ACC are used as controls. Thereafter, 150 mL of 0.1% 2,4-dinitrophenylhydrazine in 2 N HCl is added to 1 mL of the supernatant. The mixtures are reacted at 27 °C for 30 min, amended with 1 mL of 2 N NaOH and assayed for αKB at 540 nm (Honma and Shimomura 1978). The protein content of cell extracts is determined using the procedure of Bradford (1976).

3.3.2 Indole Acetic Acid (IAA) Production

Bacterial IAA production is determined according to Bric et al. (1991). Briefly, the bacterial strain is cultured for 5 days in flasks containing 20 mL of LB medium amended with 0.5 g mL-1 of L-tryptophan. After incubation, 2 mL of supernatant is transferred into a tube and mixed vigorously with 100 mL of 10 mM orthophosphoric acid and 4 mL of Salkowski reagent (1 mL of 0.5 M $FeCl_3$ in 50 mL of 35% $HClO_4$). After color development, read the absorbance values at 530 nm. The IAA concentration in culture is determined using a calibration curve of pure IAA as a standard following the linear regression analysis.

3.3.3 Phosphate Solubilization

Bacterial strain is grown in modified Pikovskayas medium [1% glucose; 0.5% $Ca(H_2PO_4)_2$; $MgCl_2$·0.5% $6H_2O$; 0.025% $MgSO_4 \cdot 7H_2O$; 0.02% KCl; $(NH_4)_2SO_4$ 0.01%] with 0.5% tricalcium phosphate at 27 °C for a week at 175 rpm. The culture supernatants are collected by centrifugation at 10,000 rpm for 15 min. The soluble phosphate in the culture supernatant is quantitatively estimated as described by Fiske and Subbarow (1925).

3.3.4 Siderophore Production

Bacterial siderophore production is qualitatively examined by a qualitative chromogenic assay using chrome azurol S (CAS) agar (Schwyn and Neilands 1987). As a highly sensitive chemical method, it is based on the affinity of siderophore for iron (Fe^{3+}). As a siderophore removes Fe from the dye, its color turns from blue to orange. Briefly, 50 μL of each exponential bacterial culture previously grown in casamino acids (CAA) medium are spotted onto CAS agar. The siderophore levels produced by bacterial strain are recorded as the diameter of orange halo. The productions of catechol and hydroxamate siderophores in culture supernatants obtained from bacteria grown in CAA medium are quantitatively determined by using 2,3-dihydroxybenzoic acid (Arnow 1937) and desferrioxamine mesylate (Atkin et al. 1970) as standards, respectively.

3.3.5 Ammonia (NH_3) Production

Bacterial NH_3 production is examined in peptone water. Briefly, bacterial strain is cultured in each tube containing 10 mL peptone water and incubated at 27 °C for 2 days. Then, 0.5 mL of Nessler's reagent (10% Hg_2Cl_2, 7% KI, 16% NaOH) is added in each tube. Development of brown to yellow color indicates NH_3 production (Cappuccino and Sherman 1992).

3.3.6 Hydrogen Cyanide (HCN) Production

Bacterial HCN production is determined according to Lorck (1948). Briefly, bacterial strain is streaked on nutrient agar plate amended with 44% glycine. A Whatman filter paper no. 1 soaked in 2% sodium carbonate in 0.5% picric acid solution is placed in the top of the plate. Plates are sealed and incubated at 27 °C for 4 days. Development of orange to red color indicates HCN production.

3.3.7 Phytagar Assay

Phytagar assay is designed to confirm bacterial plant growth promoting properties (Ma et al. 2011). The growth media are prepared using 0.5% phytagar and one-quarter strength Hoagland's nutrient solution. The surface sterilized seeds of model plants (e.g. Brassicaceae species) are inoculated with bacteria as detailed in roll towel assay and placed in 150 mL test tubes containing 25 mL of phytagar. The tubes are closed, placed in a controlled-environment growth room. After 2 months, the plants are removed from the tubes and rinsed thoroughly with SDW to remove adhering agar. Growth parameters such as plant shoot and root length, fresh and dry weights are measured.

The use of PGPB is considered as an eco-friendly approach and biotechnological tool for sustainable agriculture or environmental applications, due to their competitive colonization and abilities to stimulate plant growth. Therefore, it is important to establish the optimized methodologies for isolation and characterization of PGPB, which can be used to search beneficial PGPB strains for the development of new and efficient bioinoculants. This chapter presents the systematic methods for isolating and biochemically characterizing PGPB for biotechnological applications, which will provide further guidance on research basics aiming at undergraduate and postgraduate research.

4 An Optimized Procedure for Staining Roots of Legumes and Detecting Arbuscular Mycorrhizas

4.1 Introduction

Arbuscular mycorrhizas are the most widespread and common underground symbiotic associations and are formed between arbuscular mycorrhizal fungi and the roots of the majority of terrestrial plant species (Smith and Read 2008; van der Heijden et al. 2015). This association is generally considered to be mutualistic: the host plant supplies photosynthates to the fungal symbiont while the fungus, in return, translocates nutrients from soil to the plant (Maherali et al. 2016; Oliveira et al. 2016). Research studies on arbuscular mycorrhizas often require the observation of arbuscular mycorrhizal fungal structures in the roots of target plants. This is commonly done with the aim of assessing arbuscular mycorrhizal colonization, which is one of the most widely reported fungal parameters in arbuscular mycorrhiza research (Varga 2015).

Several techniques for root staining and detection of arbuscular mycorrhizas have been published [subject reviewed by Vierheilig et al. (2005)]. Despite the variety of existing techniques, staining with trypan blue remains one of the most commonly used to visualize arbuscular mycorrhizas (Sun and Tang 2012). Nevertheless, root staining with trypan blue can be a time consuming process, especially when large numbers of samples are to be analyzed. Therefore, even slight improvements of existing root staining methods can be of value in arbuscular mycorrhiza research.

Here we present an optimized procedure for staining roots of legumes with trypan blue for detecting arbuscular mycorrhizas. The described method is simple, reliable and relatively fast, making it suitable for dealing with large numbers of samples.

4.2 Procedure

The procedure presented below is adapted from the methods of Phillips and Hayman (1970), Koske and Gemma (1989) and Dalpé and Séguin (2013) and is optimized for colonized roots of cowpea [*Vigna unguiculata* (L.) Walp.], chickpea (*Cicer arietinum* L.) and white clover (*Trifolium repens* L.).

Roots are obtained as follows: 1 dm^3 pots are filled with a sandy loam, which had been autoclaved twice (121 °C for 25 min) on consecutive days. Seeds of cowpea, chickpea and white clover are surface sterilized with 0.5% (v/v) sodium hypochlorite for 20 min and germinated on moist paper towels at 20 °C in the dark. After germination, seedlings are transplanted singly into each pot. Six treatments are prepared: cowpea inoculated with an arbuscular mycorrhizal fungus, non-inoculated control cowpea, chickpea inoculated with an arbuscular mycorrhizal fungus, non-inoculated control chickpea, white clover inoculated with an arbuscular mycorrhizal fungus and non-inoculated control white clover. Each treatment is replicated 10 times. Each pot from the mycorrhizal treatments receives 10 g of inoculum (Symbiom Ltd., Czech Republic) of *Rhizophagus irregularis* BEG140 consisting of colonized root fragments, hyphae and spores in the mixture of zeolite and expanded clay, placed 2 cm below the root system. Every pot from the non-mycorrhizal treatments receives 10 g of inoculum autoclaved twice (121 °C for 25 min) on consecutive days. Plants are grown in a greenhouse under natural light with an average photoperiod of 12 h. Temperature and relative humidity ranges were 12–42 °C and 55–85%, respectively. After 3 months, plants are removed from the pots, the root system is separated from the shoot and gently washed to remove adhered soil. Fresh roots are cut into 1-cm pieces and stained according to the procedure described below.

1. Root pieces are placed in 20 mL glass vials and covered with 10% KOH (w/v).
2. Vials are heated at 100 °C for 30 min.
3. Root pieces are strained and the KOH solution discarded.
4. Root pieces are again placed inside the vials and thoroughly washed with tap water.
5. 5% HCl is added to the vials until the roots are completely submerged.
6. After 1 h the HCL solution is discarded.
7. Root pieces are placed inside the vials and covered with 0.1% trypan blue staining solution. The solution has the following composition: 793 mL 90% lactic acid +143 mL 99.5% glycerol +64 mL deionized water +1 g trypan blue.
8. Vials are placed in a household microwave oven (Thor, TM 17.2, 2450 MHz, 1050 W) (maximum ten vials per batch) during 15 s at maximum power.
9. Step 8 is repeated.
10. Root pieces are strained and the trypan blue staining solution discarded.
11. Root pieces are placed inside the vials and covered with 50% glycerol (v/v).

Stained root pieces are mounted on glass slides and examined with a compound microscope (Leica DM 5000-D, Germany) (×100–400) to assess the presence of arbuscular mycorrhizal structures in the roots.

No mycorrhizal colonization is observed in any control plant. Arbuscular mycorrhizas of inoculated cowpea, chickpea and white clover are well stained with trypan blue and show very good contrast. All typical arbuscular mycorrhizal fungal structures (arbuscules, vesicles and hyphae) are clearly observed. The used method is simple, reliable and relatively fast, making it suitable for dealing with large numbers of samples.

5 Sampling and Extraction of Nematodes

5.1 Introduction

Nematodes are among soil biota. Global studies of the distribution of soil nematode species show that most are endemic to a place or region, and only a small fraction are cosmopolitan. As soil nematode community composition depends on several factors (e.g. vegetation present, soil type, season, soil moisture level, amount of soil organic matter), they are useful as soil environment condition bioindicators. Nematodes are a key group for regulating biogeochemical cycling and ecosystem processes that include mineralization and decomposition in the soil system. In fact, free-living nematodes are very important and beneficial in the decomposition of organic material and the recycling of nutrients in soil. Bacterivores and fungivores nematodes do not feed directly on soil organic matter, but on the bacteria and fungi, which decompose organic matter. The presence and feeding of these nematodes accelerate the decomposition process. Their feeding recycles minerals and other nutrients from bacteria, fungi, and other substrates and returns them to the soil where they are accessible to plant roots (McSorley 2016; Orgiazzi et al. 2016).

Considering the enormous biodiversity of soil biota, it is easily understandable that there is a wide range of techniques to collect and identify the organisms present in the soil. To collect organisms, the first challenge is the separation of the organism from the soil matrix, which can be hampered by the nature of soil minerals and organic constituents, and the physical nature of the soil matrix. Means of separation by elutriation, centrifugation or density gradients is achieved due to the considerable differences in density between soil mineral constituents and organisms. Mobile organisms can be collected by boosting movement away from the soil matrix, to entrapment and collection vessels, by a combination of gravity and differential application of heat, light or water. Identification of soil organisms, historically, was chiefly based on morphological features visual observation, which can be extremely subtle and requiring considerable experience and expertise to carry out. However, currently, the traditional taxonomic tools of microscopes and systematic keys are gradually being replaced by genetic analysis of DNA derived from the organisms, despite such approaches currently require advanced laboratory equipment (Orgiazzi et al. 2016).

It is undeniable the importance of identification and/or quantification of groups of soil organisms. Nonetheless, appropriately organized sampling designs are required in order to obtain a collection of samples sufficiently representative of the area or ecosystem under consideration. Statistically robust sampling is necessary, since it allows an estimation of the likely variation in any metrics and, consequently, how accurate they are. Soil biota sampling is typically based on ex situ and in situ techniques, the former involving the removal of recommended volumes of soil, by coring or excavation, and generally transporting them back to the laboratory for assay; and the later commonly depends on the movement of the organism(s) to a collecting device (Orgiazzi et al. 2016).

When we look specifically for soil nematodes, we can consider those that are in the soil and others present in the plant roots. Soil nematodes can be directly collected from soil samples and those which are plant parasitic nematodes can be obtained from plant roots.

We can suspect that plants are infected with nematodes when they show some typical symptoms as: injuries, discoloration, and deformations and, sometimes, complete devastation in the penetration and feeding areas. Plants attacked by nematodes lose vigor, size and quality of fruits and vegetables are reduced and in extreme cases, they can die. On the other hand, the wounds left by the stylet are openings for fungal or bacterial invasion, increasing damage on plants (Esser n.d.; Krueger and McSorley 2014).

5.2 Sampling of Nematodes: Some Considerations

Sampling nematodes can have different goals. The main purposes include general surveys, population estimation for research, advisory or predictive programs, and disease diagnoses. The aims of sampling for nematodes differ with each type of experiments and then, specific objectives of sampling should be determined before a sampling program is planned. For example, the primary goals for research programs may be the characterization and understanding of nematode population dynamics. When it comes to integrated pest management programs, the final goal of sampling is to relate numbers and kinds of nematodes to crop performance and also evaluation and selection of management tactics. If quarantine programs are in mind, the goal of sampling is the detection of any nematodes present and spread prevention (Barker 1985; Mekete and Crow 2015).

Regardless of the sampling purpose, spatial and temporal patterns of nematodes should be taken into account. There are several factors that affect significantly the vertical spatial patterns of nematodes as crop, soil type, and the nematodes species involved. The horizontal spatial patterns of plant-parasitic nematodes are typically patchy or contagious, making sampling difficult. As nematodes feed on plant roots, samples should always be taken from among roots of the plants for which diagnosis is needed. It is also important to consider that temporal patterns change with crops and nematode species. Nematode populations fluctuate throughout the year and may be undetectable during the winter and early spring, however often increase to a very high population density in the early fall, before crop harvest when living roots are present. However, if there is a high initial population, the maximum population may occur before the midseason and then decline as a result of severe damage to host. After harvest, nematode populations may decline abruptly. Accordingly, taking samples at the appropriate time for a crop helps to avoid or reduce the potential of nematode problems in the future, as the risk of failing to detect a damaging nematode species is decreased (Barker 1985; Mekete and Crow 2015).

A sample should be obtained from a mixture of 10 to 20 "cores" of soil that are easily taken with a soil sampling tube, auger, shovel, or trowel. With a shovel, cut approximately 2.5 cm thick slice of soil through the soil profile. Then, collect a 2.5 to 5 cm vertical band from the slice. It is advisable to discard the top 2.5 cm from each core, as nematode numbers may be very low there. Soil sample should be taken only when soil moisture is appropriate for working the field avoiding extremely dry or wet soil conditions (Mekete and Crow 2015).

There are some specific guidelines for collecting samples from different types of symptomatic plants. For annual crops, like most vegetables, annual ornamentals, and field crops, soil should be taken from root zones of 10 to 20 affected plants that are not yet dead and "feeder" or fine roots from several of them should be includes. Soil cores are taken 15 to 20 cm deep, after remove 2.5 cm surface of soil. For fruit and nut trees, perennial shrubs and trees, if many plants are affected, cores from several of them should be included in the sample. If only one or a few are affected, several cores from around each plant should be taken. Using a shovel to dig within the "drip-line" (the area covered by the branches) find fine feeder roots. On most trees and shrubs, cores 20 cm deep should be sufficient. However, for burrowing nematode of citrus, roots and soil from below 30 cm deep should be collected. Each "core" should consist of a few fine roots and soil from immediately around them. Discard the top 2.5 cm soil and roots. For turfgrasses, 10 to 20 cores of 7.5 to 10 cm deep near the desired plant species, from areas of declining but not yet dead turf should be collected, avoiding bare spots and weeds. The sample must consist of mostly soil with a few roots discarding foliage (Mekete and Crow 2015).

All cores from a sampled area should be placed into a properly labeled plastic bag, using different plastic bags for each sampling area. A properly collected sample should have approximately 0.5 to 1 L of soil. The sample bag should be sealed, to reduce moisture loss. Water should not be added to the sample, even if it seems dry. The soil should be handled carefully since rough handling will crush nematodes living among soil particles. Nematodes will die from overheating, freezing, or drying. Then, samples should be transported in insulated containers and not be transported exposed to sunlight or in a hot car trunk or on the dashboard. In fact, soil samples for nematodes assays must be considered perishable and handled accordingly (Esser n.d.; Barker 1985; van Bezooijen 2006; Mekete and Crow 2015).

After soil sample collection, nematodes should be extracted within a week. Nevertheless, if delay is necessary, samples may be stored at 10 to 15 °C, which extends recoverability of nematodes (Barker 1985).

When the whole sample cannot be processed for nematode extraction, suitable mixing of composite soil samples before removal of an aliquant, usually 100 to 1000 cm^3, is also important (Barker 1985).

5.3 Extraction Methods

Nematode's extraction methods should be chosen in order to obtain the most consistent percentage of the nematode species from given soil and or plant samples. The actual selection of extraction methods depends on several factors as: kind and number of nematodes present, the host plant, equipment available, edaphic factors, and population dynamics. Extraction efficiency depends on several factors, including the method, soil type, nematode specie and size, and the laboratory (Barker 1985; van Bezooijen 2006).

Nematodes can be easily isolated from their host material by submerging the sample in water and select the nematodes under a microscope. But, this is a dull and hard job and it can only be done with very small samples. Consequently, most of the extraction methods are indirect, and they are based in some properties to separate nematodes from the surrounding medium that are mentioned below, being many extraction methods based on a combination of these characteristics.

(a) Weight and rate of settling: in water, nematodes are separated from particles that settle faster and can subsequently be poured off (decanted). This behavior is the basis of a number of applications, such as the use of an undercurrent that keeps nematodes afloat while other particles settle (elutriation) and use of a liquid with higher specific gravity than nematodes, which keeps them submerged whereas other particles, with a higher specific gravity than the liquid, sink to the bottom. This is applied in centrifuge floatation techniques, used to extract dried cysts from soil samples.
(b) Size and shape: due to their size and elongated shape, nematodes can be separated from soil particles by using a set of sieves with different mesh size.
(c) Mobility: since living nematodes are mobile, when samples are placed on a sieve with a moist filter paper, positioned in a shallow water-filled tray, nematodes will crawl from the sample into the water where they can be collected as a clear suspension (van Bezooijen 2006).

Methods to extract nematodes from plant material are usually based on nematode mobility. Firstly roots should be gently washed to remove adhering soil and then placed in a sealed jar for at least 3 days at room temperature. After this time, nematodes that start to leave the roots are washed with water, which is collected to a beaker. The water with the nematodes is poured on a fine sieve, which catches the nematodes. They are then washed into a Petri dish and identified. These methods vary depending on whether or not the samples have been cut in smaller pieces to speed up crawling out of nematodes. The blender centrifugal flotation method is the only method, apart from picking nematodes under a microscope, which does not make use of nematode mobility, making it suitable for the extraction of swollen endoparasitic stages and eggs (van Bezooijen 2006; Esser n.d.).

Methods usually used for soil nematode extraction are also applied for sediments, rock wool, manure, and other substrates, where nematodes can be present, and can be washed into suspension (Table 7.1). As said above, most methods use a combination of different principles. There are major differences in terms of extraction efficiency, size of the sample that can be handled, and costs. The centrifugal flotation method is the only method, apart from picking nematodes under a microscope that allows isolation of active as well as slow-moving and inactive nematodes (van Bezooijen 2006). Soil sample is placed in a container, which is half filled with water and vigorously stirred up. Then, the water is poured into another container

Table 7.1 Extraction methods for plant and soil samples (Adapted from van Bezooijen 2006)

Sample/Suspension Cleaning	Type of extraction method	Specific extraction method
Plant	Direct	Microscopic observation after dissection (apply a staining technique first, if desired)
	Indirect	Active nematodes
		Baermann funnel
		Funnel spray method
		Blender nematode filter method
		Inactive + active nematodes
		Blender centrifugal flotation method
Soil	Direct	Water microscopic observation
	Indirect	Decanting method
		Cobb's method cleaning of the suspension
		Erlenmeyer method
		Oostenbrink funnel (elutriator)
Suspension		Active nematodes
		Nematode filters
		Inactive + active nematodes
		Centrifugal flotation method

leaving the heavy soil sediments behind. As nematodes are very light, slowly settle to the bottom. Then the soil is passed through a series of fine sieves that separate the larger soil particles and the trash from the nematodes. The nematodes are then removed from the finest screen and placed in a Petri dish for examination (Esser n.d.).

5.4 *Optimized Extraction Procedures for* Globodera *and* Meloidogyne

Plant parasites are probably the most studied nematodes, taking into account the significant yield crops reduction they cause worldwide. Among the most economically important plant parasites, *Globodera* and *Meloidogyne* stand out. They are both sedentary endoparasites and can be recognized either in soil or in plant samples. In the following, optimized methods are described in more detail: the Fenwick's method which is used to extract *Globodera* spp. cysts from soil samples; the Whitehead and Hemming tray method, widely use to extract nematodes from soil including *Meloidogyne* juveniles; and the sodium hypoclorite acid fuchsin method, to stain nematodes in plant tissue, including *Globodera* and *Meloidogyne* nematodes.

5.4.1 Fenwick's Method

The Fenwick can be used to extract cysts from dried soil (max. 300 g). The method is based on floating properties of dried cysts and of difference in size between them and other fractions of the sample. In this method, coarse sample material is retained on the sieve, heavy particles passing through sink to the bottom of the can, and fine and light particles, like cysts, keep afloat. When the can overflows, the floating cysts are carried off over the overflow collar, and drop on a sieve with a pore size smaller than the cyst diameter. Large samples can be handled in a standardized way however, the method requires large amounts of water (van Bezooijen 2006).

Procedure

1. Dry a soil sample (maximum weight of 300 g).
2. Clean the can with water.
3. Close the outlet and fill the can to the rim with water. Place a 175 μm sieve under the outlet of the overflow collar, so water runs on the slightly inclined sieve, to facilitate the water running through.
4. Wash the sample through the top sieve using the spray nozzle. Continue spraying for about 5 min.
5. Carefully rinse the funnel and the collar with water to ensure that all cysts are washed on the sieve. After this the outlet can be unplugged and the can rinsed with water.
6. The cysts in the debris on the sieve can be isolated in different ways, e.g.: dry the debris, wash it into a white bowl and retrieve the floating cysts, or wash the debris onto a filter paper and pick the cysts under a dissecting microscope.

5.4.2 Whitehead and Hemming Tray Method Modified by Abrantes et al. (1976)

The Whitehead and Hemming tray method (1965) modified by Abrantes et al. (1976) is the most used method to extract small sized nematodes from soil samples to evaluate population density. There are several advantages that can be pointed out to this method: it allows the extraction of nematodes from significantly great soil samples (300–500 cm^3); it is a method of easy execution and of great effectiveness for Meloidogyne juveniles extraction; the obtained suspensions are clear, facilitating their microscopic observation (Abrantes et al. 1976).

Procedure

1. Place a net coated with tissue paper inside a plastic tray (23 × 34 cm approx).
2. Spread uniformly 500 cm^3 of soil on the tissue paper.
3. Pour water carefully into the tray until the soil is moistened and in contact with the surface of the water.
4. Do not move the tray from the initial position to obtain clear suspensions.
5. After 7 days, remove the net carefully.

6. Pour the nematode water suspension from the tray into a glass beaker (1–1.5 L).
7. Let the nematode suspension settle for at least 4 h.
8. Remove the supernatant water until 100–200 mL of suspension; the suspension can also be sieved through a 20 μm sieve and then transfer the residues retained in the sieve to a 500 mL glass beaker.
9. Record the exact volume of the suspension.
10. Observe small volume replicates and count nematodes present in each of them, under a stereomicroscope.
11. Calculate the number of nematodes/500 cm^3 of soil.

5.4.3 Sodium Hypochlorite Acid Fuchsin Staining Method

Staining simplifies microscopic detection of nematode in plant tissues. Nematodes are stained red and the plant tissue remains more or less unstained, with the exception of the meristematic regions. This method for clearing and staining nematode-infected root tissues has several advantages: it eliminates exposure of personnel to toxic substances, used in other methods; as root tissue is cleared with sodium hypochlorite prior to staining with acid fuchsin, it does not become heavily stained and consequently, the destaining time is shortened, being frequently avoided (Byrd et al. 1983).

Procedure

1. Carefully wash infected root system with water.
2. Cut the roots into segments and place them in a 150 mL beaker.
3. Clear the roots by adding 50 mL of tap water plus a suitable (depending on the root tissue age: 10 mL for young roots; 20 mL for moderate age roots; 30 mL for older or more ligneous roots) amount of sodium hypochlorite (5.25% NaOCl).
4. Soak roots for 4 min in the NaOCl solution and agitate occasionally.
5. Rinse roots for approximately 45 sec in running tap water then soak them in tap water for 15 min to remove any residual NaOCl which may affect acid fuchsin staining.
6. Drain the water transfer roots to a beaker with 30–50 mL of tap water.
7. Add 1 mL of stock acid fuchsin stain solution (prepared by dissolving 3.5 g acid fuchsin in 250 mL of acetic acid and 750 mL distilled water) to the water.
8. Boil the solution for about 30 s on a hot plate.
9. Cool the solution to room temperature, drain it from the roots, and rinse the roots in running tap water.
10. Place roots in 20–30 mL of glycerine acidified with a few drops of 5 N HCl and heat to boiling for destaining.
11. After the destaining process, roots may be either stored in acidified glycerine or observed under a stereomicroscope.

Acknowledgements J. Costa, R.S. Oliveira, Y. Ma acknowledge the support of Fundação para a Ciência e a Tecnologia (FCT) through the research grants SFRH/BPD/112157/2015, SFRH/BPD/85008/2012 and SFRH/BPD/76028/2011, Fundo Social Europeu (FSE) and Programa Operacional do Capital Humano (POCH). I. Tiago acknowledges an Investigator contract reference IF/01061/2014. This work was financed by Portuguese national funds through Programa Operacional Competitividade e Internacionalização (POCI), Project 3599 – Promover a Produção Científica e Desenvolvimento Tecnológico e a Constituição de Redes Temáticas (3599-PPCDT) and Fundo Europeu de Desenvolvimento Regional (FEDER) under Project POCI-01-0145-FEDER-016801 and by FCT under Project PTDC/AGR-TEC/1140/2014. This work was financed by the Interreg Sudoe programme through the European Regional Development Fund under Project PhytoSUDOE – SOE1/P5/E0189. J. Costa and P. Castro acknowledge the project ReNATURE - Valorization of the Natural Endogenous Resources of the Centro Region (Centro 2020, Centro-01-0145-FEDER-000007).

References

Abdul-Baki AA, Anderson JD (1973) Vigour determination in soya bean seed by multiple criteria. Crop Sci 13:630–633

Abrantes IM de O, de MMMN, Paiva IMP de FR, Santos MSN de A (1976) Análise nematológica de solos e plantas. Ciência Biológica 1:139–155

Ahmad I, Ahmad F, Pichtel J (2011) Microbes and microbial technology: agricultural and environmental applications. Microbes Microb Technol Agric Environ Appl:1–516. https://doi.org/10.1007/978-1-4419-7931-5

Alori ET, Glick BR, Babalola OO (2017) Microbial phosphorus solubilization and its potential for use in sustainable agriculture. Front Microbiol 8:971. https://doi.org/10.3389/fmicb.2017.00971.eCollection2017.

Arnow E (1937) Colorimetric determination of the components of 3,4-dihydroxyphenylalanine-tyrosine mixtures. J Biol Chem 118:531–537

Atkin CL, Neilands JB, Phaff HJ (1970) Rhodotorulic acid from species of *Leucosporidium*, *Rhodosporidium*, *Rhodotorula*, *Sporidiobolus*, and *Sporobolomyces*, and a new alanine-containing ferrichrome from *Cryptococcus melibiosum*. J Bacteriol 103:722–733

Barker KR (1985) Sampling nematode communities. In: Sasser JN, Carter JN, Carter CC (eds) An advanced treatise on Meloidogyne. Raleigh, North Carolina USA, A Cooperative Publication of the department of Plant Pathology and the United States Agency For International development, North Carolina State University Graphics, Vol. II, pp 3–17

Beck L, Römbke J, Breure AM, Mulder C (2005) Considerations for the use of soil ecological classification and assessment concepts in soil protection. Ecotoxicol Environ Safe 62(2 SPEC. ISS):189–200

Belimov AA, Dodd IC, Hontzeas N et al (2009) Rhizosphere bacteria containing 1-aminocyclopro pane-1-carboxylate deaminase increase yield of plants grown in drying soil via both local and systemic hormone signalling. New Phytol 181:413–423

Berlec A (2012) Novel techniques and findings in the study of plant microbiota: search for plant probiotics. Plant Sci 193–194:96–102

Birtel J, Walser J-C, Pichon S, Bürgmann H, Matthews B (2015) Estimating bacterial diversity for ecological studies: methods, metrics, and assumptions. PLoS One 10:e0125356. https://doi.org/10.1371/journal.pone.0125356

Blaser MJ, Cardon ZG, Cho MK, Dangl JL, Donohue TJ, Green JL, Knight R, Maxon ME, Northen TR, Pollard KS, Brodie EL (2016) Toward a predictive understanding of earth' s microbiomes to address 21st century challenges. Am Soc Microbiol 7:1–16. https://doi.org/10.1128/mBio.00714-16

Bradford MM (1976) A rapid and sensitive method for the quantitation of microgram quantities of protein utilizing the principle of protein-dye binding. Anal Biochem 72:248–254

Bric JM, Bostock RM, Silversone SE (1991) Rapid in situ assay for indole acetic acid production by bacteria immobilization on a nitrocellulose membrane. Appl Environ Microbiol 57:535–538

Brussaard L (2012) Ecosystem services provided by the soil biota. In: Wall DH, Bardgett RD, Behan-Pelletier V et al (eds) Soil ecology and ecosystem services. Oxford University Press, Oxford, pp 45–58

Byrd DW Jr, Kirkpatrick T, Barker KR (1983) An improved technique for clearing and staining plant tissue for detection of nematodes. J Nematol 15:142–143

Cappuccino JC, Sherman N (1992) Microbiology: a laboratory manual, 3rd edn. Benjamin/cummings Pub. Co, New York, pp 125–179

Dalpé Y, Séguin SM (2013) Microwave-assisted technology for the clearing and staining of arbuscular mycorrhizal fungi in roots. Mycorrhiza 23:333–340

de Castro AP, Gr F, Franco OL (2014) Insights into novel antimicrobial compounds and antibiotic resistance genes from soil metagenomes. Front Microbiol 5:1–9. https://doi.org/10.3389/fmicb.2014.00489

Delmont TO, Robe P, Cecillon S et al (2011) Accessing the soil metagenome for studies of microbial diversity. Appl Environ Microbiol 77:1315–1324

Delmont TO, Eren AM, Maccario L, Prestat E, Esen ÖC, Pelletier E, Le Paslier D, Simonet P, Vogel TM (2015) Reconstructing rare soil microbial genomes using in situ enrichments and metagenomics. Front Microbiol 6:358

Dolinšek J, Goldschmidt F, Johnson DR (2016) Synthetic microbial ecology and the dynamic interplay between microbial genotypes. FEMS Microbiol Rev 40:961–979

Dworkin M, Foster J (1958) Experiments with some microorganisms which utilize ethane and hydrogen. J Bacteriol 75:592–601

Esser RP (n.d.) What are nematodes? Accessed 7th January 2018, http://www.ontaweb.org/photosandlinks/whatarenematodes/. This article was prepared as an introduction to nematodes, particularly plant-parasitic nematodes. Websites listed on the ONTA homepage may be consulted for more advanced information on plant-parasitic nematodes

European Commission (2006) Communication from the Commission to the Council, the European Parliament, the Economic and Social Committee and the Committee of the Regions. Thematic Strategy for Soil Protection. COM(2006)231 final. Brussels

Fakruddin M, Mannan KS, Bin M (2013) Methods for analyzing diversity of microbial communities in natural environments. Ceylon J Sci (Biological Sci) 42:19–33

Ferrer M, Golyshina O, Beloqui A, Golyshin PN (2007) Mining enzymes from extreme environments. Curr Opin Microbiol 10:207–214

Fiske CH, Subbarow Y (1925) A colorimetric determination of phosphorus. J Biol Chem 66:375–400

Frostegård A, Courtois S, Ramisse V, Clerc S, Bernillon D, Le Gall F, Jeannin P, Nesme X, Simonet P (1999) Quantification of bias related to the extraction of DNA directly from soils. Appl Environ Microbiol 65:5409–5420

Ghannoum M (2016) Cooperative evolutionary strategy between the bacteriome and mycobiome. mBio 15, 7(6). https://doi.org/10.1128/mBio.01951-16

Gkorezis P, Daghio M, Franzetti A, Van Hamme JD, Sillen W, Vangronsveld J (2016) The interaction between plants and Bacteria in the remediation of petroleum hydrocarbons: an environmental perspective. Front Microbiol 7:1–27

Glick B (2012) Plant growth-promoting bacteria: mechanisms and applications. Scientifica 2012:1–15

Goodwin S, McPherson JD, McCombie WR (2016) Coming of age: ten years of next-generation sequencing technologies. Nat Rev Genet 17:333–351

Hahn AS, Konwar KM, Louca S, Hanson NW, Hallam SJ (2016) The information science of microbial ecology. Curr Opin Microbiol 31:209–216

Handelsman J, Rondon MR, Brady SF, Clardy J, Goodman RM (1998) Molecular biological access to the chemistry of unknown soil microbes: a new frontier for natural products. Chem Biol 5:R245–R249

Hawver LA, Giulietti JM, Baleja JD, Ng W-L (2016) Quorum sensing coordinates cooperative expression of pyruvate metabolism genes to maintain a sustainable environment for population stability. MBio 7:e01863-16

Heaven MW, Benheim D (2016) Soil microbial metabolomics. Beale DJ, Kouremenos KA, Palombo EA, Microbial metabolomics: applications in clinical, environmental, and industrial microbiology. Springer International Publishing, Cham, pp 147–198

Hiraoka S, Yang C, Iwasaki W (2016) Metagenomics and bioinformatics in microbial ecology: current status and beyond. Microbes Environ 31:204–212

Honma M, Shimomura T (1978) Metabolism of 1-aminocyclopropane-1-carboxylic acid. Agr Biol Chem 42:1825–1831

ISTA (1966) International rules for seed testing. Proc Int Seed Test Assoc 31:1–152

Jones OAH, Sdepanian S, Lofts S, Svendsen C, Spurgeon DJ, Maguire ML, Griffin JL (2014) Metabolomic analysis of soil communities can be used for pollution assessment. Environ Toxicol Chem 33:61–64

Kang HS (2016) Phylogeny-guided (meta)genome mining approach for the targeted discovery of new microbial natural products. J Ind Microbiol Biotechnol 44:285–293

Keisam S, Romi W, Ahmed G, Jeyaram K (2016) Quantifying the biases in metagenome mining for realistic assessment of microbial ecology of naturally fermented foods. Sci Rep 6:34155

Kim ES, Liu Y, Kim J-B, Chang I, Choi J (2015) Biological fixed film. Water Environ Res 87:974–999

Koske RE, Gemma JN (1989) A modified procedure for staining roots to detect VA mycorrhizas. Mycol Res 92:486–505

Krakat N, Anjum R, Demirel B, Schröder P (2016) Methodological flaws introduce strong bias into molecular analysis of microbial populations. J Appl Microbiol 49

Krause S, Le Roux X, Niklaus PA, Van Bodegom PM, Lennon JT, Bertilsson S, Grossart HP, Philippot L, Bodelier PLE (2014) Trait-based approaches for understanding microbial biodiversity and ecosystem functioning. Front Microbiol 5:1–10

Krueger R, McSorley R (2014) Nematode management in organic agriculture. http://edis.ifas.ufl.edu/. Accessed in 7th January 2018. This document is ENY-058 (NG047), one of a series of the Entomology & Nematology Department, UF/IFAS Extension. First published: January 2008. Reviewed June 2014. For more publications related to horticulture/agriculture, please visit the EDIS Website at http://edis.ifas.ufl.edu/

Levy SE, Myers RM (2016) Advancements in next-generation sequencing. Annu Rev Genomics Hum Genet 17:95–115

Lim NYN, Roco CA, Frostegård Å (2016) Transparent DNA/RNA co-extraction workflow protocol suitable for inhibitor-rich environmental samples that focuses on complete DNA removal for transcriptomic analyses. Front Microbiol 7:1–15

Ling LL, Schneider T, Peoples AJ, Spoering AL1, Engels I, Conlon BP, Mueller A, Schäberle TF, Hughes DE, Epstein S, Jones M, Lazarides L, Steadman VA, Cohen DR, Felix CR, Fetterman KA, Millett WP, Nitti AG, Zullo AM, Chen C, Lewis K (2015) A new antibiotic kills pathogens without detectable resistance. Nature 517:455–459

Lipiec J, Frac M, Brzezinska M, Turski M, Oszust K (2016) Linking microbial enzymatic activities and functional diversity of soil around earthworm burrows and casts. Front Microbiol 7:1–9

Lok C (2015) Mining the microbial dark matter. Nature 522:270–273

Lorck H (1948) Production of hydrocyanic acid by bacteria. Physiol Plant 1:142–146

Ma Y, Rajkumar M, Freitas H (2009) Isolation and characterization of Ni mobilizing PGPB from serpentine soils and their potential in promoting plant growth and Ni accumulation by *Brassica* spp. Chemosphere 75:719–725

Ma Y, Rajkumar M, Luo YM, Freitas H (2011) Inoculation of endophytic bacteria on host and non-host plants – effects on plant growth and Ni uptake. J Hazard Mater 196:230–237

Maherali H, Oberle B, Stevens PF, Cornwell WK, McGlinn DJ (2016) Mutualism persistence and abandonment during the evolution of the mycorrhizal symbiosis. Am Nat 188:E113–E125

Mahmoudi N, Slater GF, Fulthorpe RR (2011) Comparison of commercial DNA extraction kits for isolation and purification of bacterial and eukaryotic DNA from PAH-contaminated soils. Can J Microbiol 57:623–628

Mardis ER (2013) Next-generation sequencing platforms. Annu Rev Anal Chem 6:287–303

Masson-Boivin C, Sachs JL (2017) Symbiotic nitrogen fixation by rhizobia-the roots of a success story. Curr Opin Plant Biol 44:7–15

McSorley R (2016) Soil-inhabiting nematodes, Phylum Nematoda. http://edis.ifas.ufl.edu. Accessed in 7th January 2018. This document is EENY-012, one of a series of the Entomology and Nematology Department, UF/IFAS Extension. Original publication date July 1997. Revised July 2007. Reviewed April 2016. Visit the EDIS website at http://edis.ifas.ufl.edu

Mekete T, Crow WT (2015) Sampling instructions for nematode assays. http://edis.ifas.ufl.edu/. Accessed in 7th January 2018. This document is ENY-027 (SR011), is one of a series of the entomology and nematology department, UF/IFAS extension. Published September 2001. Revised March 2007, March 2012, and July 2015. Please visit the EDIS website at http://edis.ifas.ufl.edu/

Mendes LW, Tsai SM, Navarrete AA, de Hollander M, van Veen JA, Kuramae EE (2015) Soil-borne microbiome: linking diversity to function. Microb Ecol 70:255–265

Mendes S, Azul AM, Castro P, Röembke J, Sousa JP (2016) Chapter 16: Protecting soil biodiversity and soil functions: current status and future challenges. In: Castro P, Azeiteiro UM, Bacelar Nicolau P, Leal Filho W, Azul AM (eds) Biodiversity and education for sustainable development, World sustainability series. Springer, Cham

Morris CE, Bardin M, Berge O, Frey-Klett P, Fromin N, Girardin H, Guinebretiere MH, Lebaron P, Thiery JM, Troussellier M (2002) Microbial biodiversity: approaches to experimental design and hypothesis testing in primary scientific literature from 1975 to 1999. Microbiol Mol Biol Rev 66:592–616

Müller T, Ruppel S (2014) Progress in cultivation-independent phyllosphere microbiology. FEMS Microbiol Ecol 87:2–17

Nemergut D, Shade A, Violle C (2014) When, where and how does microbial community composition matter? Front Microbiol 5:2012–2014

Oliveira RS, Ma Y, Rocha I, Carvalho MF, Vosátka M, Freitas H (2016) Arbuscular mycorrhizal fungi are an alternative to the application of chemical fertilizer in the production of the medicinal and aromatic plant *Coriandrum sativum* L. J Toxicol Environ Health A 79:320–328

Orgiazzi A, Bardgett RD, Barrios E, Behan-Pelletier V, MJI B, Chotte JL, De Deyn GB, Eggleton P, Fierer N, Fraser T, Hedlund K, Jeffery S, Johnson NC, Jones A et al (eds) (2016) Global soil biodiversity atlas. European Commission, Publications Office of the European Union, Luxembourg 176 pp

Pajak M, Błońska E, Frac M, Oszust K (2016) Functional diversity and microbial activity of forest soils that are heavily contaminated by lead and zinc. Water Air Soil Pollut 227:348

Pajares S, Bohannan BJ (2016) Ecology of nitrogen fixing, nitrifying, and denitrifying microorganisms in tropical forest soils. Front Microbiol 7:1045

Papadopoulou A, Taberlet P, Zinger L (2015) Metagenome skimming for phylogenetic community ecology: a new era in biodiversity research. Mol Ecol 24:3515–3517

Pepper IL, Gerba CP (2009) Environmental sample collection and processing. Elsevier Inc. doi:https://doi.org/10.1016/B978-0-12-394626-3.00008-9

Phillips JM, Hayman DS (1970) Improved procedures for clearing and staining parasitic and vesicular-arbuscular mycorrhizal fungi for rapid assessment of infection. Trans Br Mycol Soc 55:158–161

Prosser JI (2015) Dispersing misconceptions and identifying opportunities for the use of "omics" in soil microbial ecology. Nat Rev Microbiol 13:439–446

Rinke C, Schwientek P, Sczyrba A, Ivanova NN, Anderson IJ, Cheng JF, Darling A, Malfatti S, Swan BK, Gies EA, Dodsworth JA, Hedlund BP, Tsiamis G (2013) Insights into the phylogeny and coding potential of microbial dark matter. Nature 499:431–437

Saia S, Ruisi P, Fileccia V, Di Miceli G, Amato G, Martinelli F (2015) Metabolomics suggests that soil inoculation with arbuscular mycorrhizal fungi decreased free amino acid content in roots of durum wheat grown under N-limited, P-rich field conditions. PLoS One 10:1–15

Saleh SS, Glick BR (2001) Involvement of gasS and pros in enhancement of the plant growth-promoting capabilities of *Enterobacter cloacae* CAL2 and UW4. Can J Microbiol 47:698–705

Schloss PD, Jenior ML, Koumpouras CC, Westcott SL, Highlander SK (2015) Sequencing 16S rRNA gene fragments using the PacBio SMRT DNA sequencing system. Peer J 4:e1869

Schwyn B, Neilands JB (1987) Universal chemical assay for the detection and determination of siderophores. Anal Biochem 160:47–56

Sharpton TJ (2014) An introduction to the analysis of shotgun metagenomic data. Front Plant Sci 5:209

Siggins A, Gunnigle E, Abram F (2012) Exploring mixed microbial community functioning: recent advances in metaproteomics. FEMS Microbiol Ecol 80:265–280

Smith SE, Read DJ (2008) Mycorrhizal symbiosis, 3rd edn. Academic, London

Solden L, Lloyd K, Wrighton K (2016) The bright side of microbial dark matter: lessons learned from the uncultivated majority. Curr Opin Microbiol 31:217–226

Staley JT, Konopka A (1985) Measurement of in situ activities of nonphotosynthetic microorganisms in aquatic and terrestrial habitats. Annu Rev Microbiol 39:321–346

Stubbendieck RM, Vargas-Bautista C, Straight PD (2016) Bacterial communities: interactions to scale. Front Microbiol 7:1–19

Sun X, Tang M (2012) Comparison of four routinely used methods for assessing root colonization by arbuscular mycorrhizal fungi. Botany 90:1073–1083

Tremblay J, Singh K, Fern A, Kirton ES, He S, Woyke T, Lee J, Chen F, Dangl JL, Tringe SG (2015) Primer and platform effects on 16S rRNA tag sequencing. Front Microbiol 6:1–15

van Bezooijen J (2006) Sampling. In: Bezooijen J van (Ed), Methods and techniques for nematology. Revised version 2006 2–10

van Elsas JD, Boersma FGH (2011) A review of molecular methods to study the microbiota of soil and the mycosphere. Eur J Soil Biol 47:77–87

Varga S (2015) On the importance of details in arbuscular mycorrhizal research. Appl Soil Ecol 87:87–90

Vierheilig H, Schweiger P, Brundrett M (2005) An overview of methods for the detection and observation of arbuscular mycorrhizal fungi in roots. Physiol Plant 125: 393–404

Violle C, Reich PB, Pacala SW (2014) The emergence and promise of functional biogeography. Proc Natl Acad Sci 111:13690–13696

Whitehead AG, Hemming JR (1965) A comparison of some quantitative methods of extracting small vermiform nematodes from soil. Ann Appl Biol 55:25–28

Wu D, Hugenholtz P, Mavromatis K et al (2009) A phylogeny-driven genomic encyclopedia of Bacteria and archaea. Nature 462:1056–1060

Xu Z, Hansen MA, Hansen LH, Jacquiod S, Sørensen SJ (2014) Bioinformatic approaches reveal metagenomic characterization of soil microbial community. PLoS One 9:e93445

Young JM, Rawlence NJ, Weyrich LS, Cooper A (2014) Limitations and recommendations for successful DNA extraction from forensic soil samples: a review. Sci Justice 54:238–244

Chapter 28
Computational Approach to Study Ecophysiology

Bibhuti Prasad Barik and Amarendra Narayan Mishra

1 Introduction

Ecophysiology is the study of how the environmental cues affect the functional aspects of an organism. This aspect is vital for the adaptation of living organisms to the ever changing surrounding, and regulates the distribution and richness of the organisms in the natural habitat. This interaction of ecosystem with an organism and its physiological status are governed by the genomic structure and time period of environmental impact. The duration and dose or intensity of the environmental impact determines the functional and genomic stability of organisms, both at individual and community level. Genome performs as unified system displaying intricate and dynamic behavior (Zhu et al. 2008). Ecophysiological genomics deciphers the alterations in gene structure and function in a specific environment. The study of such alterations in the genome, which are pivotal for the functional integrity of an organism, can be well documented through bioinformatics tools (Aubin-Horth and Renn 2009). In silico experimental strategies can ease sighting of core genomic elements, regulatory networks and conserved sequences across species and the variations in biotic and abiotic components of the environmental (McCarroll et al. 2004; Ragland et al. 2010). These studies will bring out comparative schemes didactic to the variations arising from ecological variations, leading to adaptation, speciation and evolution, as such.

B. P. Barik (✉)
Department of Zoology, Khallikote Autonomous College, Berhampur, Odisha, India

A. N. Mishra (✉)
Khallikote Cluster University, Berhampur, Odisha, India

Centre for Life Science, Central University of Jharkhand, Ranchi, Jharkhand, India

© Springer International Publishing AG, part of Springer Nature 2018
A. M. Sánchez-Moreiras, M. J. Reigosa (eds.), *Advances in Plant Ecophysiology Techniques*, https://doi.org/10.1007/978-3-319-93233-0_28

2 Ecophysiological Informatics

Ecophysiology research and analysis gradually shifted from adaptation or stress studies toward mathematical and systems model focusing on estimation, distribution and diversity of organisms with associated abiotic data (Real and Brown 1991). Computational protocols are planned to assess a set of hypothesis and the simulations and collected data and their formats differ in analysis. These make ecophysiological data extremely diverse. These data types deal with entities like numbers, genotype and ecophysiological barcodes along with the functional aspects such as the level of competition, food chain, synergism/antagonism, damage/repair etc. The heterogeneity of ecophysiological data occur in numeric, textual, imageries, movies and policies to deal with non-categorical or non-digitized environmental impacts at the functional ecosystem level (Green et al. 2005). The current demand in this systems ecology study is to access, authenticate and secure composite data types spanning from genome to biome. This is possible by developing computing algorithms have evolved to analyze by process and pattern modeling and further enabling cross platform observations. But still huge amount of ecophysiological data persist unreachable (Palmer et al. 2005). The development of big data concept will enable the allied data driven studies to augment our knowledge about ecosystems and management of natural resources, individual and ecosystem function and genomics.

Ecophysiological informatics may take advantage of the large genomic assets of the classical experimental models to understand phenotypic variations and ecotypes. Detailed genomic database archives are accessible for numerous species with molecular characterization and genomic architecture controlling variation and gene expression patterns (Kvitek et al. 2008; Smith and Kruglyak 2008).

Despite periodic or stochastic variations, species react to climate change through movement, acclimation, or adaptation (Parmesan 2006) by showing physiological plasticity. Movement, acclimation and adaptation are considered in ecological modeling. Computational programs are already developed for addressing thermal maxima, ocean warming and risk factor assessment (Berkelmans and Willis 1999). In many cases, the importance of adaptation is not clear, but the generated data specify mechanisms of acclimation to environmental stress with gene flow buffering population life styles. The functional genomics of fungi receptive to diverse environmental fluctuations (Gasch et al. 2000) is studied widely as a model system to know the genomic response in different niches with distinctive or similar physiology. Some marine animals, e.g. sea urchin, are sensitive to a rise of atmospheric carbon dioxide and consequently acidification of the seawater (Hofmann et al. 2010; O'Donnell et al. 2010). An extensive study by hybridizing heterologous microarrays within gene-genealogy models displayed compromised response to heat shock, impending synergistic impressions of ecological changes and citing a possibility to know underlying genomic engagements among co-stressors.

Interesting ecophysiological adaptations and species distribution in extreme environments were done in polar regions. Eastman (2005) studied Antarctic notothenioid fishes, which have undergone radiation pre-treatment to survive in cold niches. Ecophysiological and genomic database for the same was developed by Near et al. (2004). This computational scheme is used to know genomic mechanisms controlling trophic status and eco-morpho-physiological evolution (Podrabsky and Somero 2006; Albertson et al. 2010).

Comparative phylogenetic analysis of sculpin (superfamily Cottoidea) inhabiting varied environmental niches showing hypoxia tolerance limits have evolved ecophysiological adaptation. DNA barcoding and phylogenetic comparisons will highlight ecophysiological niche partitioning and adaptive radiation (Mandic et al. 2009). Further advanced computational comparative systems may be used to explore underlying genomic mechanisms leading to success of invasive and allied species. For example weedy and non-weedy sunflowers populations showed lineage specific genome expression variation specifying independent and recurrent evolution of invasiveness (Lai et al. 2008). Interaction of environment and species facilitate further understanding in to ecogenomic mechanisms driving minute physiological divergences. Likewise, a customized microarray study has confirmed substitution of a native intertidal congener with that of an invasive mussel responding to temperature and salinity (Lockwood and Somero 2011). Other models may be chosen which can be distinctively placed to address specific climatic responses so that computational algorithms could be developed.

First of all a suitable study system for a particular ecophysiological question is chosen and then maneuvering the genomic tools, at present which is possible through Next Generation Sequencing (NGS) to have huge genomic archives. As more and more number of complete genome accessible species is available, database sharing has opened chances of developing mixed models. Non-traditional models are becoming obsolete as analytical tools are increasingly being used for validation of gene ontology and function. Also studies on more number of wild and endemic species with diverse ecological challenges will definitely speed up alternative models for gene predictions (Colbourne et al. 2011). Computational algorithms are coming up along with generation, organization, storage, retrieval and analysis of ecogenomic data and its functional enrichment and systems biology simulations (Kim et al. 2010). These suggest clear perceptions about genome evolution and phoneme plasticity.

3 Ecophysiological Adaptation and Bioinformatics

The past decade has witnessed lot of advancement of experimental and analytical techniques, which have led to many statistical algorithms dedicated to handle gene expression and protein interactions. Computational biology has completely changed the way biological data are analyzed. With the rapid data explosion and cloud computing the demand for super specialized techniques has been on the rise to handle and analyze complex data types. Machine learning, data mining and intelligent data analysis has led to many innovative programs for quick analysis of such data. In bioinformatics, clustering similar group of genes having common expression profiles and further using this has allowed to predict the function of unknown genes. Clustering has helped to treat similar data types as single modules to reduce the number of variables while building predictive models. Grouping of disease outcome and methods to identify feature selection based biomarkers is also becoming common. Cell-cycle data types are becoming popular using time series microarray data modeling. Graph based models and genetic regulatory network approaches have led to know complexities of large-scale gene interaction towards systems modeling.

Environment circumstantial swings necessitate compensatory alteration from inhabitant species. Flexible phenotypes facilitate environmental compensation within physiological timescales. This can be inscribed on the functional genome by alteration of transcription or translation. The phenotypic plasticity directed by geno-proteomic mechanisms shall determine the acclimatization range of the species. Consequently, biogeographical distribution of species can be understood by experimental and mechanistic model approaches. Structural changes in the genomes and proteomes may be the outcome of accumulated environmental changes. Adaptive and evolutionary pressure determines allelic frequencies within species across generations.

There is a chance of overlapping of genomic elements facilitating physiological plasticity and the genomic targets of adaptive evolutionary processes. The acclimation regulated genes in show patterns of adaptive divergences compared to that of the other genes, e.g. in killfish (Whitehead et al. 2012). However, there are reports of wide variations in gene expression within a population and it is plausible that this process play significant role in evolution (Rees et al. 2011).

4 Computational Ecophysiology – A Comparative Approach

Species-specific details including gene number, genome size, patterns of sequence duplication, transposable elements etc. have been extensively reported in living organisms. But, knowledge about the evolutionary dynamics of genome architecture and their impact on ecological lineage is meagre. These problems can be solved once we have a high throughput whole genome sequences.

Researchers have taken the advantage of genomic database and programs to know functional attributes of genes in the evolutionary process. Intraspecific genome sequences reveal levels of single nucleotide polymorphisms (SNPs) confirming the signatures of adaptive radiation (Nielsen 2001). Selected candidate loci could be verified by random sampling of more number of species. Using mapping of linkage disequilibrium, naturally occurring genome wide polymorphisms are explored (Hagenblad et al. 2004). With known information regarding candidate genes, genomic variation data can provide the necessary controls for phenotypic association studies. Over the few years, it has been accepted that gene products are associated with large-scale interaction networks. It has become essential to know the evolutionary dynamics of molecular genetic networks (Cork and Purugganan 2004). Expression profiling microarrays and expressed sequence tags offer a feasible way permitting us to detect the interacting candidate genes of genetic networks and examine how the patterns of interactions change across species. The need is framing tangible for network architecture linking evolution.

Ecophysiology peruses to know the working mechanisms of complex biological systems.

Relative methods of comparison involving large number of taxa provide more insights at the fundamental level. The comparisons are made to understand the nature of biodiversity, more specifically the universal mechanisms and uniquely evolved traits. For instance, certain group of molecules acting as intracellular organic osmolytes by virtue of their physical properties stabilizes protein structure (Somero 2000). On the contrary lineage specific acceptance of urea as an osmolyte have a tendency to destabilize proteins leading to emergence of parallel lineage-specific use of methyl ammonium compounds to counter act the protein-destabilizing effects of urea. Thus the comparative study enabled detection of biological lineage-specific uniqueness.

Multiple comparative designs either universal or lineage specific are to be customized depending on the case studies. If the objective is on knowing the mechanisms of a universal phenotype then inter-specific shared traits are investigated. In contrast, if the focus is on novel lineage specific traits then more premeditated designs are considered. The objective of such studies is to know the traits that vary between taxa and adaptive for particular habitat. An alternative evolutionary elucidation for trait differences among species is done by statistical models (Harvey and Pagel 1991). These models reconstruct phylogenies based on historical divergences (Felsenstein 1985). Phylogenies offer evolutionary time scale of particular traits of ecophysiological importance with possible common ancestry.

5 Experimental Design Considerations for Comparative Ecophysiological Research

Robust comparative physiological experiments need choice of model species and pre-defined comparative framework, sequence data, statistical analysis and interpretation. Manipulation by comparison of contrasting responses of numerous taxa shall give enough opportunities to develop hypothesis in the given environmental conditions (Schlichting and Pigliucci 1998). This is repeatedly accomplished by allowing species within controlled settings subjecting them in manipulated temperature, salinity, oxygen, pollutants or symbionts. These experiments offer careful control over environmental dynamics of interest, but induced genome responses may fluctuate from those expressed in normal conditions. A substitute strategy is field transplant experiments, where organisms from diverse life styles may be transplanted to alternate environments, and transformations in genome expression is measured. These controlled studies are usually cumbersome and offer less control over environmental variables, many of which may vary, but they may more ecophysiologically convincing (Cheviron et al. 2008).

Comparisons are made between evolutionarily adaptive against neutral under strong constraints. The variance in the traits is mostly directed by interaction of neutral genetic drift and natural selection. More number of species are considered simultaneously and the neutral genetic distance tethering all sets of species can be used as a matrix of likely covariances, where species with shortest genetic distances would be expected to share more similarity for any given trait (Felsenstein 1985), and polarity of the traits can be derived. Against this neutral framework of expected covariances, one may assess whether observed trait variance among taxa matches the neutral anticipation, or discards it in support of adaptive radiations. The phylogenetic comparative approach can be used to test, which genes contribute to ecophysiologically adaptive morphotypic disagreement among taxa (Whitehead and Crawford 2006).

6 Systems Ecology

Systems ecology endeavors to simplify the structure and function of ecosystems using mathematical models and computer simulations. The observables are dimensions of the performance of the organisms, extending from phenotype to minute metabolic sketching. Complex issue of efficiently integrating diverse data types is addressed by linking nucleotide sequence, gene expression, protein-protein interactions to reach biological inferences. More emphasis should be on establishment of coherent validated data sets with universal formats for pathway analysis to define complex interactions with bio text mining tools. Ecophysiological systems biology targets to improve open-source customized platforms for easy accessible pathway models.

Ecophysiological systems are poorly defined and enough samples are to be taken to an accepted time and space in signaling and regulatory networks. The ultimate goal is to create a complete in silico model of a complete ecosystem. The existing computational efforts shall definitely integrate this information from a wide variety of platforms.

Systems ecology measure both taxonomic and functional biodiversity by using information and algorithms from all areas of bioinformatics. The dawn of next generation sequencing in molecular biology is currently serving microbial ecology to become steadfastly attached as a core sub-discipline within ecophysiology, and to test general theories about hierarchical biodiversity and life styles functioning. In addition, all systems are studied in one complete complex system vital biogeochemical progressions by linking micro and macroscopic ecospheres. Ecophysiological system is a complex phenomenon presenting evolving properties. It focuses on interfaces and connections within and between bio systems, and is exclusively concerned with the functioning of ecosystems. It reflects thermodynamics principles and cultivates explanations of complex systems.

Microorganisms are likely to involve in cycles of environmental transitions over different periodicities and these processes are reflected in the structure, genetic selection and community function. Emphasis was given on manageable microbial communities in exciting environments where genome sequence other geochemical measurements offer proof of functional attributes associated with spatiotemporal fluctuations in environmental variables and metabolic dynamics. Nucleotide sequences of important ecotypes have supported a basis for questioning exhaustive spatiotemporal facets of thermophilic and phototrophic communities, including micro sensor analysis of their physical and chemical microenvironment.

The metabolic pairing amongst populations adapted to spatial and/or temporal states is a collective theme in natural environments, and this was revealed in detail using different light-adapted ecotypes of cyanobacteria. Controlled experiments using pure cultures displayed the significance of precise nutrient demands, and how the exoproteome of a bacterium is controlled by the level of cellular oxidation. Ultimately, a clear knowledge of the complex network of abiotic and biotic interactions will also entail detailed information of the regulatory and ecophysiological events governing intron-exon boundaries, chain elongation and post-translational adaptations.

7 Mechanistic Approach to Ecophysiology Research

Current efforts strive to develop a mechanistic approach to ecophysiology, where deployment of physiochemical models encompasses our perception about how organisms intermingle with environment (Denny and Gaylord 2010). Genoproteomic approaches may be used to include our mechanistic accepting of how organisms interplay with environment. Physiological, behavioral and biomechanical attributes are the immediate providers to defining ability and delimiting the fundamental and recognized niches of species (Kearney et al. 2012). Comparative genomic methodologies are applicable for editing of biological organization. Phenotypic response is an important feature of genome structure. Interestingly organism and genome-scale information are reciprocal to each other. Although plotting genotype to phenotype is not insignificant, it is a great challenge (Schwenk et al. 2009) as genoproteomic investigations are realistic methods to know mechanisms simplifying ecologically pertinent phenotypes and reactions to environmental changes.

8 Forest Ecophysiology

Genomes of numerous forest trees are freely available which has taken forest genetics to higher level. In the coming days many draft genome sequences of forest trees will be readily online. The main plan is to use these data to fix genomic attributes affecting phenotypic attributes of trees growing in different altitudes and varied ecophysiological zonations. The complete genome database can be analytically hunted for genetic elements and concerned allelic variants. This information will expedite the well-organized tree improvement programs through genome-breeding. For the preparation of process models, knowledge of tree growth dynamics is essential to manage forest ecosystems. Improvement of ideotypes and bioinformatic approaches will also help in genetic improvement (Nelson and Johnsen 2008).

Genomic research into the ecophysiological processes of forest growth assures to deliver new visions. In addition, this technical know-how shall facilitate advances in process models. Effective process models will encourage breeders to better fix the ideotypes for specific combinations of environment and product end use. Genetic engineering offers opportunities for tree improvement and forest management.

9 Eco-Modeling and Climatic Adaptation

Genotypes, which are adapted locally, show much fitness in their native site when compared with foreign genotypes. It has been studied that both local adaptation and genomic associations with range of climatic variables are also feasible. Conditional neutrality of genomic patterns is very common as evident from SNP-environment correlations and ecophysiological pathways possibly depicted in regional adaptation. Genomic connotation approaches easily translate to non-model systems, and genetically unequivocal climate envelope models shall determine future species scattering under changing climatic conditions. The models do not include all relevant demographic processes (Weinig et al. 2014).

10 Integrated Ecophysiology and Drought Tolerance in Crops

One of the serious challenges world agriculture face is drought and it can badly hit the economy with anticipated climate change. It is crucial to understand and know the mechanisms of tolerance limits exhibited by plants to water stress. Current molecular approaches with precise genomic tools are likely to efficiently uncover genes and metabolic pathways conferring drought tolerance in crops. Importance has been given to molecular examination of tolerance by linkage, QTL cloning, gene prediction and transcriptonomics. Marker assisted back crossing, recurrent genome wide selection have been attached to molecular farming to ensure food security in changing climate by developing drought tolerant cultivars. Precise phenotyping is carried out to screen and map suitable QTL and candidate genes. A number of phenomic methodologies are accessible. Phenomics combined with modeling quickly evaluate the importance of certain traits on species performance. Models are configured to know a gene-phenotype affiliation for a complete viable system (Edmeades et al. 2004). To meet the task of enhanced productivity, functional genomics and systems biology are incorporated to bridge the gap between genes and phenomes (Yin et al. 2004). Crop models are increasingly contributing best integration of ecophysiological progressions to composite cropping patterns. Latest genomic adventures speed up large-scale barcode genotyping concomitant with stress acceptance. The need of the time is to have a combined approach to develop next generation crops in changing climatic conditions. It is expected to have integrated ecophysiological systems genomics to address food security concerns (Mir et al. 2012).

11 Ecophysiological Genomic Approaches and Salinity Stress Tolerance

Salinity is one of the most important issues reducing crop production. The complex nature of genes responding to salinity demands immediate attention. Difficulty is due to the mutagenic nature of these gene complexes. Ecophysiological genomics has improved enormously and played a very important role in generating indispensable facts for crop modeling. It has been possible to locate and characterize genes responsible for salinity stress response, figure out signaling pathways. Gene pyramiding and genome editing (for example Genome editing technologies such as Zinc finger nucleases, TALENs and CRISPR/Cas9) deliver fresher and quicker opportunities for biologists to produce engineered crops in precision (Nongpiur et al. 2016).

Salt Overly Sensitive (SOS) pathway involving SOS1, SOS2 and SOS3 genes has been documented as one of the important mechanisms governing ion homeostasis under salinity stress. SOS2 component of this pathway encrypts a serine/threonine protein kinase that together with SOS3 triggers downstream Na (+)/H (+) anti porter SOS1, reconstructing cellular ion homeostasis under salinity stress. It has been observed that the transcript levels of $BjSOS_2$ are prompted in response to various abiotic stresses. Kaur and coworkers isolated a 713 bp promoter region of SOS_2 gene from *Brassica juncea* to study the regulation of $BjSOS_2$ under various abiotic stress conditions and to survey efficacy of the cloned upstream region in controlled simulations. $BjSOS_2$ promoter retains robust multi-stress inducible nature, indicating its connection in many stress signaling hot spots. Concurrent occurrence of multiple abiotic stress settings under natural states is really a perplexing threat to crop yield; further research work may employ the $BjSOS_2$ promoter to get-up-and-go stress-inducible expression of genes reporting tolerance to multiple stresses (Kaur et al. 2015).

12 Computational Proteomics and Behavioural Ecophysiology

Proteomics is the information of the all protein complements articulated by a genome and intends to know expression, regulation, function, and interactions in a given ecophysiological state. Computational proteomics gives a fair picture of the proteins possibly connected with or accountable for precise activities at molecular level. Proteomic studies are divided into three domains, (i) Protein identification, quantification, and differential expression in different ecophysiological state; (ii) examining protein functions and protein interaction networks with other biomolecules; and (iii) structural proteomics, studying, visualizing (Barik 2013a) and predicting structure using composite algorithms. Proteomic outcomes pertinent to behavioral ecology originate from all interdisciplinary fields. Proteomics is highly multipurpose to altered experimental set ups and endorses hypothesis-motivated studies as well as exploratory assessments in which signature patterns can be tagged with phenomes without prior conventions of the essential mechanisms (van Helden 2013).

Proteomics is thus a very powerful tool for the documentation of candidate genes and corresponding proteins liable for expression patterns (Diz et al. 2012). Behavioral ecophysiology witness and quantity diverse behaviors. Computational proteomics along with behavioral ecophysiology offer natural information and ease access to the neighboring unknown mechanisms. Hypothesis engineering nature of proteomic tactics is principally apparent in both plant and animal behavioral ecophysiology. An elucidation of the huge amount of proteomic data in the behavioral ecophysiological context is still absent, though a fair amount of proteomic data is presented for toxins from many species. Proteomics has established that at least few organisms have the geno-physiological wealth to adapt in changing environments. Additionally, in vitro and in silico proteomic explorations could deliver convincing indication for the biochemical optimization. The wide-ranging uses of ecophysiological proteomics have provided a great number of computational tools and database extremely applicable for eco-behavioral biology. In reproductive ecology, metabolic cataloging has been used to evaluate the dynamics of germinal proteome and its turnover (Claydon and Beynon 2012). This particular methodology assures to deliver a quantity plasticity of protein manufacturing and distribution in response to changing ecosystems. In psychiatry, computational proteomics has been magnificently used to review behavior-changing maladies. For instance, a proteomic comparison of mice cell lines selected for nervousness-related performance exposed role of mitochondria in controlling such behavior. Maximum reviews in behavioral ecophysiology target non-model species in the absence of complete genome data. In more cases, homology-sequence-structure driven proteomics, de novo protein sequencing is employed for better results. In case of availability of only sequence information from phylogenetically diverse organisms, or in the case of swiftly evolving proteins sharing low homology, the effective proteome analysis is transcriptome sequencing followed by secondary and tertiary structure prediction. Next generation sequencing technologies have made genomic and transcriptomic data progressively reachable and cheaper (Ekblom and Galindo 2011), and it is anticipated that convergence of genomic, transcriptomic, and proteomic exploration hard works shall bring fine-tuned innovations. With the latest sci-fi improvements in mass spectrophotometry and bioinformatics and the contribution of ever-accumulative geno- transcriptomic data, the base have been strong for an wide-ranging usage of proteomic tools in addressing questions of behavioral ecophysiology.

13 Macroecological Phylogenies

There has been a rise in ecophylogenetic studies integrating phylogenetic information with enriched species samples, inspired by speculation that species having common ancestors should share niche conservatism. This is done by using many methods of cumulative complexity and robustness: using exclusive taxonomical classification; constructing super trees covering topological data; or construction of super matrices of molecular statistics that are used to estimate phylogenies with boot strap values. A concrete evolutionary hypothesis is the need of the time to build consistent ecophysiological analysis. It is essential to simplify the methodological concerns connected with the reconstruction of mega phylogenies with informative branch-lengths on large species sample size. This is possible as vast amounts of molecular data available from databases such as GenBank, and consensus knowledge on deep phylogenetic relationships for an increasing number of groups of organisms of diverse lifestyles (Barik 2015).

A key theory of evolutionary genetics is that natural selection affects genes or cluster of genes, but gene flow, range extension or blockages, leave genomic signatures. Single nucleotide polymorphism data of species data also contribute to the altered phylogenies. We are now a days confronted with huge amounts of genome information with which we can characterize population history and structure. Amount of intraspecific genome disparity also supports in the positioning of loci revealing patterns of variation (Luikart et al. 2003). As the level of structural variation in intraspecific genomes becomes clear, evolutionary dynamics of gene families and transposable elements should be established in ecophysiological context. A consequence of the proliferation of genome studies has been the documentation of patterns of genome variation between species, information that can be used to promote in construction of the ecophysiological tree of life. Furthermore, the wide-ranging genome coverage data can also be incorporated to explore molecular clocks representative of evolutionary lineages (Haubold and Wiehe 2004).

14 Metabolomics and Ecophysiology

Metabolomics is the enquiry of the comprehensive collection of small metabolites in a cell at any specified time. Metabolomics may prove to be principally vital due to the production of secondary metabolites. In a metabolite profiling research, metabolites are removed from tissues, separated and investigated in a high-throughput fashion. Metabolic fingerprinting distinguishes samples according to their phenotype or biological relevance. The metabolic network of any organism is multifaceted which could be fragmented in to sub systems called metabolic pathways, which are elementary functional structure of the entire network. Metabolic studies are primarily based on raw data arising from nuclear magnetic resonance, gas and liquid chromatography mass spectrometry to identify and compute metabolites from biological samples. For any species under study, genes are consigned in the referenced pathways (Galperin and Koonin 1999). Adjustments are presented to the current state of living systems and afterward variations are detected on transitional metabolites at specific time scales. The connections among components of a pathway are derived by metabolic profiling and comparison with altered ecophysiological states. Metabolic pathways are supposed to be vigorous and evolve continuously in different physiological status, energy level and redox potential. Environmental dynamics might encourage the preservation or the loss of metabolic pathways, or they might simply moderate gene expression via a signaling cascade prompting a pathway turned on or off. Keeping in mind about the large data sets, it is essential to use computational power for understanding the pathways. Algorithms are required to process, visualize and analyze the data from a variety of platforms. Programmed reaction network models support planned simulation studies. As the experimental confirmation of reaction networks remains difficult, in silico methods are still necessary to assume likely pathways. Computational methods are developed and many are underway for reconstructing combined metabolic pathways based on literature or reference-based frameworks and chemical transformations across species. Metabolic signatures are efficient markers of prime importance in health and disease diagnosis (Barik 2013b). Metabolites reveal the incorporation of gene expression, interfaces and other diverse governing processes. Metabolomics is the most universal and can be realistic to different organisms in their respective ecophysiological modifications. Secondary metabolites show specificity of their respective genera and species and accordingly react to specific pressure or conditions such as antioxidants, reactive oxygen species, scavengers, coenzymes, UV and excess radiation screen. In addition, enhanced accumulation of secondary metabolites during several abiotic stress responses and their cross-protection to biotic threats are of utmost importance for stress adaptation. The co-relation between certain metabolites with gene expression provides a clue for determining a probable accurate marker for tolerant organism during the process of natural selection (Arbona et al. 2013).

15 Conclusion

The challenges in ecophysiology are to deliver biological implication to the ever growing and well archived high-throughput database. Molecular networks with quantitative data allow simultaneous visualization of changes in the metabolome, proteome and transcriptome in a specific environment. This is the primary requirement for an interdisciplinary and composite approach of computational ecophysiology. Further, transitional variations, simulations and systems biology will help in modeling of metabolism and growth.

Research opportunities using computational ecophysiology will surely offer more exploration on the mechanisms enabling acclimation and evolutionary adaptation. Interestingly, confounding closely interrelated taxa that occupy and probably evolved in different geospheres, the analysis of genes and genomic programs are vital for establishing phenomes appropriate to determine specific niches. Computational ecophysiology can be used to document problems as well as produce supervised simulated models about organisms' response to continuously changing environment.

References

Albertson RC, Yan YL, Titus TA, Pisano E, Vacchi M, Yelick PC, Detrich HW, Postlethwait JH (2010) Molecular pedomorphism underlies craniofacial skeletal evolution in Antarctic notothenioid fishes. BMC Evol Biol 10:4

Arbona V, Manzi M, Ollas C, Gomez-Cadenas A (2013) Metabolomics as a tool to investigate abiotic stress tolerance in plants. Int J Mol Sci 14:4885–4911

Aubin-Horth N, Renn SCP (2009) Genomic reaction norms: using integrative biology to understand molecular mechanisms of phenotypic plasticity. Mol Ecol 18:3763–3780

Barik BP (2013a) Computational visualization of biomolecular structures. Int J Atoms Mol 3:576–578

Barik BP (2013b) In silico observations and analysis of metabolic pathways. Int J Sci Res 2:44–48

Barik BP (2015) Animal actin phylogeny and RNA secondary structure study. Int J Knowl Discov Bioinf 5:47–63

Berkelmans R, Willis BL (1999) Seasonal and local spatial patterns in the upper thermal limits of corals on the inshore central great barrier reef. Coral Reefs 18:219–228

Cheviron ZA, Whitehead A, Brumfield RT (2008) Transcriptomic variation and plasticity in rufous-collared sparrows (Zonotrichiacapensis) along an altitudinal gradient. Mol Ecol 17:4556–4569

Claydon AJ, Beynon R (2012) Proteome dynamics: revisiting turnover with a global perspective. Mol Cell Proteomics 11:1551–1565

Colbourne JK, Pfrender ME, Gilbert D, Thomas WK, Tucker A, Oakley TH, Tokishita S, Aerts A, Arnold GJ, Basu MK et al (2011) The ecoresponsive genome of *Daphnia pulex*. Science 331:555–561

Cork JM, Purugganan MD (2004) The evolution of molecular genetic pathways and networks. Bioassays 26:479–484

Denny MW, Gaylord B (2010) Marine ecomechanics. Annu Rev Mar Sci 2:89–114

Diz AP, Martinez-Fernandez M, Rolan-Alvarez E (2012) Proteomics in evolutionary ecology: linking the genotype with the phenotype. Mol Ecol 21:1060–1080

Eastman JT (2005) The nature of the diversity of Antarctic fishes. Polar Biol 28:93–107

Edmeades GO, McMaster GS, White JW (2004) Genomics and the physiologist: bridging the gap between genes and crop response. Field Crops Res 90:5–18

Ekblom R, Galindo J (2011) Applications of next generation sequencing in molecular ecology of non-model organisms. Heredity (Edinb) 107:1–15

Felsenstein J (1985) Phylogenies and the comparative method. Am Nat 125:1–15

Galperin M, Koonin E (1999) Functional genomics, enzyme evolution, homologous and analogous enzymes encoded in microbial genomes. Genetica 106:159–170

Gasch AP, Spellman PT, Kao CM, Carmel-Harel O, Eisen MB, Storz G, Botstein D, Brown PO (2000) Genomic expression programs in the response of yeast cells to environmental changes. Mol Biol Cell 11:4241–4257

Green JL, Hastings A, Arzberger P, Ayala FJ, Cottingham KL et al (2005) Complexity in ecology and conservation: mathematical, statistical, and computational challenges. Bioscience 55:501–510

Hagenblad J, Tang CL, Molitor J, Werner J, Zhao K, Zheng HG, Marjoram P, Weigel D, Nordborg M (2004) Haplotype structure and phenotypic associations in the chromosomal regions surrounding two *Arabidopsis thaliana* flowering time loci. Genetics 168:1627–1638

Harvey PH, Pagel MD (1991) The comparative method in evolutionary biology. Oxford University Press, Oxford/New York

Haubold B, Wiehe T (2004) Comparative genomics: methods and applications. Naturwissenschaften 91:405–421

Hofmann GE, Barry JP, Edmunds PJ, Gates RD, Hutchins DA, Klinger T, Sewell MA (2010) The effect of ocean acidification on calcifying organisms in marine ecosystems: an organism-to-ecosystem perspective. Annu Rev Ecol Evol Syst 41:127–147

Kaur C, Kumar G, Kaur S, Ansari MW, Pareek A, Sopory SK, Singla-Pareek SL (2015) Molecular cloning and characterization of salt overly sensitive gene promoter from *Brassica juncea* (BjSOS2). Mol Biol Rep 42:1139–1148

Kearney MR, Helmuth B, Matzelle A (2012) Biomechanics meets the ecological niche: the importance of temporal data resolution. J Exp Biol 215:922–933

Kim TY, Kim HU, Lee SY (2010) Data integration and analysis of biological networks. Curr Opin Biotechnol 21:78–84

Kvitek DJ, Will JL, Gasch AP (2008) Variations in stress sensitivity and genomic expression in diverse *S. cerevisiae* isolates. PLoS Genet 4:e1000223

Lai Z, Kane NC, Zou Y, Rieseberg LH (2008) Natural variation in gene expression between wild and weedy populations of *Helianthus annuus*. Genetics 179:1881–1890

Lockwood BL, Somero GN (2011) Transcriptomic responses to salinity stress in invasive and native blue mussels (genus *Mytilus*). Mol Ecol 20:517–529

Luikart G, England PR, Tallmon D, Jordan S, Taberlet P (2003) The power and promise of population genomics: from genotyping to genome typing. Nat Rev Genet 4:981–994

Mandic M, Sloman KA, Richards JG (2009) Escaping to the surface: a phylogenetically independent analysis of hypoxia-induced respiratory behaviors in sculpins. Physiol Biochem Zool 82:730–738

McCarroll SA, Murphy CT, Zou SG, Pletcher SD, Chin CS, Jan YN, Kenyon C, Bargmann CI, Li H (2004) Comparing genomic expression patterns across species identifies shared transcriptional profile in aging. Nat Genet 36:197–204

Mir RR, Zaman-Allah M, Sreenivasulu N, Trethowan R, Varshney RK (2012) Integrated genomics, physiology and breeding approaches for improving drought tolerance in crops. TAG Theor Appl Genet 125:625–645

Near TJ, Pesavento JJ, Cheng CHC (2004) Phylogenetic investigations of Antarctic notothenioid fishes (Perciformes: Notothenioidei) using complete gene sequences of the mitochondrial encoded 16S rRNA. Mol Phylogenet Evol 32:881–891

Nelson CD, Johnsen KH (2008) Genomic and physiological approaches to advancing forest tree improvement. Tree Physiol 28:1135–1143

Nielsen R (2001) Statistical tests of selective neutrality in the age of genomics. Heredity 86:641–647

Nongpiur RC, Singla-Pareek SL, Pareek A (2016) Genomics approaches for improving salinity stress tolerance in crop plants. Curr Genomics 17:343–357

O'Donnell MJ, Todgham AE, Sewell MA, Hammond LM, Ruggiero K, Fangue NA, Zippay ML, Hofmann GE (2010) Ocean acidification alters skeletogenesis and gene expression in larval sea urchins. Mar Ecol Prog Ser 398:157–171

Palmer M, Bernhardt ES, Chornesky EA, Collins SL, Dobson AP et al (2005) Ecological science and sustainability for the 21st century. Front Ecol Environ 3:4–11

Parmesan C (2006) Ecological and evolutionary responses to recent climate change. Annu Rev Ecol Evol Syst 37:637–669

Podrabsky JE, Somero GN (2006) Inducible heat tolerance in Antarctic notothenioid fishes. Polar Biol 30:39–43

Ragland GJ, Denlinger DL, Hahn DA (2010) Mechanisms of suspended animation are revealed by transcript profiling of diapause in the flesh fly. Proc Natl Acad Sci U S A 107:14909–14914

Real LA, Brown JH (eds) (1991) Foundations of Ecology: Classic Papers with Commentaries. University of Chicago Press, Chicago, USA, pp. 920

Rees BB, Andacht T, Skripnikova E, Crawford DL (2011) Population proteomics: quantitative variation within and among populations in cardiac protein expression. Mol Biol Evol 28: 1271–1279

Schlichting C, Pigliucci M (1998) Phenotypic evolution: a reaction norm perspective. Sinauer Associates, Sunderland

Schwenk K, Padilla DK, Bakken GS, Full RJ (2009) Grand challenges in organismal biology. Integr Comp Biol 49:7–14

Smith EN, Kruglyak L (2008) Gene-environment interaction in yeast gene expression. PLoS Biol 6:810–824

Somero GN (2000) Unity in diversity: a perspective on the methods, contributions, and future of comparative physiology. Annu Rev Physiol 62:927–937

van Helden P (2013) Data-driven hypotheses. EMBO Rep 14:104

Weinig C, Ewers BE, Welch SM (2014) Ecological genomics and process modeling of local adaptation to climate. Curr Opin Plant Biol 18:66–72

Whitehead A, Crawford DL (2006) Neutral and adaptive variation in gene expression. Proc Natl Acad Sci U S A 103:5425–5430

Whitehead A, Pilcher W, Champlin D, Nacci D (2012) Common mechanism underlies repeated evolution of extreme pollution tolerance. Proc R Soc B 279:427–433

Yin X, Struik PC, Kropff MJ (2004) Role of crop physiology in predicting gene-to-phenotype relationships. Trends Plant Sci 9:426–432

Zhu J, Zhang B, Smith EN, Drees B, Brem RB, Kruglyak L, Bumgarner RE, Schadt EE (2008) Integrating large-scale functional genomic data to dissect the complexity of yeast regulatory networks. Nat Genet 40:854–861

MIX
Papier aus verantwortungsvollen Quellen
Paper from responsible sources
FSC® C105338

If you have any concerns about our products,
you can contact us on
ProductSafety@springernature.com

In case Publisher is established outside the EU,
the EU authorized representative is:
Springer Nature Customer Service Center GmbH
Europaplatz 3, 69115 Heidelberg, Germany

Printed by Libri Plureos GmbH
in Hamburg, Germany